Understanding the Chemistry of the Environment

Understanding the Chemistry of the Environment

FRANCISCO G. CALVO-FLORES
University of Granada
Granada, Spain

Registered Offices
John Wiley & Sons, Inc., 111 River Street, Hoboken, NJ 07030, USA
John Wiley & Sons Ltd, New Era House, 8 Oldlands Way, Bognor Regis, West Sussex, PO22 9NQ, UK

For details of our global editorial offices, customer services, and more information about Wiley products visit us at www.wiley.com.

Wiley also publishes its books in a variety of electronic formats and by print-on-demand. Some content that appears in standard print versions of this book may not be available in other formats.

Library of Congress Cataloging-in-Publication Data applied for:
Paperback ISBN: 9781119568636

Cover Design: Wiley
Cover Image: © tommyvideo/Pixabay

Set in 10/12pt STIXTwoText by Straive, Pondicherry, India
SKY10094310_121924

This book is dedicated to my dear daughters, Irene and Sofia, with all my love. They are passionate about the intrinsic value of nature, and both have shown me, with their lifestyles, that every living creature and every corner of our planet has a purpose and deserves respect.

Contents

Presentation

This book, entitled ***Understanding the Chemistry of the Environment***, delves into the intricate relationship between chemistry and the environment, exploring how chemical substances and processes impact the world we live in. As our planet faces unprecedented environmental challenges, it is imperative that we understand the underlying chemistry driving these issues and work towards sustainable solutions.

Environmental chemistry is a multidisciplinary field that draws from various branches of science, including chemistry, biology, geology, physics, ecology and engineering. It seeks to unravel the complex interactions between chemicals and the environment, from the molecular to the global scale. This book provides a comprehensive overview of the fundamental principles and concepts of environmental chemistry while also delving into the latest research and practical applications.

I have always had an affinity for teaching environmental chemistry, so I am delighted to introduce this book to prospective readers. This book is the culmination of my experience as an educator, researcher and practitioner in the field of environmental chemistry, and it is my hope that it will serve as a valuable resource for students, researchers and professionals alike.

Throughout my career, I have had the privilege of working with diverse groups of students and engaging in thought-provoking discussions about the chemistry of the environment. These interactions have shaped my understanding of the challenges and complexities of environmental chemistry and have inspired me to share my knowledge and insights through this book. I have also been fortunate to collaborate with esteemed colleagues of the Grupo de Modelización y Diseño Molecular of the University of Granada in the field of environmental sciences and chemistry, whose expertise and contributions have enriched the content of this book.

In writing this book, I have strived to provide a comprehensive and accessible overview of environmental chemistry, tailored to the needs of students and professionals seeking to deepen their understanding of this fascinating field for a better understanding of the chemistry of the environment. I should be very satisfied if it could also serve as a valuable resource for people involved in environmental management and decision-making.

The content is organised in a logical and progressive manner, starting with the description of the environmental structure, always from the point of view of chemistry, and building up to more complex subjects while also incorporating real-world examples and case studies to illustrate the relevance of environmental chemistry in our daily lives and how science is a fundamental part of providing solutions to environmental problems.

The book covers a wide range of topics, including the sources and fate of pollutants in different environmental compartments, the chemistry of natural processes such as biogeochemical cycles, the impacts of pollutants on human health and ecosystems, and the role of chemistry in addressing environmental challenges such as climate change and pollution remediation.

As a teacher, my goal is to inspire and empower learners to become environmentally literate citizens who can critically analyse and address complex environmental issues. I have therefore included learning objectives, topics of interest and practical examples throughout the book to facilitate active learning and stimulate critical thinking. I have also strived to emphasise the importance of interdisciplinary approaches and the need for sustainable solutions to environmental problems.

I am deeply grateful to my students, colleagues, mentors and the wider scientific community for their continuous support and inspiration in my journey as an environmental chemistry educator.

I hope that **Understanding the Chemistry of the Environment** serves as a valuable resource for your learning and inspires you to appreciate the beauty and complexity of environmental chemistry. Together, let us strive towards a better understanding of our environment and work towards a more sustainable and healthier planet.

Credits

Figures and illustrations have been created by Sofía García-Ruiz, graduate in art.

Acknowledgements

I extend my heartfelt gratitude to my esteemed colleagues within the *Grupo de Modelización y Diseño Molecular (GMDM)*. Together, we have delved into the intricate realms of molecular structures, forging pathways to new understanding and innovation, and especially thanks to Dr. José Dobado for his invaluable assistance in reviewing and correcting the manuscript, which greatly enhanced the quality of the work. His insights and suggestions have undoubtedly improved the clarity and coherence of the content. I also thank him for his unwavering support and camaraderie throughout our collaborative endeavours. Likewise, I want to express my gratitude to my colleague, Dr. Joaquín Isac-García for his friendship, invaluable assistance and unwavering support.

I extend my deepest gratitude to every member of my family, with special recognition to my mother, Magdalena. She stood out as a beacon of inspiration, being the first woman to pursue the study of chemistry at the University of Granada. Her fervent passion for this captivating subject ignited my curiosity, propelling me forward on this academic journey.

To my esteemed professors and mentors, I am indebted for their guidance, wisdom and encouragement during my period as an undergraduate student. Their mentorship has been instrumental in shaping my academic and professional trajectory, instilling in me a fervent commitment to the pursuit of knowledge and excellence.

I express my heartfelt appreciation to Sofia Garcia-Ruiz, a talented graduate in art, for her work in crafting the figures and illustrations featured in this text. Sofia's creativity, attention to detail and artistic flair have breathed life into the scientific concepts presented within these pages, elevating the visual representation of the text to new heights. I am also thankful to the team at Wiley for their collaboration and expertise in bringing this book to fruition throughout the process of manuscript preparation and meticulous editing.

In closing, I am profoundly thankful to every individual who has contributed, whether in a significant or subtle way, not only to the achievement of this book but also to every aspect of my personal life. Special appreciation goes to Mane, who has been by my side throughout these recent years on this thrilling journey called life.

Introduction

Understanding the Chemistry of the Environment is a comprehensive book that delves into the complex relationship between the natural world and chemical processes. This book provides a unique perspective on the various spheres of nature and the biogeochemical cycles through the lens of chemistry, exploring the concept of pollution and its different types, with particular emphasis on chemical pollution, and delving into the toxic nature of many chemical substances.

The first part of the book focuses on describing the different spheres of the environment and the biogeochemical cycles from a chemical standpoint. It explores how these cycles are affected by human activities and their subsequent impact on the environment. Additionally, it addresses the concept of pollution, examining its various forms and highlighting the significance of chemical pollution. The toxicological aspects of chemical substances and their potential harm to living organisms are thoroughly discussed.

In the second part of the book, a series of chapters are dedicated to studying the major inorganic and organic substances that play a role as pollutants. The latter group, the organic pollutants, receives more extensive coverage due to the abundance of organic chemicals that may interact with the environment, both of natural and synthetic origin. These chapters provide in-depth analysis and understanding of these pollutants, their sources, behavior, and potential ecological implications.

Moving on to the third part of the book, it explores the most common pollutants found in the context of soil, water, and the atmosphere. Descriptions of these pollutants are provided in such spheres, with an exploration of the scientific and technical methods available for their control and, ideally, their elimination. The book highlights the importance of remediation techniques for minimizing environmental contamination and preserving the well-being of ecosystems.

Finally, the book dedicates a section to the greenhouse effect and so-called climate change, encompassing both natural and anthropogenic aspects. It examines the causes and effects of this phenomenon, with a particular emphasis on its role in climate change. The interplay between greenhouse gases, global warming, and the resulting climate shifts is thoroughly explored, providing readers with a comprehensive understanding of this pressing environmental issue.

Understanding the Chemistry of the Environment has been developed as a resource for students, researchers, and professionals seeking a comprehensive overview of the chemical dynamics in the environment. By elucidating the interconnections between chemistry, pollution, and environmental impact, this book equips readers with the knowledge and tools necessary to address the challenges facing our planet today.

CHAPTER 1

The Environment

Part I: Basic Concepts Related to the Environment: Environmental Spheres

1.1 Environmental Chemistry

Environmental chemistry is a branch of chemistry that focuses on the study of chemical processes occurring in the environment, including air, water and soil. It draws upon principles and concepts from various fields such as chemistry, biology, geology, physics, ecology and engineering. Environmental chemistry is concerned with the chemistry of natural processes as well as pollutants and contaminants and their impact on the environment and human health.

Key areas of study in environmental chemistry include atmospheric chemistry, aquatic chemistry, soil chemistry, environmental toxicology and environmental analytical chemistry. Environmental chemists employ a variety of techniques to study chemical processes in the environment, including laboratory experiments, field studies and computer modelling.

The goal of environmental chemistry is to understand the chemical processes occurring in the environment and the impact of human activities on these processes in order to develop strategies to mitigate environmental problems and protect nature and human health.

1.2 The Environment

The environment can be defined as the surroundings or conditions in which an organism, person or group operates. It includes all the physical, biological and social factors that influence the lives of living organisms.

The environment encompasses both natural and human-made elements, such as air, water, land, industry, agricultural and livestock activities, human settlements, infrastructures, etc., constituting a complex system that is constantly changing and interacting with the living and non-living components within it.

Understanding the Chemistry of the Environment, First Edition. Francisco G. Calvo-Flores.
© 2025 John Wiley & Sons Ltd. Published 2025 by John Wiley & Sons Ltd.

The environment is structured into ecosystems. An ecosystem refers to the community of living and non-living things that interact with each other in a specific area. It includes all of the organisms in a particular habitat, along with their physical surroundings, such as air, water and soil. Ecosystems can be small, such as a pond or a forest, or they can be large, such as a desert or an ocean. All living organisms in an ecosystem are interconnected and depend on each other for survival. For example, plants provide oxygen and food for animals, while animals help pollinate plants and disperse seeds. Bacteria and fungi are also important components of ecosystems, breaking down dead matter and recycling nutrients back into the environment.

Human activities can have a significant impact on ecosystems. Pollution, deforestation and climate change are just a few of the ways that human actions can disrupt the delicate balance of an ecosystem. Protecting and preserving ecosystems is important for the health of the planet and all of its inhabitants.

The different parts of the environment that make up the Earth's system are named *spheres*. These spheres interact with each other and influence the planet's climate, weather patterns and ecosystems.

The environment can be broadly categorised into five spheres, each of which has its own unique characteristics and dynamics:

1. **Atmosphere:** This is the layer of gases that surrounds the Earth. It is composed of different gases, including nitrogen, oxygen and carbon dioxide (CO_2), and it protects the Earth from harmful radiation from the sun.

2. **Hydrosphere:** This is the part of the Earth that consists of all water on and under the surface, including oceans, lakes, rivers, groundwater and glaciers.

3. **Lithosphere:** This is the solid part of the Earth's surface, including the continents, rocks and soil.

4. **Biosphere:** This is the part of the Earth that includes all living organisms, including plants, animals and microorganisms.

5. **Anthroposphere:** This refers to the human-made environment, including buildings, cities and infrastructure, and the impact of human activities on other environmental spheres.

1.3 The Atmosphere

The atmosphere (etymology: ἀτμός (atmos), 'vapor' and σφαῖρα (sphaira), 'sphere') is the layer of gases that surrounds the Earth and is held in place by the planet's gravitational field.

The Earth's atmosphere is a complex, heterogeneous system that plays several crucial functions. Some of the main functions of the atmosphere include:

1. **Protection:** The atmosphere protects life on Earth from harmful radiation and solar wind from the sun. It also shields the planet from meteoroids, which would otherwise cause widespread destruction upon impact.

2. **Regulation of temperature:** The atmosphere helps regulate the temperature of the planet by trapping heat from the sun and preventing it from escaping

into space. This helps keep the planet's temperature within a range that is suitable for life.

3. **Transportation of heat, moisture and gases:** The atmosphere is responsible for transporting heat from the equator to the poles, which helps regulate the global climate. It also transports moisture and gases around the planet, which are essential for life.

4. **Provision of gases for respiration and photosynthesis:** The atmosphere provides oxygen for respiration and carbon dioxide for photosynthesis which are essential for life. It also contains other gases such as nitrogen, which are important for the planet's ecology and the survival of many organisms.

5. **Support of weather patterns:** The atmosphere is the stage on which weather patterns occur. It is responsible for creating and sustaining atmospheric pressure systems, which drive the movement of air masses and cause weather patterns such as hurricanes, tornadoes and thunderstorms.

6. **Communication:** The atmosphere allows for the transmission of radio and television signals, as well as other forms of wireless communication.

The Earth's atmosphere is composed of several major and minor components (Table 1.1).

a) **Major components:** nitrogen (N_2), oxygen (O_2), argon (Ar).

b) **Minor components:** carbon dioxide (CO_2), neon (Ne), helium (He), methane (CH_4), krypton (Kr), xenon (Xe), ozone (O_3).

The percentage of every single gas represents the volume of this gas in the air. For example, the percentage of 20.95 for oxygen means that for every 100 units of air, 21 units are oxygen.

Water vapour is also a significant component of the atmosphere, but its concentration varies greatly depending on location and weather conditions. It can range from less than 1% in cold, dry regions to over 4% in humid, tropical areas.

TABLE 1.1 **Components of the atmosphere and their respective percentages**

Gas	Percentage
Nitrogen (N_2)	78.08%
Oxygen (O_2)	20.95%
Argon (Ar)	0.93%
Carbon dioxide (CO_2)	0.04%
Neon (Ne)	0.0018
Helium (He)	0.0005%
Methane (CH_4)	0.0002%
Krypton (Kr)	0.0001%
Xenon (Xe)	Trace amounts
Ozone (O_3)	Trace amounts

Additionally, the atmosphere also contains various pollutants and aerosols, which can have significant impacts on air quality and climate.

Topics of Interest: The Atmospheres of the Solar System's Planets

The atmospheres of the planets in the solar system can vary greatly depending on a number of factors, such as distance from the sun, composition of the planet and the planet's history. Here's a brief overview of the atmospheres of the major planets in our solar system:

- **Mercury** has a very thin atmosphere consisting mostly of helium and traces of sodium, potassium and oxygen. Its close proximity to the sun causes the solar wind to strip away its atmosphere.
- **Venus** has a thick, toxic atmosphere composed mainly of carbon dioxide and sulfuric[1] acid. The pressure at the surface of Venus is about 90 times that of Earth's atmosphere, which is equivalent to being about 900 m underwater on Earth.
- **Earth's** atmosphere is composed mainly of nitrogen, oxygen and trace amounts of other gases such as carbon dioxide, methane and O_3. This combination of gases is what allows life to thrive on Earth.
- **Mars:** Mars has a thin atmosphere composed mainly of carbon dioxide. It also has traces of nitrogen and argon, but the atmospheric pressure is only about 1% of that on Earth.
- **Jupiter:** Jupiter has a very thick atmosphere composed mainly of hydrogen and helium. It also has traces of methane, ammonia and water vapour. Jupiter's atmosphere is known for its swirling bands of clouds and the famous Great Red Spot.
- **Saturn's** atmosphere is similar to Jupiter's, with hydrogen and helium making up the majority of the atmosphere. Saturn's atmosphere also contains traces of methane, ammonia and water vapour.
- **Uranus** has a much colder atmosphere than the gas giants Jupiter and Saturn. Its atmosphere is composed mainly of hydrogen and helium, with traces of methane. The methane in Uranus' atmosphere gives it a blue-green colour.
- **Neptune's** atmosphere is similar to Uranus, composed mainly of hydrogen and helium with traces of methane. Neptune's atmosphere is also known for its dark, swirling storms, such as the Great Dark Spot.

1.4 Residence Time and Half-Life of Atmospheric Components

Residence Time

The gaseous components of the atmosphere are constantly formed and eliminated, maintaining a dynamic equilibrium that keeps their concentrations relatively constant within certain limits. If each molecule of the atmosphere could be labelled,

---

[1] Sulfur is the preferred spelling of the International Union of Pure and Applied Chemistry (IUPAC) since 1990, for the element and their derivatives and is the default form employed by many scientific journals .The alternative spelling sulfur with "ph" may still be found in common use in the UK and Commonwealth countries.

it would have a permanence time from the moment it is emitted until it disappears. This is what is called residence time (Table 1.2).

Hence, residence time denotes the average duration a gas molecule remains in the atmosphere prior to its elimination through a sink or conversion into a different compound. The residence time can vary widely, depending on the gas in question. For example, carbon dioxide has a residence time of several decades to centuries, while methane has a much shorter residence time of around a decade.

Half-Life

On the other hand, half-life is another parameter about the time it takes for half of a given quantity of a substance to be removed or transformed. This is often used to describe the rate of decay of radioactive isotopes, but it can also be useful to describe the rate of removal or transformation of gases in the atmosphere.

It's worth noting that the residence time and half-life of gases in the atmosphere are not the same. While the residence time gives an idea of how long a gas molecule stays in the atmosphere, the half-life provides information about how quickly the gas

TABLE 1.2 **Estimated residence time and half-life for some common atmospheric gases**

Gas	Residence time (years)	Half-life (years)	Notes
Water vapour	A few days to weeks	N/A	Highly variable; influenced by weather conditions
Carbon dioxide (CO_2)	4–5 years	N/A	Varies with exchange between the atmosphere and oceans
Methane (CH_4)	9–15 years	12 years	Influenced by chemical reactions and removal processes
Nitrous oxide (N_2O)	114 years	121 years	Removed through atmospheric and soil processes
Ozone (O_3)	Days to months	Minutes to months	Highly variable; influenced by atmospheric dynamics
Carbon monoxide (CO)	A few months	1–2 months	Removed through oxidation and deposition
Nitrogen (N_2)	Forever (effectively)	N/A	Chemically inert; does not participate in atmospheric reactions
Oxygen (O_2)	Thousands of years	N/A	Reacts slowly with other elements; largely constant

is removed or transformed. Both parameters are important for understanding the behaviour and impact of different gases in the atmosphere.

The distribution of these gases varies with altitude due to differences in their molecular weight and behaviour under different atmospheric conditions. Near the Earth's surface, the concentration of nitrogen and oxygen is relatively uniform, while other gases are present in trace amounts. As altitude increases, the concentration of all gases decreases, but the rate of decrease varies depending on the gas.

For example, the concentration of oxygen decreases rapidly with altitude, while the concentration also decreases with increasing altitude in the Earth's atmosphere due to a combination of factors, including sources and sinks of carbon dioxide at the Earth's surface and its relatively heavier molecular weight. The concentration of water vapour also varies greatly with altitude, ranging from nearly 0% at high altitudes to as much as 5% by volume near the Earth's surface.

1.5 Structure of the Atmosphere

The atmosphere is composed of gases that surround the Earth and is held in place by gravity. It has a structure divided by layers according to their pressure, temperature and composition.

Atmospheric pressure arises due to the gravitational force exerted by the air mass above a specific area. Consequently, as one ascends through the atmosphere, the quantity of air above diminishes, leading to a decrease in pressure. The exact rate at which pressure decreases with height depends on various factors, including temperature and humidity. The vertical profile of the atmospheric pressure is given in Figure 1.1.

The barometric formula describes the mathematical relationship known as the dependence of atmospheric pressure on altitude (or height). The barometric formula states that atmospheric pressure decreases exponentially with increasing altitude. Equation 1.1 provides a frequently employed barometric expression.

$$P = P_o \left[\frac{1 - (L \times h)}{T_o} \right]^{(g/R \times L)} \tag{1.1}$$

where:

- P is the pressure at a given altitude
- P_0 is the pressure at a reference altitude (usually sea level)
- L is the temperature lapse rate, which represents the rate at which temperature decreases with altitude
- h is the height or altitude above the reference altitude
- T_0 is the temperature at the reference altitude
- g is the acceleration due to gravity
- R is the specific gas constant for dry air

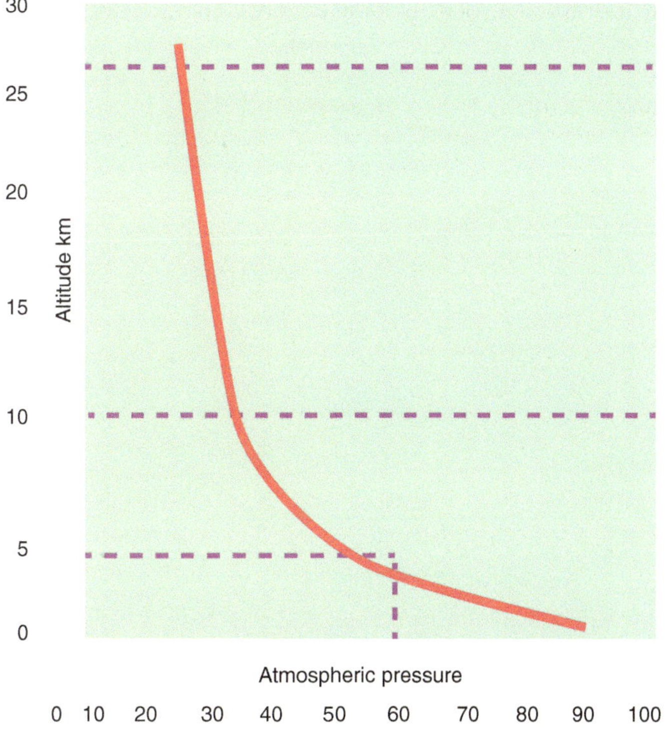

FIGURE 1.1   Pressure/altitude diagram

This formula indicates that as the altitude (h) increases, the pressure (P) decreases. The rate of decrease depends on the values of the other parameters. In reality, the temperature lapse rate varies with different atmospheric conditions, so the exact relationship between pressure and altitude can be more complex.

The atmosphere is structured into a set of layers (Figure 1.2). Every layer is characterised by its temperature range and average gas composition. Boundaries between the different atmospheric layers show a strong dependence on latitude and season, so the limit of every layer is not clear. Between every layer, there is a transition zone that is nominated with the suffix **pause**. From the Earth's surface upwards, these layers are: troposphere, stratosphere, mesosphere, thermosphere and exosphere.

Troposphere

The troposphere is the lowest layer of the Earth's atmosphere, extending from the Earth's surface up to an altitude of about 7–20 km, depending on the location and season. It contains approximately 75% of the Earth's atmospheric mass, and almost all of its weather occurs within this layer. The troposphere is characterised by a decrease in temperature with increasing altitude, which is caused by the decrease in air pressure and the absorption of solar radiation by the Earth's surface. This decrease

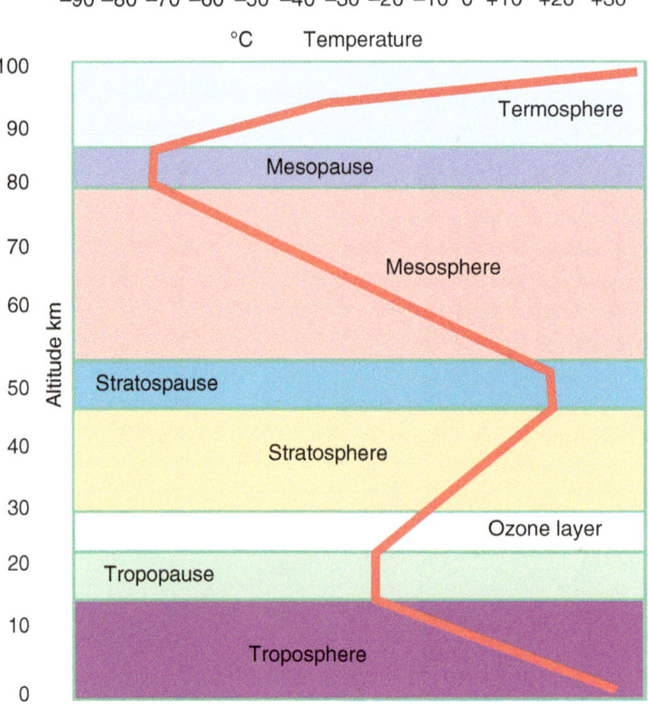

FIGURE 1.2 Structure of atmosphere

in temperature is known as the *lapse rate* and is responsible for the formation of clouds, precipitation and other weather phenomena. The troposphere also contains a mixture of gases, primarily nitrogen (about 78%) and oxygen (about 21%), as well as trace amounts of other gases such as carbon dioxide, water vapour and methane. The troposphere plays a crucial role in regulating the Earth's climate and weather patterns, as it is where most of the atmospheric processes that affect the Earth's climate occur. Changes in the troposphere, such as increases in greenhouse gas concentrations or changes in temperature, can have significant impacts on global climate and weather patterns.

The tropopause is the boundary layer between the troposphere and the next layer, the stratosphere. It is defined as the layer of the atmosphere where the temperature stops decreasing with increasing altitude and begins to increase instead. The tropopause typically lies at an altitude of about 8–18 km above sea level, depending on latitude and season. The tropopause is an important boundary in atmospheric science because it marks the transition from the layer where most weather occurs (the troposphere) to the layer where the O_3 layer is located (the stratosphere). It is also an important boundary for aviation, as it marks the upper limit of most weather-related turbulence and the lower limit of the jet stream.

Stratosphere

The stratosphere is the layer of Earth's atmosphere that extends from about 10–50 km above the planet's surface. It lies above the troposphere, which is the layer closest to the surface where weather occurs. The stratosphere is characterised by a gradual increase in temperature with altitude due to the presence of O_3, which absorbs ultraviolet radiation from the sun and heats the surrounding air. This layer also contains much less water vapour than the troposphere, which makes it much drier. One of the most important features of the stratosphere is the O_3 layer, which plays a critical role in protecting life on Earth from the harmful effects of the sun's ultraviolet radiation. In the past few years, human-made chemicals like chlorofluorocarbons (CFCs) have been endangering the O_3 layer by destroying O_3 molecules. The stratosphere is also home to several important weather phenomena, such as the polar vortex, which is a large-scale circulation pattern that forms over the poles during the winter months. This circulation helps keep cold air trapped over the poles and prevents it from moving southward.

Topics of Interest: The Molecule of Ozone

The history of O_3 traces back to the early 19th century, marked by its revelation by Christian Friedrich Schönbein, a chemist of German-Swiss origin.

Schönbein discovered O_3 in 1839 while he was experimenting with electricity and gases. He noticed a peculiar odour, similar to the smell of the air after a thunderstorm, which he attributed to a new gas he called O_3, from the Greek word *ozein* meaning 'to smell'.

Schönbein went on to study the properties of O_3 and discovered that it was a highly reactive gas with powerful oxidising properties. He also found that it had a bleaching effect on certain substances, such as indigo, and that it could kill bacteria and other microorganisms.

O_3 quickly gained attention for its potential applications, and by the late 19th century, it was used for a variety of purposes, including disinfecting water and sterilising medical equipment.

O_3 is a gas composed of three oxygen atoms; therefore, its chemical formula is O_3. The molecule can be represented in Lewis representation as a resonance hybrid with two contributing structures. In each contributing structure, there is a single bond on one side and a double bond on the other side. The resulting arrangement gives rise to an overall bond order of 1.5 for both bonds (Figure 1.3).

The resonance between these structures results in the delocalisation of electrons. This means that the electrons involved in the double bonds are not fixed between specific pairs of atoms but are distributed over all three oxygen atoms. The delocalisation of electrons gives O_3 its unique properties.

Resonance stabilisation in O_3 helps explain its reactivity. The delocalised electrons make the molecule more susceptible to chemical reactions, especially with other molecules that can accept or donate electrons.

O_3's reactivity is crucial for its role in the Earth's O_3 layer and in various chemical reactions in the atmosphere.

In the early 20th century, scientists discovered that the O_3 layer in the Earth's upper atmosphere played a crucial role in protecting life on Earth from harmful ultraviolet radiation from the sun. However, in the mid-20th century, it was found that human activities, such as the use of some halogenated compounds, were depleting the O_3 layer.

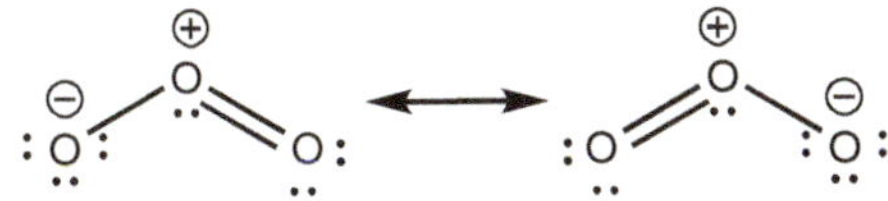

FIGURE 1.3   Lewis's representation of ozone (O_3)

Mesosphere

The mesosphere is the third layer of the Earth's atmosphere, located above the stratosphere and below the thermosphere. It extends from about 50–85 km above the Earth's surface. The temperature in the mesosphere decreases with altitude, reaching as low as −100 °C at the upper boundary. It is also the layer of the atmosphere where the air pressure is lowest. Because of the thinness of the air at this altitude, the mesosphere is not able to efficiently transfer heat by convection, and so it can be thought of as a 'cold trap' for incoming meteoroids.

The mesosphere is also the layer of the atmosphere where noctilucent clouds form, which are rare and beautiful cloud formations that can be seen at high latitudes during the summer months.

In terms of human activity, the mesosphere is too high to be easily accessible by aircraft or balloons, but it is studied using rockets and satellites. The mesosphere is above the stratosphere and extends to about 85 km in altitude. In this layer, the temperature decreases with altitude, and it is where meteors burn up upon entry into the Earth's atmosphere.

Thermosphere

The thermosphere is the uppermost layer of the Earth's atmosphere, located between the mesosphere and the exosphere. It extends from about 80 km to about 600 km above the Earth's surface. In the thermosphere, the air is extremely thin, and the temperatures can reach as high as 1,727 °C due to the absorption of solar radiation.

The thermosphere is also the layer of the atmosphere where many satellites and other spacecraft orbit the Earth. The high temperatures and ionisation in the thermosphere can affect the behaviour of these objects and cause drag, which can gradually cause them to lose altitude and eventually re-enter the Earth's atmosphere.

The boundary between the mesosphere and the thermosphere is the mesopause, located at an altitude of about 80–85 km above the Earth's surface. It is the coldest point in the Earth's atmosphere, with temperatures reaching as low as −100 °C or even colder, depending on latitude and season. The temperature at the mesopause is so low because this region is where the Earth's atmosphere is the least dense, and as a result, it cannot absorb and retain as much heat. In addition, the cooling effect of outgoing radiation from the Earth's surface and the cooling of the air due to rising and falling motions in the atmosphere also contribute to the low temperatures at the mesopause. The mesopause is an important region for studying the Earth's atmosphere and its interactions with the sun, as it is where many interesting phenomena occur, such as noctilucent clouds, airglow and gravity waves.

Exosphere

The exosphere is the outermost layer of the Earth's atmosphere, which gradually transitions into the vacuum of space. It starts at an altitude of about 500 km and extends to the edge of space, which is typically considered to be about 10,000 km above the Earth's surface. The exosphere is a very thin layer of gas with a density that decreases exponentially with altitude. It contains very few atoms and molecules, mostly low-mass ones such as hydrogen or helium. However, the density of these particles is so low that they rarely collide with each other, and their behaviour is mostly determined by the Earth's magnetic field and the solar wind[2].

1.6 Hydrosphere

The hydrosphere is the part of the Earth's surface where water exists, including all the oceans, rivers, lakes, groundwater and even the water in the atmosphere. It is one of the four major components of the Earth's system, along with the atmosphere, lithosphere and biosphere. The hydrosphere occupies approximately 70% of the Earth's surface and constitutes the habitat of many animals and plants.

A remarkable fact is that water in nature is not a pure chemical compound. Water is able to dissolve gases and solids and drag particles. Depending on the source of water bodies, the chemical composition of natural waters is variable.

Water Bodies

Water bodies can be categorised into different types based on their composition, size, depth and location.

1. **Oceans:** They are the largest water bodies on Earth, covering more than 70% of the planet's surface. They include the Atlantic, Pacific, Indian, Southern and Arctic oceans.

2. **Seas:** Smaller than oceans but still large bodies of saltwater, seas are usually partially enclosed by land. Examples include the Mediterranean Sea, the Red Sea and the Caribbean Sea.

3. **Lakes:** Lakes are bodies of freshwater that are surrounded by land. They can be small or large and are usually fed by rivers and streams. Examples include Lake Superior, Lake Victoria and Lake Baikal.

4. **Rivers:** Rivers are flowing bodies of freshwater that move from higher elevations to lower elevations, usually ending in oceans or lakes. Examples include the Amazon, the Nile, the Yenisei, the Mississippi and the Danube.

5. **Waterfalls:** Waterfalls are formed when a river or stream flows over a steep drop in elevation, creating a dramatic cascade of water. Examples include Niagara Falls, Victoria Falls and Angel Falls.

6. **Streams:** Streams are smaller, flowing bodies of freshwater that feed into larger rivers or lakes. They are often found in mountainous regions or forests.

7. **Ponds:** Ponds are small bodies of freshwater that are surrounded by land. They are usually fed by underground springs or rainfall and are home to a variety of aquatic plants and animals.

---

[2] Solar wind is a stream of charged particles, mostly protons and electrons, that are continuously emitted from the outer layer of the Sun's atmosphere, travelling at speeds ranging from 300 to 900 km/s and extending outward from the sun to the edge of the solar system.

8. **Wetlands:** Wetlands are areas of land that are saturated with water, either seasonally or permanently. They include marshes, swamps and bogs and are home to many species of birds, insects and other wildlife.

9. **Groundwater:** It refers to the water that is located beneath the Earth's surface, in spaces between rock, soil and sediment. It is an important source of freshwater for many communities. Groundwater is recharged through precipitation and surface water that seeps down into the ground, and it typically moves slowly through underground layers of soil and rock.

Water is a key component for the development of life on Earth, and it is always on our planet in continuous movement through the so-called water cycle.

1.7 The Molecule of Water

Water is an angular molecule formed by one atom of oxygen and two atoms of hydrogen linked by a covalent bond. Figure 1.4 shows different forms to represent water molecules. Oxygen shows sp^3 hybridisation; two hybrid orbitals are used to form covalent bonds with hydrogen atoms, and the other two orbitals each contain a non-bonding electron pair.

Hydrogen Bonds

Hydrogen bonds are non-covalent bonds that occur when a hydrogen atom is covalently bonded to a highly electronegative atom such as oxygen, nitrogen or fluorine. They are a type of dipole–dipole interaction, meaning that they occur between molecules with partially charged ends. Such bonds are present in both inorganic and organic molecules such as water, ammonia, hydrogen fluoride, alcohols, and carboxylic acids, showing an important role in biological molecules such as DNA and proteins, where they help stabilise the three-dimensional structure of the molecules.

Hydrogen bonds condition the behaviour of water and other solvents, as they can influence the solubility and reactivity of many substances.

The three main characteristics of hydrogen bonds are:

1. **Strength:** Hydrogen bonds are stronger than typical dipole–dipole interactions but weaker than covalent bonds. They are usually about 5–10% as strong as covalent bonds.

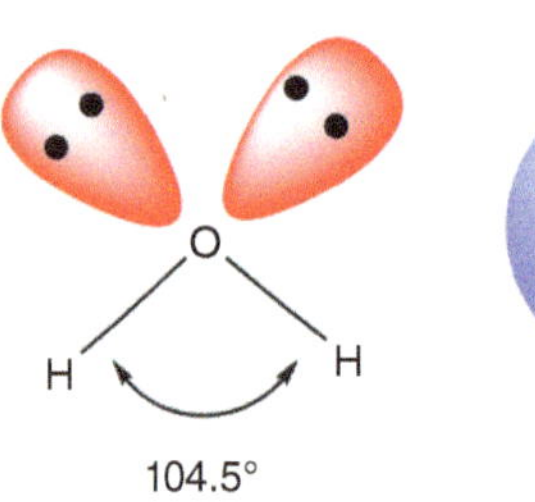

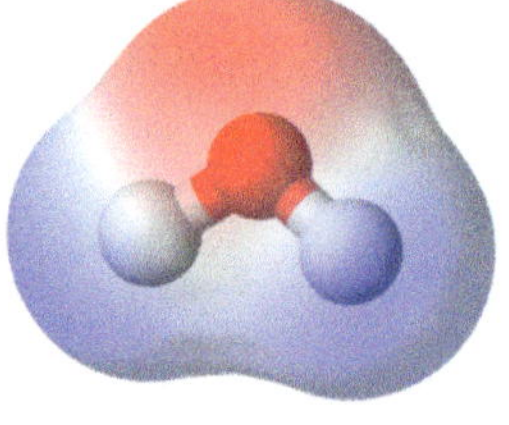
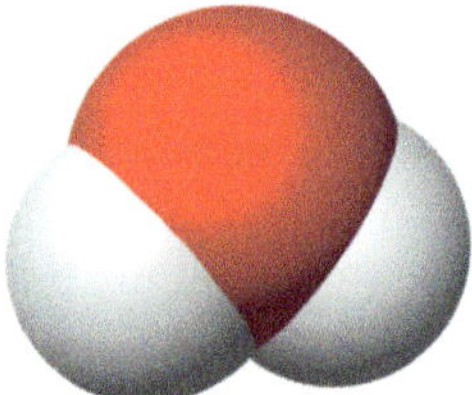

FIGURE 1.4 Molecule of water

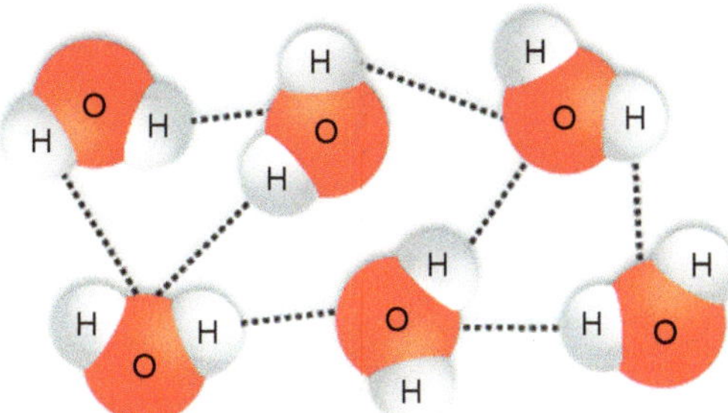

FIGURE 1.5   Representation of hydrogen bonds in water

2. **Bond length:** Hydrogen bonds are typically longer than covalent bonds, with bond lengths ranging from 1.5 to 2.5 Å.

3. **Directionality:** Hydrogen bonds have directionality, meaning that the bond is stronger when the atoms involved are aligned in a straight line.

Water molecules are polar because the oxygen atom is more electronegative than the hydrogen atom, which means that it attracts electrons more strongly. As a result, a partial negative charge ($\delta-$) is formed on the oxygen atom and a partial positive charge ($\delta+$) on the hydrogen atoms. The hydrogen bonds in water molecules arise from the attraction between the partially negative oxygen atom of one water molecule and the partially positive hydrogen atoms of adjacent water molecules (Figure 1.5). These bonds are constantly breaking and reforming as water molecules move and interact with each other.

Hydrogen bonds are responsible for many of the unique properties of water, such as its high boiling point, high surface tension, high heat capacity and ability to dissolve many substances.

1.8 Chemical–Physical Properties of Water with Environmental Transcendence

Density of Pure Water, Freshwater and Seawater

The density of pure water is approximately 1 gram per cubic centimetre (g/cm³) at standard temperature and pressure (STP)[3]. However, the density of water can vary slightly depending on its temperature and pressure. At 4 °C, the temperature at which water has its maximum density, the density of water is approximately 0.99997 g/cm³. At higher temperatures, the density of water decreases, and at lower temperatures also, it decreases (Figure 1.6).

---

[3] STP is defined as 1 atmosphere (atm) of pressure and a temperature of 25 °C.

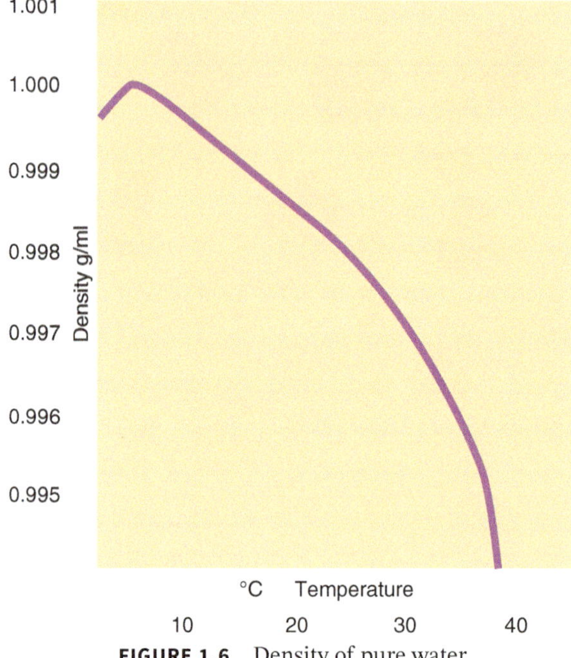

FIGURE 1.6   Density of pure water

In addition, at higher pressures, such as those found in the deep ocean, the density of water can increase significantly. Water in nature is not a pure chemical substance, containing several components, mainly through dissolution.

The densities of freshwater and seawater are different because of differences in their composition and temperature. The density of freshwater varies with temperature, but the density of seawater also varies with salinity.

At STP, the density of freshwater is about $1\,g/cm^3$. However, as the temperature of freshwater changes, its density also changes. For example, at $4\,°C$, the density of freshwater is about $1{,}000\,g/cm^3$, but at $25\,°C$, the density decreases to about $0.997\,g/cm^3$.

On the other hand, the density of seawater is about $1{,}025\,g/cm^3$ at STP, which is higher than the density of freshwater due to the presence of dissolved salts. The salinity of seawater can vary, but as it increases, the density of seawater also increases.

Density of Ice in Nature

Ice is a solid state of water that forms when water molecules slow down and bond together in a regular, crystalline pattern. The density of ice formed from pure water is about $0.916\,g/mL$ at $0\,°C$ at standard atmospheric pressure. In nature, ice is formed in a different way depending on the water body: lakes, rivers or seawater.

Freshwater and pure water become less dense as they near the freezing point. This difference in density explains why ice cubes float in a glass of water. When it is very cold, low-density water stays at the surface of lakes and rivers, forming an ice layer on top. This layer of ice on the surface blocks the exchange of energy between the cold air in the surroundings and the warm water below; therefore, cooling continues at the surface, but instead of dropping the temperature of the water below, the heat losses will be manifested in the production of ice.

In contrast to freshwater from lakes or rivers, the salts dissolved in ocean water cause the density of the water to increase as it nears the freezing point, and very cold ocean water tends to sink. As a consequence, sea ice forms quite slowly compared to freshwater ice because saltwater sinks away from the cold surface before it cools enough to become ice. The reported values of ice density vary over a wide range from 0.84 to 0.94 g/mL, with an average of approximately 0.91 g/mL[4].

The crystalline structure of ice is known as hexagonal close packing, which is composed of a repeating unit cell with a hexagonal shape. The unit cell contains two interpenetrating sub-lattices, one of oxygen atoms and the other of hydrogen atoms.

Topics of Interest: Ice Crystals in the Atmosphere

Ice crystals in the atmosphere are tiny particles of ice that form when the air temperature is below freezing point and there is enough moisture in the air. These ice crystals can take on a variety of shapes and sizes, including needles, columns, pyramids, bullets, hexagon plates, sectored hexagon plates, stars and highly branched dendrites, depending on temperature and humidity (Figure 1.7).

Snowflakes are perhaps the most well-known form of ice crystals in nature. Each snowflake is a unique and intricate pattern of ice crystals that forms in the clouds when water vapour freezes onto dust or other particles.

Ice crystals are an important component of many atmospheric phenomena, such as halos around the sun or moon, sundogs and diamond dust. They also play a critical role in the formation of clouds, which are made up of billions of tiny ice crystals or water droplets suspended in the air.

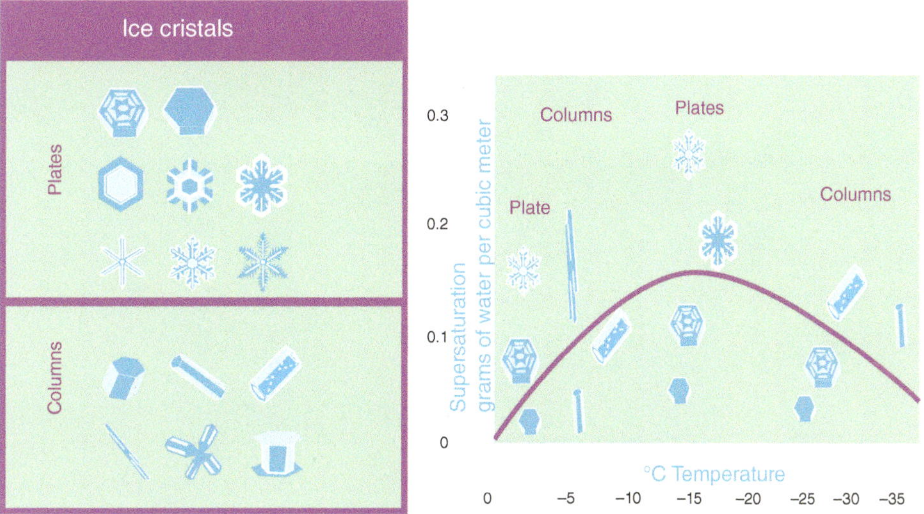

FIGURE 1.7   Ice crystal shapes

[4] Timco, G.W. and Frederking, R.M.W. (1996). A review of sea ice density. *Cold Regions Scienceand Technology* 2: 41.

Melting and Boiling Points of Water

Water is an extraordinary substance that can exist in nature in three states: solid, liquid and gas.

The melting point of pure water at a standard atmospheric pressure of 1 atm is 0 °C. Comparing the melting points of binary chemical substances formed by hydrogen and other elements surrounding oxygen in the periodic table, water shows higher values than expected. Similar conclusions may be drawn for the melting point.

These anomalously high values are a consequence of the hydrogen bonds between water molecules. In pure water at room temperature, the average energy of a hydrogen bond is around 4 kcal/mol (17 kJ/mol). However, in more complex environments, such as biomolecules, the strength of hydrogen bonds can vary widely depending on the specific structural and chemical context.

It is worth noting that the energy of hydrogen bonds in water is relatively weak compared to covalent bonds, which typically have energies in the range of 50–100 kcal/mol (200–400 kJ/mol). However, the large number of hydrogen bonds in water collectively contribute to its high melting and boiling points and other properties such as its high surface tension and ability to dissolve many substances.

In Figure 1.8, the values of boiling points for binary compounds of hydrogen and elements of groups 14–17 of the periodic table are dropped.

The boiling point increases with molecular mass for every group, with the exception of those compounds showing hydrogen bonds. Water molecules have the biggest deviation from the expected values without considering hydrogen bonds.

The values of the melting point and boiling point of water play critical roles in shaping many natural processes on Earth. The melting point of water, which is 0 °C, is important for several reasons.

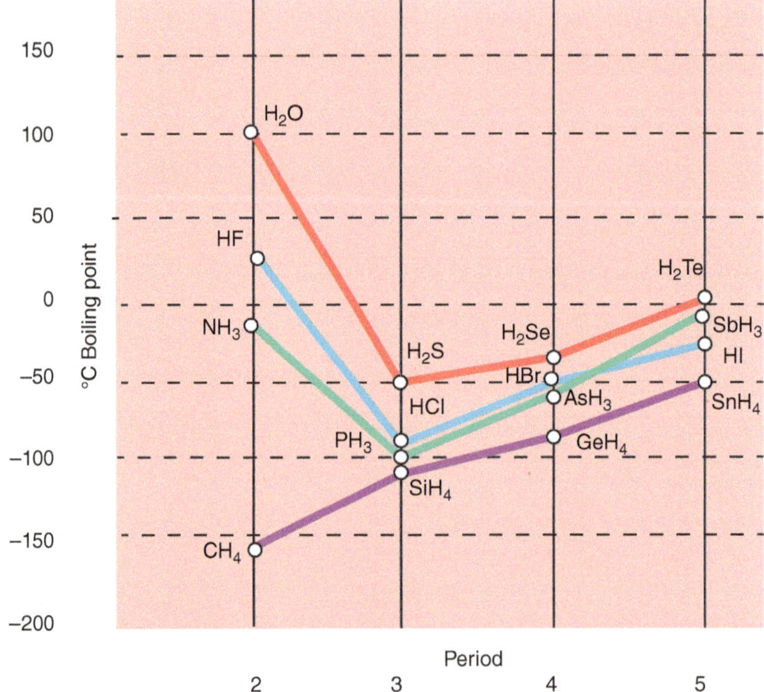

FIGURE 1.8 Boiling points of binary hydrogen compounds

First, it is the temperature at which water transitions from a solid to a liquid state, which allows for the formation of ice and snow, which are critical components of Earth's climate system. Additionally, the expansion that occurs when water freezes creates unique habitats for aquatic life and helps regulate ocean currents.

The melting of water also has important consequences for individual organisms. For example, many animals have adaptations that allow them to survive in freezing water, such as antifreeze proteins or heat shock proteins. The properties of water also affect the way that nutrients and other essential molecules are transported within and between cells, which is critical for the survival of all living organisms.

The boiling point of water, which is 100 °C at standard atmospheric pressure, is also extraordinarily high if compared to boiling points of similar substances.

Heat Capacity and Vapourisation Heat of Water

Those are physical properties related to the absorption of energy and changes of state in water. The heat capacity of water (C) is defined as the amount of heat required to change the temperature of a mass unit by one degree. It can be determined when pressure is constant (C_p) or when volume is constant (C_V).

- Water (liquid) $C_p = 4,185.5\,\mathrm{J/(kg\,K)}$ (15 °C, 101.325 kPa)
- Water (liquid) $C_V = 74.539\,\mathrm{J/mol\,K}$ (25 °C)

Both values are unusually high and imply that water stores a lot of heat energy, which makes it a good medium for spreading the planet's heat.

These high heat capacity values are one of the reasons why water is such an important substance for regulating temperature in the natural world.

Because of its high heat capacity, water can absorb a lot of energy without undergoing a large temperature change. This means that water can help stabilise temperatures in the environment, which is important for many living organisms.

The vapourisation heat of water, also known as the enthalpy of vapourisation of water (enthalpy of vapourisation, ΔH_{vap}), is the amount of energy required to convert one unit of mass of liquid water at a given temperature into water vapour at the same temperature.

At standard atmospheric pressure (1 atm), the vapourisation heat of water is approximately 40.7 kJ/mol or 2,260 J/g. However, this value can vary depending on the temperature and pressure at which the water is vapourised.

The vapourisation heat of water is a critical factor in weather phenomena such as evaporation and condensation, which play a significant role in the Earth's water cycle. The evaporation of water from the oceans, lakes and rivers is driven by solar energy and helps regulate the Earth's climate. In addition to these broader effects, it is also a critical factor in many engineering applications, such as power generation, refrigeration and heating and cooling systems.

Surface Tension of Water

Surface tension is a property of liquids that results from the cohesive forces between the molecules in the liquid. It is the tendency of the surface of a liquid to resist an external force due to the attractive forces between its molecules. The surface tension

FIGURE 1.9 Surface tension of water

of water at 20 °C is approximately 72.8 mN/m. This means that it takes a force of 72.8 mN to break a surface film of water that is 1 m long. Water has a relatively high surface tension compared to other liquids, such as ethanol or acetone, which have surface tensions of approximately 22.3 and 23.9 mN/m, respectively. This is because water molecules are strongly attracted to each other due to their hydrogen bonds, which results in the formation of a strong cohesive network at the surface of the liquid (Figure 1.9).

This property of water is important in many natural processes, such as the ability of small insects like water striders to walk on water. The high surface tension of water allows the insects to distribute their weight over a larger area, enabling them to stay afloat.

In addition, surface tension in water is why water droplets tend to form spherical shapes. When a water droplet forms, the surface tension causes the water molecules to pull together tightly, minimising the surface area of the droplet. This results in a shape that has the smallest possible surface area for the given volume, which is a sphere.

The surface tension of water also allows water droplets to adhere to surfaces, which is why they can stick to leaves, spider webs and other surfaces. This is because the cohesive forces between the water molecules are stronger than the adhesive forces between the water and the surface.

Capillary action is the ability of a liquid, such as water, to be drawn upwards by a narrow tube or porous material against an external force, such as gravity

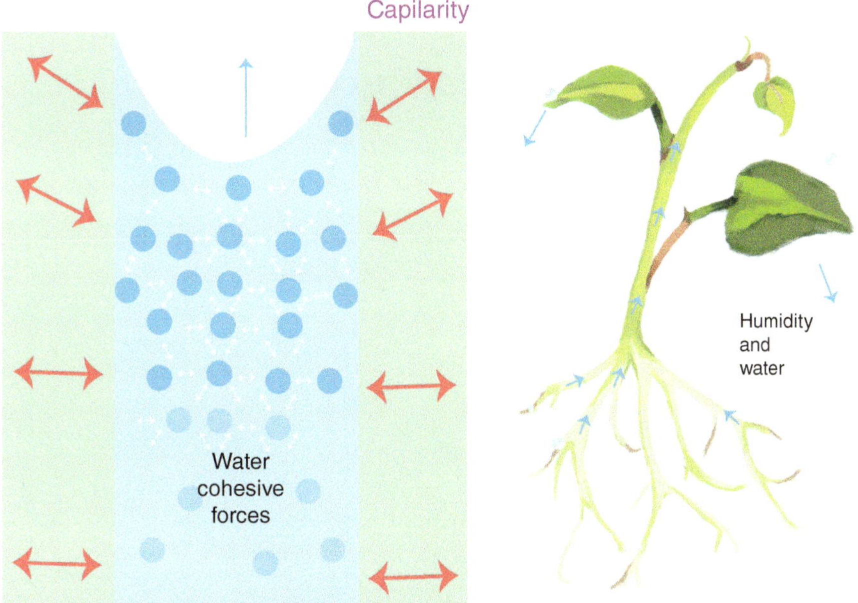

FIGURE 1.10   Capillary example

(Figure 1.10). Capillary action is a spontaneous phenomenon resulting from two attractive forces:

1. Cohesive forces between water molecules caused by surface tension.
2. Adhesive forces are forces between water molecules and the surface of a narrow tube or porous material.

The narrower the tube, the greater the height at which the liquid arrives. In the case of water, the adhesive forces between the water molecules and the tube or material surface are greater than the cohesive forces between the water molecules, allowing water to 'climb' up the tube or material. It is an essential phenomenon in plants. They put down their roots into the soil, carrying up water and dissolved nutrients into the plant's tissues.

Topics of Interest: Lake Stratification

Thermal stratification is the phenomenon by which, in temperate areas, lake water is divided into layers based on their different density and temperature that behave like immiscible layers. Depending on climate conditions, lakes are divided into three categories:

1. **Amictic lakes:** The surface is covered with ice throughout the year. This prevents the mixing of the waters beneath, and therefore they exhibit inverse cold water stratification in such a way that temperature increases with the increase in depth. They are located in extremely cold regions.
2. **Holomictic lakes** are arranged in layers. Each layer has a specific density and temperature, regardless of depth, during a particular period of the year. The upper layer is known as the epilimnion, the middle layer is the metalimnion (or thermocline), and the bottom layer is the hypolimnion (Figure 1.11).

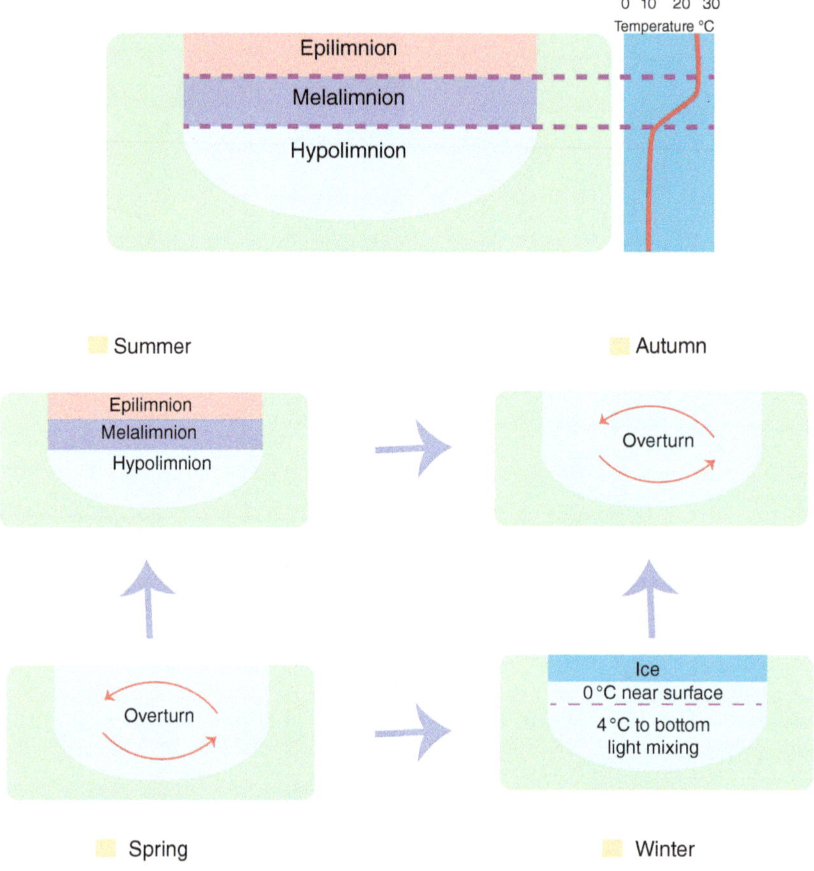

FIGURE 1.11   Lake stratification in temperate areas

In a cyclic way, it takes place through the formation of layers and mixing periods.

Most of the lakes in the world are holomictic. They may be subdivided into three different types:

  a. Monomictics: a single mixing period per year

  b. Dimictics: two mixing periods per year

  c. Polymictics: several mixing periods per year

3. **Meromictic lakes:** They have layers of water that do not intermingle due to a lack of vertical circulation, usually due to salinity. They are characterised by the absence of water intermixes and the fact that the lower layer of the lake does not contain dissolved oxygen and, therefore, is practically devoid of life, except for purple sulfur bacteria. There are only a few meromictic lakes in the world compared to holomictic lakes. The nature of thermal stratification in lakes is a determinant of the types of living organisms inhabiting them.

1.9   Water as Solvent in Nature

Water is a very common solvent in nature, particularly in aqueous environments like oceans, rivers, lakes and underground water systems. There are several reasons for this:

1. **Abundance:** Water is one of the most abundant substances on earth, making up over 70% of the earth's surface and 60% of the human body.

2. **Polarity:** Water is a polar molecule, which means it has a partial positive charge on one end (the hydrogen atoms) and a partial negative charge on the other end (the oxygen atom). This polarity allows water to dissolve other polar substances, such as salts and sugars, and to form hydrogen bonds with other polar molecules, like itself.

3. **Versatility:** Water is an excellent solvent for most ionic compounds and many other polar molecules. Because it dissolves more substances than any other liquid, water used to be called the universal solvent, although strictly speaking, this is not correct because many apolar molecules are not soluble in water. This versatility makes it an ideal solvent for many biological processes.

4. **Stability:** Water is a very stable substance, meaning it does not easily break down or react chemically with other substances. This stability makes it an ideal solvent for long-term processes, such as those involved in biological systems.

In nature, water plays a key role in the transportation of many substances, both in hydrologic and biological cycles.

The amount of solute that can be dissolved in water is variable and is described by the concept of solubility. In chemistry, solubility is defined as the upper concentration a compound reaches in a solution. Solubility is often expressed in terms of the mass of solute (in grams, milligrams or mol) per amount of solvent (expressed as volume or mass).

The composition and properties of water bodies in nature depend strongly on the species that are dissolved in water, solids or gases.

Solubility of Gases in Water

Gases may be dissolved in liquids to form solutions. The solubility of a gas in a specific solvent depends on temperature and pressure in such a way that it decreases with temperature and increases with pressure.

Water is capable of dissolving gases from the atmosphere while remaining in contact with air, and oxygen, nitrogen and carbon dioxide are the more relevant natural gases present in natural waters.

At a given temperature, the solubility of a gas in a liquid is directly proportional to the partial pressure of the gas above the liquid. This fact is known as Henry's Law:

At a constant temperature, the amount of a given gas that dissolves in a given type and volume of liquid is directly proportional to the partial pressure of that gas in equilibrium with that liquid.

Solubilities of gases are typically reported in g/L or mol/L, and Henry's law can be expressed in two ways:

1. $S_1/P_1 = S_2/P_2 = k$
2. $C = kP_p$

1. S_1 and P_1 are the solubility and the pressure at an initial set of conditions, S_2 and P_2 are the solubility and pressure at another changed set of conditions and k is Henry's law constant with units depending on those used for solubility and pressure.

2. The concentration of the dissolved gas, C, is proportional to partial pressure of the gas (Figure 1.12).

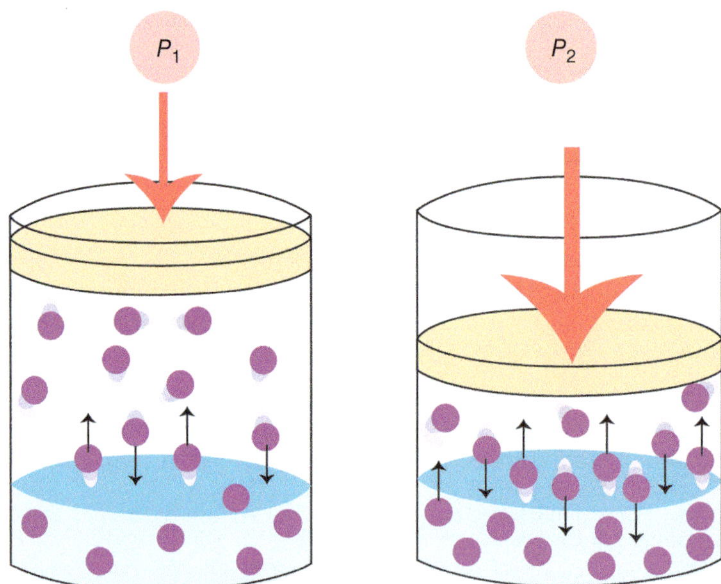

FIGURE 1.12 Henry's law representation

Topics of Interest: Units of Concentration

In chemistry, the measure of how much of a given substance is mixed with another substance Such relationships can be expressed in several ways. When reactions occur, or in the case of stoichiometric calculus, concentrations based on moles are more frequent. Such units used to be named *chemical units of concentration* (Table 1.3).

For solutions and mixtures in which there are no chemical changes, concentrations are expressed in terms of relative percentages of mass and/or volumes of solutes and solvents depending on averages of solution components (*physical unit concentrations*) (Table 1.4).

Both ppm_m and ppb_m are units frequently used for reporting the concentrations of trace substances in water and soils.

One ppm_m is equivalent to 1 milligram of substance per litre of water (mg/L), assuming that a litre of water weighs approximately 1,000 g, but actually, it is 1 milligram of substance per kilogram of solution (mg/kg). For ppb_m, the values are related to μg of substance in solution.

Concentrations in gas systems are computed on a volume-per-volume ratio, and they are accurately termed ppm_v and ppb_v. Such units are used for reporting concentrations of gases in the atmosphere. The values remain unchanged as gases are compressed or expanded. The

TABLE 1.3   Concentrations with moles

Concept	Units
Molarity	Mol solute/L solution
Molality	mol solute/kg solvent
Mole fraction	$Moles_A / \Sigma\ moles\ A + B + C \ldots$

TABLE 1.4 **Concentrations with mass and volume**

Concept	Units
Mass/weight percentage	g of solute per 100 g of solution.
Volume percentage	mL of solute per 100 mL solution
Mass by volume percentage	g of solute per 100 mL of solution.
Parts per million in mass (ppm_m)	Mass of solute/mass of solution $\times 10^6$ (mg/L)
Parts per million in volume (ppm_v)	Volume of solute/volume of solution: 10^6 (μL/L)
Parts per billion in mass (ppb_m)	Mass of solute/mass of solution $\times 10^9$ (μg/m^3)
Parts per billion in volume (ppb_v)	Volume of solute/volume of solution $\times 10^9$ (μg/m^3)

relationship between ppmv and ppb_v with mg/m^3 and μg/m^3, respectively, depends on their pressure, temperature and molecular weight. But a good approximation, for a generic gas A in a mixture of 1 ppm_v, implies that for every 1,000,000 molecule constituent of the mixture, about one molecule is of *A*.

The concentration of dissolved gases in water is of prime importance when considering the quality of the water. Oxygen concentration is a determinant for the development of living beings in water bodies. Oxygen solubility in pure water at 25 °C and a pressure of 1 bar is 40 mg/L. In air with a normal composition, the oxygen partial pressure is 0.2 atm, so the solubility of oxygen in water that comes in contact with air is: $40 \times 0.2 = 8$ mg O_2/L.

The solubility of oxygen in seawater and freshwater differs slightly, and as with any other gas, it depends on temperature. Three major considerations can be made:

1. The solubility of oxygen decreases as temperature increases. This means that warmer surface water requires less dissolved oxygen to reach 100% air saturation than does deeper, cooler water.

2. Dissolved oxygen decreases exponentially as salt levels increase. At the same pressure and temperature, seawater holds about 20% less dissolved oxygen than freshwater (Figure 1.13).

3. Dissolved oxygen increases as pressure increases. Water at lower altitudes can dissolve less oxygen than water at higher altitudes.

Colder, deeper freshwaters have the capability to hold higher concentrations of dissolved oxygen.

In addition to oxygen solubilising according to Henry's law, there are other factors that can contribute to obtaining values above or below the average of oxygen in water. Aquatic respiration and decomposition of algae and other living beings provide lower dissolved oxygen concentrations, while rapid aeration, for example, by wind, and photosynthesis of aquatic plants may produce the phenomenon of supersaturation.

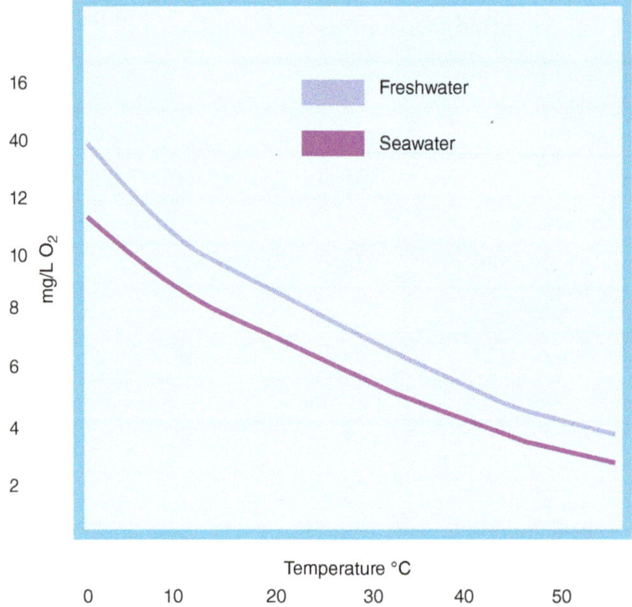

FIGURE 1.13 Oxygen solubility in fresh and seawater

Solubility of Salts in Water

Water is able to solubilise solid substances such as ionic compounds and polar molecules. Two parameters describe solubilisation of substances in water:

1. **Concentration** is defined as the amount of solute expressed in grams, milligrams, mols, etc., dissolved in a given amount of solution expressed in mass or volume.

2. **Solubility:** It is the maximum amount of solute dissolved in a given volume of solvent at equilibrium at a specific temperature. The resulting solution is called a saturated solution.

Most solids increase their solubility with temperature. Figure 1.14 shows the solubility dependence of some common salts on temperature.

Salt solubility can be predicted using the information in Tables 1.5–1.7, which summarise the so-called solubility rules. Salts that can dissolve in water are listed in Table 1.5.

Total dissolved solids (TDS) comprise inorganic soluble salts and some small amounts of organic matter that may be dissolved in water. The measurements of total dissolved solids are made by gravimetric analysis[5] and conductivity[6]. TDS is given in parts per million (ppm). One ppm indicates 1 milligram of dissolved solids per kilogram of water.

---

[5] The amount of solute dissolved is calculated by weighing the remaining solids from a sample of water after the water has evaporated.

[6] Conductivity is a measure of water's capability to pass electrical flow, usually measured in micro- or millisiemens per centimetre (μS/cm or mS/cm).

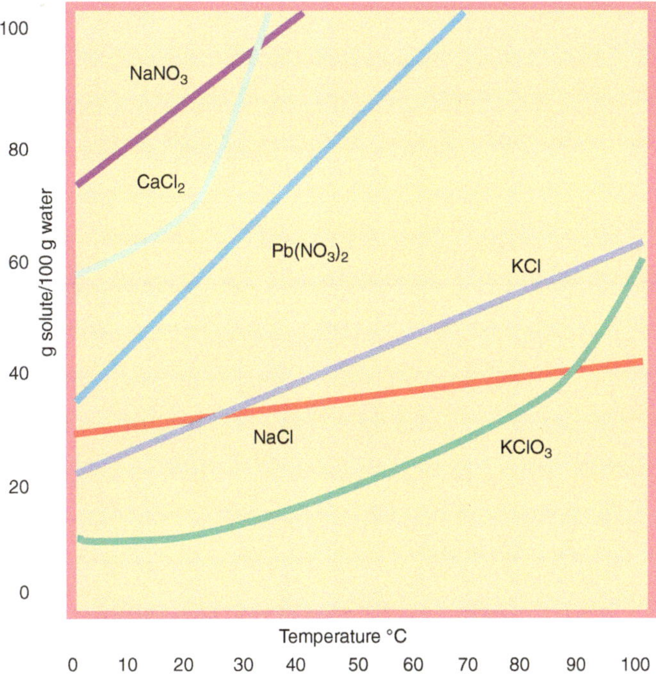

FIGURE 1.14   Dependence of solubility of salts on temperature

TABLE 1.5   **Soluble salts in water**

Ions	Exception
Alkaline metals. Na^+, K^+, etc.	
Ammonium ions (NH_4^+)	
Nitrates, acetates, chlorates and perchlorate NO_3^-, CH_3COO^-, ClO_3^-, ClO_4^-	
Binary compounds of halogens (chloride, bromide, iodide, etc.) with metals	F^-, Ag^+, Pb^{2+}* and Hg^{2+} *Lead halides are soluble in hot water.
SO_4^{2-}	Ba^{2+}, Sr^{2+}, Ca^{2+}, Pb^{2+}, Ag^+ and Hg^{2+}

TABLE 1.6   **Slightly soluble salts**

Ions	Exception
SO_4^{2-} with Pb^{2+}, Ag^+ and Hg^{2+}	*Lead sulfate is poorly soluble.
Hydroxides of Ca^{2+}, Sr^{2+}, etc.	Ba^{2+}

TABLE 1.7   Insoluble salts

Ions	Exceptions
S^{2-}	Ca^{2+}, Ba^{2+}, Sr^{2+}, Mg^{2+}, Na^+, K^+ and NH_4^+
OH^-	Na^+, K^+, etc. Al^{3+}, NH_4^+
CO_3^{2-}, $C_2O_4^{2-}$, CrO_4^{2-} and PO_4^{3-}	Na^+, K^+, etc. NH_4^+ *Lithium phosphate is poorly soluble

According to TDS levels, water can be classified into:

- **Freshwater:** TDS = 500 ppm
- **Brackish water:** TDS = 500–30,000 ppm
- **Saline water:** TDS = 30,000–40,000 ppm
- **Hypersaline:** TDS greater than 40,000 ppm

1.10 Metal Ions in Natural Waters

Metal ions are commonly found in natural waters and can come from a variety of sources, such as the weathering of rocks, leaching from soils, industrial discharges and atmospheric deposition. Some of the most common metal ions found in natural waters include calcium (Ca^{2+}), magnesium (Mg^{2+}), sodium (Na^+), potassium (K^+), iron (Fe^{2+} and Fe^{3+}), manganese (Mn^{2+}), copper (Cu^{2+}) and zinc (Zn^{2+}). The concentration of these metal ions can vary depending on the location and type of water body. For example, rivers and streams that flow through areas with high agricultural or industrial activities may have higher concentrations of metal ions compared to pristine mountain streams.

In natural waters, metal ions can exist in different forms: as solvated ions, complex ions and organometallic compounds:

1) **Solvated ions:** In natural waters, metal ions are often found in a solvated state, meaning they are surrounded by a shell of water molecules. The solvation of metal ions in water is driven by the electrostatic attraction between the metal ion and the partial charges on the water molecules. This attraction can vary depending on the charge and size of the metal ion, as well as the temperature and pressure of the water. For example, iron (Fe^{2+} and Fe^{3+}) can have different coordination numbers depending on their oxidation states and the ligands present in the solution. In general, iron(II) (Fe^{2+}) ions are coordinated by six water molecules, while iron(III) (Fe^{3+}) ions are coordinated by either six or seven water molecules, depending on the ligands present (Figure 1.15).

FIGURE 1.15   Representation of solvated Fe(III) ion

The specific reactions that solvated metal ions undergo in natural waters depend on a variety of factors, including the concentration and speciation of the metal ion, the pH and temperature of the water, and the presence of other chemical species. Understanding these reactions is important for predicting their reactivity and bioavailability and assessing their potential impacts on the environment.

Solvated metal ions in natural waters can undergo a variety of chemical reactions that affect their behaviour, hydrolysis and properties. Some of the most common reactions include precipitation, complexation, redox reactions and hydrolysis.

a) **Hydrolysis** occurs when solvated metal ions undergo stepwise hydrolysis, in which they deliver one or more H^+ to form hydrolysis products. For example, for hydrated Fe^{3+}, the first step is represented by Eq. 1.2.

$$\left[Fe\left(H_2O \right)_6 \right]^{3+} + H_2O \rightarrow \left[Fe\left(H_2O \right)_5 OH \right]^{2+} + H_3O^+ \tag{1.2}$$

b) Each step in hydrolysis reduces the charge as hydroxy cations are formed. Precipitation occurs when the solubility of a metal ion in water is exceeded and the metal ion forms a solid phase, such as the metal hydroxide $Fe(OH)_3$, or with other chemical species in the water. This can occur due to changes in pH, temperature or the presence of other ions that can form insoluble compounds with the metal ion. This set of reactions depends on the pH of the water and can also affect the bioavailability and mobility of the metal ions.

c) **Redox reactions** involving solvated ions are common in natural waters. One common redox reaction in natural waters involves the oxidation of dissolved ferrous ions (Fe^{2+}) to ferric ions (Fe^{3+}). This reaction can occur in the presence of dissolved oxygen, which acts as an oxidising agent (Eq. 1.3).

$$Fe\left(H_2O \right)_6^{2+} \rightarrow Fe\left(OH \right)_3 + 3H_2O + e - + 3H^+ \tag{1.3}$$

2) **Complex ions:** Complex ions are formed when a central metal ion is surrounded by a group of ligands. Such ligands are molecules or ions that donate a pair of electrons to the metal ion. The formation of these complexes can affect the solubility, reactivity and mobility of metal ions in the environment. In natural waters, metal ions can form complexes with a variety of ligands, both inorganic and organic.

Some examples of inorganic ligands found in natural waters include:

a) **Carbonate (CO_3^{2-}):** Carbonate is a common inorganic ligand in natural waters. It can form complexes with metal ions such as calcium, magnesium and iron.

b) **Sulfate (SO_4^{2-}):** Sulfate is another inorganic ligand commonly found in natural waters. It can form complexes with metal ions such as calcium and magnesium.

c) **Chloride (Cl^-):** Chloride is a simple inorganic ligand that can bind to metal ions such as silver and mercury.

d) Nitrate (NO_3^-): Nitrate is a common ligand found in natural waters that can form complexes with metal ions such as copper and iron.

e) Phosphate (PO_4^{3-}): Phosphate is another common inorganic ligand found in natural waters. It can form complexes with metal ions such as calcium and magnesium.

Organic ligands are naturally occurring substances in natural waters and are important in regulating the behaviour and bioavailability of metals in aquatic environments. In natural waters, various organic molecules can form complexes with metals through a process called complexation. These organic molecules have functional groups that can coordinate with metal ions, forming stable complexes. The complexation of metals by organic ligands is an essential process in aquatic chemistry and has implications for metal speciation, bioavailability and transport.

Some examples of these organic molecules that can form complexes with metals in natural waters are:

1. **Fulvic and humic acids:** Fulvic and humic acids are complex mixtures of organic compounds derived from decaying plant matter (Figure 1.16). They are abundant in soil and natural waters and their functional groups, such as carboxyl and phenolic groups, can bind to metal ions, forming complexes (Figure 1.17).

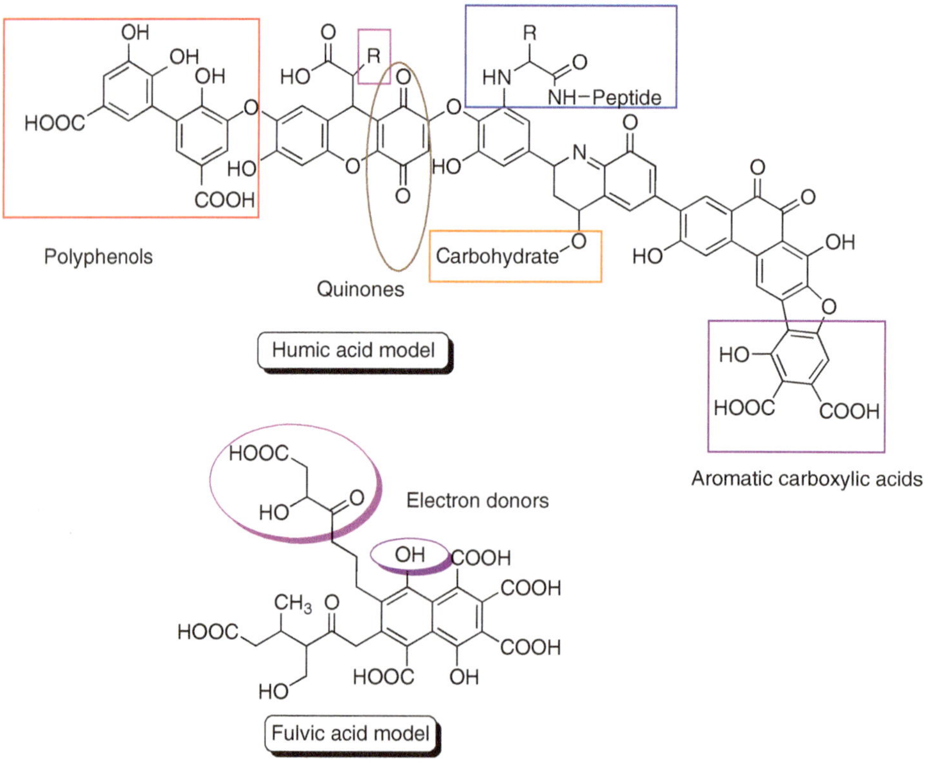

FIGURE 1.16   Schematic representation of humic and fulvic acids

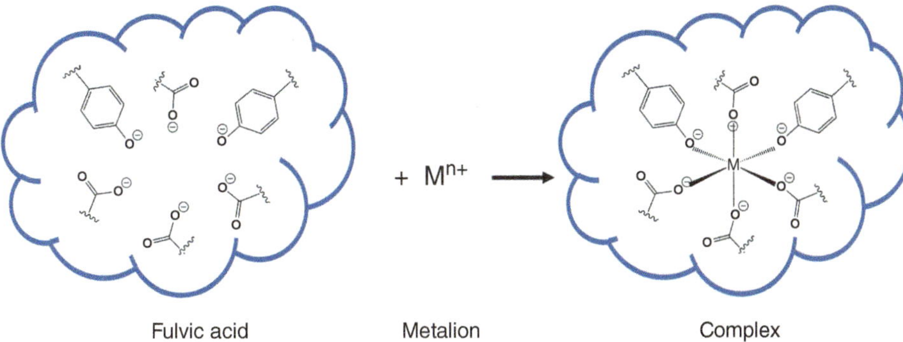

FIGURE 1.17   Schematic representation of metal complexes with humic substances

2. **Amino acids and peptides:** Amino acids and peptides, the building blocks of proteins, can also form complexes with metals. They have functional groups like amino and carboxyl groups that can coordinate with metal ions.

3. **Organic acids:** Various organic acids, such as acetic acid, citric acid and oxalic acid, can form complexes with metals. These acids contain carboxyl groups that can chelate with metal ions.

4. **Carbohydrates:** Some carbohydrates, like glucose and gluconic acid, can bind with metals. Hydroxyl groups present in these compounds can coordinate with metal ions.

5. **Siderophores:** Siderophores are specialised organic molecules produced by microorganisms to scavenge and transport essential metals like iron. They have high affinity for metal ions and form complexes to facilitate metal uptake. The structures of siderophores can vary significantly, but they typically contain polar functional groups that form strong complexes with iron(III) ions, such as carboxylate, phenolate, catecholate and hydroxamate, groups (Figure 1.18).

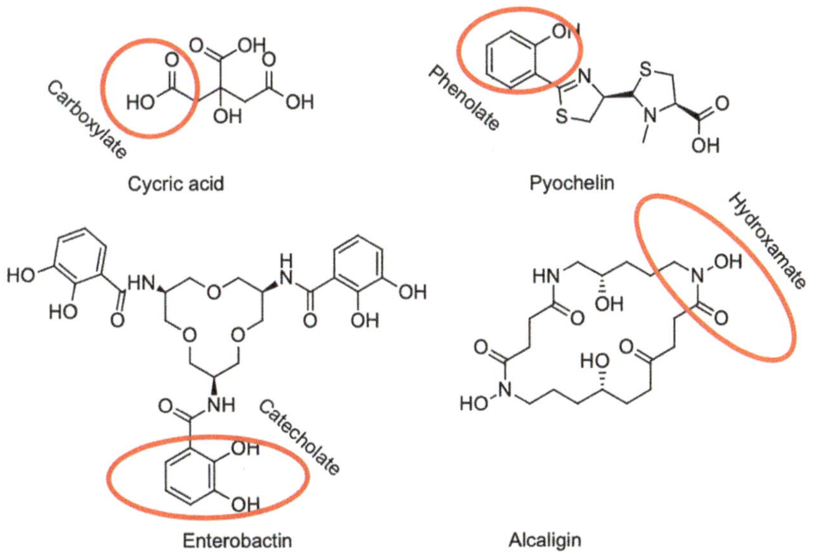

FIGURE 1.18   Examples of siderophores

6. **Phytochelatins:** Phytochelatins are peptides synthesised by plants and some microorganisms in response to heavy metal stress. They bind tightly to metal ions, particularly heavy metals like cadmium, copper and zinc, forming stable complexes.

   Some organic ligands can increase the solubility and mobility of metal ions, making them more readily available for uptake by organisms, or, on the contrary, decrease their solubility, forming less soluble complexes. They can also reduce the toxicity of certain metals by forming stable complexes that are less toxic than the free metal ions.

   For example, the complexation of copper ions with organic matter can increase their solubility and mobility in soil and water, making them more bioavailable to aquatic organisms. On the other hand, the complexation of iron ions with hydroxide ions can cause them to precipitate out of solution, reducing their bioavailability.

   The presence of complex metal ions in natural waters can also affect water quality and human health. For example, high levels of complex mercury in fish can pose a health risk to humans who consume them. Complexation occurs when the solvated metal ion forms a complex with other chemical species in the water, such as organic matter or other metal ions. This can influence the bioavailability and toxicity of the metal ion and also affect its mobility and reactivity in the water.

3) **Organometallic compounds:** These compounds contain at least one chemical bond between a carbon atom and a metal atom. Organometallic compounds can be found in natural waters through both natural and anthropogenic sources. These compounds are formed when organic matter, such as decaying plants or animal waste, reacts with metal ions in the water. Some examples of natural organometallic compounds include:

   a) **Methylmercury:** This is a naturally occurring organometallic compound that is formed when inorganic mercury is methylated by certain bacteria in aquatic sediments (Figure 1.19). It can accumulate in fish and other aquatic organisms and pose a potential risk to human health when consumed.

   b) **Organotin compounds:** Examples of organic compounds that contain metal atoms and are introduced into natural waters through human activities are organotin compounds. The carbon-tin bond in such compounds is relatively stable, and they may remain in water for long periods of time. Tributyltin (Figure 1.20) and other organotin compounds have been used as biocides in antifouling paints, but they can also be toxic to marine organisms.

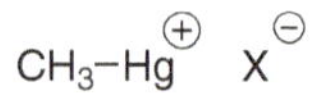

$$CH_3\!-\!\overset{\oplus}{Hg} \quad \overset{\ominus}{X}$$

FIGURE 1.19 Structure of methyl mercury

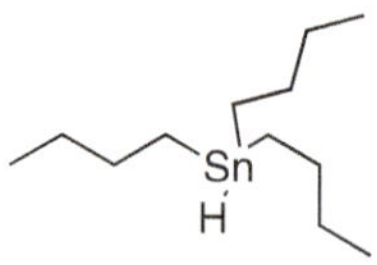

FIGURE 1.20 Structure of tributyltin

Topics of Interest: pH Scale

A measure of the acidity or alkalinity of water is described in terms of pH (pH stands for the so-called hydrogen *potential*). pH is a logarithmic scale, from 1 to 14, so it is not linear. The mathematical expression of pH is as follows (Eq. 1.4):

$$pH = -\log_{10}\left[H^+\right] = \log_{10} 1/\left[H^+\right] \tag{1.4}$$

- Seven corresponds to neutral solutions
- Values below seven indicate acidity because $[H^+] > [OH^-]$, which increases as the number decreases.
- Values above seven indicate alkalinity, which increases as the number increases, with fourteen being the most alkaline.
- Scheme examples include popular items that are acidic or basic (Figure 1.21).

In nature, the pH values of water are a consequence of the dissolved species: acids, bases or hydrolyzable salts.

In nature, water shows pH values different from neutrality:

- Seawater has a pH range between 7.5 and 8.5, depending on the local conditions.
- Freshwater lakes, ponds and streams usually have a pH of between 6 and 8, depending on the surrounding soil and bedrock
- Rainwater in the absence of pollutants has a pH of about 5.6 because of the interaction of raindrops with carbon dioxide molecules in the atmosphere according to the equilibrium represented in Eq. 1.5:

$$CO_2 + H_2O \rightleftharpoons H_2CO_3 \tag{1.5}$$

Carbonic acid is a weak acid that produces protons in water.

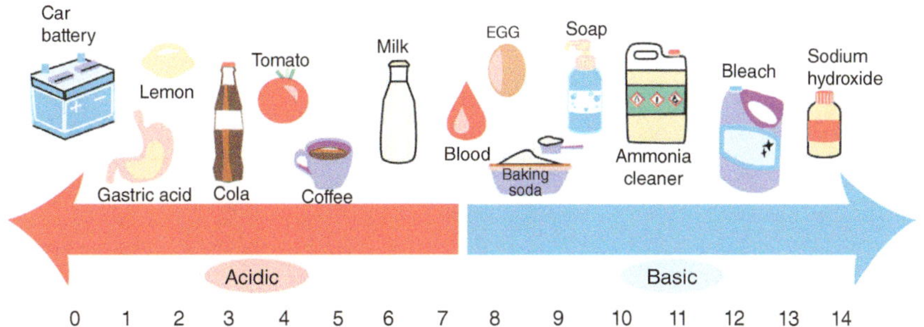

FIGURE 1.21   Acidic and basic items

1.11   Colloids in Natural Waters

A colloid is a mixture of at least two substances, and, as in solutions, all combinations of the phases (solid, liquid and gas) can occur. The major component is nominated as a *dispersed phase* and the minor component or components as a *dispersant phase*. They are also known as colloidal dispersions.

The dispersant phase is formed by particles ranging from 1 to 1,000 nm in diameter and is dispersed or suspended over the dispersant phase. If the size of the particles is smaller, the mixture is considered a solution, and if they are larger, the mixture is considered a suspension.

TABLE 1.8   Types of binary colloidal dispersions

Dispersed phase	Dispersant phase	Denomination	Example
Liquid	Gas	Liquid aerosol	Mist, spray
Solid	Gas	Solid aerosol	Smoke
Gas	Liquid	Foam	Foam on beer, bubbles in sparkling wine, whipped cream
Liquid	Liquid	Liquid emulsion	Mayonnaise, milk and vinaigrette
Solid	Liquid	Sol/gel	Whitewash, skimmed milk, paints, puddings, jelly and mud
Gas	Solid	Solid foam	Meringue, ice cream, bread and sponge cake
Liquid	Solid	Solid emulsion	Butter, cheese
Solid	Solid	Solid sol or solid solution	Some metal alloys and gemstones

Colloids are classified into four groups: Sols, Emulsions, Foams, and Aerosols There are only eight possible phase combinations (Table 1.8).

Key properties of colloids are:

- They are a heterogeneous mixture.
- The size of the particles in a colloid is too small to be detected by the naked eye.
- Colloids are big enough to scatter a beam of light passing through them and make its path visible.
- Particles are not able to settle down by gravity.
- The mixture of phases cannot be separated by conventional filtration, but centrifugation or separation through certain membranes can be used to isolate the colloidal particles.

Colloids can be commonly found in natural waters, including rivers, lakes, oceans and groundwater. Some examples of colloids in natural waters are given in the following list:

1. **Suspended inorganic particles:** Rivers and streams often carry suspended particles, including certain minerals such as iron, manganese and silica and sediments eroded from the land, transported by rivers and deposited in lakes or oceans, which can exist in colloidal form in natural waters, affecting water clarity and light penetration and having ecological consequences.

2. **Phytoplankton:** Phytoplankton are microscopic, single-celled algae that inhabit marine and freshwater environments. They are colloidal in nature and are responsible for the greenish appearance of water bodies during algal blooms. Phytoplankton are an essential part of the aquatic food chain.

3. **Organic matter (OM):** Natural waters may contain organic matter, formed by a complex mixture of organic compounds derived from decaying plants, microorganisms and other organic sources. It consists of colloidal-sized particles that contribute to the colour and overall composition of water.

4. **Colloidal contaminants:** Natural waters can also contain colloidal contaminants, such as heavy metals, organic pollutants and pathogens. These contaminants can attach to colloidal particles and be transported in water bodies, posing risks to aquatic ecosystems and human health.

The presence of colloids in natural waters has implications for water quality, aquatic ecosystems and various environmental processes. Understanding their behaviour and interactions is essential for studying water pollution, sediment transport, nutrient cycling and overall ecosystem dynamics.

Figure 1.22 shows a comparison of sizes of soluble and insoluble waterborne particles.

Many biological fluids are colloids and natural waters are able to form colloids. These mixtures play an important role in the transfer of many substances, such as nutrients from plants and pollutants in the environment, over short and long distances.

The chemical-physical properties of water with environmental transcendence are summarised in Table 1.9.

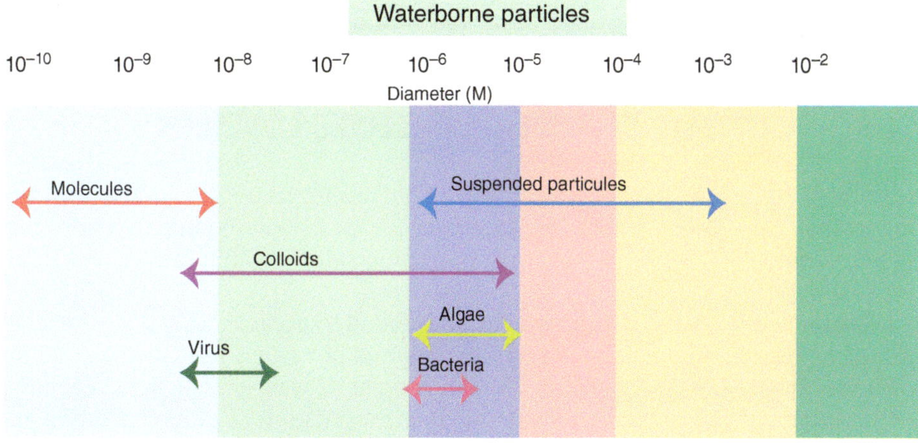

FIGURE 1.22   Waterborne particles

TABLE 1.9   **Chemical-physical properties of water with environmental transcendence**

Property	Highlights	Environmental consequences
Density	Varies with temperature but not linearly, showing the higher value at 3.98 °C, not in the melting point	Stational stratification in lakes takes place
Boiling and melting points	The melting point of ice is 0 °C (273 K) at standard pressure; however, pure liquid water can be supercooled well below that temperature without freezing if the liquid is not mechanically disturbed. It can remain in a fluid state down to its homogeneous nucleation point of about 231 K (−42 °C)	Water remains liquid in most of ecosystems

(Continued)

TABLE 1.9 **(Continued)**

Property	Highlights	Environmental consequences
Heat capacity	Shows the highest value with the exception of ammonia	It is able to moderate temperature changes close to water bodies
Evaporation heat	One or the higher Kwon value	It is a determining factor in energy transfer procedures in atmosphere, lakes and oceans
Surface tension	High value	Water is responsible for the shapes of water drops and dew
Solvent properties	Excellent solvent for most ionic compounds and polar molecules	It is a key factor in the transportation of many substances in hydrologic and biological cycles

1.12 Composition of Natural Waters

Natural waters refer to bodies of water that occur naturally in the environment, including lakes, rivers, streams and oceans. These bodies of water contain a variety of different substances, including salts, organic matter and living organisms.

Seawater is a type of natural water that is found in oceans and seas. It is characterised by its high salt content, which is composed of various ions such as sodium, chloride, magnesium and calcium. The salt content of seawater can vary depending on location and environmental factors such as evaporation rates, precipitation and freshwater inputs from rivers and streams. Additionally, there may be other trace elements and compounds present in seawater in smaller quantities. Typical concentrations of inorganic ions in oceans are shown in Table 1.10. The values given in tables

TABLE 1.10 **Typical concentrations of inorganic ions in oceans in mg/L**

Anions	Concentration	Cations	Concentration
Chloride (Cl^-)	19,000–20,000	Sodium (Na^+)	10,500–11,000
Sulfate (SO_4^{2-})	2,600–2,800	Magnesium (Mg^{2+})	1,250–1,350
Bicarbonate (HCO_3^-)	140–150	Calcium (Ca^{2+})	400–420
Carbonate (CO_3^{2-})	1–20	Potassium (K^+)	380–390
Bromide (Br^-)	65–70	Strontium (Sr^{2+})	8–9
Fluoride (F^-)	1.3–1.4	Barium (Ba^{2+})	0.01–0.06
Nitrate (NO_3^-)	0.05–0.15		

TABLE 1.11	Estimation of the total salinity (in mg/L) of several seas and oceans	
Sea/ocean	**Total salinity (mg/L)**
Baltic Sea	7,000–8,000
Black Sea	18,000–22,000
Mediterranean Sea	38,000–40,000
Red Sea	40,000–41,000
Persian Gulf	40,000–45,000
Arabian Sea	35,000–41,000
Indian Ocean	34,000–39,000
Atlantic Ocean	35,000–37,000
Southern Ocean	34,000–37,000
Pacific Ocean	34,000–37,000

are approximate and can vary depending on location and other factors. Table 1.11 compares the total salinity (in mg/L) of several seas and oceans.

Freshwater

Freshwater is a term used to describe any naturally occurring water that has a low salt content and is suitable for drinking, irrigation and other uses. It is typically found in lakes, rivers, streams and groundwater aquifers, and it is essential for the survival of many plants and animals, including humans. The major cations in typical freshwater are calcium (Ca^{2+}), magnesium (Mg^{2+}), sodium (Na^+), ammonium (NH_4^+) and hydrogen (H^+). The most important of the negatively charged ions (or anions) are sulfate (SO_4^{2+}), chloride (Cl^-) and nitrate (NO_3^-).

Most of the water present on our planet corresponds to oceans and seas, so 97.5% of the water is saltwater, and only 2.5% of the total amount corresponds to freshwater (Figure 1.23). However, the largest reserves of freshwater are glaciers and ice sheets,

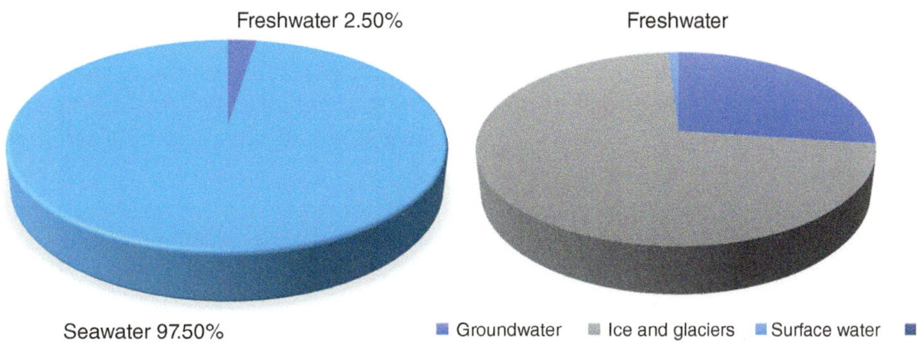

FIGURE 1.23 Freshwater and saltwater averages

with more than 24 million km³, 68.6% of the total ice mass, though this is only 1.74% of Earth's water.

Groundwater is the second-largest freshwater reserve on the planet, though only 45% of it is freshwater, whereas the other 55% is more or less salty water. Still, groundwater accounts for 30% of global freshwater reserves. The lakes and wetlands account for only 0.014% of the total water on Earth, and nearly half their volume is saltwater, so the volume of all freshwater lakes on Earth is only around 0.29% of the reserves.

1.13   Lithosphere: The Soil

Soil is the thin layer of material covering the earth's surface that is composed of mineral particles, organic matter, water, air and living organisms. It is a complex mixture of inorganic and organic materials that support plant growth and provide habitat for a wide range of organisms. Soil is formed through the weathering of rocks and minerals, and its composition and characteristics vary depending on factors such as climate, topography, vegetation and time. Soil serves as a critical resource for agriculture, forestry, construction and many other human activities and plays a vital role in the Earth's ecosystem by supporting biodiversity, regulating the water cycle and storing carbon.

Soil can be characterised by a variety of physicochemical properties such as composition, texture, structure, density, porosity, consistency, temperature, colour, resistivity and acidity.

1.14   Composition of Soil

Soil is a complex, heterogeneous material with great diversity, formed by solid particles, water and air. Solid particles can be inorganic or organic. However, most soils show some common major components: minerals, water, gases and organic material.

1. **Mineral particles:** Mineral particles in soil are formed by the disintegration and decomposition of rocks. Minerals constitute the largest component of the soil, forming about 45%. They are formed by silicates, oxides and hydroxides of iron, aluminium, manganese, etc., forming sand, silt and clay.

2. **Water:** It is the second basic component of the soil, with a variability of 2–50%. Soil shows pores between solid particles, constituting a thick and extended network that allows water to move in the ground. Soil water usually contains dissolved salts and other chemical substances.

3. **Gases:** Soil has many cavities and holes capable of retaining gases. The average amount of gases retained in soil is very variable (2–50%), and they are a consequence of the contact with atmosphere and reactions of plant roots, microbes and other chemicals in the soil.

4. **Residual organic matter:** It is produced by the decay of plant residues, animal remains and microbial tissues and forms approximately 1–5% of soil. It has a very high water-holding capacity.

5. **Living organisms:** Both as big as worms and as small as insects and microorganisms.

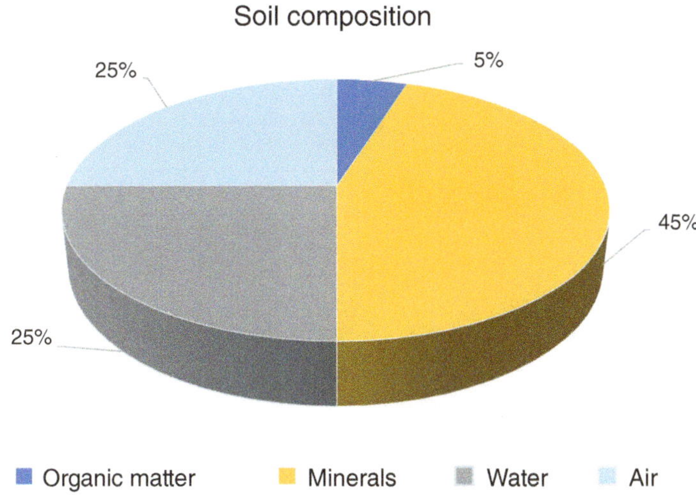

FIGURE 1.24 Soil composition

The sum of the organic material (residual organic matter and living organisms) in soil is called humus. Humus is a basic component of soil that improves soil structure and provides plants with water and nutrients.

The average of each component depends on several factors, such as the quantity of vegetation, soil compaction and the amount of water available. To have healthy soil, the amount and proportion of every component must be equilibrated and sufficient to promote and sustain plant and microorganism life. A good approximation of healthy soil is described in Figure 1.24.

1.15 Soil Texture

Soil texture refers to the composition of soil based on the sizes of the mineral particles that make it up. There are three main types of mineral particles that make up soil: sand, silt and clay.

1. **Sand particles** are the largest of the three, with diameters ranging from 0.05 mm to 2.0 mm.

2. **Silt particles** are smaller, ranging from 0.002 to 0.05 mm in diameter.

3. **Clay particles** are the smallest, with diameters less than 0.002 mm.

The relative proportions of sand, silt and clay particles in a soil determine its texture. Soils can be classified as sandy, loamy or clayey, depending on the relative proportions of these particles. A soil's texture can have a significant impact on its properties, such as water-holding capacity, nutrient retention and aeration, and can therefore affect plant growth and overall soil health.

Soils may be described according to the proportion of these three components using a triangular diagram (Figure 1.25).

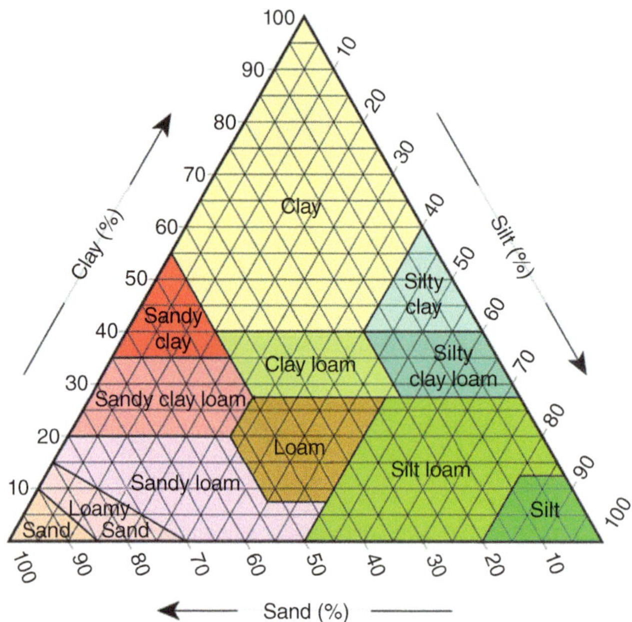

FIGURE 1.25   Soil classification

1.16   Soil Profiles and Soil Horizons

Soil is not a homogeneous material. It is formed by different layers, from the surface to the so-called underlying rock. The so-called profile gives information about the different layers. It is obtained by digging a hole from the surface to the underlying rock. The soil profile is a fingerprint that differs from other soil samples based on factors like its colour, texture, structure and thickness, as well as its chemical composition. Each layer of a soil profile is referred to as a soil horizon, identified by a capital letter. *O* corresponds to **organic matter**, and *A* is the closer layer to the surface (Top soil). Moving deeper into the layers, horizons are nominated as *B (Subsoil)* and *C (Parent material)*. *A*, *B* and *C* represent the three main horizons (Figure 1.26).

1.17   Chemical Properties of Soil

According to FAO[7] criteria, the main chemical properties of soils are as follows:

A. **pH** is a measure of hydrogen ion concentration in an aqueous solution. In the case of soils, it ranges from 3.5 (very acid) to 9.5 (very alkaline), depending on soil composition[8]. pH is a clue about the presence or availability of certain ions. For example, very acidic soils (<5.5) tend to have toxic amounts of aluminium and manganese. In cases of high alkalinity (>8.5), it tends to disperse.

---

[7] The Food and Agriculture Organisation (FAO) is a specialised agency of the United Nations that leads international efforts to defeat hunger.

[8] A common procedure is to prepare mixtures of soil-water (1, 1) and soil-$CaCl_2$ solutions and use a chemical kit or a pH metre.

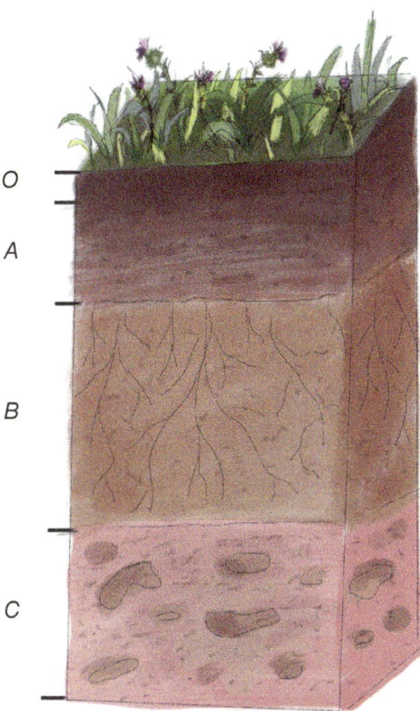

FIGURE 1.26   Soil horizon

B. **Cation exchange capacity (CEC):** This depends on pH and is a measure of the maximum quantity of total cations, such as Ca^{2+}, Mg^{2+}, Na^+, K^+, NH_4^+ etc., that soil is capable of holding and is available for exchange with ions dissolved. CEC is a measure of fertility, nutrient retention capacity and the capacity to protect groundwater from cation contamination. It is expressed as centimol of hydrogen per kg (cmolc/kg or 100 meqc/100 g). Most of the soil's CEC occurs in clay and humus.

C. **Base saturation percentage:** Some cations and anions are responsible for pH values (named acid and base-forming). The fraction of the base-forming cations that occupy positions on the soil colloids is called the *base saturation percentage*. In neutral soil (pH 7), base saturation is 100%, and there are no hydrogen ions available on the colloids. Base saturation is almost in direct proportion to pH, and with the exception of calculating the amount of lime needed to neutralise an acid soil, it is of little use.

D. **Plant nutrients:** There are sixteen essential nutrients for plant growth and living organisms in the soil. They are classified into two different categories, but this classification does not mean that one nutrient element is more important than another, just that they are required in different quantities.

   a. Macronutrients are those elements required in greater quantities. They include carbon (C), oxygen (O), hydrogen (H), nitrogen (N), phosphorus (P), potassium (K), calcium (Ca), magnesium (Mg) and sulfur (S).

   b. **Micronutrients:** They are needed in smaller amounts than macronutrients, but they are still crucial for plants. They include iron (Fe), zinc (Zn), manganese (Mn), boron (B), copper (Cu), molybdenum (Mo) and chlorine (Cl). These elements are in soil as cation or anion forms, and they are taken up from the dissolved soil.

E. **Organic soil carbon** is produced by organic matter decay from roots, leaves, etc. Some carbon is transferred again to the atmosphere as carbon dioxide by aerobic decomposition, respiration of microorganisms and anaerobic decomposition of organic matter as methane. Soil organic carbon improves the mechanical properties of the soil, increases the CEC and the water-holding capacity, and contributes to the structural stability of clay soils. Soil organic matter improves soil fertility because it is a source of carbon and holds a great proportion of nutrients necessary for plant growth. On the other hand, organic matter prevents nutrient leaching, is integral to the organic acids that make inorganic compounds available to plants and also protects soil from strong changes in pH because of its buffer capacity. Organic soil carbon is a major factor in its overall health.

F. **Soil nitrogen:** Nitrogen is the most important nutrient for plants. The two major nitrogen species in soil are the cation ammonium, NH_4^+ and the anion nitrate, NO_3^-, but they are also available in gas compounds in the soil as nitrogen oxides (N_2O, NO and NO_2), ammonia (NH_3) and molecular nitrogen (N_2). However, these quantities are very small and difficult to detect. Nitrogen content is a key data point for soil fertility.

G. **Soil salinity:** Corresponds to the total salt content of soil. The most common salts are formed by cations such as sodium, calcium, magnesium and potassium, with the corresponding anions chloride, sulfate and carbonate, to maintain electroneutrality. When concentration of salts is too high, salinisation occurs naturally in cases of seawater intrusion or is promoted by irrigation practices that are carried out without due attention to drainage and leaching of the salts. As soil salinity increases, salt effects can result in degradation of soils and vegetation.

H. **Calcium carbonate content:** Calcium carbonate ($CaCO_3$) concentrations are not very soluble salts that may occur in various forms and concentrations in soils. $CaCO_3$ in moderate amounts is favourable for soil structure and is often used to correct the pH of acidic soils. However, when the concentration of calcium in the soil exceeds the capacity of the soil to absorb it, it binds with other elements and forms insoluble compounds that are difficult for plants to absorb. Excess amounts of calcium may restrict the availability of other nutrients such as phosphorus, boron and iron to plants.

I. **Calcium sulfate (gypsum) content:** Calcium sulfate ($CaSO_4.2H_2O$) is a material commonly found in layered sedimentary deposits. In soils, it is very common in the driest climates. In case of soils with high quantities of gypsum, plant growth may be interfered with. The natural vegetation on soils with high gypsum used to be quite poor and dominated by xerophytic shrubs and trees and/or ephemeral grasses.

1.18 Biosphere

The term biosphere comes from Greek *bios*, life and *sphaira*, sphere.

The biosphere is the part of the Earth's surface and atmosphere where life exists. It includes all living organisms and the environments they inhabit, such as forests, oceans, grasslands and deserts. The biosphere is interconnected with the geosphere, hydrosphere and atmosphere, and the interactions between these spheres shape the Earth's climate, weather patterns and ecosystems.

The biosphere is home to a vast array of living organisms, from microscopic bacteria to towering trees and whales. These organisms interact with each other and with their environment in complex ways, forming ecosystems that are adapted to their specific habitats. These ecosystems provide important services to humans, such as food, clean air, water and raw materials. The specific study of the biosphere is known as ecology, and it is a critical field for understanding how the Earth's ecosystems function and how they are impacted by human activities.

1.19 Anthroposphere

The anthroposphere refers to the part of the Earth's environment that has been created or significantly altered by human activities. It includes all human-made structures such as buildings, roads, cities and other infrastructure, as well as the social, economic and cultural systems that humans have developed. The anthroposphere is one of the four major components of the Earth's system, with strong interactions with the atmosphere, hydrosphere and biosphere. These interactions can have both positive and negative effects on nature.

a) Positive interactions between the anthroposphere and nature include conservation efforts such as protected areas and restoration of degraded landscapes, sustainable agriculture and renewable energy sources. These efforts aim to mitigate the negative impacts of human activities on the environment and promote the conservation and sustainable use of natural resources.

b) Negative interactions between the anthroposphere and nature are also widespread. Examples include deforestation, urbanisation, pollution, overfishing and climate change. These activities have a detrimental impact on the natural environment, leading to habitat loss, biodiversity loss, water and air pollution and other negative impacts on ecosystem functioning. As human populations continue to grow and develop, the anthroposphere has become increasingly important and has had significant impacts on other components of the Earth's system, leading to environmental and ecological challenges.

Part II: Biogeochemical Cycles

1.20 Biogeochemical Cycles

Biogeochemical cycles are the natural processes by which essential elements and nutrients, such as carbon, nitrogen, phosphorus and sulfur, are exchanged between the living and non-living components of the Earth's ecosystems. These cycles involve the movement of substances through the biotic and abiotic components of the Earth's systems, including the atmosphere, hydrosphere, lithosphere and biosphere.

There are several biogeochemical cycles that operate on Earth, including:

1. **The carbon cycle:** The carbon cycle describes the processes by which carbon moves through the atmosphere, oceans and terrestrial ecosystems. This cycle includes processes such as photosynthesis, respiration, decomposition and fossil fuel combustion.

2. **The nitrogen cycle:** The nitrogen cycle describes the processes by which nitrogen is transformed and recycled in the biosphere. This cycle includes processes such as nitrogen fixation, nitrification, denitrification and assimilation.

3. **The phosphorus cycle:** The phosphorus cycle describes the movement of phosphorus through the lithosphere, hydrosphere and biosphere. This cycle includes processes such as weathering, erosion and assimilation.

4. **The sulfur cycle:** The sulfur cycle describes the movement of this element through the atmosphere, hydrosphere and biosphere. This cycle includes processes such as volcanic activity, weathering and biological assimilation.

These biogeochemical cycles are crucial for maintaining the balance of nutrients and elements required for life on Earth. Human activities, such as deforestation, pollution and climate change, can disrupt these cycles and have significant impacts on the Earth's ecosystems.

1.21　The Carbon Cycles

Carbon is the sixth most abundant element in the universe, and it is the backbone element for life on Earth. It is present in all living organisms, and it is part of the carbon-bearing reservoirs. Most of Earth's carbon is stored in rocks, and it is present in oceans, the atmosphere, soil and the living beings that they contain.

The carbon cycle consists basically of a series of cyclic movements of carbon compounds associated with chemical transformations between each reservoir. Scientists have considered two carbon cycles depending on whether they have slow or fast components.

Short-Term Carbon Cycle

The short-term carbon cycle (Figure 1.27) describes the flux of carbon between soil, atmosphere, oceans and living organisms in the short term if compared with geological transformations. The fast carbon cycle shows fluxes of carbon compounds between different spheres, with values estimated in gigatons of carbon per year.

In the short-term carbon cycle, the displacement of this element across reservoirs takes a relatively short time (from minutes, hours, days, months, to several years) and involves the changes that occur within the atmosphere, terrestrial ecosystems and marine ecosystems. In contrast, a long-term carbon cycle may take thousands to millions of years to occur. The excess carbons from the short-term cycling are stored in the 'long-term' reservoir until they are removed after a long time.

An essential part of the fast carbon cycle corresponds to biochemical transformations in which carbon dioxide is involved. Carbon dioxide regulates the key movement of carbon between the ocean, land and atmosphere.

The carbon cycle is divided into the following four steps:

1. **Entry of carbon into the atmosphere as carbon dioxide:** The carbon cycle starts with the incorporation of carbon dioxide and volatile organic compounds into the atmosphere. Carbon dioxide is directly emitted by respiration, aerobic organic matter decomposition, combustion or the transformation of volatile organic compounds by photochemical oxidations and converted into carbon dioxide in the atmosphere.

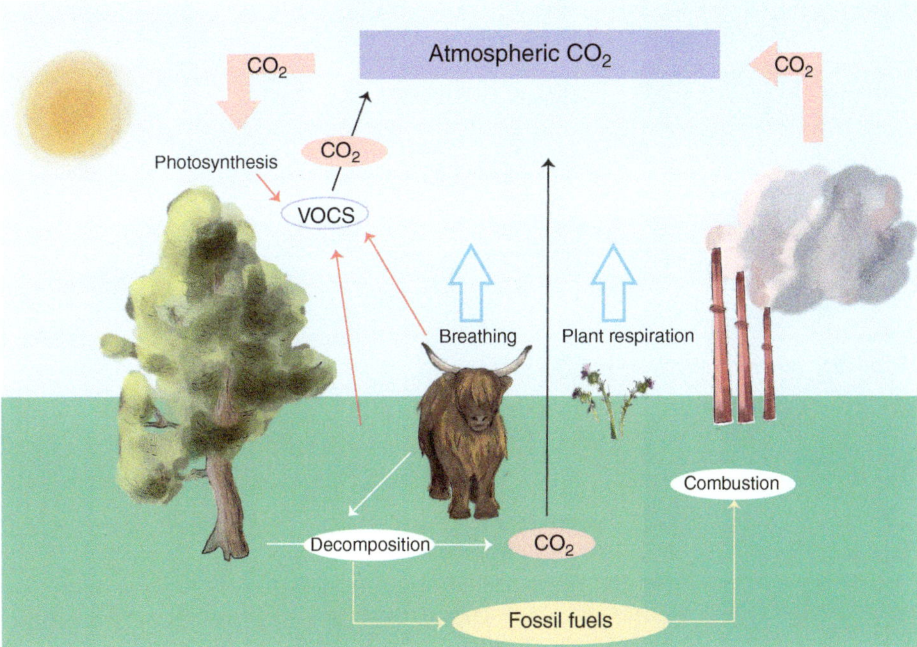

FIGURE 1.27   Fast carbon cycle

2. **Carbon dioxide absorption by producers:** The next step is the absorption of carbon dioxide by plants and algae (*producers*) in the photosynthetic process to transform carbon dioxide and water into carbohydrates and oxygen, as shown in Eq. 1.6.

$$CO_2 + H_2O + energy \rightarrow Carbohydrates + O_2 \tag{1.6}$$

3. **Passing of the carbon organic compounds in the food chain:** Following the absorption by producers, carbon compounds from plants are incorporated into the food chain when animals consume these plants and other animals.

4. **Return of carbon to the atmosphere:** The last step is the return of carbon to the atmosphere:

   a. The *decomposers* (bacteria and fungi) provide the degradation of the dead bodies of plants and animals. This is because the process of decomposition gives off carbon dioxide as a by-product.

   b. By spontaneous combustion processes of biomass or volcanic activity.

When both production and consumption are balanced, an equilibrium situation takes place. In the case of a supplementary addition of carbon dioxide, the balance is broken down, for example, by fossil fuel burning and supplementary carbon dioxide rises in the atmosphere due to anthropogenic causes.

Topics of Interest: Long-Term Carbon Cycle

It takes place over hundreds of millions of years and involves five key steps in the movement of carbon around the environment (Figure 1.28).

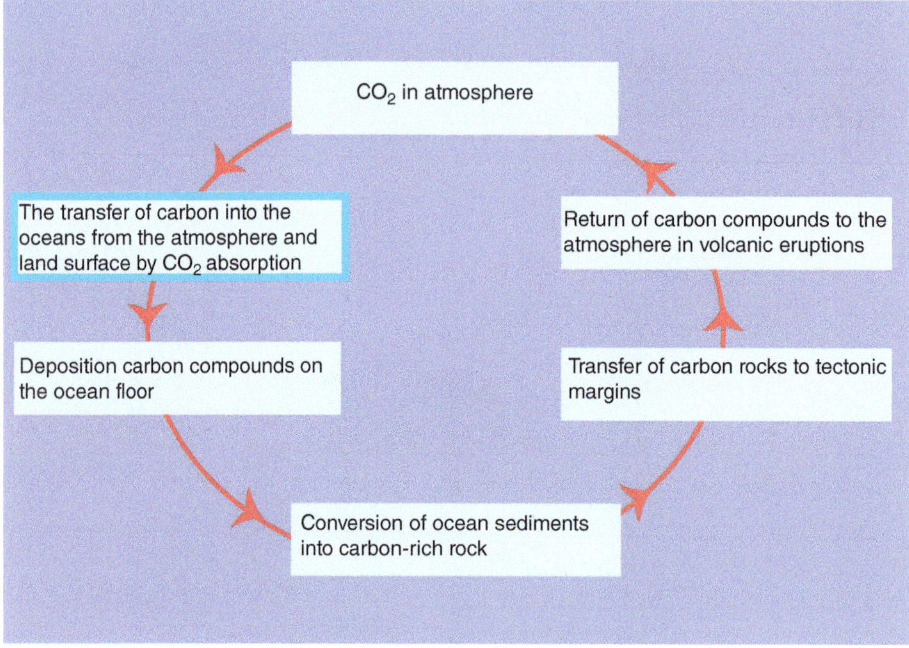

FIGURE 1.28 Long-term carbon cycle

1. **The transfer of carbon into the oceans from the atmosphere and land surface by direct carbon dioxide absorption:** As part of the atmosphere-ocean exchange, it is supplemented by the erosion of carbon-rich terrestrial surfaces as naturally acidic rainfall dissolves surface rocks and transfers soluble bicarbonate compounds, via rivers, to the sea.

2. **Deposition of carbon compounds on the ocean floor:** Marine plants (including phytoplankton) absorb carbon dioxide, and marine creatures take in carbon to construct skeletons and shells. Phytoplankton are consumed by zooplankton, and their carbon-rich excrement falls to the ocean floor. The skeletal and shell remains of marine creatures also fall into the seabed.

3. **Conversion of ocean sediments into carbon-rich rock:** On continental shelves, carbon-rich accumulations of deposits may be converted into carbon-rich rocks (such as chalk and limestone) or become contained as concentrations within sandstones and shales to form organic deposits, some of which become fossil-fuel reserves in time. This process of sedimentary rock formation is called lithification.

4. **Transfer of carbon rocks to tectonic margins:** As sedimentary rocks are created by heat and pressure over millions of years, they are also moved in the direction their crustal plate is moving. If they eventually become collision margins, they may be uplifted to become surface mountain ranges (as in the Himalayas). The carbon-rich strata may then be exposed to weathering and erosion and return to the ocean as eroded carbonate rocks.

5. **Return of carbon compounds to the atmosphere in volcanic eruptions:** At subduction zones, carbon-rich rocks may be ejected at the surface from volcanic eruptions, usually in the form of gaseous compounds, into the atmosphere. Here, carbon dioxide contributes to the formation of carbonic acid in clouds, which then begins the process of the solution of surface rocks and the start of the terrestrial component again or being absorbed by ocean surfaces for the marine component.

1.22 The Oxygen Cycle

Oxygen is one of the most abundant elements on Earth and represents a large portion of each main reservoir. It is present as an element or combined with other elements to form water, silicates, carbonates, organic compounds, etc. The oxygen cycle (Figure 1.29) describes oxygen displacement through the atmosphere, hydrosphere, biosphere and soil. The largest reservoir of oxygen is the lithosphere. This cycle is very closely related to the carbon cycle.

The oxygen cycle has three steps:

1. **Oxygen generation:** It is generated in two ways:
   a. Green plants, through photosynthesis, produce carbohydrates from carbon dioxide, water and sunlight. Oxygen molecules are formed as a by-product of this process.
   b. **Photochemical production:** Some oxygen gas is produced both when sunlight reacts with water vapour in the atmosphere and by photolysis of O_3.

2. **Oxygen consumption:** Oxygen is consumed
   a. **Respiration:** Aerobic organisms consume oxygen in respiration from the atmosphere and exhale carbon dioxide back into the atmosphere.
   b. **Organic matter aerobic decomposition:** The dead living beings decay into the ground, and the organic matter formed by carbon, oxygen, water and other elements is returned back into the soil and air. This process is carried out by decomposers (fungi, bacteria and some insects). The entire process requires oxygen and produces carbon dioxide.

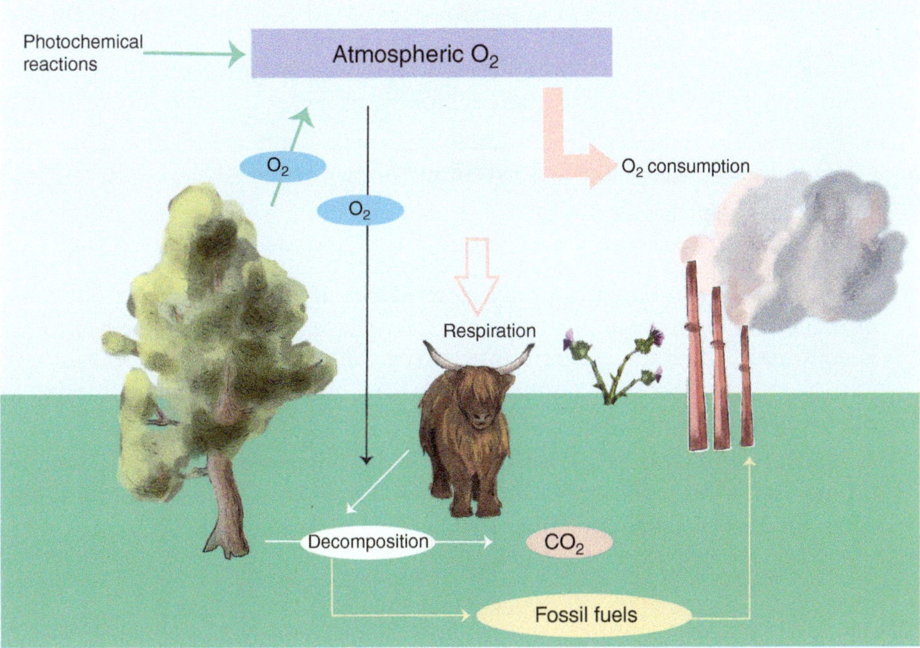

FIGURE 1.29 The oxygen cycle

   c. Combustion: It occurs when any of the organic materials, such as biomass, fossil fuels and other materials, are burned. The presence of oxygen is required and releases carbon dioxide into the atmosphere.

   d. Oxidation and rusting: Many substances are able to spontaneously combine with oxygen in the environment (oxidation reactions). In this process, metals like iron or alloys rust when they are exposed to moisture and oxygen for a long period of time, and new compounds of oxides are formed by the combination of oxygen with the metal.

3. **Return to atmosphere:** Oxygen goes back to atmosphere combined with carbon as carbon dioxide by respiration or combustion processes; so, it is again used by the plants during photosynthesis.

1.23   The Nitrogen Cycle

Nitrogen is probably the chemical element that shows the greatest diversity in nature, forming organic and inorganic compounds in the atmosphere, soil and living organisms. It is the most abundant element in the atmosphere as a diatomic molecule (N_2), a component of biomolecules such as proteins or nucleic acids and an essential nutrient for plants. Atmospheric nitrogen can barely be used directly by plants, so it must be transformed for biological use.

The nitrogen cycle (Figure 1.30) is the biogeochemical cycle by which nitrogen is transformed into several chemical compounds while it moves in the atmosphere and marine and terrestrial ecosystems. But activities such as fossil fuel combustion, the massive use of nitrogen fertilisers and the release of nitrogen compounds in wastewater have altered the global nitrogen cycle.

The nitrogen cycle consists of four main steps:

1. **Nitrogen fixation**[9]: The atmospheric nitrogen, N_2, is converted into ammonia or ammonia salts (NH_4^+) by certain microorganisms:

   a. Bacteria like Azotobacter and Rhizobium have symbiotic relationships with leguminous plants which are harboured in their roots

   b. Free anaerobic bacteria

   c. Certain Algae

2. **Nitrification:** Ammonia can only be used directly by some plants. For most of them, it is highly toxic. Nitrification takes place in soil when fixed nitrogen is converted by certain bacteria into nitrite ion (NO_2^-) and then into nitrate ion (NO_3^-). The bacteria responsible for that transformation are known as nitrifying bacteria.

3. **Assimilation:** Plants take up from soil several nitrogen compounds such as nitrate, nitrite, ammonia and ammonium and transform them into other molecules, for example, plant proteins.

---

[9]This situation has dramatically changed since the Haber-Bosch procedure, which is able to obtain ammonia from molecular nitrogen by catalytic transformation. About 30% of the total fixed nitrogen is produced industrially using this reaction as a platform to obtain artificial nitrogen fertilisers.

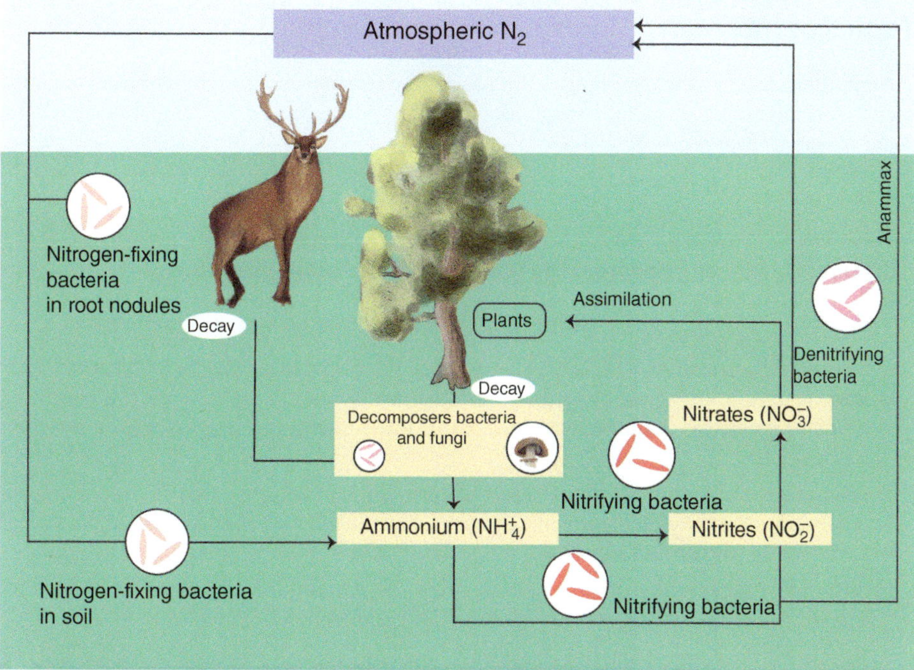

FIGURE 1.30   Nitrogen cycle

4. **Ammonification or mineralisation:** When plants or animals die or produce waste, nitrogen is part of organic molecules. Immediately, bacteria or fungi convert organic nitrogen into ammonium (NH_4^+).

5. **Denitrification:** This is the reverse of the nitrification process. Denitrification takes place in the deep layers of wet soil in an environment with very low concentrations of oxygen. Specific soil bacteria, such as Pseudomonas and Clostridium, in anaerobic conditions use the nitrate as an electron acceptor instead of oxygen during respiration to convert the nitrate ion (NO_3^-) into N_2 and other nitrogen gaseous compounds like NO_2 that are emitted into the atmosphere.

6. **Anaerobic ammonia oxidation:** This process is the major nitrogen conversion in the oceans. It is nominated as *anammox*[10], an abbreviation for anaerobic ammonium oxidation. The reaction is mediated by a specialised group of planctomycete-like bacteria and takes place in many natural environments. The balanced reaction for this transformation is given in Eq. 1.7.

$$NH_4^+ + NO_2^- \rightarrow N_2 + 2H_2O\left(\Delta G^\circ = -357\,kJ\,/\,mol\right) \tag{1.7}$$

[10] Anammox is also a registered procedure for removing ammonium compounds in wastewater based on such concept.

Topics of Interest: Marine Nitrogen Cycle

The nitrogen cycle in oceans (Figure 1.31) has a close relationship with other biogeochemical cycles of elements like carbon, phosphorus or sulfur. Nitrogen and its compounds enter the water in three ways:

A. Precipitation

B. Runoff

C. Gas exchange of gases from the atmosphere (the main procedure)

In seawater, the nitrogen cycle is driven by complex microbial transformations.

1. **Nitrogen fixation:** As it occurs on land, molecular nitrogen cannot be utilised directly by phytoplankton, so it must undergo a previous transformation. In this case, it is performed predominantly by cyanobacteria that transform molecular nitrogen into ammonium ions (NH_4^+).

2. **Assimilation:** Ammonium ions are consumed almost as fast as they are produced and incorporated into living organisms. But some marine microbes consume another form of assimilation (nitrite and nitrate).

3. **Nitrification:** It is the process of converting the ammonium ion to nitrite (NO_2^-) and then nitrite to nitrate (NO_3^-). It is carried out by some microorganisms.

4. **Remineralisation:** A host of organisms consume organic nitrogen particles and dissolved organic nitrogen to obtain ammonium ions.

5. **Anammox:** This process is the major nitrogen conversion in the oceans. It is an abbreviation for anaerobic ammonium oxidation, and it is considered by scientists that this process may be responsible for 30–50% of the molecular nitrogen produced in the oceans. The bacteria that produce this transformation were identified in 1999, and it

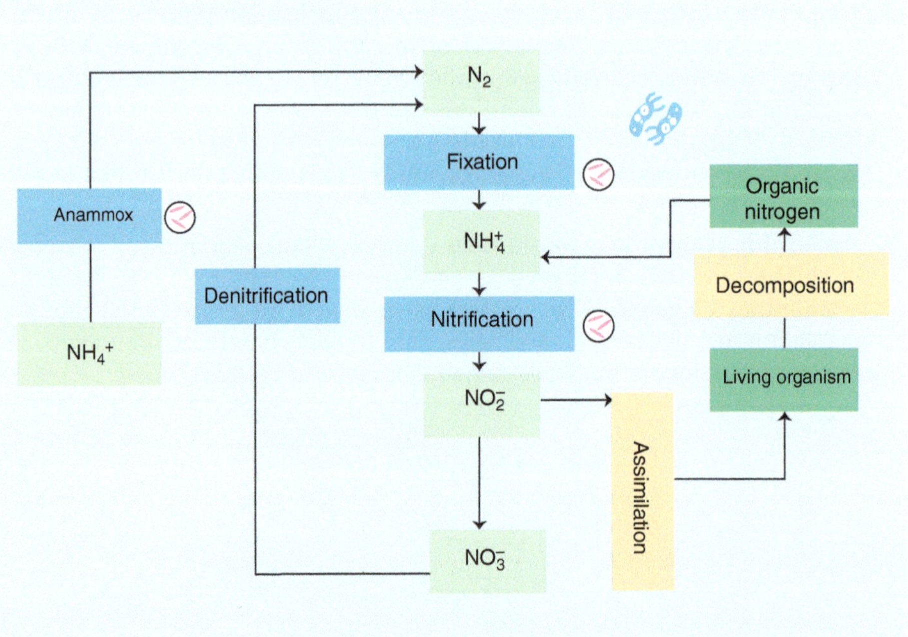

FIGURE 1.31   Marine nitrogen cycle

was actually a great surprise for the scientific community. The balanced reaction for this transformation is given in Eq. 1.8.

$$NH_4^+ + NO_2^- \rightarrow N_2 + 2H_2O \left(\Delta G° = -357\,kJ/mol \right) \tag{1.8}$$

6. **Denitrification:** To complete the complex cycle, other microorganisms convert nitrate to nitrogen gas.

1.24   The Phosphorus Cycle

Phosphorus is an essential element for plant and animal growth, and a key element for the health of microbes inhabiting the soil, but is gradually depleted from the soil over time. Phosphorus is necessary for the formation of biomolecules as ATP (adenosine triphosphate) that stores energy for biochemical transformations, nucleotides, which comprise DNA and RNA, or phospholipids, components of membranes of cells. Inorganic derivatives of phosphorus as calcium phosphate is also an essential component of mammalian bones and teeth or insect exoskeletons. In agriculture, an insufficient amount of phosphorus in the soil produces a decreased crop yield.

The phosphorus cycle (Figure 1.32) is the biogeochemical cycle that describes the displacement of phosphorus through the lithosphere, hydrosphere and biosphere. It is a slow process which involves five key steps:

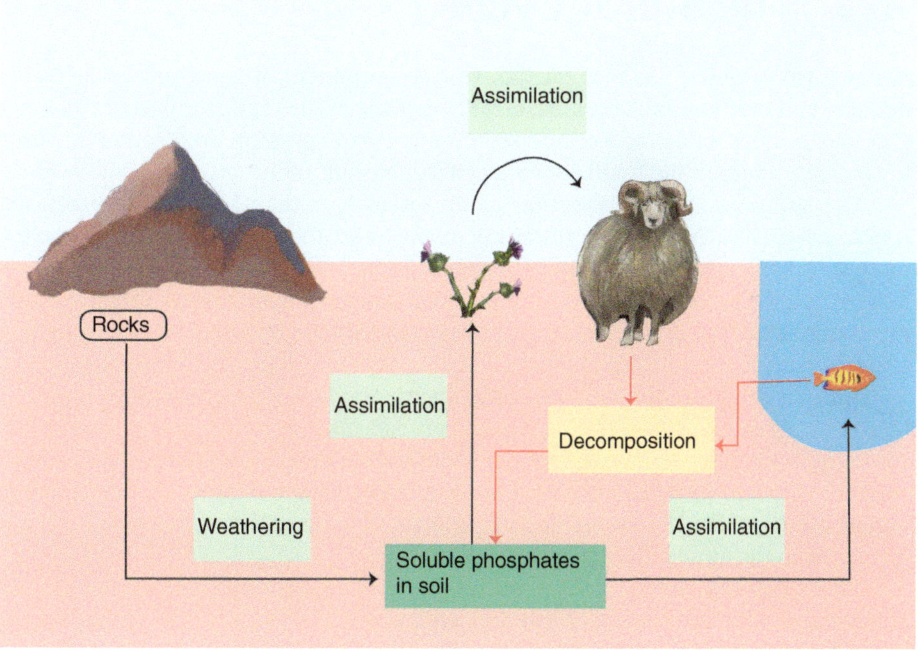

FIGURE 1.32   Phosphorus cycle

1. **Weathering:** Since the main sources of phosphorus in nature are rocks and minerals, phosphorus is obtained from reservoirs by weathering. Weather events, like rain, runoff, wind or any other cause erosion, provide the transport of phosphorus as phosphate salts into the soil.

2. **Absorption by plants from soil:** Plants absorb phosphorus salts dissolved in soil water. These salts come from weathering (step 1) or are formed in soil from organic compounds of phosphate that are transformed by bacteria into inorganic forms of phosphorus. This process is known as mineralisation. Despite the quantities of phosphorus in the soil being normally small, it is one of the main factors that undermine plant growth. In case of aquatic plants phosphorus salts are absorbed from water and the bottom layer of water bodies.

3. **Absorption by animals:** Animals obtain phosphorus from three sources:

   a. Drinking water

   b. Eating plants for herbivores

   c. Plant-eating animals for carnivores.

   The organic compounds of phosphorus obtained by these three ways are transferred to the next level of consumers.

4. **Return to the environment via decomposition:** When plants and animals die, decomposition processes return phosphorus back to the water and soil and plants and animals can use this phosphorus again, and step 2 of the cycle is repeated.

Human activity has caused major changes to the global cycle mainly through shipping of minerals and the use of fertilisers.

1.25 The Sulfur Cycle

Sulfur is an essential element for life. It is a component of some amino acids like cysteine and methionine, vitamins, etc. Most of the earth's sulfur is part of rocks and minerals or buried deep in oceanic sediments. Pyrite, gypsum and elemental sulfur in volcanic areas are the main sources of available sulfur.

The sulfur cycle is a collection of processes by which sulfur moves between rocks, waterways, living systems, water bodies and the atmosphere. This cycle is more complex than other biogeochemical cycles because of the different redox states in which sulfur can occur, both in organic and inorganic compounds, integrating the metabolic activity of diverse microorganisms, and abiotic transformations in soil and atmosphere.

Sulfur is in soil as inorganic salts and organic and inorganic compounds produced by living organisms when they decay, but it can also be found in the atmosphere combined with other elements, forming volatile compounds from natural and anthropogenic sources. In addition, volcanic eruptions, bacterial processes or decaying organisms from land or water bodies produce volatile sulfur compounds that are emitted into the atmosphere (Figure 1.33).

There are two scopes of the sulfur cycle. The first one takes place in the atmosphere with interactions with soil and water bodies, and the second one is basically performed by microorganisms in soil that transforms organic sulfur into inorganic species and vice versa (Figure 1.34).

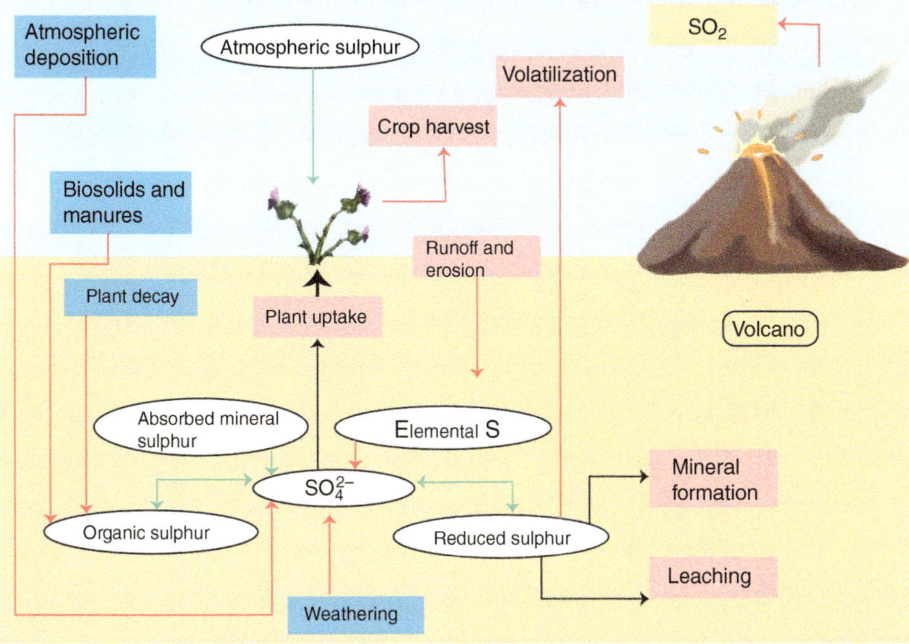

FIGURE 1.33   Sulfur cycle

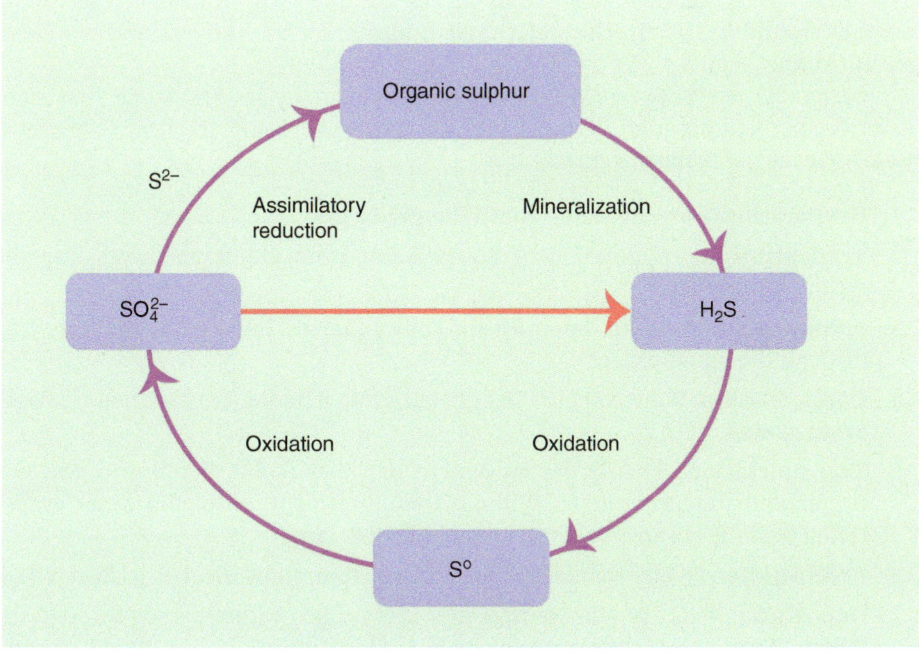

FIGURE 1.34   Microbial sulfur cycle

The main four steps of the sulfur cycle performed by microorganisms are:

1. **Mineralisation of organic sulfur** into inorganic compounds (hydrogen sulfide, inorganic sulfides and elemental sulfur).
2. **Oxidation of hydrogen sulfide, sulfides and elemental sulfur** to ion sulfate (SO_4^{2-}).
3. **Reduction of sulfate to sulfite.**
4. **Incorporation of sulfur into organic compounds** (including metal-containing derivatives).

These steps can be grouped into two parts.

- Dissimilatory
- Assimilatory

The dissimilatory side of the cycle includes reductive processes of sulfate and sulfur reduction as well as oxidative processes such as phototrophic and chemotrophic sulfide and sulfur transformations.

The assimilatory side includes sulfate and sulfide assimilation by bacteria, fungi and plants and is mainly transformed into amino acid sulfur obtained from living and dead organic matter by excretion and decomposition. Animals, including humans, are unable to reduce sulfate for this purpose and therefore have to rely on sulfur-supplying food.

1.26 The Water Cycle

Water is the only substance that can be found in nature in all three states of matter. The water cycle describes how water moves continuously from one reservoir to another around the Earth. The water cycle is also known as the hydrologic cycle or the H_2O cycle (Figure 1.35).

Generally, ten steps are described for the water cycle. The three first steps describe the incorporation of water into the atmosphere and the next seven, how water returns to the Earth's surface:

1. **Evaporation:** Turning water from a liquid to a gas.
2. **Transpiration:** Evaporation of liquid water from plants and trees into the Earth's atmosphere.
3. **Sublimation:** The solid water (ice and snow) changes into a gas without moving through the liquid phase.
4. **Condensation:** Water vapour changes back into a liquid. This is when we begin to see clouds.
5. **Transportation:** This is the movement of water (solid, liquid and gaseous) throughout the atmosphere. Without this movement, all of the water evaporating over the ocean would not precipitate over land.
6. **Precipitation:** Water that falls to the earth as rain, snow, sleet, drizzle and hail.
7. **Deposition:** Water vapour changes into ice (a solid) without going through the liquid phase. This is the reverse of stage 3. This is most often seen on clear and cold nights when frost forms on the ground.

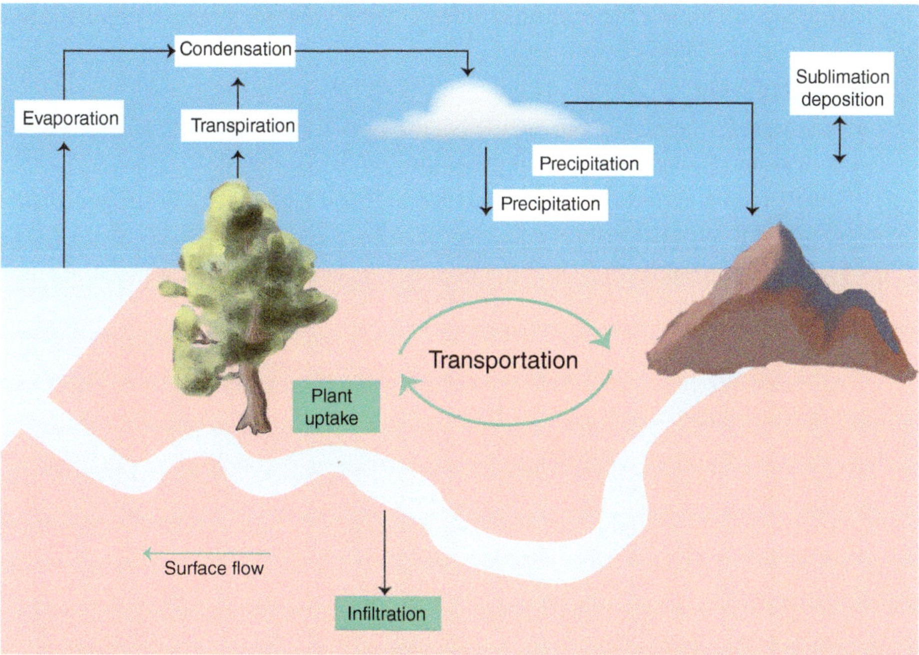

FIGURE 1.35 The water cycle

8. **Infiltration:** Water moves into the ground from the surface.

9. **Surface flow:** Water moves to oceans mainly by streams and rivers.

10. **Plant uptake:** Water is taken from the groundwater flow and soil moisture.

The water cycle involves the exchange of energy, which leads to temperature changes and temperature regulation on the earth's surface. For instance, when water evaporates, it takes up energy from its surroundings and cools the environment. When it condenses, it releases energy and warms the environment. These heat exchanges strongly influence the climate.

1.27 Biogeochemical Cycles and Human Influence

Since the Industrial Revolution, biogeochemical cycles have been strongly influenced by human activities.

Main human influences and consequences on biogeochemical cycles are as follows:

1. **Carbon cycle:**
   - **Deforestation:** Reduces the number of trees available to absorb CO_2 through photosynthesis.
   - **Fossil fuels:** Burning coal, oil and natural gas releases stored carbon into the atmosphere as CO_2.

- **Impact:** Enhanced greenhouse effect, global warming, ocean acidification and climate change.

2. **Nitrogen cycle:**
   - **Fertilisers:** Excessive use of nitrogen-based fertilisers leads to runoff into water bodies, causing nutrient pollution.
   - **Fossil fuels:** Emissions from vehicles and industry release nitrogen oxides (NOx) into the atmosphere.
   - **Impact:** Eutrophication of water bodies, loss of biodiversity, soil acidification and contribution to smog and acid rain.

3. **Phosphorus cycle:**
   - **Agriculture:** Use of phosphate fertilisers increases phosphorus runoff into rivers and lakes.
   - **Waste:** Improper disposal of animal waste and sewage contributes to phosphorus pollution.
   - **Impact:** Algal blooms, hypoxia (low oxygen levels), fish kills and disruption of aquatic ecosystems.

4. **Sulfur cycle:**
   - **Industrial emissions:** Release of sulfur dioxide (SO_2) from burning fossil fuels and industrial processes.
   - **Impact:** Formation of sulfuric acid in the atmosphere, leading to acid rain, which harms ecosystems, corrodes buildings and affects human health.

5. **Water cycle:**
   - **Urbanisation:** Increases surface runoff and reduces groundwater recharge due to impermeable surfaces.
   - **Deforestation:** Reduces transpiration and alters precipitation patterns.
   - **Impact:** Altered hydrological cycles, increased flood risk, reduced water quality and habitat loss.

Strategies for mitigation of the impact of human activities on biogeochemical cycles, which must be based on more sustainable practices. They can be summarised on the following points:

1. **Carbon cycle:**
   - **Reforestation:** Planting trees to absorb CO_2 to improve carbon balance
   - **Renewable energy:** Transitioning to solar, wind and other renewable energy sources to reduce CO_2 emissions.

2. **Nitrogen cycle:**
   - **Precision agriculture:** Using fertilisers more efficiently to minimise soil accumulation and runoff.
   - **Alternative energy:** Reducing fossil fuel use to lower NOx emissions.

3. **Phosphorus cycle:**
   - **Sustainable agriculture:** Managing fertiliser application to prevent excess phosphates on soil and runoff.
   - **Waste management:** Proper treatment of animal waste and sewage to reduce phosphorus pollution.

4. **Sulfur cycle:**
   - **Cleaner technologies:** Using methods and devices in industrial processes to reduce SO_2 emissions.
   - **Alternative fuels:** Shifting to low-sulfur fuels and renewable energy sources.
5. **Water cycle:**
   - **Water management:** Implementing efficient water use practices in agriculture, industry and households and improving the treatment methods of wastewater.
   - **Green infrastructure:** Using permeable surfaces, green roofs and rain gardens to enhance groundwater recharge and reduce runoff.

CHAPTER 2

Chemical Pollution and Toxicity of Chemical Compounds

2.1 Introduction

The environment is exposed to a constant interchange of matter and energy between the different spheres, as described in Chapter 1. Hundreds of organic and inorganic substances are produced in nature, forming parts of rocks, soils, air, waterbodies or living beings and are transformed and emitted into the surroundings. Human activities have dramatically improved the number and amounts of substances that have been incorporated into the environment in a direct or indirect manner.

According to the Chemical Abstracts Service[1], there are more than 127 million substances. Some of them are *natural products*, meaning substances that can be found in Nature, but many others are *synthetic products* that Nature is not able to produce[2]. This amount is increasing every day and most of them are organic compounds.

Many of newly registered compounds in Chemical Abstract Service have only an academic interest, coming from research work of scientists and industrialists and, for now, they barely show relevance in the environment, but many others are used massively because of their application in many branches of human activity, such as solvents, drugs, personal care products, phytochemicals, building materials, polymers and food additives.

---

[1] www.cas.org.

[2] There is a remarkable group of natural products that can be prepared by chemical synthesis and they are structurally identical to the compounds found in nature. They are named as synthetic natural products.

Understanding the Chemistry of the Environment, First Edition. Francisco G. Calvo-Flores.
© 2025 John Wiley & Sons Ltd. Published 2025 by John Wiley & Sons Ltd.

When a substance, natural or synthetic, comes into contact with the environment and is able to produce any kind of damage, it is considered a pollutant and produces a ***Chemical Pollution*** episode. Chemical pollution is not the only cause of possible damage to the environment. Other physical or biological phenomena such as temperature, radiation, noise or pathogens are capable of disturbing the natural development of living beings, including humans and the common ecosystems. Such phenomena are also considered pollution.

There are many types of pollution and pollutants and it is not easy to give a strict definition for both terms.

In the case of pollution, a reasonable option should be:

The phenomenon by which soil, water, air or any other part of the environment is damaged by any kind of external factor that makes the environment unsafe or unsuitable for living beings.

Pollution can be classified according to two major criteria:

a. Types of pollution

b. Extension of area affected

According to type, there are three main groups of pollution based on the nature of the phenomenon:

1. **Chemical pollution:** It may be produced naturally by processes carried out in the environment that increase the amount of chemical substances in higher concentrations than the considered safe values. Volcanic eruptions or wildfires are one of the most extensive natural sources of chemical pollution. But in our contemporary world, the main origin for the introduction of undesirable chemicals that are found in Nature is caused by human activity. Branches such as transport, industry, mining, agriculture or livestock are responsible for most of the chemical pollution.Chemical pollution is the most common and extended pollution phenomenon and probably produces the most major concern in society.

2. **Physical pollution:** Refers to the presence of harmful episodes such as vibrations, thermal alterations or electromagnetic radiation that cause damage to the environment and are considered normal processes.

3. **Biological pollution:** It is produced by the presence of exogenous species of living organisms or microorganisms and/or substances or materials produced by them which disturb ecosystem development, such as insects, acarus, algae, pollen and toxins.

Depending on the extent of the affected area, pollution can be classified into three groups:

1. **Local:** It refers to when pollution phenomena affect a restricted area.

2. **Regional:** When an extended area is affected, for example, a continent, an island, several countries, etc.

3. **Global:** When pollution episodes affect the whole planet.

2.2   Contamination and Pollution

Sometimes the term *contamination* is used as a synonym for pollution, but it might cause some confusion because the differences between both terms remain elusive. In 2007, Chapman[3] introduced an interesting and valuable remark:

> *"Contamination is simply the presence of a substance where it should not be or at concentrations above background. Pollution is contamination that presents or can present adverse biological effects to resident communities. All pollutants are contaminants, but not all contaminants are pollutants".*

2.3   Sources of Pollution

Classifying pollution sources involves categorising them based on various characteristics or properties: natural or manmade, if they are produced by fixed or not fixed locations or facilities and according to the environmental sphere affected: air, water or soil.

The first criteria involve two broad types: natural and anthropogenic.

1. **Natural sources of pollution:** This type of pollution occurs in the environment without any direct human intervention. They include:
   a. **Volcanic eruptions:** Volcanic eruptions release large amounts of ash, dust, gases and other pollutants into the atmosphere, which can affect air quality and contribute to global air pollution.
   b. **Forest fires:** Forest fires, either caused by lightning or natural combustion, can release smoke, particulate matter and other pollutants into the air, leading to air pollution.
   c. **Dust storms:** Dust storms, which are natural occurrences in arid and semi-arid regions, can generate large amounts of dust and particulate matter, causing air pollution.
   d. **Biological processes:** Natural processes like decomposition of organic matter, decay of vegetation and natural emissions from plants and animals can release gases like methane, ammonia and volatile organic compounds (VOCs) into the atmosphere, contributing to air pollution.

2. **Anthropogenic sources of pollution** are the types of pollution that are directly or indirectly caused by human activities. These include:
   a. **Industrial activities:** Industries and manufacturing processes release large amounts of pollutants, such as particulate matter, sulfur dioxide (SO_2), nitrogen oxides (NO_x), carbon monoxide (CO) and VOCs, into the air, water and soil, leading to air, water and soil pollution.

---

[3] Chapman, P.M. (2007). Environment International 33 (4): 492–501.

b. **Transportation:** Emissions from vehicles, such as cars, trucks, boats and aeroplanes, are a significant source of pollution, releasing pollutants like carbon dioxide (CO_2), NO_x, particulate matter and VOCs into the atmosphere.

c. **Energy production:** The burning of fossil fuels, such as coal, oil and natural gas, for energy production in power plants and other facilities releases pollutants like CO_2, SO_2, NO_x and particulate matter, contributing to air and water pollution.

d. **Agricultural activities:** The use of fertilisers, pesticides and herbicides in agriculture can lead to water and soil pollution, as these chemicals can contaminate water bodies and soil, affecting aquatic ecosystems and agriculture.

e. **Waste disposal:** Improper waste disposal, including dumping of solid waste, hazardous waste and sewage into the environment, can result in pollution of land, water and air, causing environmental degradation and health risks.

f. **Deforestation:** Deforestation, which involves clearing of forests for agricultural expansion, logging and infrastructure development, can result in soil erosion, loss of biodiversity and release of carbon dioxide into the atmosphere, contributing to air and water pollution.

g. **Construction:** Construction activities, including excavation, demolition and dust generated from construction sites, can release particulate matter, dust and other pollutants into the air, contributing to air pollution.

The second criterion for categorising the source of pollution is to consider if the point of emission is located in a fixed area or if it moves over an interval of time, as follows:

a. **Stationary sources:** Stationary sources of pollution refer to fixed or stationary facilities that emit pollutants into the environment. These sources are typically located at a specific location and can include natural sources such as volcanoes and wildfires and anthropogenic sources such as industrial facilities, power plants, factories, refineries, incinerators and other similar installations. Stationary sources of pollution can release pollutants into the air, water or soil, depending on the type of facility and its operations. Anthropogenic stationary sources are often subject to environmental regulations and permits that set emission limits and require monitoring and control measures to reduce the impact of their emissions on the environment and public health.

b. **Mobile sources:** Mobile sources of pollution refer to movable sources that emit pollutants as a result of their operation. These sources are typically vehicles that move from one location to another, such as cars, trucks, buses, motorcycles, aeroplanes, ships, trains and other mobile equipment. Mobile sources of pollution can emit pollutants into the air, including exhaust emissions from engines, evaporative emissions from fuel systems and dust from tyres and brakes. Mobile sources are also subject to environmental regulations and standards that aim to reduce their emissions, such as emission standards for vehicles, fuel quality standards and regulations for transportation systems and infrastructure.

And finally, sources of pollution can be categorised according to the environmental sphere affected: air, water or soil.

Sources of Air Pollution

In case of air pollution, the main sources are

1. **Natural sources**, including volcanic eruptions, windblown dust, sea-salt spray wildfires, dust storms, aerosols formed from volatile substances from plants and trees, emissions of gaseous and particulate matter from wild animals, etc.
2. **Burning of fossil fuels** used in power generation, transport, industry and households
3. **Emissions of gases**, aerosols and dusts produced in industrial processes
4. **Mining industries**
5. **Domestic and industrial solvent use**
6. **Agriculture**
7. **Waste treatment**

Sources of Water Pollution

According to the type of source, pollution is classified into:

1. **Natural sources:**
   a. Putrefaction of organic matter by microorganisms
   b. **Pathogens and disease-causing organisms:** Animals that live on water in large numbers, such as aquatic birds and put manure directly into the water cause pollution that can contaminate the water with disease-causing organisms or some aquatic organisms that produce toxins for themselves
   c. Sediments, powder, sands and other solid particles related to large storm events, landslides and floods
2. **Industrial waste:** Such pollution is very diverse and already specific to every industrial branch. Pollutants such as lead, mercury, sulfur, asbestos, nitrates and many other harmful chemicals use to be present in industrial waste. Many industries do not have a proper waste management system and drain the waste into freshwater, which goes into rivers, canals and later into the sea. The toxic chemicals have the capability to change the colour of water, increase the number of minerals (also known as eutrophication), change the temperature of water and pose a serious hazard to aquatic organisms.
3. **Sewage and wastewater:** The sewage and wastewater that are produced by each household are chemically treated and released into the sea with freshwater. The sewage water carries harmful bacteria and chemicals that can cause serious health problems. Pathogens are known as common water pollutants. The sewers of cities house several pathogens and, thereby, diseases. Microorganisms in water are known to be causes of some very deadly diseases and become breeding grounds for other creatures that act as carriers. These carriers inflict these diseases via various forms of contact on an individual. A very common example of this process is malaria.

4. **Mining activities:** Mining is the process of crushing rock and extracting coal and other minerals from underground. These elements, when extracted in their raw form, contain harmful chemicals and can increase the number of toxic elements when mixed with water, which may result in health problems. Mining activities emit a large amount of metal waste and sulfides from the rocks, which are harmful to the water.

5. **Marine dumping:** The garbage produced by each household in the form of paper, aluminium, rubber, glass, plastic and food is collected and deposited into the sea in some countries. These items take from 2 weeks to 200 years to decompose. When such items enter the sea, they not only cause water pollution but also harm animals in the sea.

6. **Accidental oil leakage:** An oil spill poses a huge concern as a large amount of oil enters the sea and does not dissolve with water, thereby opening problems for local marine wildlife such as fish, birds and sea otters. For example, a ship carrying a large quantity of oil may spill oil if it meets with an accident, which can cause varying damage to species in the ocean depending on the quantity of the spill, the size of the ocean and the toxicity of pollutants.

7. **The burning of fossil fuels:** Fossil fuels like coal and oil, when burned, produce a substantial amount of ash in the atmosphere. The particles, which contain toxic chemicals when mixed with water vapour, result in acid rain. Also, carbon dioxide is released from the burning of fossil fuels, which results in global warming.

8. **Chemical fertilisers and pesticides:** Chemical fertilisers and pesticides are used by farmers to protect crops from insects and bacteria. They are useful for the plant's growth. However, when these chemicals are mixed with water, they are harmful for plants and animals. Also, when it rains, the chemicals mix up with rainwater and flow down into rivers and canals, which pose serious damage to aquatic animals.

Sources of Soil Pollution

For soil pollution, the major sources are as follows:

1. **Natural sources** include some metal elements present in the original minerals and some rocks, volcanic eruptions, materials dragged away in floods or other natural processes.

2. **Effluents from industrial and commercial services**, mining, oil extraction and production, power plants, etc.

3. **Domestic and municipal wastes and wastewater**

4. **Agriculture and livestock:** Fertilisers and agrochemicals; livestock health products; manures; irrigation with saline or untreated water.

5. **Transport spills on land**

6. **Deficient storage of oil, chemicals and industrial residues**

2.4 Pollutant Classification

There are many points of view on how to categorise pollutants, which are actually complementary to understanding the complexity of chemical pollution.

Pollutants can be classified according to their:

1. Chemical composition or structure
2. Physical state
3. If produced from natural sources or not
4. The concentrations that appear in the environment
5. How they evolve in the environment once they are emitted
6. The regulatory normative

1. **According to chemical composition or structure:**

   Pollutants can be classified based on their chemical composition, such as organic pollutants, inorganic pollutants and their chemical structure. Considering chemical structure, it is a hard task to carry out a systematic classification. As has been commented on before, there are more than 127 million registered compounds. Other data that strengthens this fact is that the World Health Organization (WHO) has reported more than 100,000 chemicals that are released into the global environment every year as a consequence of their production, use and disposal.

2. **According to the physical state:**

   It's important to note that pollutants can often have multiple physical states depending on the conditions and context in which they are released, transported and transformed. The common classification of pollutants according to their physical state includes:

   a. **Gaseous pollutants:** These pollutants exist in the form of gases, mainly in the atmosphere. They can be released from various sources, such as industrial emissions, vehicle exhaust and natural processes. Examples of gaseous pollutants include carbon dioxide (CO_2), sulfur dioxide (SO_2, NO_x), ozone (O_3) and VOCs.

   b. **Liquid pollutants:** These pollutants are in liquid form and can contaminate water bodies such as rivers, lakes and oceans. They can come from various sources, such as industrial discharges, agricultural runoff and sewage treatment plants. Examples of liquid pollutants include some chemical practices and oils.

   c. **Solid pollutants:** These pollutants are in solid form and can come from activities such as mining, construction and waste disposal. Solid pollutants can include various types of waste, debris and hazardous materials that can have adverse impacts on the environment and human health.

   d. **Particulate pollutants:** These pollutants are tiny solid or liquid particles suspended in the air. They can be released from activities such as burning fossil fuels, construction activities and agricultural practices. Particulate pollutants can have various sizes and compositions and they can have detrimental effects on human health and the environment. Examples of particulate pollutants include dust, soot, smoke and aerosols.

3. **According to their origin:**

Pollutants can be classified into two categories based on whether they are produced by human activities or not, as follows:

1. **Natural pollutants:** These are substances of natural origin. Examples of these pollutants are gases and dust produced by volcanoes, sea salt particles, photochemically formed ozone or material from plant decay.

2. **Anthropogenic pollutants** are substances produced by human activity: industry, transportation, agriculture, livestock, human settlements, etc. There are two groups of substances belonging to this class.

   a. Pollutants produced exclusively by human activity are facts foreign to the biosphere and its natural transformations. Such chemical substances may be found within ecosystems and individual organisms, but they are not naturally produced or expected to be present within the organisms. Anthropogenic pollutants in this group are nominated as xenobiotics. They include inorganic compounds such as heavy metals and other non-natural salts, acids and bases and organic compounds such as synthetic polymers, pesticides, surfactants, halogenated derivatives and many others.

   b. Pollutants produced by anthropogenic and natural sources: A typical example of such a pollutant is carbon dioxide. Carbon dioxide is produced in combustion of coal or crude derivatives, but at the same time, carbon dioxide is produced in respiration processes, volcanic eruptions or organic matter decay by aerobic decomposition.

4. **According to concentrations found in the environment:**

Based on quantities and concentrations present in the environment, pollutants may be classified into:

1. **Quantitative pollutants:** These are substances normally occurring in the environment, but they may acquire the status of a pollutant when their concentration is increased due to human activities. Again, carbon dioxide can be considered in this group. It is present in the atmosphere, but activities such as transportation and industries are increasing the concentration; therefore, it is classified as a quantitative pollutant.

2. **Qualitative pollutants:** They are substances that do not normally occur in nature but are introduced by man, for example, some polymers (plastics) or phytosanitaries.

5. **Based on evolution in the environment:**

Pollutants can be classified into different categories based on their behaviour and evolution in the environment. Here are some common classifications:

a. **Non-persistent pollutants:** These pollutants are less stable and tend to degrade relatively quickly in the environment through natural processes, so they do not tend to accumulate in the environment or in organisms. One key process for degradation is so-called biodegradation. Biodegradation is the term used to describe the general process of the breakdown of organic chemicals by the biological action of microorganisms. In these processes, many organic molecules, pollutants or not, are transformed into other species that are considerably less harmful.

b. **Persistent pollutants:** These pollutants are highly stable and do not easily break down in the environment. They can persist in the environment for a long time, leading to long-term accumulation and potential bioaccumulation in the food chain.

c. **Bioaccumulative pollutants:** These pollutants have the ability to accumulate in living organisms, often in fatty tissues, as they are absorbed from the environment or through food intake. They can bioaccumulate in the food chain, leading to higher concentrations in organisms at higher trophic levels. Examples of bioaccumulative pollutants include certain pesticides, polychlorinated biphenyls (PCBs) and some heavy metals.

d. **Non-bioaccumulative pollutants:** These pollutants do not tend to accumulate in living organisms and are more readily excreted or eliminated from the body. They do not biomagnify in the food chain. Examples of non-bioaccumulative pollutants include some water-soluble chemicals, some gases and some metals that are not readily absorbed by organisms.

e. **Mobile pollutants:** These pollutants can easily move and spread in the environment through natural processes such as air currents, water currents or the migration of organisms. Examples of mobile pollutants include some air pollutants such as particulate matter and gases, oil spills and invasive species.

f. **Localised pollutants:** These pollutants are confined to a specific area or location and do not spread extensively in the environment. Examples of localised pollutants include spills of hazardous substances, localised industrial emissions and waste disposal sites.

g. **Global pollutants:** These pollutants have the potential to spread globally and can have widespread impacts on the environment and human health. Examples of global pollutants include greenhouse gases such as carbon dioxide and methane, which contribute to climate change and ozone-depleting substances such as chlorofluorocarbons (CFCs), which affect the ozone layer.

h. **Primary pollutants** are substances that are directly emitted into the atmosphere as a result of human activities or natural processes.

i. **Secondary pollutants** are formed through chemical reactions involving primary pollutants and other components. This second group is usually more difficult to handle. They are produced by chemical or biological processes such as photochemical reactions initiated by ultraviolet (UV) light in the atmosphere or metabolised by living organisms from former substances that used to be toxic or potentially toxic and were first emitted into the environment before their transformation. It is very frequent that both the original molecules and the transformed molecules are considered pollutants.

It's important to note that many pollutants can exhibit characteristics of multiple categories and the behaviour of pollutants in the environment can vary depending on factors such as their chemical properties, environmental conditions and human activities. Proper identification and classification of pollutants are essential for effective pollution management and mitigation strategies.

6. **According to regulations:**

Another important aspect of pollutants is related to regulations and norms. According to such criteria, pollutants may be classified into two major categories:

1. **Pollutants under regulation (priority pollutants):**

   They are a set of chemical substances that have been selected based on their known or suspected nocive effects, such as carcinogenicity, mutagenicity, teratogenicity or high acute toxicity, for which there are well-defined analytical test methods in soil, water or air. Most countries and supranational organisations agree about the type and permitted levels of these substances. For example, the Environmental Protection Agency (EPA) of the United States defines these types of pollutants as *priority pollutants*. They are a group of regulated chemical substances that, in the case of water, include a set of 126 **priority pollutants**. Most of these priority pollutants are subject to regulation by rules and laws not only in the United States but also in individual countries or supranational agencies. In Europe, the levels of priority pollutants in waters are regulated according to the 2008/105 EU directive.

2. **Emerging pollutants:**

   This is a new class of pollutants that are currently not regulated, but because of their increasing concern, they are candidates for future regulation. Emerging pollutants are not necessarily new chemicals. Many of them have been present in the environment for decades, but they have been detected recently because of the improvement of analytical methods.

Topics of Interest: Environmental Regulatory Organisms

European Environment Agency

The European Environment Agency (EEA, https://www.eea.europa.eu/en) is an agency of the European Union (EU) that provides information on the environment. It was established to support sustainable development and to help achieve significant and measurable improvement in Europe's environment through the provision of timely, targeted, relevant and reliable information to policymakers and the public.

The EEA was established by European Economic Community (EEC) Regulation 1210/1990 (amended by EEC Regulation 933/1999 and EC Regulation 401/2009) and became operational in 1994. It is headquartered in Copenhagen, Denmark and is formed by the member states of the EU to *provide sound, independent information on the environment. It operates as a major information source for those involved in developing, adopting, implementing and evaluating environmental policy, as well as the general public.*

- **Mission:** The EEA's mission is to help achieve significant and measurable improvement in Europe's environment through the provision of timely, targeted, relevant and reliable information to policymakers and the public.
- **Information gathering and analysis:** The agency collects data on various environmental issues, such as air and water quality, biodiversity, climate change and more. It analyses and synthesises this information to provide comprehensive assessments of the state of Europe's environment.

- **Reporting:** The EEA produces regular reports and indicators, including the European Environment State and Outlook Report (SOER), which provides an integrated assessment of Europe's environment and the challenges it faces.
- **Coordination:** The agency collaborates with its member countries, international organisations and other stakeholders to ensure the exchange of environmental information and best practices.
- **Assessment and monitoring:** The EEA monitors and assesses the implementation of environmental policies and measures, providing feedback on their effectiveness and suggesting improvements.
- **Networks and partnerships:** The EEA works closely with a network of national environmental agencies and other partners across Europe. This collaboration enhances the sharing of environmental data and expertise.
- **Communication:** The agency communicates its findings to policymakers, the public and the scientific community to raise awareness and foster informed decision-making on environmental issues.

The EEA plays a crucial role in supporting the EU's environmental policies and initiatives. It contributes to the development and implementation of these policies by providing the necessary knowledge and information for effective decision-making and public engagement.

EEA is dependent on the European Commission that has recently approved the EU Action Plan titled 'Towards Zero Pollution for Air, Water and Soil,' marking a pivotal milestone in the implementation of the European Green Deal. This plan, unveiled during this year's EU Green Week, represents a comprehensive strategy aimed at achieving a vision for 2050 in which pollution is mitigated to levels that pose no threat to human health and the well-being of natural ecosystems.

At its core, the action plan integrates various EU policies designed to address and prevent pollution, placing a particular emphasis on leveraging digital solutions as instrumental tools in the fight against pollution. The initiative recognises the need for a holistic approach, interlinking diverse policy frameworks to effectively combat pollution across different environmental domains.

A key aspect of the plan involves thorough reviews of pertinent EU legislation to identify any existing gaps and areas requiring enhanced implementation. This process is crucial for ensuring that the legal obligations outlined in EU frameworks are met comprehensively. By scrutinising and refining relevant legislation, the European Commission aims to strengthen the regulatory framework and enhance its capacity to address the multifaceted challenges associated with pollution.

In summary, the EU Action Plan signifies a commitment to a pollution-free future by 2050, aligning with the overarching goals of the European Green Deal. The plan's emphasis on digital solutions and the comprehensive review of existing legislation reflects a forward-looking strategy, demonstrating the EU's dedication to robust environmental policies and sustainable practices.

Environmental Protection Agency (EPA)

The United States Environmental Protection Agency (EPA, https://www.epa.gov) is a federal agency responsible for protecting human health and the environment. It was established on December 2, 1970 in response to growing concerns about environmental pollution and its impact on public health. The EPA's mission is to enforce regulations and laws that address various environmental issues, including air and water quality, hazardous waste management, chemical safety and pollution prevention.

Key responsibilities of the EPA include:

- **Air quality:** The EPA sets and enforces national air quality standards to protect public health and the environment. This includes regulations on emissions from vehicles, industrial facilities and other sources.

- **Water quality:** The EPA works to ensure safe and clean water by regulating discharges into water bodies, setting water quality standards and addressing contamination issues.
- **Toxic substances:** The EPA regulates the production, use and disposal of toxic substances to protect human health and the environment. This includes overseeing the implementation of laws such as the Toxic Substances Control Act (TSCA).
- **Hazardous waste:** The EPA manages hazardous waste through the Resource Conservation and Recovery Act (RCRA), which establishes a comprehensive program for the safe management of hazardous waste from generation to disposal.
- **Chemical safety:** The EPA evaluates and regulates the use of chemicals to ensure they do not pose unreasonable risks to human health or the environment.
- **Environmental justice:** The EPA addresses environmental justice concerns by working to ensure that all communities, regardless of race or income, have the same degree of protection from environmental and health hazards.
- **Climate change:** While not originally part of its mandate, the EPA has taken on a role in addressing climate change by regulating greenhouse gas emissions and implementing programs to reduce their impact.
- **Research and education:** The EPA conducts scientific research to inform environmental policies and provides educational resources to the public.

The EPA collaborates with state and tribal governments, industry, environmental groups and the public to implement and enforce environmental laws. The agency plays a crucial role in safeguarding the nation's natural resources and public health.

The EPA is a federal government agency of the United States that was established in December 1970. The key objective is to protect human and environmental health and it is a world reference in terms of environmental protection. Headquartered in Washington, DC, the *EPA is responsible for creating standards and laws promoting the health of individuals and the environment.*

2.5 Pollutant Pathways in Environment

Once a pollutant is emitted, it remains for a period in the environment. A key question related to the activity and possible hazards of a pollutant is how much it remains in the environment.

Residence time and half-life are two parameters used in environmental science and chemistry to describe the behaviour and fate of pollutants in various media, such as air, water and soil.

1. Residence time

Residence time refers to the average amount of time a pollutant stays in a particular environmental compartment before being removed or transformed. It can be calculated using the following expression:

$$\text{Residence time} = \frac{\text{Amount of pollutant in the compartment}}{\text{Rate of input or output of pollutant from the compartment}} \tag{2.1}$$

Residence time is typically expressed in units of time, such as hours, days or years and it can vary depending on the specific pollutant, environmental compartment and environmental conditions. In figure, it is plotted the residence time in the atmosphere of some pollutants is plotted against their environmental impact, from local to global. In general, for a pollutant with a longer residence time, a greater and more extensive environmental impact is produced.

2. **Half-life**

The half-life is the time it takes for half of the initial amount of a pollutant to decay or degrade. It is a measure of the rate of decay or degradation of a pollutant and it is often used to estimate the persistence or longevity of pollutants in the environment. The half-life of a pollutant can be determined through laboratory experiments or field measurements and it can vary widely depending on the type of pollutant and the environmental conditions.

The relationship between residence time and half-life can be described mathematically as follows (Eq. 2.2):

$$\text{Half} - \text{life} = 0.693 \times \text{Residence time} \tag{2.2}$$

Where 0.693 is the natural logarithm of 2 and this relationship assumes a first-order decay or degradation process.

Usually, a pollutant breaks down to 50% of its original amount after a half-life period. After two half-lives, the amount drops to 25% and to about 12% after three half-lives. This process continues until the amount remaining is nearly zero (Figure 2.1).

Both residence time and half-life are important parameters for understanding the behaviour of pollutants in the environment, as they can provide insights into their persistence, transport and potential impacts on ecosystems and human health. They are commonly used in environmental risk assessments and management strategies for controlling pollution.

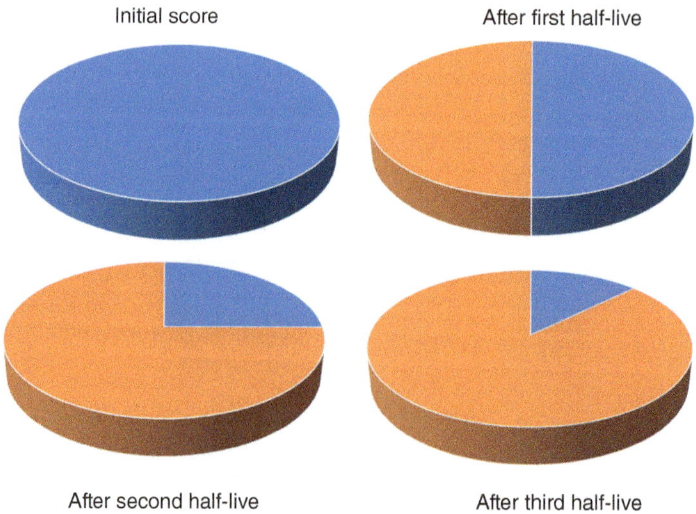

FIGURE 2.1   Half-live representation

The distribution, behaviour, impact and fate of a pollutant can vary and are dependent on numerous other factors, such as its physical properties and ability to undergo chemical transformations.

Physical Properties that Condition Behaviour of Pollutants

Several physical properties can influence the behaviour of pollutants in the environment. Some of the key physical properties that can condition the behaviour of pollutants include:

1. **Solubility:** The solubility of a pollutant refers to its ability to dissolve in a particular medium, such as water, air or soil. Highly soluble pollutants, such as salts or certain organic compounds, tend to dissolve easily in water and may be transported through surface runoff, leaching or atmospheric deposition. Insoluble pollutants, such as certain heavy metals or hydrophobic organic compounds, may accumulate in soil or sediments and persist for longer periods of time.

2. **Hydrophobicity/hydrophilicity:** The hydrophobicity or hydrophilicity of a pollutant refers to its affinity or repulsion towards water. Hydrophobic pollutants, such as some organic compounds, tend to repel water and may accumulate in fatty tissues or sediments, while hydrophilic pollutants, such as salts or certain organic compounds, readily dissolve in water and can be transported through water bodies.

3. **Volatility** is the property of how readily a substance vaporises at a given temperature and pressure. It is quantified with vapour pressure data. From a thermodynamic standpoint, this can be viewed from the well-known Clausius-Clapeyron expression (Eq. 2.3):

$$\ln \frac{P_2}{P_1} = \frac{\Delta H_{vap}}{R}\left(\frac{1}{T_2} - \frac{1}{T_1}\right) \tag{2.3}$$

   where P_1 and P_2 are the vapour pressures at two temperatures, T_1 and T_2, ΔH_{vap} is the heat of vaporisation and R is the gas constant. ΔH_{vap} is proportional to the energy needed to break intermolecular bonds in the pure condensed liquid. It is one of the major predictors of the environmental fate and transport of chemicals.

4. **Density:** The density of a pollutant refers to its mass per unit volume. Pollutants with higher densities, such as heavy metals, tend to settle or sink in water or soil due to gravity, which can result in their accumulation in sediments or lower layers of soil. Pollutants with lower densities, such as certain organic compounds or gases, may remain in the atmosphere or float on water surfaces.

5. **Diffusion** is the net movement of a pollutant from a zone of water, air or soil of higher concentration to a region of lower concentration. Diffusion is driven by a gradient in concentration.

6. **Leaching** is the process by which soluble or partially soluble pollutants become detached from or extracted from a dump, land field, asphalt, etc., by water.

7. **Adsorption** is a mass transfer in which a porous solid, such as soil, meets a liquid or gaseous material to deposit (adsorb) them onto the solid, generally by non-covalent forces. The opposite process is called desorption.

Chemical Transformations of Pollutants

Once pollutants are emitted, they can undergo different types of transformations. Chemical transformations of pollutants in the environment, for example, may occur through natural processes. These transformations can affect the fate, transport and toxicity of pollutants and can take place in different environmental compartments, such as air, water, soil and biota[4]. Chemical transformations include reactions leading to the formation of new, simpler compounds.

In the environment, there are two ways in which pollutants can be transformed:

1. **Abiotic transformations:**

   Abiotic transformations of pollutants refer to chemical and physical processes that occur in the environment without the involvement of living organisms. These processes can affect the fate and behaviour of pollutants and may either enhance or reduce their toxicity.

   Some examples of abiotic transformations of pollutants include **photolysis, hydrolysis, oxidation and reduction**.

2. **Biotic transformations:**

   Biotic transformations of pollutants are the processes by which living organisms, such as bacteria and fungi, break down or transform harmful substances into less harmful or non-toxic forms through enzymatic degradation of organic molecules. That may occur in soil and water bodies, where enzymes produced by microorganisms break down complex organic compounds into simpler molecules that can be absorbed by plants or recycled in the ecosystem. Such processes can be carried out in aerobic or anaerobic conditions. These transformations are known as the biodegradation of organic substances. Many organic pollutants are suitable for such transformations.

In soil, enzymes such as hydrolases, oxidoreductases and dehydrogenases play a crucial role in the degradation of organic pollutants.

Hydrolases are enzymes that catalyse the hydrolysis of chemical bonds using water, leading to the breakdown of complex organic compounds into smaller molecules.

Oxidoreductases and dehydrogenases are enzymes that facilitate redox reactions, which involve the transfer of electrons between molecules. Dehydrogenases are

---

[4] Biota refers to all the living organisms, including plants, animals, fungi, bacteria and other microorganisms, in a particular region or ecosystem.

responsible for the removal of hydrogen atoms from organic molecules, which is an essential step in the breakdown of pollutants. Dehydrogenases facilitate the transfer of electrons during redox reactions, where one molecule is oxidised (loses electrons) and another is reduced (gains electrons). These enzymes are produced by a variety of microorganisms, including bacteria.

In water, enzymes such as peroxidases, laccases and phenol oxidases can degrade pollutants such as polycyclic aromatic hydrocarbons (PAHs), pesticides and herbicides. Microorganisms carry out biodegradation by *endoenzymes* and *exoenzymes*. Endoenzymes provide degradation within the cell. The molecules must be soluble in cell fluids for efficient degradation. Exoenzymes are excreted out of the cell and then interact with substrates and produce degradation.

Enzymatic degradation is a promising method for the bioremediation of contaminated sites, as it is an environmentally friendly and cost-effective solution. However, the effectiveness of enzymatic degradation depends on various factors, including the type and concentration of pollutants, the environmental conditions and the presence of suitable microorganisms. Biodegradation rates are difficult to predict and to formulate a generally applicable predictive theory.

Attenuation of Pollutants

The concept of attenuation refers to a set of natural processes such as dilution, dispersion, precipitation, sorption, biodegradation, bioaccumulation, volatilisation and chemical and biochemical stabilisation of pollutants in environment, which effectively reduce pollutant mobility, bioavailability, toxicity or concentration to lower levels that are not overly harmful to human health and ecosystems.

2.6 Degradation Reactions of Pollutants

Pollutants are broken down or transformed in the environment by several reactions, mediated or not by microorganisms. Significant degradation reactions of pollutants in the environment include photolysis, redox reactions and hydrolysis.

Photochemical Transformations

Photolysis of pollutants typically occurs when certain molecules absorb light energy, which leads to the breaking of chemical bonds and the formation of reactive species that can further react with other compounds or undergo degradation. The specific mechanism and rate of photolysis depend on the nature of the pollutant, the type of light energy (e.g., ultraviolet, visible or infrared) and the environmental conditions (e.g., temperature, humidity and presence of other chemicals). Such reactions are defined as *photolysis*.

One of the most common examples of photolysis of pollutants is the degradation of air pollutants, such as VOCs, NO_x and O_3, in the atmosphere by sunlight. When these pollutants are exposed to UV radiation from sunlight, they can undergo photolysis, leading to the formation of reactive radicals or other products that are usually less harmful or more easily removed from the atmosphere.

Photolysis can also occur in water bodies and soil, where pollutants such as pesticides, herbicides and pharmaceuticals can be broken down by exposure to sunlight. In this case, the radiation can directly break the chemical bonds in the pollutants or generate reactive species that can react with the pollutants and degrade them into less toxic or more biodegradable forms.

There are two types of spontaneous photochemical reactions in environment (see Figure 2.2):

1. **Direct photolysis:**

   Direct photolysis occurs when sunlight energy is absorbed by the pollutant molecule, causing it to break down into smaller molecules or atoms. This process can be important in the removal of pollutants from the atmosphere and the surface of water bodies.

   It takes place when photons are absorbed by a molecule (X) without any other intermediary to give an excited species (X*) that is transformed. The primary step of a photolysis reaction is the formation of an electronically excited state of the starting molecule X (Eq. 2.4)

$$X + h\nu \rightarrow X^{*} \tag{2.4}$$

   X* may be evolve by several ways

   a. **Dissociation:** $X^{*} \rightarrow Y + Z$

   b. **Isomerisation:** $X^{*} \rightarrow X'$

   c. **Rearrangement:** $X^{*} \rightarrow Y$

   d. **Ionisation:** $X^{*} \rightarrow X^{+} + 1e^{-}$

   e. **Direct reaction:** $X^{*} + Y \rightarrow Z_{1} + Z_{2}$

   f. **Collisional deactivation:** $X^{*} + M \rightarrow X + M$

   Photolysis is a first-order reaction. The three factors on which direct photolysis depends are:

   a. The structure and electronic absorption spectrum of the molecule

   b. The quantum yield (Ø). This is a magnitude defined as the fraction of the amount of molecules decomposed (or product formed) for every photon absorbed, since not all photons absorbed produce an effective transformation. The typical quantum yield will be less than 1 (Eq. 2.5).

$$\Phi = \frac{\text{Molecules decomposed}}{\text{Photons absorbed}} \tag{2.5}$$

2. **Indirect photolysis:**

   Indirect photolysis consists of the degradation or transformation of pollutants in the environment through chemical reactions initiated by the absorption of light energy by other substances different from the pollutants, such as molecules or particles, that are present in the environment. This process can occur through several mechanisms, including sensitised photolysis, electron transfer reactions and radical-mediated reactions.

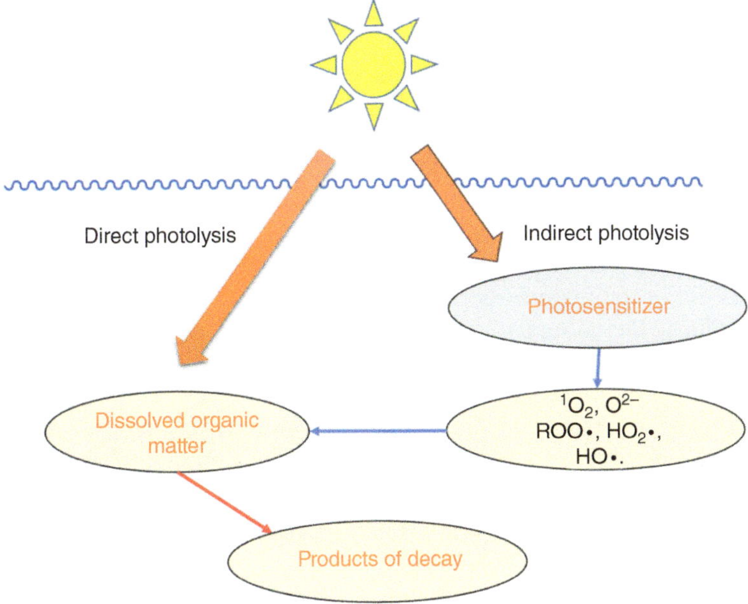

FIGURE 2.2   Types of photolysis in water bodies

Sensitised photolysis involves the presence of a *photosensitiser*, a chemical species that absorbs light energy and then transfers it to the pollutant molecule, causing it to break down. Natural organic matter (NOM) or dissolved organic matter (DOM) in water bodies can act as sensitisers when they absorb light energy, they may form various reactive species, such as singlet oxygen ($1O_2$), superoxide anion (O_2-) or radicals such as organic peroxides (ROO˙), hydroperoxyl (HO_2˙) and hydroxyl (HO˙) radicals. These reactive species provide effective oxidation of other molecules quite refractory to oxidation, such as aliphatic hydrocarbons in pharmaceuticals, especially those drugs that do not appreciably absorb light above 290 nm. This is the reason that the presence of a great variety of oxygenated compounds in seawater after an oil spill can be detected, such as aldehydes, alcohols, ketones, sulfoxides, sulfones and phenols. This process can occur in natural environments, such as rivers, lakes and oceans, where sunlight can penetrate the water surface and trigger sensitised photolysis reactions, resulting in the degradation of pollutants.

Another example of indirect photolysis is the electron transfer reactions that can occur between pollutants and other molecules or particles in the environment. For instance, pollutants can undergo electron transfer reactions with minerals, such as iron or manganese oxides, present in soil or sediment, leading to their transformation or degradation upon exposure to sunlight.

Indirect photolysis of pollutants plays an important role in the natural attenuation and degradation of pollutants in the environment and it is influenced by various factors, including the properties of the pollutants, the presence of other substances in the environment and the intensity and wavelength of the light source. It is an important process to consider in the fate and transport of pollutants in the environment and can be harnessed for environmental remediation strategies.

Oxidation

Oxidation reactions involve the transfer of electrons from a pollutant to an oxidising agent, resulting in the formation of more oxidised products. Oxidation is the main degradation transformation for most organic compounds in the troposphere and for many substances in surface waters. Many of these reactions are related to indirect photolysis reactions because the reactive species described above are actually oxidants. In Table 2.1 examples of the differences in reactivity of various organic chemicals for gas-phase oxidation in the troposphere are given.

The oxidation of hydrocarbons in the atmosphere leads to the formation of various products, including carbon dioxide, H_2O, formaldehyde (HCHO) and other substances.

It's important to note that the specific products of hydrocarbon oxidation in the atmosphere can vary depending on a variety of factors, including the type of hydro-carbon, the availability of O_2 and other reactive species, temperature and atmospheric conditions. Atmospheric chemistry is a complex and dynamic process that involves the interaction of multiple species and the final products of hydrocarbon oxidation can be influenced by a wide range of factors.

TABLE 2.1 **Half-lives (days) for some organic compounds caused by oxidation reactions**

Family of compounds	Half-lives (days) for tropospheric oxidation
Alkanes	1–10
Alkenes	0.06–1
Aromatic hydrocarbons	1–10
Alcohols	1–3
Halogenated derivatives[a]	100–4700

[a] Higher time-live values correspond to chlorofluorohydrocarbons (CFCs).

Topics of Interest: Oxidation of Inorganic Compounds

The oxidation of inorganic compounds may be performed both chemically or by a class of microorganisms named *chemolithotrophs*, which obtain energy for their vital processes through the transformation of inorganic materials. Some inorganic forms of reduced sulfur, such as sulfide (H_2S/HS^-) and elemental sulfur (S_0), can be oxidised in this way. Such reactions used to be coupled to the reduction of O_2 or nitrate (NO_3^-).

In soil and water bodies for inorganic species, most redox reactions are pH-dependent. For example, the oxidation of ferrous to ferric iron takes place by the reduction of O_2 in the presence of water (Eq. 2.6).

$$O_2 + 4Fe^{2+} + 4H^+ \rightleftarrows 4Fe^{3+} + 2H_2O \tag{2.6}$$

Since most redox reactions are pH-dependent, a convenient way of depicting these reactions is to use the Pourbaix diagrams. They are graphical representations that illustrate the thermodynamic stability of chemical species over a range of pH and electrode potential for a particular chemical system, typically involving a metal or other electrochemically active species. They are commonly used in electrochemistry and corrosion science to predict the predominant forms of a chemical species under different conditions of pH and electrode potential.

In a Pourbaix diagram, the x-axis represents the electrode potential (usually in volts) and the y-axis represents the pH of the solution. The diagram is divided into regions corresponding to different chemical species or phases that may exist under given conditions. These regions are delineated by lines called equilibrium lines, which separate the domains of stability for different species.

The primary species depicted in a Pourbaix diagram are usually the metal (or other central element) in different oxidation states, as well as various forms of water (e.g., H_2O, H^+, OH^-, H_2, O_2). By examining the Pourbaix diagram for a particular metal, it is possible to determine the conditions under which different forms of that metal are stable, for example, whether it exists as a metal, an oxide, a hydroxide or in some other chemical form.

Pourbaix diagrams for metals in natural waters provide valuable insights into their behaviour in aquatic environments, which is crucial for understanding corrosion processes, environmental fate and even remediation strategies.

For instance, iron can exist in various oxidation states in natural waters, primarily Fe(II) and Fe(III). The Pourbaix diagram for iron shows the stability regions of different iron species as a function of pH and electrode potential. In oxygenated waters, Fe(III) species such as $Fe(OH)_3$ and Fe_2O_3 are stable at high pH, while Fe(II) species like $Fe(OH)_2$ are predominant at low pH (Figure 2.3).

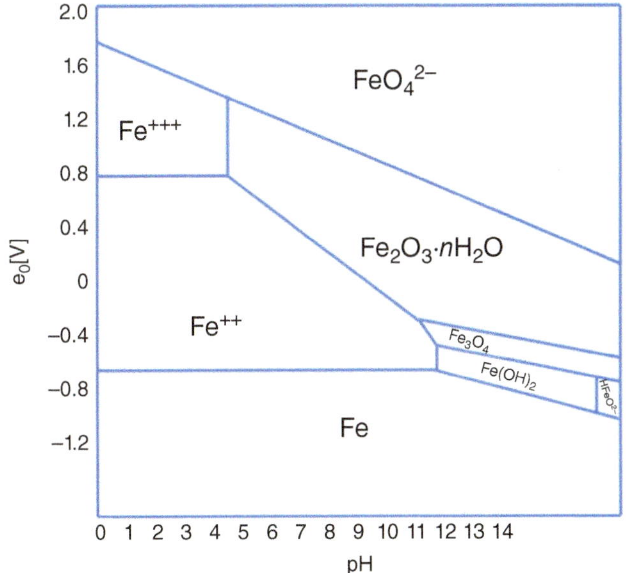

FIGURE 2.3   Pourbaix diagram for iron species

In soil and water bodies, organic matter and single organic molecules suffer oxidation reactions performed by microorganisms by their enzymatic systems constituted by enzymes such as laccases, monooxygenases and dioxygenases.

Laccases are a type of enzyme that belongs to the class of multicopper oxidases. They are found in many living organisms, including bacteria, fungi, plants and insects. Laccases are known for their ability to oxidise a wide range of organic and inorganic compounds, particularly phenols and aromatic compounds, using molecular O_2 as a co-substrate.

Mono-oxygenases and dioxygenases are types of enzymes that are involved in oxygenation reactions in living organisms. They both catalyse the incorporation of O_2 atoms into various organic molecules, but they differ in the number of O_2 atoms they utilise and the type of reactions they perform.

Mono-oxygenases and dioxygenases are specifically involved in the oxidation of pollutants, including pesticides, pharmaceuticals, aromatic compounds, PAHs and chlorinated compounds.

These enzymes play a significant role in maintaining the global carbon cycle through either the transformation or complete mineralisation of organic molecules, including organic pollutants.

Reduction

Reduction reactions involve the addition of electrons to a pollutant, resulting in the formation of more reduced products. For example, certain pollutants, such as heavy metals, can be reduced to less toxic forms through reactions with reducing agents, such as sulfides or organic matter, leading to the precipitation or immobilisation of the pollutants. Also, substances such as aromatic hydrocarbons, azo and nitro compounds or halogenated derivatives (aliphatic and aromatic) may be reduced under certain conditions in the environment. Electrons from the reduced organic molecule are transferred to other molecules, for example, via porphyrins that are formed in the decomposition of biological materials or via $Fe(II)/Fe(III)$ or $HS-/S_0$ systems.

For example, reduction reactions are the major mechanism for microbial reduction of pentachloronitrobenzene (Figure 2.4), which is reduced to pentachloroaniline or, in case of DDT, converted into DDD (Figure 2.5).

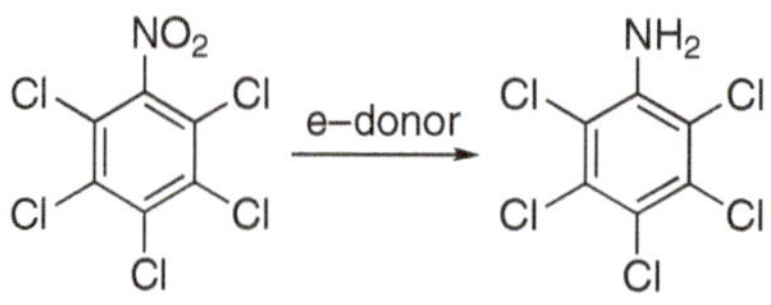

FIGURE 2.4   Nitro group reduction

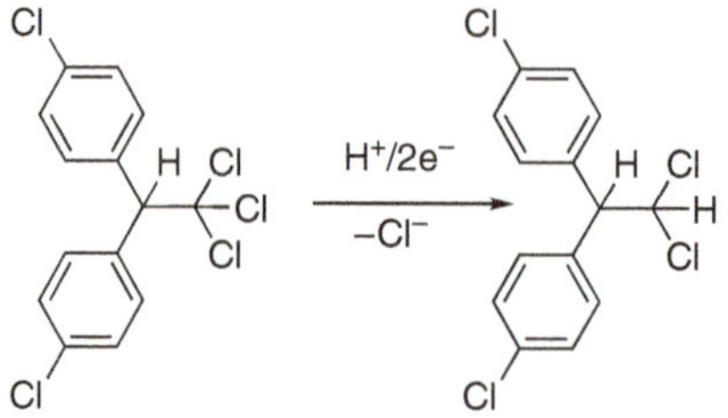

FIGURE 2.5   DDT biotic reduction

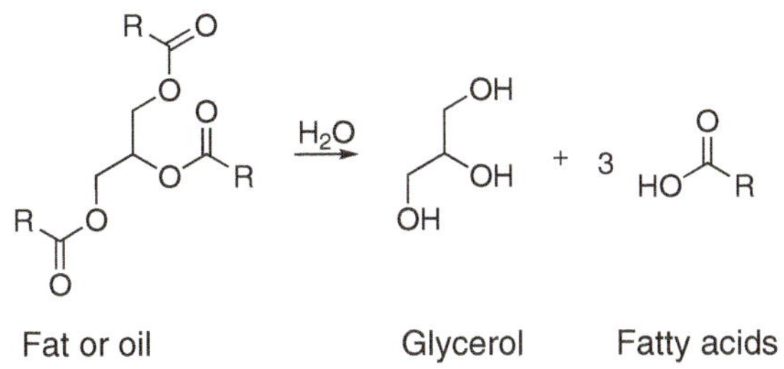

FIGURE 2.6   Hydrolysis of fats and oils

Reduction reaction pathways contribute significantly to the removal of organic molecules in sewage sludge, anaerobic biological systems, saturated soil systems and anoxic sediments. The reduction rate of specific compounds depends on factors such as temperature, pH, the prevailing redox potential and the physical and chemical properties of the substance to be reduced. As in hydrolysis reactions, usually more polar products are formed from the parent compound, which makes them more susceptible to further chemical attack and less likely to accumulate.

Hydrolysis

Hydrolysis reactions involve the breakdown of pollutants through reactions with water. For example, certain organic pollutants such as esters, amides or halogenated compounds can undergo hydrolysis in the environment by abiotic or biotic processes. Abiotic hydrolysis reaction rates can be a function of chemical parameters such as pH values, DOM, dissolved metal ions or many other compounds. In the case of biotic hydrolysis, the rate of reaction is often proportional to bacterial biomass and/ or specific enzyme concentrations. For example, fats and oils, as they are known, are esters formed by a molecule of glycerol and three fatty acids and the hydrolysis gives the parent molecules (Figure 2.6).

The hydrolysis reactions used to give more polar products than the parent molecules; consequently, when hydrolysis takes place, new molecules that are more water-soluble and less lipophilic are formed. Hydrolysis reactions are commonly catalysed in acidic or basic media and the rate of hydrolysis directly depends on the pH. Hydrolysis rate constants, kh, generally obey pseudo- first-order kinetics. The study of hydrolysis reactions used to be measured experimentally in laboratory tests, starting with the known quantity of the compound in a solution of fixed pH and following the disappearance of the compound over time. The concentration of the chemical typically declines exponentially with increasing time.

Topics of Interest: How to Predict Biodegradation of Organic Pollutants

Predicting the biodegradation of organic compounds can be challenging as it depends on a variety of factors, such as the chemical structure of the compound, environmental conditions and the microbial population present. However, there are several methods and tools that can be used to predict the biodegradability of organic compounds:

1. **Quantitative structure–activity relationship (QSAR) models:** These models use mathematical equations to predict the biodegradability of organic compounds based on their structural properties, such as molecular weight, functional groups and atom connectivity.

2. **Biodegradation prediction software:** There are several software programmes available that can predict the biodegradability of organic compounds based on their chemical structure and other factors such as temperature, pH and nutrient availability. There are several software programmes available that can predict the biodegradability of organic compounds. Here are some examples:

    a. **BIOWIN (Biodegradation Probability Program):** Developed by the US EPA, BIOWIN is a set of six models that use the structure of the compound to predict its biodegradability. Each model focuses on a different biodegradation pathway, such as primary biodegradation or ultimate biodegradation.

    b. **CATABOL (Computer-Automated Biodegradation Evaluation Tool):** Developed by the European Chemical Industry Council (CEFIC), CATABOL predicts the biodegradability of organic compounds based on their chemical structure and functional groups.

    c. **TOPKAT (Toxicity Prediction by Komputer-Assisted Technology):** Developed by Accelrys, TOPKAT is a suite of software tools that includes a module for predicting biodegradation. The module uses a database of biodegradation data to predict the likelihood of a compound being biodegraded.

    d. **ECOSAR (Ecological Structure Activity Relationships):** Developed by the US EPA, ECOSAR predicts the biodegradability and other environmental fate properties of organic compounds based on their chemical structure.

    e. **EPI Suite:** Developed by the US EPA, EPI Suite is a suite of software tools that includes modules for predicting the biodegradability, as well as other environmental fate properties, of organic compounds.

        It's important to note that the accuracy of these software programmes may vary depending on the specific compound being evaluated and the conditions of its exposure. Therefore, it's important to use multiple prediction methods and to validate the results with laboratory testing where possible.

3. **Biodegradation tests:** Laboratory-based tests can be used to determine the biodegradability of organic compounds. These tests typically involve exposing the compound to a microbial population under controlled conditions and monitoring its degradation over time. Biodegradation tests involve exposing the compound to a microbial population under controlled conditions and monitoring its degradation over time. Here are some examples of biodegradation tests:

    a. **OECD 301 ready biodegradability test:** This is a standard test method developed by the Organisation for Economic Cooperation and Development (OECD) for measuring the biodegradability of organic compounds in water. The test involves exposing the compound to a microbial population under aerobic conditions and measuring the amount of carbon dioxide produced over a 28-day period.

    b. **Modified sturm test:** This test is used to evaluate the aerobic biodegradation of organic compounds in soil. The test involves adding the compound to a soil sample and monitoring its degradation over time.

    c. **Closed bottle test:** This is a simple method for evaluating the biodegradability of organic compounds in water. The test involves adding the compound to a sealed bottle of water along with a microbial population and monitoring the amount of carbon dioxide produced over time.

4. **Microcosm studies:** Microcosm studies involve creating a small-scale ecosystem in the laboratory that mimics natural conditions. The compound is added to the microcosm

along with a microbial population and the degradation is monitored over time. This method allows for the evaluation of the biodegradation of compounds under conditions that are more like those found in the environment.

5. **Bioreactors:** Bioreactors are larger-scale systems that allow for the controlled study of biodegradation under different conditions. They are often used to evaluate the effectiveness of bioremediation strategies for the cleanup of contaminated soil or water.

   It's important to select a biodegradation test that is appropriate for the specific compound being evaluated and to consider the conditions of exposure that are relevant to the environment where the compound may be released.

6. **Literature review:** A literature review can provide valuable information on the biodegradability of organic compounds that have been previously studied. This can include information on the microbial populations involved, environmental conditions and degradation pathways.

2.7   Toxicity of Chemicals

Pollutants are present in the environment in the air, soil and water. Most known pollutants are anthropogenic, but many others have a natural source and they may affect plants, animals and humans because they interfere with or modify many biochemical and biological processes when living beings are exposed to them. Almost all known chemical substances cause or are suspected of causing damage depending on their exposure time or amount.

Concept of Toxicity

The toxicity of a substance is defined as its capacity to cause injury to a living organism. Toxicity[5] is produced by:

a. Acute exposure occurs when a single exposure to a substance may cause severe biological harm or death, usually characterised as lasting no longer than a day.

b. Chronic exposure is defined as continuous exposure to a toxic substance over an extended period of time (months or years) that can cause irreversible side effects.

Toxicity is a relative concept that must be referred to in four keyways:

1. Type and severity of injury and the time needed to produce that injury. This factor is a consequence of the so-called *intrinsic toxicity* of a substance. Objectively, there are substances with similar effects on living beings that, in the same conditions, produce several grades of injuries.

2. The quantity of a substance is known as a **dose**. That is a key subject to determine the effect on a living being of a substance.

---

[5] Ecotoxicity is a complementary concept to classic toxicity. This concept arises from observations that some chemicals can exert toxic effects in an ecosystem in several places, even far away from their initial emission point of introduction into the environment. Pollutants are part of a series of disturbing agents named stressors, defined as chemical or biological agents or any other external stimulus causing damage to an organism.

3. The way in which this quantity has been distributed, administered or absorbed: inhalation, ingestion, injection, etc. The effect and consequences of a toxic substance are very dependent on the way that a living being has been exposed to it.

4. Exposition time (short- or long-term expositions) and single doses or repeated doses.

How Is Toxicity Evaluated?

To estimate and evaluate the toxicity of chemical substances to living beings, it has been defined as a group of parameters based on laboratory tests (Table 2.2). They are performed with chemically pure substances; mixtures are rarely studied.

1. **NEC** (*No Effect Concentration*) is defined as the concentration of a substance that will not harm the species involved with respect to the effect that is studied. It is often the starting point for environmental policy and regulations.

2. **NOEC** (*No Observed Effect Concentration*) is the highest tested concentration for which there is no statistically significant difference in the effect produced by a substance when compared to the control group over the long term. NOEC is highly dependent on dose group setting and some authorities prefer the use of the ECx parameter as a chronic endpoint. ECx is defined as *the effect concentration corresponding to x-percent response*. The advantage of the estimation of ECx is that information from the whole concentration–effect relationship is taken into account and that confidence intervals can be calculated.

3. **NOAEL** (*No Observed Adverse Effect Level*) is the highest dose level that does not produce a significant increase in adverse effects in comparison to the control group. Animal toxicology studies of three types are used to determine the

TABLE 2.2   **Toxicity parameter' summary**

Acronym	Name	Typical units
NEC	No effect concentration	ppm or milligrams per litre (mg/L), in water or air mg/kg in soil or sediment studies
NOEC	No observed effect concentration	ppm or milligrams per litre (mg/L), in water or air mg/kg in soil or sediment studies
NOAEL	No observed adverse effect level	ppm or milligrams per litre (mg/L), in water or air mg/kg in soil or sediment studies
LD50	Lethal dose for the 50% of the population	mass per kilogram of body weight (e.g., milligrams per kilogram or mg/kg).
LC50	Lethal concentration for the 50% of the population	(ppm) or milligrams per litre (mg/L) for water or milligrams per cubic metre (mg/m^3) for air.

NOAEL. These are studies that detect:

i. Overt toxicity, such as lesions

ii. Biomarkers such as levels of liver enzymes in the blood serum

iii. Exaggerated pharmacodynamic effects

4. **LD50:** Corresponds to the abbreviation for *lethal dose, 50%*. The value of LD50 for a substance is the *dose required to* **kill half the members of a tested population after a specified test duration.** Usually, it is expressed as the mass of substance administered per unit mass of the test subject. The typical units are milligrams of substance per kilogram of body mass for bigger test subjects (mg/kg) or per 100 g for smaller animals (mg/100 g). For extremely toxic substances, LD50 is sometimes expressed as micrograms or nanograms of substance per unit mass of the test subject, generally kilograms of body mass (µg/kg or ng/kg, respectively). Lower LD50 values imply higher toxicities. Theoretically, tests may be performed with many kinds of animals, but in order to normalise and make results reproducible, most of the tests are done with rats and mice. On the other hand, tests must specify the route of entry or administration.

   a. Dermal, applied to the skin.

   b. Oral, given by mouth.

   c. Injected by intravenous, intramuscular or intraperitoneal (into the abdominal cavity) injection.

   Dermal and oral are the most common administration methods. In the last few years, other alternative tests for several substances have been approved without animals[6]. LD50 tests are very useful but somewhat unreliable because results may vary greatly between testing facilities due to multiple factors such as animal species and their genetic characteristics, mode of administration and environmental factors.

5. **LC50:** Corresponds to the abbreviation for lethal concentration, 50%. It is defined as a **lethal concentration of a substance able to kill half the members of a tested population after a specified test duration**. As in LD50, it is very important to clearly define how tests have been carried out and under what criteria.

With toxicity parameters, summarised in table, it is possible to perform a hazards guide of substances,

2.8 Classification of Substances According to Toxicity Parameters

Aquatic Toxicity

Aquatic toxicity refers to the harmful effects of substances on aquatic organisms, including fish, invertebrates, algae and other aquatic plants and is determined by organisms representing the three trophic levels:

---

[6] For example in 2011, the U.S. Food and Drug Administration approved alternative methods to LD50 for testing substances used in cosmetics without animal tests.

TABLE 2.3 **GHS classification for aquatic toxicity**

Exposure	Category 1	Category 2	Category 3	Category 4
Acute	≤1.0 mg/L	≤10 mg/L	≤100 mg/L	—
Chronic	≤1.0 mg/L	≤10 mg/L	≤100 mg/L	Long-lasting harmful effects

1. Algae or plants, representing 'primary producers'.
2. Invertebrates (e.g., crustaceans such as Daphnia spp.), representing 'primary consumers/secondary producers'
3. Vertebrates (usually fish) represent 'secondary consumers'.

In general, there are acute and chronic endpoints in aquatic toxicity.

Under GHS[7] criteria for aquatic toxicity measures, both acute and chronic exposure are divided into categories (Table 2.3), three for acute toxicity and four for chronic toxicity. Categories 1–3 for chronic toxicity are based on combining the criteria for Categories 1–3 for acute toxicity with criteria for persistence in the aquatic environment. Category 4 chronic toxicity may cause long-lasting harmful effects on aquatic life.

Subject key indicator species are exposed to concentrations of a substance for a determined period. For example, to determine acute toxicity measures of LC50, fish are exposed to toxics for 96 hours, while crustaceans are exposed for 48 hours.

a. Vertebrates (fish)
   i. Acute toxicity to fish (96 hours, LC50 in mg/L)
   ii. Long-term toxicity (28 days, NOEC in mg/L)
b. Invertebrates (crustaceans such as Daphnia)
   i. Acute toxicity to Daphnia (48 hours, EC50 in mg/L)
   ii. Long-term toxicity to Daphnia (21 days; NOEC in mg/L)
c. Algae
   i. Acute toxicity to algae (72–96 hours, EC50 in mg/L)
   ii. Chronic toxicity (long-term exposure for the determination of NOEC LOEC or ECx values)

Toxicity in Non-Aquatic Media

In the case of air, toxicity may be produced by gases, vapours, dust or mists, a typical experiment involves groups of animals exposed to a concentration or series of concentrations of a substance for a set period of time (usually four hours). The animals are clinically observed for up to 14 days. Other durations of exposure other than the traditional four hours may apply, depending on specific laws.

According to LD50 and LC50 data, substances are categorised into five groups (Table 2.4). Category 1 requires the least amount of exposure to be lethal and

[7]The Globally Harmonised System of Classification and Labelling of Chemicals (GHS) is an internationally agreed-upon standard managed by the United Nations that was set up to replace the assortment of hazardous material classification and labelling schemes previously used around the world.

TABLE 2.4   **Toxicity categories according to LD50 and LC50 values**

Method of administration	Category 1	Category 2	Category 3	Category 4	Category 5
Oral: LD50 measured in mg/kg of bodyweight	5	50	300	2,000	5,000
Dermal: LD50 measured in mg/kg of bodyweight	50	200	1,000	2,000	5,000
Gas inhalation: LC50 measured in ppmV	100	500	2,500	20,000	Undefined
Vapour inhalation: LC50 measured in mg/L	0.5	2.0	10	20	Undefined
Dust and mist inhalation: LC50 measured in mg/L	0.05	0.5	1.0	5.0	Undefined

Category 5 requires the most exposure to be lethal. GHS Category 5 corresponds to substances with the lowest concern about effects and covers chemicals with toxicities expected to fall in the range of oral and dermal LD50s of 2000–5000 mg/kg or equivalent doses for inhalation toxicity.

Topic of Interest: History of the LD50 Parameter

In 1927, J.W. Trevan attempted to find a way to estimate the relative poisoning potency of drugs and medicines used at that time. He developed the LD50 test because the use of death as a 'target' allows for comparisons between chemicals that poison the body in very different ways. Since Trevan's early work, other scientists have developed different approaches for more direct, faster methods of determining the LD50.

In the case of drugs, In pharmacology, an effective dose (ED) or effective concentration (EC) is a dose or concentration of a drug that produces a biological response. The term effective dose is used when measurements are taken in vivo, while the term effective concentration is used when the measurements are taken in vitro.

Examples of Estimated Toxicity Values

In Table 2.5, LD50 values are listed for a set of substances in descending order. It is remarkable that many common substances, even those included in the human diet, may show toxicity if a determined dose is overcome. This agrees with an idea developed before that toxicity is a relative concept.

TABLE 2.5 **Examples of toxicity values**

Substance	Animal, route	LD50
Water	Rat, oral	90,000 mg/kg
Sucrose	Rat, oral	29,700 mg/kg
Glucose	Rat, oral	25,800 mg/kg
Vitamin C	Rat, oral	11,900 mg/kg
Glyphosate	Rat, oral	10,537 mg/kg
Aspartame	Mice, oral	10,000 mg/kg
Urea	Rat, oral	8,471 mg/kg
Ethanol	Rat, oral	7,060 mg/kg
Methanol	Human, oral	810 mg/kg
Sodium chloride	Rat, oral	3,000 mg/kg
Paracetamol	Rat, oral	1,944 mg/kg
Ibuprofen	Rat, oral	636 mg/kg
Formaldehyde	Rat, oral	600–800 mg/kg
Hydrochloric acid	Rat, oral	238–277 mg/kg
Ketamine	Rat, intraperitoneal	229 mg/kg
Aspirin	Rat, oral	200 mg/kg
Caffeine	Rat, oral	192 mg/kg
Dichlorodiphenyltrichloroethane (DDT)	Mouse, oral	135 mg/kg
Cocaine	Mouse, oral	96 mg/kg
Thiopental sodium (used in lethal injection)	Rat, oral	64 mg/kg
Methamphetamine	Rat, intraperitoneal	57 mg/kg
Nicotine	Rat, oral	50 mg/kg
Heroin	Mouse, intravenous	21.8 mg/kg
Lysergic acid diethylamide (LSD)	Rat, intravenous	16.5 mg/kg
Arsenic trioxide	Rat, oral	14 mg/kg
Metallic arsenic	Rat, intraperitoneal	13 mg/kg
Sodium cyanide	Rat, oral	6.4 mg/kg
Hydrogen cyanide	Mouse, oral	3.7 mg/kg
Nicotine	Mice, oral	3.3 mg/kg
Mercury(II) chloride	Rat, oral	1 mg/kg
Nicotine	Human, oral	0.8 mg/kg (estimated)

TABLE 2.5   (Continued)

Substance	Animal, route	LD50
Sarin gas	Mouse, subcutaneous injection	172 µg/kg
VX (nerve agent)	Human, oral, inhalation, absorption through skin or eyes	140 µg/kg (estimated)
2,3,7,8-Tetrachlorodibenzodioxin (TCDD, in agent orange)	Rat, oral	20 µg/kg
Latrotoxin (from widow spider)	Mice	4.3 µg/kg
Batrachotoxin (from poison dart frog)	Human, subcutaneous injection	2–7 µg/kg (estimated)
Tetanospasmin (tetanus toxin)	Mice	2 ng/kg
Botulinum toxin (botox)	Human, oral, injection, inhalation	1 ng/kg (estimated)

2.9 European and USA Regulations

REACH Regulation

The REACH[8] regulation is a normative enacted by the EU in 2006 and its aim is to ensure a high level of protection of human health and the environment from the risks posed by chemicals.

The REACH regulation requires companies that manufacture, import or use chemicals in the EU to register those chemicals with the European Chemicals Register Agency (ECHA). The registration process involves providing information on the properties and hazards of the chemicals, as well as their intended uses.

The regulation also requires that chemicals that are of particular concern be evaluated by the ECHA and, if necessary, subject to authorisation or restriction. Authorisation is required for substances that are of very high concern, such as carcinogens, mutagens and reproductive toxins and are restricted or banned from use unless the manufacturer can demonstrate that the benefits of using the substance outweigh the risks. Restriction is used to limit or ban the use of certain hazardous chemicals in specific applications.

EPA Regulations

The EPA classifies chemicals based on their potential to cause harm to human health and the environment. This classification system helps the EPA identify which chemicals may require further regulation or management.

---

[8] REACH is the acronym for Registration, Evaluation, Authorisation and Restriction of Chemicals. A European Union regulation that requires companies to register and report information on the production, usage and hazards of chemical substances.

There are several different classifications used by the EPA, including:

1. **Hazardous Air Pollutants (HAPs):** These are pollutants that can cause or contribute to serious health problems such as cancer, respiratory problems and birth defects. The EPA has identified a list of HAPs that are subject to regulation under the Clean Air Act.

In addition to these classifications, the EPA also uses other criteria, such as the chemical's physical and chemical properties, to classify chemicals and determine their potential risks.

1. **Toxic Substances Control Act (TSCA) inventory:** The TSCA inventory is a list of all chemical substances that are manufactured or imported into the United States. The EPA classifies chemicals on the TSCA inventory as either 'active' or 'inactive.' Active chemicals are those that are currently being manufactured or imported, while inactive chemicals are not.

2. **Persistent, Bioaccumulative and Toxic (PBT) chemicals:** These are chemicals that are highly toxic, do not break down easily in the environment and can accumulate in the bodies of living organisms. The EPA has identified a list of PBT chemicals that require special attention and regulation.

3. **Hazardous waste:** The EPA regulates the disposal of hazardous waste, which is waste that is dangerous or potentially harmful to human health or the environment. Hazardous waste is classified based on its characteristics, such as ignitability, corrosivity, reactivity and toxicity.

CHAPTER 3

Inorganic Pollutants

3.1 Elements and Periodic Table

The periodic table of elements is a display of the chemical elements, arranged by atomic number (Figure 3.1). The vertical columns in the periodic table are nominated *groups* and contain elements with similar chemical behaviours. Horizontal rows are named *periods* and they are numbered 1–7 on the left-hand side of the table. The atomic numbers increase from up to down and left to right. Elements are classified by metals, nonmetals and metalloids.

Of the elements that have been discovered, there are 90 that occur in nature in appreciable amounts. There is another set of elements that occur in nature as a result of the radioactive decay of other elements, so the grand total of natural elements is 94. Elements such as neptunium, plutonium, americium, curium, berkelium and californium have been artificially synthesised and show radioactivity and short life spans, so they have little practical interest.

Some of these groups have accepted names. In Table 3.1, the more significant groups are:

Nature is constituted by chemical elements, there is nothing out of the Periodic table.

The abundance of every element on the periodic table is very diverse but, in all cases, limited.

The European Chemical Society sponsored in 2019, the international year of the periodic table, a representation of the periodic table that shows the scarcity of the 90 natural elements[1] that make up everything, in which the area assigned to every element is proportional to their abundance, paying special attention to elements used in the fabrication of an object of daily use in our society: the smartphone (Figure 3.2).

It is striking that about a dozen elements are under serious threat of scarcity in the next century if current consumption is maintained.

---

[1] https://www.euchems.eu/euchems-periodic-table/.

Understanding the Chemistry of the Environment, First Edition. Francisco G. Calvo-Flores.
© 2025 John Wiley & Sons Ltd. Published 2025 by John Wiley & Sons Ltd.

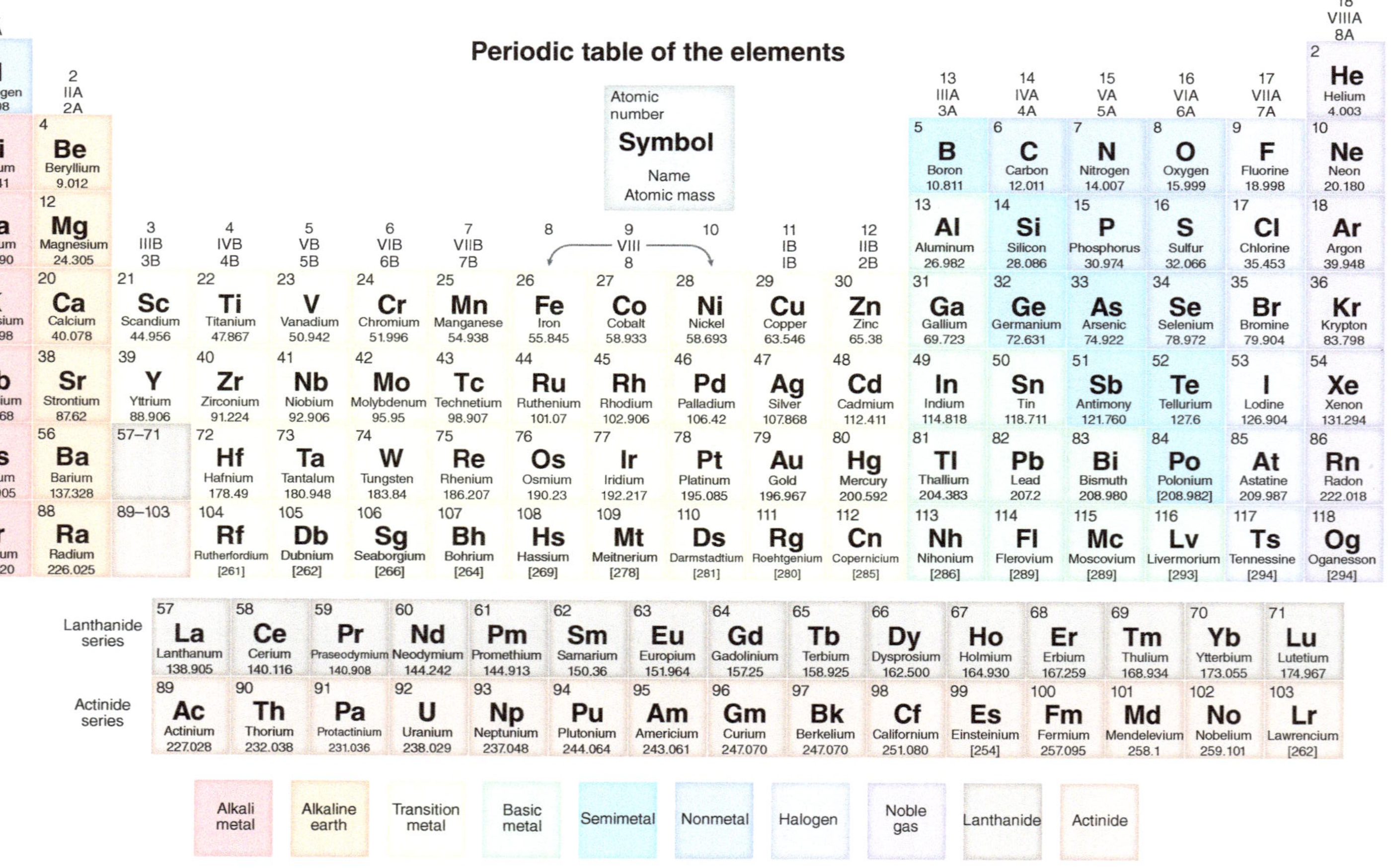

FIGURE 3.1 Periodic table

TABLE 3.1   Periodic table groups

Group	Name
Group 1	Alkaline metals
Group 2	Alkaline earth metals
Group 15	Pnictogens
Group 16	Chalcogens
Group 17	Halogens
Group 18	Noble gases
Groups 3–12	Transition metals
Groups 57–71	Lanthanides
Groups 89–103	Actinides

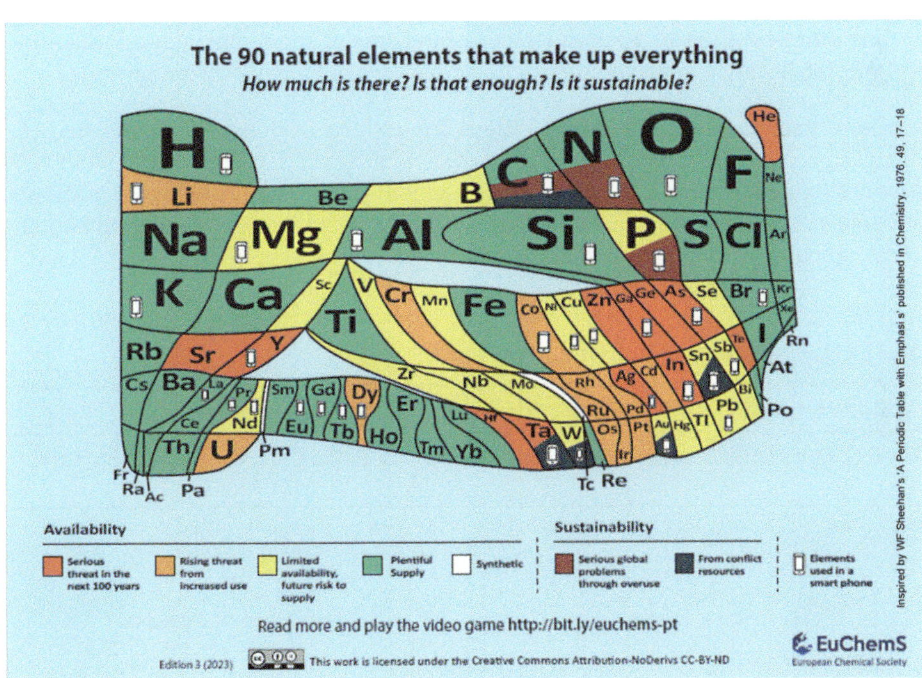

FIGURE 3.2   Natural elements and occurrences. *Source: EuChemS / CC BY ND*

3.2   Chemical Elements and Living Beings

All the elements of the periodic table are not present in living beings. Some of them have a key role in living beings, forming part of their structure or participating in diverse biological processes.

By contrast, other elements have no nutritional properties and, upon intake, show serious toxic effects. According to this perspective, elements are classified into two groups:

1. **Essential elements:** These are elements considered indispensable for the normal development of living beings. Deficiencies of essential elements cause abnormal development, severe malfunctions or even death.

2. **Non-essential elements** can be defined as elements that don't play any role in living organisms. The absence of non-essential elements does not affect the body of an organism.

Living organisms contain about 25 different elements that are classified into three groups (Table 3.2).

1. **Major elements:** They are a set of six elements (carbon, hydrogen, oxygen, nitrogen, phosphorus and sulfur) that make up about 98% of the mass of any living organism. The acronym CHNOPS is frequently used for these elements.

2. **Minor elements** are present in living organisms in lower proportions, but they also have an important role in the correct functioning of organisms. These elements include sodium, chlorine, fluorine, potassium, calcium, iron and magnesium.

3. **Trace elements:** They are present in living beings in very low concentrations or any other measure of amount (traces), but some of them may play an essential role, for instance, when bound to the active sites of enzymes or coenzymes.

Both essential and nonessential elements induce responses in living beings. When concentration or dose is plotted versus biological response, a *dose–response curve* is obtained (Figure 3.3). For essential elements, the dose–response curve shows four zones up and down from a hypothetical line corresponding to harm and benefit doses: deficient, normal, toxic or lethal.

TABLE 3.2 **Elements in living organisms**

Major elements	Minor elements	Trace elements
C, H, N, O, P and S	Na, Ca, Mg, Cl, F and Fe	B, Co, Cu, I, Fe, Mn, Mo, Zn, Cr, F, Ni, l, Se and V

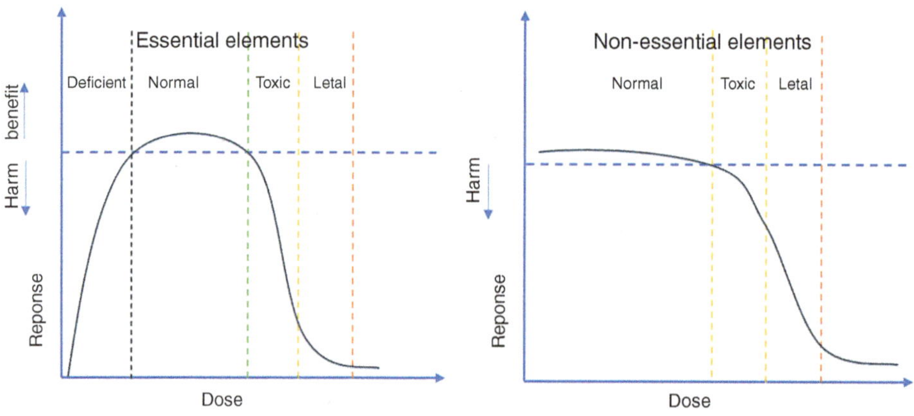

FIGURE 3.3 Biological response of essential and non-essential elements

In the case of non-essential elements, the response is slightly different. For non-essential elements, there exists a range of doses at which living beings show a certain tolerance, but soon toxic effects appear and finally death occurs.

If the contribution to the diet of an essential element is deficient, negative effects on the normal development of living beings are produced. No Observed Adverse Effect Level (NOAEL) occurs for a range of concentrations or contributions. At increasing doses or concentrations, toxic effects arise and, finally, death occurs.

3.3 Inorganic Pollutants

Inorganic pollutants are chemical substances derived from non-living sources. They can occur in the environment by natural processes such as volcanic eruptions, wildfires, dust particles becoming airborne, metal salts dissolved in water bodies, etc., which can sometimes reach levels with adverse effects on the environment and human health. But also they can be released into the environment through various human activities such as industrial processes, transportation, agriculture, mining and waste disposal. Main inorganic pollutants are heavy metals and their derivatives, certain salts, gases from combustion processes, inorganic compound components of fertilisers, radionuclides, chlorinated compounds, etc.

3.4 Pollution with Heavy Metals

The term heavy metals refers to a variety of metals based on their density, atomic number and toxicity, even at low concentrations. The more accepted list of hazardous heavy metals is as follows: cadmium (Cd), mercury (Hg), thallium (Tl), chromium (Cr) and lead (Pb). The list also includes two metalloids[2], such as arsenic (As) and antimony (Sb), that are also able to induce toxicity at low levels of exposure.

Other metals such as iron (Fe), copper (Cu), selenium (Se) or zinc (Zn) are considered essential to maintain the metabolism of many living beings, but in very low amounts; however, at higher concentrations, these elements become toxic.

For decades, there has been an increasing concern associated with environmental pollution by heavy metals because ecosystems and human exposure have risen dramatically as a result of exponential growth in their use in agriculture and industry and for their wide technological and domestic applications. Detectable amounts of heavy metals may be found in soil, water bodies or the air as aerosols, constituting a serious pollution problem.

Heavy metals are toxic to living organisms, including humans, due to their ability to accumulate in tissues and interfere with various biological processes. The toxicity of heavy metals depends on factors such as the specific metal, its chemical form, the dose, the duration of exposure and the route of exposure. The main features of significant heavy metals such as cadmium (Cd), mercury (Hg), thallium (Tl), chromium (Cr), lead (Pb), arsenic (As) and antimony (Sb) are described below.

---

[2] Metalloids, or semimetals, are a small class of elements that show some properties of both metals and nonmetals such as B, Si, Ge, Sb, and Te.

3.5 Cadmium (Cd)

Cadmium is a naturally occurring element that is found in the Earth's crust. The highest level of cadmium compounds in the environment is accumulated in sedimentary rocks and marine phosphates contain about 15 mg of cadmium/kg. It is most commonly found in association with zinc, lead and copper ores. Cadmium is also present in some mineral deposits, such as greenockite, which is a rare mineral containing cadmium sulfide.

Cadmium is commonly used in a variety of applications, including:

- Rechargeable nickel–cadmium (Ni–Cd) batteries. These batteries are commonly used in portable electronic devices such as mobile phones, laptops and cameras.
- Fabrication of cadmium-based coatings and plating to protect steel from corrosion. These coatings are commonly used in the aerospace and automotive industries.
- Production of pigments with bright and intense colours in paints, plastics and ceramics.
- In nuclear reactors as a neutron absorber to control the rate of nuclear reactions and prevent the buildup of radioactive isotopes.
- Cadmium telluride is used as a component in the manufacture of thin-film solar cells.

Toxicity

Cadmium exposure can occur through various means, including inhalation of cadmium-containing dust, ingestion of contaminated food or water and diarrhoea or skin contact with materials containing cadmium. Acute exposure to cadmium can result in symptoms like nausea, vomiting, diarrhoea, abdominal pain and respiratory distress. However, chronic exposure to cadmium can lead to more serious health issues, such as kidney damage, lung damage, bone disease and even lung and prostate cancer.

3.6 Chromium (Cr)

Chromium is the 21st most abundant element in the Earth's crust and can be found in various minerals. The primary source of chromium is chromite ore, which is a mineral that contains chromium, iron and oxygen ($FeCr_2O_4$).

Chromium is widely used in a variety of applications, including:

- Stainless steel production: The majority of chromium produced is used in the production of stainless steel, including kitchenware, cutlery and medical instruments.
- Production of aircraft parts, such as landing gear and engine components, due to their high strength and corrosion resistance.
- Production of various alloys, including nickel/chromium alloys, which are used in dental and medical implants.

- Plating to provide a hard, durable and corrosion-resistant surface to a variety of metal objects, including car parts, plumbing fixtures and motorcycle components.
- Production of refractory materials, which are used in high-temperature applications such as kilns, furnaces and ovens.
- Pigments for paint, ink and ceramics.
- Tanning industry of leather with chromium salts.

The toxicity of chromium compounds is highly dependent on their oxidation state and solubility. Trivalent chromium compounds, like chromium (III) chloride and chromium (III) oxide, are generally less toxic compared to hexavalent chromium compounds.

Trivalent chromium is naturally present as a trace element in various foods, such as meat, shellfish, fish, eggs, whole-grain cereals, nuts and some fruits and vegetables. It may play a role in carbohydrate, lipid and protein metabolism by potentiating insulin action, but it is not considered an essential mineral because its absence or deficiency does not produce abnormalities that can be reversed by adding chromium.

Nevertheless, some people use it as a supplement, with the belief that chromium forms a compound that enhances the effects of insulin and helps lower glucose levels. This use as a supplement is somewhat controversial.

The toxicity of trivalent chromium is considered moderate when present in higher concentrations than traces, as it is poorly absorbed by organisms through any route. However, high doses of trivalent chromium sometimes may cause gastrointestinal ulcers and kidney and liver damage.

On the other hand, hexavalent chromium compounds, such as chromium (VI) oxide, potassium dichromate or other hexavalent compounds, are highly toxic. They are known to be carcinogenic, particularly for workers who come into contact with them. Prolonged exposure to hexavalent chromium increases the risk of developing lung, nasal and sinus cancer. Workers in industries like chromate production, chromate pigment and chrome electroplating in the United States who were employed before the 1980s have shown increased rates of lung cancer mortality. Animal studies have also demonstrated that certain hexavalent chromium compounds can cause lung cancer when directly placed in the lungs.

3.7 Lead (Pb)

Lead is a metal obtained from galena, the primary ore mineral, which is mainly composed of lead sulfide (PbS). It is easily extracted and converted into metallic lead and it shows a relatively low melting point (as measured by its °C), ductility and relative inertness.

Lead has been used since ancient times. The Romans used lead extensively to design many objects, such as plates, cooking utensils, urns for wine, makeup and indoor plumbing.

Currently, lead metal has a wide range of applications in various industries due to its unique properties, such as its high density, low melting point, corrosion resistance and malleability, including:

- It has been employed since the Middle Ages for the fabrication of bullets and ammunition
- Their alloys have been used in letterpress printing since its invention.
- Rechargeable lead-acid for automobiles, boats, UPS systems and solar power storage.
- Due to its high density, lead is an effective material for radiation shielding in medical, nuclear and industrial applications. It is also used to line x-ray and gamma-ray rooms.
- For decades, it has been used in plumbing systems, such as pipes, fittings and solder, due to its corrosion resistance and malleability and in roofing materials, such as flashing and gutters, due to its durability and resistance to corrosion.
- Fishing gear and belts weigh more for diving due to their density.
- Production of pigments, such as lead white, which is used in oil paints, as well as in ceramic glazes and glass production.

Toxicity

Lead is listed as one of *the ten major chemicals of public health concern by the World Health Organization (WHO)*. It is a very toxic cumulative pollutant that affects multiple organs and is particularly harmful to children. The main sources of lead exposure are inhalation of lead particles generated by burning materials containing lead, industrial activities, ingestion of lead-contaminated drinking water and direct contact with lead-based products in work spaces or homes.

Lead toxicity can occur from a variety of sources, including lead-based paint, contaminated soil, lead pipes and plumbing fixtures and certain types of consumer products.

Lead is toxic when it enters the body and it can cause a wide range of health problems. Lead toxicity is especially dangerous for children, as their developing brains and bodies are more susceptible to its effects. Exposure to high levels of lead can cause developmental delays, learning difficulties and behavioural problems. It can also damage the nervous system, kidneys and other organs.

Prevention of lead exposure is key and efforts to reduce or eliminate sources of lead are important for protecting public health.

Topics of Interest: Saturnism

Lead poisoning is known as Saturnism. It is a type of illness that occurs when a person is exposed to high levels of lead over an extended period of time. Lead can be found in various sources, including old lead-based paint, contaminated soil, drinking water from old lead pipes and some industrial activities.

Symptoms of Saturnism can vary depending on the severity of the poisoning and the duration of exposure. In children, symptoms can include developmental delays, behavioural problems, abdominal pain, anaemia and learning difficulties. In adults, symptoms can include high blood pressure, joint and muscle pain, memory loss and mood disorders.

Lead poisoning is typically diagnosed through blood tests that measure the level of lead in the bloodstream. Treatment can involve chelation therapy, which involves administering medications that bind to lead and help remove it from the body. However, the best course of action is to prevent lead exposure in the first place by avoiding sources of lead and taking measures to reduce exposure.

The three forms of lead that are an environmental concern and their sources are as follows:

a. **Elemental lead:** The three main sources of elemental lead pollution are car batteries, bullets and hunting ammunition and lead pipes and plumbing connections that have been used for decades in homes and water supply installations.

Batteries must be properly recycled to avoid environmental hazards. They are crushed in order to separate out their different components, which are employed to make new batteries.

Lead-based ammunition is an important source of lead pollution in terrestrial and aquatic environments, their inhabitants, and finally, humans. Legislative actions are being developed to substitute lead ammunition[3], for example, with steel-based shot, the most common alternative, but such ammunition is regarded with scepticism by many hunters and conflicts with the commercial interests of suppliers.

Lead pipes and other lead plumbing materials are the main sources of lead in drinking water. It is a serious problem, especially in areas where the water has high acidity or low mineral content. In both cases, lead is transformed into soluble substances that are transported by drinking water. An increase in lead concentrations in drinking water may be caused by inadequate municipal water treatment. This is the case in Washington, DC, which changed in 2001 from free chlorine to chloramines for water disinfection without first studying the potential impact. The chloramines made the water far more corrosive and tragically, extremely high lead levels pervaded the city. Authorities are still working to fix the lead problem today. In the case of slightly basic drinking water containing sulfate or carbonate salts, a patina of very insoluble lead salts protects the inner surface of pipes, avoiding lead dissolution. To address this real hazard, many municipalities added corrosion inhibitors to the water, such as zinc orthophosphate ($Zn_3(PO_4)_2$) or procured pH adjustments.

Nowadays, this problem has been minimised by replacing lead with other materials, mainly copper or PVC. Only in old buildings and installations does it persist.

b. **Inorganic derivatives:** An important source of lead pollution in homes is lead paint-based. Lead was a component of very popular paints used in fine arts, house decoration and toys. Lead compounds were intentionally added as pigments, drying agents and anti-corrosive agents. Such paints have been very popular for decades, but they have been banned in many countries due to their harmful effects on human health, particularly on children because they taste sweet. By ingestion, they can cause damage to the nervous system and kidneys and may cause delayed development in children. Other effects of these inorganic lead compounds include reproductive problems, including a decrease in sperm concentration in adult men.

---

[3] In 2017, the European Commission requested the European Chemicals Agency (ECHA) to prepare a purpose for restrictions on lead gunshot in wetlands under the EU REACH (Registration, Evaluation, Authorization, and Restriction of Chemicals) Regulation.

The United States banned the manufacture of lead-based paint for residential use in 1978 through the Residential Lead-Based Paint Hazard Reduction Act (RLPHRA), also known as Title X. This law was enacted to protect children and pregnant women from lead-based paint hazards in housing built before 1978 when lead-based paint was commonly used. The US Environmental Protection Agency (EPA) regulates lead-based paint activities, including renovation, repair and painting (RRP) in target housing and child-occupied facilities. The European Union (EU) has taken steps to regulate lead-based paint as well. In 2003, the EU banned the marketing and use of lead-based paint in newly constructed buildings and in 2010, it further extended the ban to include all consumer paints, including those used for industrial and professional purposes, through the Registration, Evaluation, Authorization and Restriction of Chemicals (REACH) Regulation. Many other countries around the world have also banned or restricted the use of lead-based paint. For example, Canada banned the use of lead-based paint in residential properties in 1991 and Australia phased out lead-based paint in 1997. Other countries, such as Japan, New Zealand and several countries in Latin America, have also implemented regulations to ban or limit the use of lead-based paint. Table 3.3 summarises common lead pigments that have been used in paints and their environmentally friendly alternatives[4].

c. **Organic lead:** Lead is able to form *organolead compounds* when it bonds to organic moieties. One example of such a compound is lead tetra acetate, $Pb(OAc)_4$, which is used in organic synthesis as an oxidising agent.

By far, the organolead compounds that show environmental relevance and concern are tetraethyl lead $(CH_3CH_2)_4Pb$ (TEL) and tetramethyl lead $(CH_3)_4Pb$ (TML). Both present Pb−C bonds and they have been used for decades as anti-knock additives for gasoline in order to boost the octane rating, increasing vehicle performance and fuel economy. But they have represented an important source of atmospheric pollution. In the engine, tetraethyl lead may be oxidised to give lead particles according to the following reactions (Eq. 3.1):

$$\left(CH_3CH_2\right)_4 Pb + 13O_2 \rightarrow 8CO_2 + 10H_2O + Pb$$
$$Pb + 1/2O_2 \rightarrow PbO \tag{3.1}$$

TABLE 3.3 **Lead pigments and alternatives**

Fórmula	Common name	Alternative
$PbCrO_4$	Chrome yellow	Cadmium, strontium and barium chromates
Pb_3O_4/Pb(II) and Pb(IV)	Red lead (Minium)	Red oxide (Fe_2O_3)
$PbCO_3$	White lead	Titanium oxide (TiO_2)

[4] A 1:1 replacement for pigments is almost impossible.

Pb and PbO would be produced in such amounts that they would quickly accumulate, damaging the engine. For this reason, halogenated lead scavengers such as 1,2-dichloroethane and 1,2-dibromoethane are added to gasoline to give the volatile lead (II) chloride ($PbCl_2$) and lead (II) bromide ($PbBr_2$), respectively (Eq. 3.2), that are emitted to the atmosphere. TML shows similar behaviour to TEL.

$$\left(CH_3CH_2\right)_4 Pb + C_2H_4Cl_2 + 16O_2 \rightarrow 10CO_2 + 12H_2O + PbCl_2 \qquad (3.2)$$

The use of leaded gasoline in cars was banned in Canada in 1990, in the United States in 1996 and in the EU in 2002.

Topics of Interest: New Additives for Gas

Lead additives are incompatible with catalytic converters and an alternative to lead compounds for octane boosters must be used. One early solution was methyl tertiary butyl ether (MTBE) and mixtures of aromatic hydrocarbons such as benzene, toluene, ethylbenzene and xylene (BTEX). However, MTBE is quite water-soluble and poses a potential risk for the contamination of drinking water. According to EPA regulations, it has been removed from commercial gasoline formulations in the United States since 2005. Other oxygenate compounds are available as additives for gasoline, such as ethanol and ethers such as ethyl *tert*-butyl ether (ETBE). However, nowadays additives for gasoline and diesel fuels are complex mixtures of substances with a multipurpose action, not only to modify octane numbers but also to stabilise the fuel over time, improve combustion, control soot and protect the engine.

3.8 Mercury (Hg)

Cinnabar is the most common mineral source of mercury and it is the main ore of mercury. Cinnabar is a bright red mineral that contains mercury sulfide (HgS).

The industrial demand for mercury has grown throughout the 20th century and began to sharply decline between 1980 and 1994 as a result of the emergence of regulatory laws. Its use in consumer products and industrial processes has been greatly reduced due to environmental and health concerns. However, it still has some specialised applications, such as laboratory and scientific devices, dental amalgams, electrical switches, industrial processes, art and decoration, etc.

Mercury is a unique metal that may be used in three forms: elemental, inorganic and organic compounds.

a. **Elemental mercury** is a liquid at room temperature that has a high vapour pressure and is released into the environment as mercury vapour. It has been used in a variety of applications:

- Mercury thermometers are widely used in laboratories and medical facilities for temperature measurement. Mercury expands when heated and contracts when cooled, allowing the temperature to be read from a scale on the thermometer.

- Dental amalgam, which is a mixture of mercury, silver, tin and copper, has been used for dental fillings for over a century. Although the use of mercury in dental fillings is controversial, it remains used due to its durability and ease of use.
- Electrical switches and relays in electrical equipment because mercury is a good conductor of electricity and does not corrode.
- Mercury vapour lamps are used in outdoor lighting, streetlights and other industrial applications. These lamps are energy-efficient and have a long lifespan.
- Production of various chemicals, such as chlorine and caustic soda.
- Art and cultural heritage: Mercury was historically used in the production of certain pigments for paintings and in gilding techniques. It is also used for conservation and restoration of cultural heritage artefacts.

b. **Inorganic compounds of mercury:** The metal shows oxidation states of +1 or +2. Mercury salts have a variety of applications in various fields, including:

- Manufacturing of various drugs and medicines. They are also used as preservatives in some vaccines.
- Fungicides and herbicides in agriculture. Some salts can be used to control fungal diseases in crops and prevent weed growth.
- Laboratory reagents in various chemical tests and experiments. For example, they can be used to test for the presence of certain elements in a sample.
- Electroplating processes to deposit metals onto surfaces to create a decorative or protective coating on a surface.
- Mercury salts as an electrode material.
- In the mining industry to extract gold from ores. This process is known as amalgamation and involves mixing mercury with the ore to form an alloy.

c. **Organomercury compounds:** They are chemical compounds that contain a carbon-mercury bond, where mercury is covalently bonded to a carbon atom in alkyl or aryl groups. Their use has been limited due to their toxicity and environmental concerns. Some of the applications of organomercury compounds are:

- Organomercury compounds were once used as fungicides and seed disinfectants in agriculture. However, due to their toxicity and environmental concerns, their use has been banned in many countries.
- As catalysts in some chemical reactions, such as the oxymercuration reaction, which is used to add alcohol groups to alkene molecules.
- It has been used in the past as an antimicrobial agent and a preservative in some vaccines. However, due to their toxicity, they are no longer used for these purposes.
- **Laboratory research:** Organomercury compounds are used in some laboratory research, such as in the preparation of organic mercury reagents for use in organic synthesis.

In nature, the most common organic mercury compound is methylmercury. It is formed as a result of the methylation of some inorganic +2 mercury compounds by algae or microorganisms to give an organometallic cation, Me-Hg$^+$, bonded to anions such as chloride (Cl$^-$), hydroxide (OH$^-$) or nitrate (NO$_3^-$). Methylmercury is particularly known for its toxic effects on the nervous system and it is often associated with environmental contamination of fish and seafood.

Toxicity

Mercury is a toxic heavy metal that can cause a variety of health problems when ingested or inhaled. Because mercury is ubiquitous in the environment, humans, plants and animals are all unable to avoid exposure to any form of mercury.

The two types of mercury compounds, organic and inorganic, differ in their degree of toxicity and in their effects on the nervous, digestive and immune systems, on the lungs, kidneys, skin and eyes. Both inorganic and organic mercury compounds are absorbed through the gastrointestinal tract and affect other systems through this route.

Elementary mercury mainly causes health effects when inhaled as a vapour, where it can be absorbed through the lungs. Workers exposed to an elemental mercury level in the air of $20\,\mu g/m^3$, may observe mild, subclinical signs of central nervous system toxicity.

Exposures to inorganic mercury derivatives are similar to exposures to elementary mercury and include damage to the skin and nervous system, such as tremors, changes in nerve responses, emotional changes (mood swings, irritability, nervousness and excessive shyness), insomnia, muscular weakness, muscle atrophy, twitching, headaches, disturbances in sensations, kidney and respiratory failures, and finally, death.

Organic mercury enters the food chain through fish, shellfish and eventually humans. It is easily absorbed through the skin and mucous membranes in tissues and organs. Common symptoms of methylmercury poisoning may include loss of peripheral vision, 'pins and needles' feelings in the hands, feet and around the mouth, lack of coordination of movements, impairment of speech, hearing, walking and muscle weakness. Children and infants are especially affected, producing severe impacts on their cognitive thinking, memory, attention, language, fine motor skills and visual-spatial skills.

According to the Agency for Toxic Substances and Disease Registry (ATSDR), mercury is not considered a human carcinogen, although the EPA classifies mercury chloride and methyl mercury as possible human carcinogens.

Topics of Interest: Mercury Poisoning in Minamata

Mercury poisoning in Japan has been a significant environmental and health issue for several decades. The most notorious case of mercury poisoning occurred in the mid-20th century in the city of Minamata, located on the southern coast of Japan. The poisoning was caused by industrial wastewater discharged into Minamata Bay by a local chemical factory, which contained high levels of methylmercury. This led to the contamination of fish and shellfish in the bay, which were then consumed by the local population.

The resulting outbreak of Minamata disease, a neurological disorder caused by mercury poisoning, affected thousands of people, resulting in deaths and serious disabilities. The incident led to a public outcry and spurred environmental and health regulations to prevent similar incidents from happening in the future.

Since then, there have been several other cases of mercury contamination in Japan, including in the town of Niigata in the 1960s and 1970s, where industrial wastewater containing mercury was discharged into the Agano River, leading to widespread contamination of fish and shellfish. More recently, in 2010, a study found that a significant proportion of the population in the coastal city of Taiji had elevated levels of mercury in their hair, likely due to the consumption of contaminated fish and shellfish.

3.9 Thallium (Tl)

Thallium is a relatively rare and toxic element that occurs in nature mainly as a trace element associated with other minerals such as heavy metal sulfide ores.

Thallium minerals are, therefore, not very common.

Thallium is a soft, grey metal that has some unique applications in various fields:

- Thallium-201 is a radioactive isotope of thallium that emits gamma rays. It is used in nuclear medicine for imaging purposes, such as cardiac imaging and tumour detection.
- In the manufacture of semiconductors as it is a good electrical conductor and has unique optical properties.
- Thallium-based glasses and crystals are used in infrared optics for applications such as thermal imaging and night vision.
- Materials that have been used to develop high-temperature superconductors, which have potential applications in energy storage and transmission.
- Used in the past as a triggering agent for nuclear weapons.
- Thallium sulfate has been widely used in the past as a rat poison.

Toxicity

The main sources of thallium exposure include the gaseous emissions of cement factories, coal-burning power plants and metal sewers. In the case of water bodies, thallium pollution is produced by the leaching of ore processing operations.

Thallium compounds are extremely toxic. In most cases, poisoning is difficult to detect because thallium usually shows nonspecific effects due to its multi-organ involvement. The initial symptoms include fever, gastrointestinal problems, psychotic behaviour with hallucinations and dementia, convulsions and coma.

Because of its high toxicity, the WHO recommended in 1973 against its use as a rodenticide.

3.10 Arsenic (As)

Arsenic is relatively abundant in the earth's crust, with an estimated concentration of 1–2 parts per million (ppm). However, its abundance varies depending on the location and geological conditions. Arsenic-rich deposits can be found in areas with volcanic activity, hydrothermal vents and sedimentary rocks in various forms, including sulfides, oxides, arsenates and arsenites. Some common arsenic minerals are arsenopyrite, realgar, orpiment and enargite.

This element has a variety of applications, but many of them are limited or restricted due to its toxicity. Here are a few of the most common applications:

- Despite its toxicity, arsenic derivatives have been used for medicinal purposes for centuries and it is still used in some traditional medicines, and in some cases, in the treatment of acute promyelocytic leukaemia (APL), a type of blood cancer.
- Arsenic containing compounds were once widely used in pesticides, but their use has declined or banned due to their toxicity.
- Manufacture of pigments in paints and dyes, but their use has declined due to their toxicity.
- Production of glass to improve its clarity and brilliance.
- As preservative in the treatment of wood to protect it from insects, fungi and decay.
- Production of semiconductors, including transistors, solar cells and light-emitting diodes (LEDs).
- Particularly during World War I, it was used in the form of hydrogen arsenide, also known as arsenic hydride or 'Lewisite', which is a colourless, oily liquid with a garlic-like odour and other organic derivatives.

Some inorganic arsenic compounds are being used for medical purposes for the treatment of various diseases, such as leukaemia, psoriasis and chronic bronchial asthma. For example, arsenic trioxide (As_2O_3) is currently used in combination with tretinoin (a class of retinoid) to treat APL, a type of cancer in which there are too many immature blood cells in the blood and bone marrow.

Arsphenamine, also known as Salvarsan (Figure 3.4), was the first effective treatment for syphilis and African trypanosomiasis that was introduced at the beginning of the 1910s.

It is considered the first antimicrobial agent that has been substituted by modern antibiotics. However, arsenicals such as melarsoprol (Figure 3.5) are still used for the treatment of trypanosomiasis despite their toxicity.

A controversial case is the use of Roxarsone as a feed additive (Figure 3.6) in poultry production. Under some conditions, the arsenic in chicken feed is converted

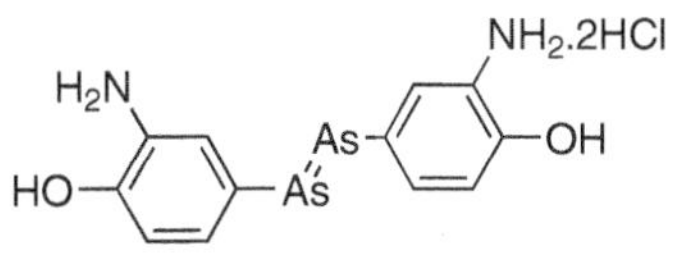

FIGURE 3.4   Salvarsan

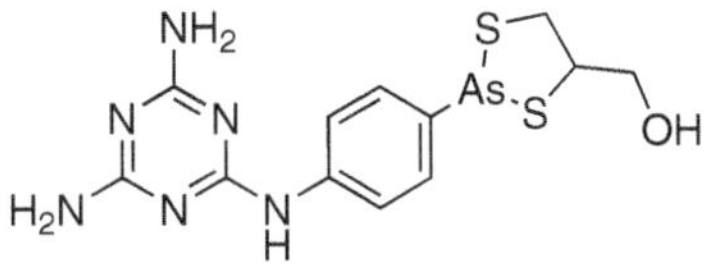

FIGURE 3.5   Melarsoprol

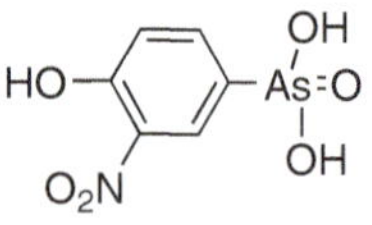

FIGURE 3.6 Roxarsone

to toxic inorganic forms at elevated levels in treated chickens. Roxarsone has been banned in the EU since 1999 and in many other countries in following years.

Arsenical pesticides have been used in agriculture for many years. Records show that arsenic sulfides have been in use since AD 900 in China. Lead arsenate was first prepared as an insecticide in 1892 in the United States. Initially, the first formulations of insecticides were prepared by farmers, and over time, they were commercially prepared as basic lead arsenate $[Pb_5OH(AsO_4)_3]$ and acid lead arsenate ($PbHAsO_4$). Lead arsenate and calcium arsenate were spread across this country's fields and orchards. Such arsenic substances have been banned since the middle of the 20th century.

Before the toxicity of arsenic compounds was widely recognised, two arsenic-based pigments have been widely used since their discovery, Paris Green and Scheele's Green, prepared by combining copper(II) acetate and arsenic trioxide and sodium carbonate arsenious oxide, respectively.

Toxicity

According to the WHO reports, arsenic is one of the 10 chemicals of major public health concern. The toxicity of arsenic compounds depends on their chemical form, with inorganic forms generally considered more toxic than organic forms.

Inorganic arsenic compounds, such as arsenic trioxide and sodium arsenite, are the most toxic forms of arsenic. These compounds are commonly found in contaminated drinking water, industrial waste and some food sources.

Acute arsenic poisoning includes vomiting, abdominal pain, diarrhoea, lesions, respiratory problems, peripheral neuropathy, cardiovascular disease and diabetes.

Long-term exposure produces skin pigmentation changes, skin lesions and hard patches on the palms and soles of the feet (hyperkeratosis). Chronic exposure to inorganic arsenic has been linked to an increased risk of cancer, including lung, bladder and skin cancer.

Arsenic is also associated with adverse pregnancy outcomes and infant mortality and many studies recognise arsenic as carcinogenic to humans.

Organic arsenic compounds, such as those found in seafood, are generally considered less toxic than inorganic forms. However, exposure to high levels of organic arsenic can still cause health problems. For example, exposure to high levels of arsenobetaine (Figure 3.7), a common organic arsenic compound found in seafood, has been linked to skin lesions and developmental problems in foetuses.

The toxic properties of arsenic compounds are known from ancient times and it has been nominated as the 'King of Poisons', because throughout history,

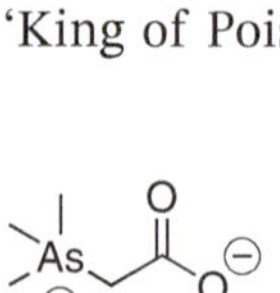

FIGURE 3.7 Arsenobetaine

TABLE 3.4 **Organoarsenic compounds used as chemical weapons**

Name and effects	Structure
Phenyldichloroarsine: vesicant	
Lewisite: vesicant and lung irritant	
Adamsite: irritation of the eyes, mucous membranes and lungs	

it has been used many times to poison royalty and thus alter who would ascend to the throne.

Arsenic chemical weapons were used during the Second World War, especially those featuring As-Cl bonds (Table 3.4).

3.11 Antimony (Sb)

Antimony is often found in minerals such as stibnite (antimony sulfide), valentinite (antimony oxide) and kermesite (antimony sulfide oxide).

Antimony and its compounds have a variety of applications in various fields, including:

- Antimony oxides are used as flame retardants in textiles, plastics and other materials. They act as a synergist with halogenated flame retardants, enhancing their effectiveness.
- In lead-acid batteries to improve their performance and reduce corrosion.
- As a dopant in the semiconductor industry to increase the conductivity of silicon.
- In the production of glass to remove bubbles and improve its refractive index.
- In the production of pigments such as antimony orange and Naples yellow.
- Some antimony compounds have been used in medicine to treat parasitic infections such as leishmaniasis and schistosomiasis.
- Used as an alloying element in various alloys, such as pewter, which is a tin-based alloy.
- Catalyst in the production of polyethylene terephthalate (PET) bottles.

Toxicity

Exposure to antimony can occur through inhalation, ingestion or skin contact, mainly due to occupational exposure or during therapy. Acute exposure to high levels of antimony can cause symptoms such as nausea, vomiting, diarrhoea, abdominal pain and headaches. Chronic exposure to lower levels of antimony can lead to more serious health effects, such as lung and heart problems, skin rashes and neurological effects. Antimony is also classified as a carcinogen, meaning that long-term exposure to high levels of the element may increase the risk of developing cancer.

3.12 Biomarkers of Heavy Metal Poisoning and Treatments

Heavy metal poisoning can lead to various health problems and it's essential to identify the condition as soon as possible to prevent long-term damage. Some biomarkers of heavy metal poisoning include:

- **Blood Lead Level (BLL):** This is a common biomarker for lead poisoning. BLL is measured in micrograms of lead per deciliter of blood (μg/dL).
- **Hair mineral analysis:** Hair mineral analysis can help identify chronic exposure to heavy metals, including lead, mercury, arsenic and cadmium. The levels of these metals can be detected and measured in hair samples.
- **Urine metal analysis:** Urine testing can detect the presence of heavy metals in the body, including arsenic, cadmium and mercury.
- **Red Blood Cell (RBC) protoporphyrin:** An elevated level of RBC protoporphyrin is indicative of lead poisoning.
- **Zinc protoporphyrin (ZPP):** Elevated levels of ZPP may indicate chronic exposure to lead or iron deficiency anaemia. Blood and urine mercury levels: These tests can detect the presence of mercury in the body, which is a common biomarker for mercury poisoning.
- **Total arsenic in urine:** This test measures the amount of total arsenic in urine, which is a common biomarker for arsenic poisoning.

3.13 Metal Poisoning Treatments

Treatments for heavy metal toxicity may include:

1. **Removal of the source of exposure:** The first step in treating heavy metal toxicity is to identify and remove the source of exposure, such as contaminated food or water, lead-based paint or industrial chemicals.
2. **Nutritional support:** A diet rich in nutrients such as calcium, magnesium and zinc can help support the body's natural detoxification processes and promote healing.
3. **Supplements:** Certain supplements, such as selenium and vitamin C, may help reduce the toxicity of heavy metals and support the immune system.
4. **Medications:** In some cases, medications may be prescribed to help reduce the symptoms of heavy metal toxicity, such as pain relievers or anti-inflammatory drugs.

5. **Antidote therapy:** Prussian blue is the antidote for thallium poisoning. It works by binding with thallium in the digestive tract and preventing it from being absorbed into the body. Prussian blue is usually given orally and the dose and duration of treatment depend on the severity of the poisoning.

6. **Supportive care:** Thallium poisoning can cause a range of symptoms, including gastrointestinal problems, neurological symptoms and renal failure. Supportive care may include intravenous fluids, medications to manage symptoms and dialysis in severe cases.

7. **Chelation therapy:** Chelation therapy is a process in which a chelating agent binds to the metal ions to form a complex that can be excreted. For example, ethylenediaminetetraacetic acid (EDTA) is used for lead poisoning and Prussian blue (potassium ferric hexacyanoferrate) in the case of thallium is administered to bind to the metal ions in the bloodstream and form a complex that can be excreted through urine. The process is done under medical supervision and can take several hours to complete. Chelation therapy may be used in severe cases of poisoning, but it is not always effective and can have serious side effects.

Topics of Interest: Exposure to Metals by Certain Pigments and Coins

Inorganic Pigments

They are pigments that are derived from minerals and are typically characterised by their high resistance to light, heat and chemicals. They are commonly used in paint formulations due to their durability and colour stability.

Some common inorganic pigments used in paint include:

- Titanium dioxide (TiO_2) is a white pigment that provides opacity and brightness to paints.
- Iron oxide pigments, including red, yellow and black varieties, provide a range of earthy and rust-coloured hues.
- Ultramarine blue is a synthetic pigment made from the mineral lapis lazuli that provides a deep blue colour.
- Chromium oxide green is a bright green pigment that is highly resistant to fading.
- Cadmium pigments, including yellow, red and orange varieties, provide bright and intense colours.
- Zinc oxide is a white pigment that is often used as a corrosion inhibitor in paints.
- Carbon black, a black pigment made from the incomplete combustion of organic materials, is used to provide black colour to paints.

These inorganic pigments are often preferred over organic pigments in industrial and exterior applications because of their greater lightfastness, chemical resistance and colour stability.

Some inorganic pigments can be toxic if they are not handled properly. For example, lead-based pigments such as lead chromate and lead oxide can be harmful if ingested or inhaled as dust. Similarly, cadmium-based pigments such as cadmium red and cadmium yellow can be toxic if they are ingested or inhaled in large amounts. Other inorganic pigments, such as TiO_2, iron oxide and chromium oxide, are generally considered safe for use in paint formulations. However, it is important to note that even these pigments can be harmful if they are not handled properly. For example, if they are inhaled as dust during manufacturing or application, they can cause respiratory problems.

Coins

There are several metals that are commonly used to produce coins. The toxicity of these metals can vary depending on the specific metal and the form in which it is present. Here are some common coin metals and their toxicity:

- **Copper:** Copper is a common metal used to produce coins. It is not toxic in small amounts, but excessive exposure to copper can lead to copper toxicity, which can cause gastrointestinal distress, liver damage and neurological symptoms.
- **Zinc:** Zinc is often used as a coating on coins to prevent corrosion. It is not toxic in small amounts, but excessive exposure to zinc can lead to zinc toxicity, which can cause nausea, vomiting and diarrhoea.
- **Nickel:** Nickel is sometimes used in coins to give them a silver colour. Nickel can cause skin irritation and allergic reactions in some people. It can also be toxic in large amounts, causing lung damage, kidney damage and cancer.
- **Silver:** Silver is a precious metal often used in commemorative coins. It is not toxic in small amounts, but excessive exposure to silver can cause a condition called argyria, which turns the skin blue-grey.
- **Gold:** Gold is a precious metal often used in commemorative coins. It is not toxic in small amounts, but excessive exposure to gold can cause skin irritation and allergic reactions in some people.

In general, the toxicity of coin metals depends on the amount of exposure and the form in which they are present. Most people are not at risk of toxicity from handling coins, but people who handle large quantities of coins or who have allergies to specific metals may be at increased risk.

3.14 Attenuation of Metal Pollutants

Metal pollutants are not biodegradable because they are not digested or destroyed by any microorganism or enzyme. Thus, these inorganic substances are chemically transformed, diluted, sorbed or excreted, but the core of them remains unalterable.

Many hazards and environmental risks produced by metal pollutants may be reduced by **attenuation**. It consists in the decrease of the pollutant concentration in the soil, water bodies, groundwater or air via natural processes without human intervention; consequently, the total amount of them available diminishes, and therefore, the exposure pathway or the toxicity of inorganic chemicals, also diminishes over time. There are two types of attenuation:

1. **Physical attenuation:** Consists of a set of physical processes that cause a net decrease in pollutant levels and therefore the hazards produced by them.

    Some common physical attenuation mechanisms for inorganic pollutants include:

    a. **Sedimentation:** Many inorganic pollutants, such as heavy metals and sediment-bound pollutants, can settle and accumulate in the sediment at the bottom of water bodies due to gravity. This process can effectively remove these pollutants from the water column and reduce their concentration.

    b. **Filtration:** Filtration is a physical process that involves passing contaminated water through a filter medium, such as sand or clay, to trap and remove inorganic pollutants.

c. **Adsorption:** Adsorption is a process where inorganic pollutants are attracted to and adhere to the surface of solid particles, such as soil particles or suspended solids in water. This can occur through various physical forces, such as van der Waals forces or electrostatic interactions.

d. **Erosion:** Erosion is a natural process that can physically attenuate inorganic pollutants by removing contaminated soil or sediment from an area. For example, runoff from rainfall or snowmelt can wash away soil particles containing inorganic pollutants, effectively reducing their concentration in the affected area.

Physical attenuation tends to depend upon the velocity of the dispersing phase (e.g., a fast-moving stream versus a slow-moving river) and its rate of mixing.

2. **Chemical attenuation:** It is caused by the chemical transformations of pollutants that decrease the pollutant concentrations in air, soil, surface waters or groundwater. Chemical attenuation is rather complicated and specific to the particular contaminant class.

Natural chemical attenuation for metal pollutants can occur through several mechanisms:

a. **Precipitation and sorption:** Metal pollutants may undergo precipitation or sorption reactions in natural environments, where they form insoluble compounds or become adsorbed onto solid particles, respectively. This can reduce their mobility and bioavailability, limiting their potential impact on the environment.

b. **Co-precipitation:** Inorganic pollutants can also be co-precipitated with minerals or other solid phases in natural environments, resulting in their immobilisation and reduced mobility. This can occur through processes such as adsorption, co-precipitation with mineral surfaces or incorporation into mineral structures.

c. **Redox reactions:** Reduction–oxidation reactions involve the transfer of electrons between chemical species and can result in the transformation of inorganic pollutants into less toxic or less mobile forms. For example, some metals can undergo reduction reactions in natural environments, converting from more toxic forms, such as hexavalent chromium ($Cr(VI)$), to less toxic trivalent chromium ($Cr(III)$).

d. **Complexation:** Inorganic pollutants can form complexes with other chemicals in the environment, which can alter their mobility and reactivity. For example, heavy metals can form complexes with organic matter or inorganic ligands in soil or water, reducing their bioavailability and potential for environmental harm.

3.15 Inorganic Carbon Pollutants

Carbon Monoxide (CO)

Carbon monoxide is a colourless, odourless, tasteless and flammable poisoning gas that is slightly less dense than air. Carbon monoxide occurs from both natural and anthropogenic sources and it is present in the atmosphere in concentrations of about 80 ppb. Carbon monoxide is formed in two ways:

a. Photochemical transformation of organic volatile compounds in troposphere

b. From volcanic emissions or incomplete combustion of biomass or fossil fuels

Carbon monoxide is significantly present in fumes produced by cars and trucks, lanterns, stoves, gas ranges and heating systems with improper maintenance or incorrect installation. In indoor spaces with a poor flow of fresh air, it is especially dangerous because people exposed to carbon monoxide can be intoxicated or even die before they have symptoms[5]. For this reason, carbon monoxide is known as the *invisible killer*. Every year, many deadly episodes occur and thousands of people are treated in hospital emergency rooms for carbon monoxide poisoning all around the world.

Carbon monoxide is produced intentionally as a component of the so-called *syngas* (Eq. 3.3), a gas mixture consisting primarily of hydrogen and carbon monoxide that can be obtained from methane (natural gas), coal, biomass or virtually any hydrocarbon feedstock that may be used as a combustible or starting material in the petrochemical industry.

$$CH_4 + H_2O \rightarrow CO + 3H_2 \tag{3.3}$$

Carbon Dioxide (CO$_2$)

Carbon dioxide is a colourless gas that plays a key role in the carbon cycle. It is a minor component of the atmosphere that is generated in the respiration of animals and employed as a carbon source by plants in the photosynthesis of carbohydrates.

When it solubilizes in water, it forms carbonic acid that is able to give HCO_3^- and $CO_3^=$ depending on pH (Eq. 3.4).

$$CO_2 + H_2O \rightleftharpoons H_2CO_3 \rightleftharpoons H^+ + HCO_3^- \rightleftharpoons H^+ + CO_3^= \tag{3.4}$$

Carbon dioxide is also formed in the combustion of carbon-containing materials, such as biomass or fossil fuels and it is produced in the aerobic fermentation of organic matter.

Carbon dioxide has multiple applications in the chemical and food industries, for example, as a raw material, a supercritical solvent or to produce carbonated soft drinks and fire extinguishers.

In indoor spaces, low concentrations of carbon dioxide show low toxicity, but at higher concentrations, it may increase respiratory rate and induce tachycardia, cardiac arrhythmias and impaired consciousness. Higher concentrations than 10% may cause convulsions, comas and death.

Carbon dioxide contributes to the natural greenhouse effect by absorbing and emitting infrared radiation at its two infrared-active vibrational frequencies. Atmosphere temperature, which makes life possible on earth.

---

[5]The initial symptoms of CO poisoning are fatigue, shortness of breath, nausea, and vomiting chest pain, confusion, and dizziness.

The current carbon dioxide concentration in the atmosphere exceeds 410 ppm, rising significantly from pre-industrial times due to human activities such as power generation, industry and transportation.

There is a broad consensus about the effect of anthropogenic carbon dioxide as a key factor in the increase in atmospheric temperature and its contribution to global climate change.

Cyanides (CN⁻)

Cyanides are a family of chemical compounds containing the cyano group (CN). Inorganic cyanides show the cyanide anion (CN^-) linked to a hydrogen or cation.

Hydrogen cyanide (HCN) is a highly volatile liquid that is produced industrially by the acidification of cyanide salts. Exposure to HCN is lethal even at low concentrations[6] and sodium and potassium salts are highly toxic too.

Cyanide is widely used in many branches of industry, for example, the paper industry, textiles, plastics, metallurgy for electroplating, metal cleaning and removing gold from its ore. Cyanide gas is used to exterminate pests and vermin on ships and buildings.

Cyanides can be found in Nature in the so-called *cyanogenic glycosides* in apricot kernels, cassava roots and bamboo shoots.

Cyanide has also been detected in the smoke of cigarettes, constituting the major source of cyanide exposure for the public in general and it may be found in the fumes of combustion of some synthetic polymers and wools in residential areas. This is a relatively frequent exposure in cases of smoke inhalation from residential or industrial fires.

Symptoms of cyanide poisoning vary widely, depending on the dose and route of exposure and may range from minor upper airway irritation to death. Cyanide inhibits oxygen metabolism and energy production in cells, killing a severely exposed individual within minutes.

3.16   Inorganic Nitrogen Pollutants

Nitrogen Oxides (NO$_x$)

The term nitrogen oxides corresponds to a set of binary combinations between nitrogen and oxygen atoms, such as nitrous oxide (N_2O), nitric oxide (NO) and nitrogen dioxide (NO_2). They are collectively referred to as NO_x. They are very reactive species in the atmosphere, for example, they oxidise volatile organic compounds to give the so-called photochemical smog.

N_2O[7] is produced in the soil during the *denitrification* process (Eq. 3.5) in which the nitrate ion is reduced to nitrogen in a series of reactions provided by microorganisms.

---

[6] Hydrogen cyanide, which was used by the Nazis in gas chambers under the name Zyklon B.

[7] Nitrous oxide, also known as *laughing gas,* is used in medical and dental procedures as a sedative.

$$NO_3^- \rightarrow NO \rightarrow N_2O \rightarrow N_2 \qquad (3.5)$$

In nature, denitrification can take place in both terrestrial and marine ecosystems and through these processes, NO_x are emitted into the atmosphere. In the last few decades, N_2O emissions have increased dramatically because of the overuse of fertilisers in agriculture. This is a matter of concern because N_2O has shown a double role: it contributes to photochemical smog by participating in several photochemical reactions and it is a potent greenhouse gas. Such emissions are relevant in rural areas.

Other sources of NO_x are combustion processes, mostly the direct combination of atmospheric oxygen and nitrogen in flames or engines or produced as a by-product in some industrial processes, for example, during the manufacturing of nitrogenated fertilisers. These emissions are relevant in urban and industrial areas, where they often appear as a brownish gas mantle.

In combustion reactions, the primary NO_x formed is NO, accompanied by a small proportion of NO_2. NO is readily oxidised to give NO_2, which, combined with water vapour, gives nitric acid (HNO_3) and nitrate aerosols, which are dissolved in cloud droplets. Both are components of acid rain.

The main route of NO_x exposure is inhalation. They are irritating to the eyes, skin, mucous membranes and respiratory tract. As it takes place in the atmosphere, on contact with moisture in the eyes, mucous membranes and respiratory tract, NO_2 forms HNO_3 and other nitric substances in situ.

Ammonia (NH$_3$)

Ammonia is a quite water-soluble gas with a characteristic odour. It is one of the most common chemicals produced by industry in the world, with numerous applications. It is the starting material for the synthesis of nitrogen compounds, including fertilisers, HNO_3 and nitrogenated organic compounds. When ammonia is dissolved in water, it is represented by the formula NH_4OH. The solution is formed mainly by water (H_2O), ammonia (NH_3) and smaller quantities of ammonium ions (NH_4^+) and hydroxide ions (OH^-). Ammonium hydroxide solutions are basic and corrosive at high concentrations.

Ammonia also has a natural occurrence. It is essential for many biological processes and serves as a precursor for biomoléculas such as amino acids and nucleotides. In the environment, ammonia is part of the nitrogen cycle. It is produced in soil by bacterial processes and organic matter decay.

Ammonia is considered an air pollutant largely emitted from industry, agriculture, waste depots and natural sources. Because of its basicity, it is able to react in the atmosphere with other acidic pollutants such as NO_x and sulfur dioxide (SO_2) to give aerosols containing ammonium ions (NH_4^+), which contribute to visibility degradation and atmospheric deposition of nitrogen pollutants in sensitive ecosystems.

The effects of direct exposure to ammonia depend on the form of ammonia: gas, liquid ammonia or ammonium hydroxide. Exposure to external sources of gas may be produced by ingestion, inhalation, or contact with the skin, mucous membranes or eyes. Exposure to low and moderate concentrations can cause coughing and nose and throat irritation. Higher concentrations of ammonia in the air cause acute, immediate burning of the eyes, nose, throat and respiratory tract and can result in blindness, lung damage or death.

Nitrates (NO_3^-)

The salts containing the polyatomic ion NO_3^- are known as nitrates. They are a naturally occurring form of nitrogen that participates in the nitrogen cycle and can be used by plants as a nutrient. Almost all nitrates are water-soluble and they constitute a chemical raw material used for the production of fertilisers, explosives, construction materials and, on a small scale, nutritional supplements.

Nitrates are produced in human and animal bodies, so they are naturally found in foods such as vegetables, dairy products and meat and they can be found extensively in manures, decaying plants and animals and other organic wastes. Nitrates are also added to processed meats and other foods as a preservative, labelled E252.

The massive use of fertilisers, careless farming practices with manure and other wastes and inadequate septic installations are the main sources of nitrate pollution. Such soluble salts are incorporated into underground waters, rivers and lakes and can finally be found in drinking water.

The main exposure to nitrates is due to the ingestion of polluted waters and foods. High levels of nitrates in humans affect oxygen transportation in the blood, especially in infants, causing *methemoglobinemia*, known as blue baby syndrome, because the skin turns a bluish colour, along with decreased blood pressure, an increased heart rate, headaches, stomach cramps, vomiting, and finally, death.

Nitrates are commonly found in food, water and the environment and they can be converted into nitroso compounds through several different pathways. One of the primary pathways involves the reduction of nitrates to nitrites (NO_2^-) by enzymes called nitrate reductases, which are present in various tissues and organs, including the gut, saliva and skin. Nitrites, in turn, can react with amines or amides present in the body to form nitroso compounds.

For example, in the stomach, nitrite can react with amines derived from dietary sources, such as proteins, to form nitrosamines, which are a type of nitroso compound. Nitrosamines have been shown to be potentially carcinogenic and can be found in certain processed foods, cured meats and tobacco products (Eq. 3.6).

$$NO_3^{\ominus} \longrightarrow NO_2^{\ominus} \longrightarrow NO^{\ominus} + R_1R_2NH \longrightarrow R_1R_2N\text{-}N\text{=}O$$

nitrate.   nitrite nitrosil ion secondary amine   nitrosamine   (3.6)

Another undesirable effect of nitrates in water bodies is so-called eutrophication. The excessive input of nitrates and phosphates[8] from human activities can lead to harmful algal blooms. The continuous decay of aquatic plants provoques a strong oxygen consumption. Such *hypoxia*[9] kills fish and other organisms and produces toxins.

Topics of Interest: Nitrogen Compounds in the Environment

Nitrogen is an essential element for life and is present in various compounds in the environment. Nitrogen compounds exhibit diverse valences in different compounds, which result from their ability to form different types and numbers of bonds with other atoms. This versatility

---

[8] Both are the main components of fertilisers.
[9] The term *hypoxia* refers to low or depleted oxygen in a water body.

TABLE 3.5 Nitrogen compounds and their valences

Compound	Valence of nitrogen
Ammonia (NH_3)	−3
Nitric oxide (NO)	+2
Nitrate (NO_3^-)	+5
Nitrite (NO_2^-)	+3
Nitrogen gas (N_2)	0
Ammonium ion (NH_4^+)	−3
Nitrogen dioxide (NO_2)	+4
Nitrous oxide (N_2O)	+1
Amides ($RCONH_2$)	−3
Amines (R_3N)	−3
Nitro compounds (RNO_2)	+3

allows nitrogen to participate in various chemical reactions and biological processes, contributing to the cycling of nitrogen in the environment.

Table 3.5 provides a snapshot of some nitrogen compounds found in nature and their respective valences.

3.17 Inorganic Sulfur Pollutants

Sulfur Oxides: SO_2 and SO_3

SO_2 and trioxide are two binary combinations of sulfur and oxygen that have both natural occurrences in volcanic emissions and anthropogenic sources[10].

SO_2 is a colourless gas that can be detected by taste and smell in the range of 1,000–3,000 micrograms per cubic metre ($\mu g/m^3$). At concentrations of 10,000 $\mu g/m^3$, it has a pungent, unpleasant odour. Exposure to SO_2 has been associated with reduced lung function, an increased incidence of respiratory damages, irritation of the eyes, nose and throat and premature mortality. It is assumed that the gas has a half-life of 6–24 hours, consequently, only about 5% of the emitted gas is present in the lower atmosphere after 1–4 days.

Sulfur trioxide (SO_3) is another oxide of sulfur that may be emitted directly into the atmosphere as a pollutant or produced from SO_2 in the atmosphere by oxidation with several atmospheric oxidants (Eq. 3.7).

$$SO_2(g) \rightarrow SO_3(g) \tag{3.7}$$

---

[10] Other sources of sulfur oxides are the oxidation of biogenic organic sulfides.

SO_2 dissolves readily in water present in the atmosphere to form sulfurous acid (H_2SO_3), but about 30% of the SO_2 is converted to the more oxidised forms of sulfate aerosols (acid aerosol), which are removed through wet or dry deposition processes.

SO_3 is a highly reactive sulfur oxide and it is rapidly converted to sulfuric acid (H_2SO_4), which, when combined with water, gives sulfuric acid (Eqs. 3.8 and 3.9).

$$SO_2(g) + H_2O(l) \rightarrow H_2SO_3(aq) \tag{3.8}$$

$$SO_3(g) + H_2O(l) \rightarrow H_2SO_4(aq) \tag{3.9}$$

Bulky amounts of SO_2 are used every year as an intermediate for the synthesis of sulfuric acid. In addition, SO_2 is frequently formed as a by-product in the case of the burning of sulfur-bearing fossil fuels such as coal or oil and in some industrial processes such as extracting metal from ore. The largest anthropogenic source of SO_2 in the atmosphere is produced in power plants, furnaces and other large industrial fuel consumers.

In the atmosphere, SO_2 is oxidised to sulfur trioxide, which, combined with water, gives sulfuric acid (Eqs. 3.10 and 3.11).

$$SO_2(g) \rightarrow SO_3(g) \tag{3.10}$$

$$SO_3(g) + H_2O(l) \rightarrow H_2SO_4(aq) \tag{3.11}$$

Sulfur oxides, sulfuric acid and sulfate aerosols formed from them contribute to acid rain. The presence of SO_2 causes damage to living beings and sensible ecosystems.

The environmental effects of sulfur oxides and their derivative, sulfuric acid are significant on plants, soil, surface waters and acid-sensitive materials such as marble.

1. **Plants:** They can have detrimental effects on plant life, especially on sensitive species. It damages plant tissues, interferes with photosynthesis and reduces chlorophyll production. This can lead to stunted growth, reduced crop yields and even the death of plants in severe cases. Forests and natural vegetation can also be negatively impacted, affecting biodiversity and ecosystem health.

2. **Soil:** Sulfur oxides can be deposited onto the soil through dry deposition or as a component of acid rain. The acidic nature of such sulfur compounds can lower the soil's pH, making it more acidic. This disrupts the soil's nutrient balance and can harm beneficial microorganisms, affecting plant growth and soil fertility.

3. **Surface waters:** Sulfur oxide emissions can lead to the formation of sulfuric acid when SO_2 combines with water, which can lower the pH of surface waters, such as rivers, lakes and streams. This altered pH can harm aquatic life, including fish, amphibians and invertebrates. Acidification of water bodies disrupts aquatic ecosystems and can lead to declines in populations of sensitive species.

4. **Impact of acid-sensitive materials:** Acidic pollutants like SO_2 and their derivatives can react with surfaces of buildings, monuments and sculptures made from acid-sensitive materials such as marble, limestone and certain types of metals. This chemical reaction results in the gradual deterioration and corrosion of these materials, leading to the loss of cultural heritage and economic costs for restoration.

In humans, short-term exposures can harm mucous membranes and the respiratory system, making breathing difficult. Patients with asthma and children are especially sensitive to these effects. Aerosols formed from SO_2 may affect the lungs and other vital organs. SO_2 can also have other harmful effects on human health. It is a major contributor to the formation of fine particulate matter (PM2.5) and can irritate the respiratory system, leading to respiratory problems, especially in people with pre-existing conditions like asthma and bronchitis.

To mitigate the environmental effects of SO_2, various measures can be taken, including the implementation of emission control technologies in industrial processes and power plants, the use of cleaner fuels and the enforcement of regulations to limit SO_2 emissions from various sources. International agreements, such as the United Nations' Convention on Long-Range Transboundary Air Pollution, have also been established to address transboundary air pollution and reduce the impacts of pollutants like SO_2 on a global scale.

To reduce the impact of sulfur oxides as pollutants, many industries have implemented technologies to control emissions, such as using scrubbers to remove SO_x from flue gases. Governments have also set emissions standards and regulations to limit the amount of SO_x that can be released into the environment. Additionally, individuals can take steps to reduce their own contribution to SO_x pollution by conserving energy and reducing their use of fossil fuels.

Reduced Sulfur Compounds

The term total reduced sulfur (TRS) refers to a gaseous mixture of pollutants containing one or more sulfur atoms in their reduced state, such as hydrogen sulfide (H_2S), carbon disulfide (CS_2), carbonyl sulfide (COS), dimethyl sulfide (CH_3-S-CH_3), methyl mercaptan (CH_3SH) and dimethyl disulfide (CH_3-S-S-CH_3). TRS compounds are naturally formed during the anaerobic decomposition of organic materials in marshes, bogs, lakes and coastal regions and emitted in other sectors related to human activities, such as petroleum refining, pulp and paper manufacturing, aluminium production, landfilling, wastewater treatment, etc. Usually, the individual quantities reported for each pollutant are converted to a single unit (tonnes of H_2S).

Hydrogen Sulfide (H_2S)

Hydrogen sulfide is a colourless gas with the characteristic odour of rotten eggs. Hydrogen sulfide occurs naturally in volcanic emissions and hot springs and can be detected in sewers, manure pits, waste depots and any place where organic matter decomposes through the bacterial breakdown of proteins containing sulfur, including the intestinal tract of mammals.

Sources of anthropogenic hydrogen sulfide are oil and gas refining, pulp and paper processing, rayon manufacturing, tanning and mining.

The H_2S emitted in the atmosphere is oxidised by atmospheric oxidants to SO_2 and finally sulfate aerosols.

Hydrogen sulfide is highly toxic and many health effects appear even at low concentrations, such as headaches or eye irritation at low concentrations and unconsciousness and death in cases of severe exposure.

Topics of Interest: Resume of Inorganic Gaseous Pollutants (Table 3.6)

TABLE 3.6   Inorganic gaseous pollutants

Gas	Residence time	Concentration	Fate
CO	~60 days	0.05–0.12 ppm	Oxidation to CO_2 (95%) or diffusion into the stratosphere.
CO_2	5–15 years[a]	>410 ppm	Photosynthesis, land, oceans
N_2O	~120 years	~330 ppb	Nitrate species
$NO + NO_2$	0.5–2 days	10 ppb	Nitrate species
NH_3	2–10 days	<5 ppm	Nitrate species
SO_2	1–4 days	10 pptv–1 ppbv	Sulfate species
H_2S	1–5 days	5–500 pptv	Sulfate species

[a] Long-term (100 years) rise in atmospheric concentration is not due to anthropogenic emissions.

3.18   Asbestos

The term *asbestos* refers to a group of six fibrous, naturally occurring silicates[11] The best-known type is chrysotile, which accounts for about 95% of all asbestos in commercial use and shows the following molecular formula: $Mg_3Si_2O_5(OH)_4$.

Asbestos has been used in a wide range of materials, including roofing shingles, siding, insulation, cement pipes, wallboard, ceiling tiles, flooring materials, protective clothing for workers in high-heat or fire-prone industries, automotive parts and many others.

These materials show great fibre strength, heat resistance and durability. They do not dissolve in water or evaporate, they are resistant to heat and fire, chemically very stable and inert to microorganisms.

During the fabrication of asbestos materials or when they are damaged, fine dust containing asbestos fibres may be produced and dispersed through the air. With exposure to low levels of asbestos fibres, most people do not develop any illness, but long-term exposure, especially in workspaces with asbestos products, produces serious health hazards.

Asbestos toxicity primarily occurs when asbestos fibres are released into the air and inhaled or ingested by the body. Once inhaled or ingested, asbestos fibres can become lodged in the lungs, where they can cause a range of health problems over time.

One of the first effects of asbestos fibres is asbestosis. It is a chronic lung disease caused by prolonged exposure to asbestos fibres. Asbestosis occurs when

[11] Fibre is a particle that is more than 5 μm in length and has a length-to-width ratio of at least 3:1.

asbestos fibres accumulate in the lungs, causing inflammation and scarring of lung tissue. Over time, this scarring can lead to breathing difficulties and decreased lung function. Symptoms of asbestosis may include shortness of breath, a persistent cough, chest pain and finger clubbing (enlarged fingertips). In severe cases, it can lead to respiratory failure and death.

Asbestos exposure has also been associated with an increased risk of several cancers, such as mesothelioma, a rare and aggressive form of cancer that affects the lining of the lungs, abdomen, heart or other organs. Mesothelioma is almost exclusively caused by asbestos exposure (Figure 3.8).

Asbestos exposure has also been linked to lung cancer, particularly in individuals who smoke or have a history of smoking. Asbestos exposure and smoking together can greatly increase the risk of lung cancer and other cancers: in organs such as the throat, oesophagus, pancreas, colon and kidneys.

It's important to note that asbestos-related diseases can have a long latency period, often taking decades to develop after exposure. Therefore, even brief or low-level exposure to asbestos can still pose a risk to health. To mitigate asbestos toxicity, strict regulations have been put in place in many countries to limit the use of asbestos and protect workers and the general public from exposure. Proper handling, removal and disposal of asbestos-containing materials are critical to preventing asbestos-related health risks. If you suspect or have been exposed to asbestos, it's important to seek medical advice and follow appropriate safety precautions.

In many countries, the manufacture, import, processing and distribution of asbestos-containing products have been banned since the 1980s. According to the WHO, occupational exposure to asbestos is estimated to cause over 107,000 deaths annually worldwide.

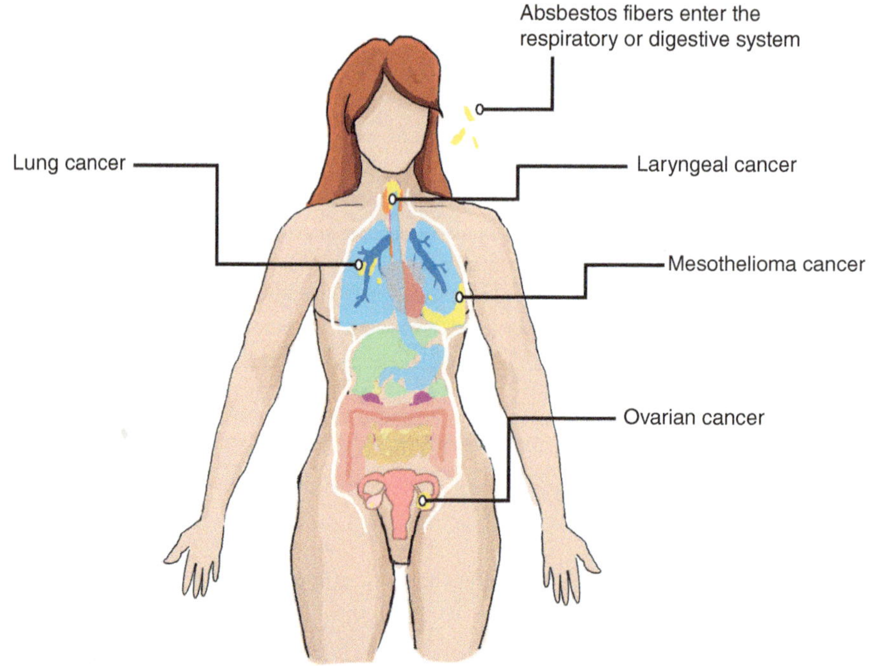

FIGURE 3.8   Asbestos damages

3.19 Radionuclides

The radionuclides are unstable forms of some elements that are spontaneously transformed into one or more new elements, emitting particles and radiation until they become stable nuclei. The successive elements that are formed during the decomposition of a parent radionuclide until they arrive at a stable nucleus are named as *radioactive series*.

To measure the decay rate of a specific radionuclide, the concept of *half-life* is used. It takes time for one-half of the nuclei of a radioactive sample to decay, emitting particles or radiation. The half-life is a characteristic property of each radionuclide and can vary widely depending on the specific radionuclide. Half-life values range from mere seconds to millions of years, depending on the specific radionuclide.

Each radionuclide emits a combination of particles and radiation with energies expressed as MeV. Table 3.7 summarises the different types of radioactive emissions. The more common radioactive emissions are constituted by alpha, beta and gamma particles or a combination of them.

The capacity of penetration of radioactive particles depends on several factors, including the type of radioactive particle, its energy level and the medium through which it is passing. In general, radioactive particles can penetrate different materials to varying extents. The common types of radiation, alpha, beta and gamma show these penetration capacities:

a. **Alpha particles (α):** Alpha particles are large, positively charged particles consisting of two protons and two neutrons, which are emitted by certain types of radioactive materials such as uranium and radium. Due to their size and charge, alpha particles have a low penetration capacity and can be stopped by a sheet of paper, a few centimetres of air or a thin layer of other materials, such as a T-shirt.

b. **Beta particles (β):** Beta particles are high-energy electrons that have a greater penetration capacity compared to alpha particles and can pass through materials such as paper, wood and plastic. However, they can be stopped by materials with

TABLE 3.7 Radioactive emissions

Emission	Description	Penetrating power
Alpha (α)	Helium nucleus $^4_2$He	Very limited ability to penetrate materials
Beta (β)	Electrons e$^-$	They can displaced a few metres in the air and penetrate skin
Gamma (γ)	Ionising electromagnetic radiation	Ability to penetrate several cm of a dense materials such lead,
Neutrons (n)	Neutral subatomic particles	Exceptional ability to penetrate other materials. They can travel great distances in the air and require very thick hydrogen-containing materials, such as concrete or water, to block them.
Neutrino (ν)	Subatomic particle with no electric charge and very little mass	The most penetrating form of radiation. It can pass through the entire Earth without interacting with anything along the way

higher atomic numbers, such as aluminium or lead or thicker layers of other materials; therefore, they can penetrate partially into the human body.

c. **Gamma rays (γ):** Gamma rays are high-energy electromagnetic radiation emitted by certain types of radioactive materials. They have the highest penetration capacity among the three common types of radioactive particles and can penetrate materials with high atomic numbers, such as lead and concrete. Gamma rays can only be effectively attenuated by very dense materials and their penetration capacity depends on their energy level.

Radioactivity can be detected using various methods, depending on the type of radiation being measured. The choice of method depends on the type of radiation being measured, the sensitivity required and the specific application. Some common methods for detecting radioactivity are:

a. **Geiger-Muller (GM) counter:** A GM counter is a device that detects ionising radiation by measuring the electrical charge generated when radiation passes through an inert gas-filled tube such as Argon (Figure 3.9). When radiation interacts with the gas in the tube, it ionises the gas atoms, producing a pulse of electrical current that can be detected and counted. GM counters are commonly used for detecting alpha, beta and gamma radiation.

b. **Scintillation detector:** A scintillation detector uses a scintillating material, which is a material that emits light when it interacts with ionising radiation. When radiation interacts with the scintillating material, it produces flashes of light that are detected by a photomultiplier tube or a photodiode. Scintillation detectors are used for detecting gamma and beta radiation and can also be configured to detect alpha radiation with the use of a thin window.

c. **Solid-state detectors:** Solid-state detectors are made of semiconductor materials, such as silicon or germanium and are used to detect ionising radiation.

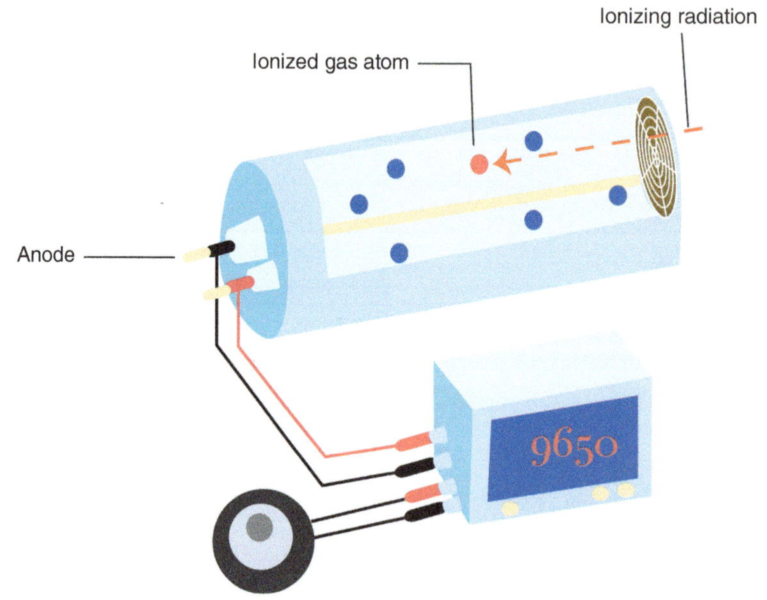

FIGURE 3.9   Geiger-Muller counter

When radiation interacts with the semiconductor material, it produces electrical charges that can be detected and measured. Solid-state detectors are commonly used for detecting gamma and X-ray radiation.

d. **Film badges:** Film badges are passive detectors that use photographic film to measure radiation exposure. The film is enclosed in a badge or a film holder and is worn by a person who may be exposed to radiation. After a certain period of time, the film is developed and the radiation exposure can be determined by the density of the film.

e. **Ionisation chambers:** Ionisation chambers are devices that measure the electrical charge produced when radiation ionises a gas inside the chamber. The charge is collected and measured and the radiation dose can be calculated based on the charge produced. Ionisation chambers are commonly used for detecting gamma and X-ray radiation in medical and industrial applications.

f. **Radiochemical methods:** Radiochemical methods involve collecting a sample and using chemical processes to separate and concentrate radioactive isotopes. The concentration of radioactive isotopes in the sample can then be measured using techniques such as liquid scintillation counting or gamma spectroscopy.

It's important to note that the penetration capacity of radioactive particles also depends on the duration of exposure. Prolonged exposure to radioactive particles can increase the risk of health effects and appropriate safety measures should always be followed when working with or around radioactive materials. Figure 3.10 shows schematically how radiation can interact with humans.

Some radionuclides occur naturally, while others are man-made. The main sources of natural radiation are uranium and thorium minerals. These two elements are called *primordial radionuclides* and trace amounts of both and their decay

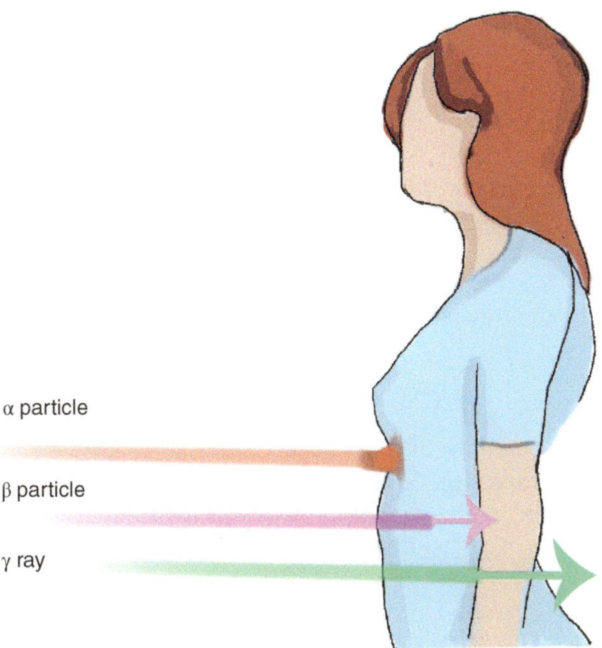

FIGURE 3.10   The penetration capacity of radioactive particles in human body

products can be found everywhere. Terrestrial radiation levels produced by uranium and thorium minerals are not equal on the Earth's surface, they vary by location depending on specific geological materials present in an area and such radiation is present mainly in soil and water bodies.

Living beings are exposed to the sum of natural radiation and other forms of so-called ionising radiation[12], including some man-made sources[13].

This amount is named as *background radiation*, a measure of the level of radiation present in a particular location that is not due to the introduction of deliberate radiation sources.

The main sources of background radiation for humans are:

a. Rocks and soils containing radionuclides and water bodies in contact with them (terrestrial background radiation)

b. Living things, such plants and animals, absorb radioactive materials from the soil or water and these pass up the food chain

c. Presence in the environment of radioactive materials from nuclear power stations, nuclear waste, nuclear weapons testing and some consumer products

d. Cosmic rays that reach the Earth's surface from space.

The amount of ionising radiation that a person is exposed to is measured in units of sieverts (Sv) or millisieverts (mSv). The risk of harmful effects from ionising radiation depends on several factors, including the amount of radiation exposure, the duration of exposure and the type of radiation. It is important to minimise exposure to ionising radiation whenever possible and to take appropriate safety measures when working with sources of ionising radiation.

To the average dose of radiation coming from natural background radiation, there has to be added radiation from medical procedures (X-ray diagnosis and radio-therapy). Ionising radiation can have both beneficial and harmful effects. However, exposure to ionising radiation can also cause damage to living cells and tissues, leading to an increased risk of cancer and other health problems.

Table 3.8 describes the common units of measurement for activity and dose exposure to radiation.

The activity of a sample is measured in terms of Becquerel units (particles or photons emitted per second). The basic unit to quantify radiation dose is the grey (Gy) and dose rates are expressed per unit time (Gy/h). The dose is calculated on the basis of how much of this energy is absorbed in 1 kg of material. Other units to express radiation dose are: absorbed dose, equivalent dose and effective dose, which take into consideration the effect of radiation on organs, tissues or the whole body.

3.20 Types of Exposure to Radionuclides

1. External exposure is produced by sources located outside of the body. The alpha and beta particles may be stopped by clothing and the outermost layers of the skin. The gamma rays may pass through the body that are responsible for the hazards.

---

[12] Ionising radiation refers to a type of radiation that has enough energy to remove electrons from atoms or molecules, resulting in the formation of ions. This type of radiation can come from natural sources such as cosmic rays as well as human-made sources such as X-rays, gamma rays, and radioactive materials.
[13] Trace amounts of radionuclides are present in the environment from radioactive materials used in industry and consumer products, as well as from nuclear weapons tests and nuclear accidents.

TABLE 3.8 Units for measurement of activity and dose exposure to radiation

Concept	Unit	Definition
Activity	Becquerel (Bq)	Particles or photons emitted per second
Exposure	Coulomb/kg	Amount of radiation travelling through the air
Absorbed dose	Gray	1 J of energy absorbed in 1 kg of material
Equivalent dose	Sievert	Amount of radiation absorbed, adjusted to account for the type of radiation and the effect on particular organs
Effective dose	Sievert	Amount of radiation absorbed, adjusted to account for the type of radiation and the effect on whole body

2. **Internal exposure** is the most dangerous. Radionuclides enter the body by inhalation, ingestion or through breaks in the skin, being transported to various sites, such as bones, marrow and organs, where they continue emitting radiation until they are removed or decay.

3. **Accidental exposure** is caused by an unexpected event. That in the case of nuclear accidents in power plants or other installations, nuclear bombs and criminal acts.

4. **Planned exposure** is a deliberate exposure to radionuclides or radiation with a diagnostic or therapeutic purpose.

It's important to note that not all radiation is harmful. For example, we are exposed to natural background radiation every day and small doses of ionising radiation can be used in medical imaging and cancer treatment with minimal risk. The harmful effects of radiation depend on the type of radiation, the amount of exposure and the type of organism being exposed.

The mechanisms of radiation damage are complex and depend on a variety of factors, including the type and dose of radiation, the tissue type and the individual's genetic susceptibility.

When ionising radiation interacts with biological tissue, it can cause a number of different types of damage, including DNA damage, cell death and mutation. The exact mechanism by which radiation produces biological damage depends on the type of radiation and the characteristics of the tissue it interacts with.

There are two mechanisms for DNA damages (Figure 3.11)

a. **Indirect effect:** One common mechanism of radiation damage is the production of free radicals, which are highly reactive molecules that can damage DNA, proteins and other cellular components. When ionising radiation passes through biological tissue, it can strip electrons from molecules and create free radicals. These free radicals can then react with nearby molecules, causing damage to the cell.

b. **Direct effect:** Another mechanism of radiation damage is the induction of DNA double-strand breaks. Ionising radiation can cause breaks in both strands of the DNA double helix, which can lead to chromosome aberrations, cell death or the activation of DNA repair mechanisms. If the DNA damage is not repaired correctly, it can lead to mutations that increase the risk of cancer.

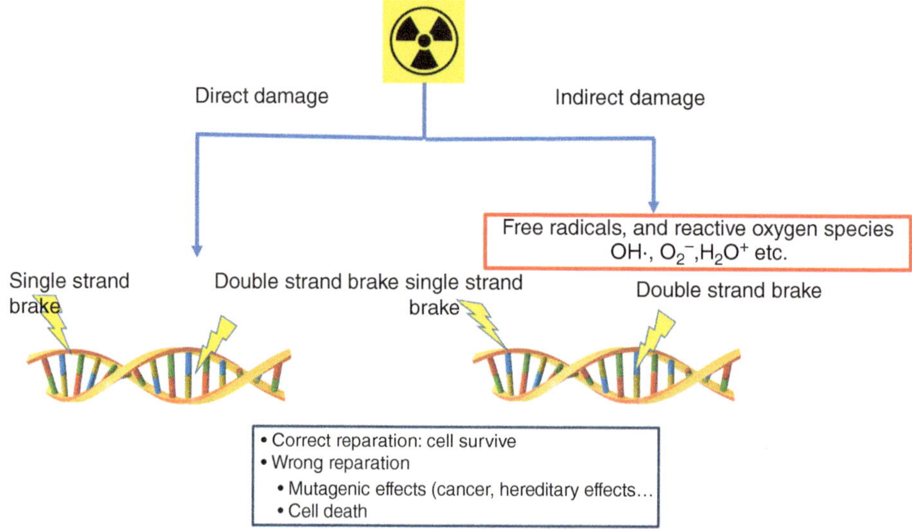

FIGURE 3.11   Schematic representation of radiation damage on DNA

Radiation can also cause damage to cellular membranes and organelles, leading to dysfunction and cell death. Additionally, radiation can induce changes in gene expression and signalling pathways, which can have long-term effects on cellular function and the risk of cancer.

Radiation can have harmful effects on living beings, depending on the type of radiation and the amount of exposure.

Here are some of the effects that radiation can have on living beings:

1. **Radiation sickness:** High doses of radiation can cause radiation sickness, which can lead to nausea, vomiting, diarrhoea and even death.

2. **Cancer:** Exposure to ionising radiation can increase the risk of cancer, particularly leukaemia, thyroid cancer and breast cancer.

3. **Mutations:** Radiation can damage DNA, which can lead to mutations in cells that can cause genetic disorders and birth defects.

4. **Infertility:** Radiation can damage reproductive cells and cause infertility.

5. **Skin damage:** Radiation can cause skin burns and other damage, particularly with exposure to high levels of ultraviolet (UV) radiation from the sun or tanning beds.

6. **Cataracts:** Exposure to ionising radiation can increase the risk of cataracts, which can lead to blindness.

Symptoms of acute radiation exposure include nausea, vomiting, diarrhoea, headache, dizziness, weakness, fatigue, bleeding, hair loss, swelling, itching, redness or other skin problems, and finally, death in the case of very large doses.

Topics of Interest: Chernobyl Accident

An example of dramatic accidental exposure to radionuclides and their consequences took place after the Chernobyl accident in 1986 with iodine 131. It decays with a half-life of 8.02 days with beta and gamma emissions to give the stable nucleus xenon-131 (Eq. 3.12).

$$^{131}_{53}\text{I} \rightarrow\ ^{131}_{54}\text{Xe} +\ ^{0}_{-1}\text{e} \tag{3.12}$$

Due to beta decay, iodine-131 causes mutation and death in the cells that it penetrates and other surrounding cells. In Chernobyl, radioactive iodine 131 was released into the air and significant amounts of it contaminated fresh products such as milk, meat and vegetables, which were eventually ingested by the population. Iodine is an element that attaches itself very easily to the thyroid and the radioactive isotope 131 is not an exception. Health consequences of the Chernobyl accident have been reported, showing a significant increase in thyroid cancer, with special significance among children and adolescents living near the reactor.

3.21 Radon Exposure

Radon-222 is a naturally occurring colourless, odourless and tasteless radioactive gas produced from the decay of uranium nuclei, with a half-life of 3.8 days.

Radon can be found in rocks and soils containing uranium minerals, which is the initial element of the series and may be present in natural waters and in the atmosphere in trace amounts. The radon decay chain is a continuous process and each isotope transforms into its daughter isotope until a stable lead isotope is reached. Radon-222, in particular, is of concern due to its potential to accumulate in indoor environments, posing health risks if inhaled in significant quantities when escaping from the ground into the air and decaying into the radioactive nuclei polonium-218 and lead-214.

Outdoor concentrations of radon are very low and generally they don't cause a health problem. The average radon level in outdoor spaces varies from 5 to 15 Bq/m^3.

Indoor radon levels can vary substantially from 10 to more than 10,000 Bq/m^3, depending on the materials and local geology.

Radon is trapped in indoor spaces, entering buildings through cracks and other holes in the foundation. Ineffective ventilation produces the accumulation of radon in buildings, but depending on the specific design, the accumulation of radon can substantially vary between nearby buildings, therefore, most radon exposure occurs in affected areas inside homes, schools and workplaces.

Concern about radon exposure has been growing since the 1970s, when research about radon effects on human health established the relationship between radon exposure and lung cancer. In fact, most human radiation exposure is caused by radon (Figure 3.12).

When radon is breathed in, it can cause damage to lung tissue caused by its decay, which can lead to lung cancer. The WHO estimates that radon causes between 6% and 15% of lung cancers worldwide and is the leading cause of lung cancer death for nonsmokers. In the case of smokers, the risk of developing lung cancer is significantly higher than for nonsmokers[14].

---

[14] Stimations consider that more than 85% of radon-induced lung cancer deaths are among smokers.

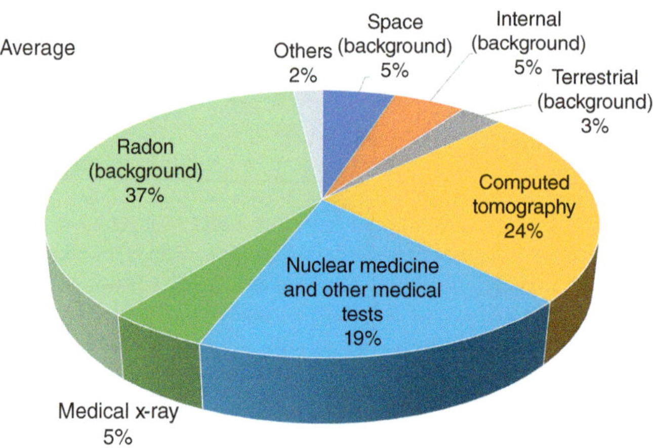

FIGURE 3.12   Sources of radiation exposure

Radon detectors are devices used to measure the amount of radon gas in indoor air. There are two main types of radon detectors: short-term and long-term.

a. Short-term detectors measure radon levels for a period of several days to several months. Short-term detectors are often used for the initial screening of a home or building to determine if further action is needed. These detectors can be placed in the lowest livable area of the building for a minimum of 48 hours and up to 90 days. They can be either passive or active. Passive detectors do not require power to operate and include charcoal canisters, alpha track detectors and electret ion chamber detectors. Active detectors require power and include continuous radon monitors and pulsed ionisation detectors.

b. Long-term detectors measure levels for a period of several months to a year or more. They are used to get a more accurate and reliable measurement of radon levels over a longer period of time. These detectors are typically passive and include alpha track detectors, electret ion chamber detectors and charcoal canisters. They can be left in place for a period of several months, a year or more.

In the event that high levels of radon are detected in a building, some measures can be taken to reduce radon exposure. The most common method is called radon mitigation, which involves installing a ventilation system to draw radon out of the building and vent it outside. Other methods include sealing cracks and openings in the foundation and improving ventilation.

To minimise radon exposure, it is recommended to increase the rate of air exchanges, use an underground ventilation system in the building and keep the equipment or the building tested, either with detection equipment or without. For example, the EPA suggests mitigating radon levels in buildings if they are at or above 148 Bq/m.

Radon is the major component of the radiation dose received by humans annually from various sources[15]. Figure 3.12 shows an estimation of the percentage of the different sources contributing to such a background radiation dose.

---

[15] https://19january2017snapshot.epa.gov/radiation/radiation-sources-and-doses_.html.

3.22 Nuclear Waste Management

Nuclear waste is the radioactive material that is generated as a byproduct of nuclear power generation, nuclear medicine and various industrial and research activities that involve nuclear materials. It can remain hazardous for thousands of years and pose significant risks to human health and the environment if not properly handled and disposed of.

Nuclear waste is typically classified into two categories based on its level of radioactivity and potential hazard: low-level waste and high-level waste.

1. Low-level waste (LLW) includes materials that have relatively low levels of radioactivity and can be safely disposed of in specialised landfills. This type of waste can include items such as clothing, tools and medical equipment that have been contaminated with low levels of radioactive material. Low-intensity waste has relatively low levels of radioactivity and can be safely disposed of in specialised landfills, while high-intensity waste is highly radioactive and can remain hazardous for thousands or even millions of years (Figure 3.13)

2. High-level radioactive waste, on the other hand, contains highly radioactive materials that can remain hazardous for thousands of years. This type of waste is typically generated by nuclear power plants and nuclear weapons production facilities. HLW includes materials such as spent nuclear fuel, which is highly radioactive and contains a variety of isotopes, some of which remain dangerous for millions of years.

   a. **Interim storage:** This involves storing the waste at a designated site until a permanent solution can be found. Interim storage facilities are typically designed to safely store nuclear waste for up to 100 years (Figure 3.14). This option provides flexibility and allows for further research and development of disposal technologies. It is designed for short-term storage of waste materials

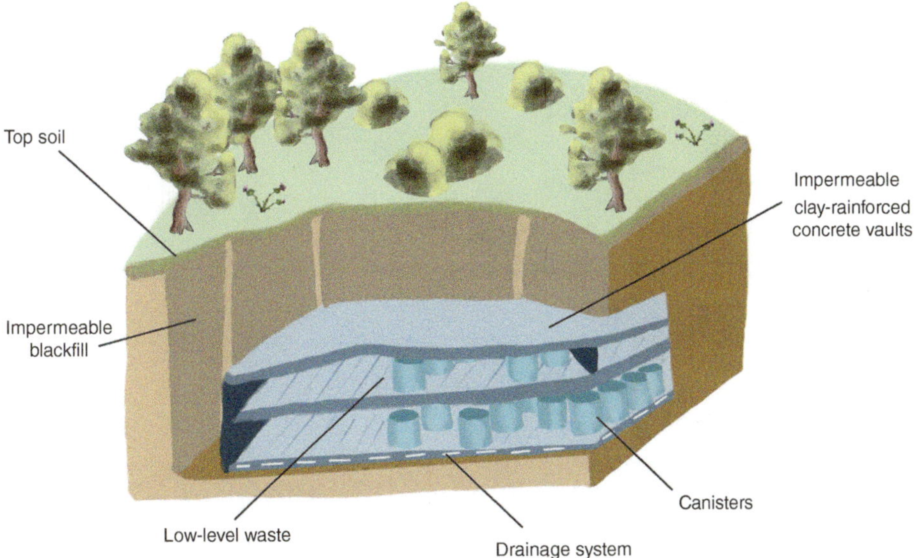

FIGURE 3.13 Low-level radioactive storage

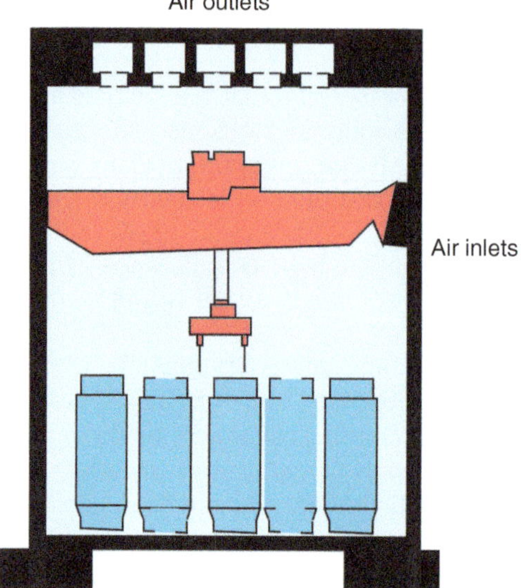

FIGURE 3.14   Interim storage of high-level radioactive waste

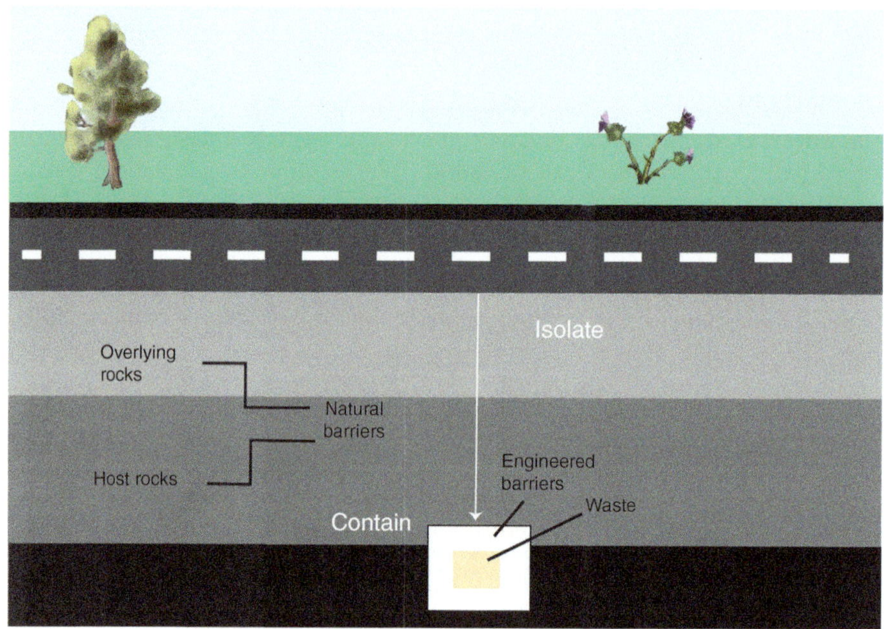

FIGURE 3.15   Deep geological repository of high-level radioactive waste

generated by industrial, commercial or other activities. These residues may include hazardous materials, such as chemicals or radioactive waste or non-hazardous materials, like scrap metal or construction debris.

b. High-Level Deep Geological Repository (DGR): This involves burying the waste in geological formations that are stable and have low ground-water movement (Figure 3.15). The waste is then sealed in multiple layers of protective barriers, including thick layers of concrete, metal and plastic.

The goal is to ensure that the waste is isolated from the environment for tens of thousands of years until it has decayed to a level that is safe for human exposure. The residues stored in a DGR include the waste that is left over after the nuclear fuel has been used and the waste that is generated during the operation of the facility itself. This waste can include a variety of materials, such as spent fuel rods, contaminated clothing and tools and other materials that have come into contact with radioactive materials.

The selection of a disposal method depends on factors such as the type of waste, the quantity of waste and the characteristics of the site. It is important to note that nuclear waste disposal is a highly regulated process and the design and operation of disposal facilities must comply with strict safety and environmental standards to minimise the risks associated with nuclear waste.

Topics of Interest: Dumping Nuclear Waste into the Sea

The practice of dumping nuclear waste into the sea occurred in the past, particularly during the 1950s and 1970s, when there was little regulation and awareness of the potential environmental and health impacts.

The United States and the Soviet Union were among the countries that disposed of nuclear waste in the ocean during this period. The United States Navy admitted to dumping over 90,000 barrels of radioactive waste in the Atlantic and Pacific Oceans between 1946 and 1970. The Soviet Union also dumped a significant amount of nuclear waste into the Arctic and Pacific Oceans, including liquid and solid waste from nuclear power plants and nuclear weapons production facilities[16].

The environmental and health impacts of these practices became increasingly evident over time and international pressure led to the adoption of the London Convention in 1972, which established a framework for regulating the dumping of these practices into the ocean.

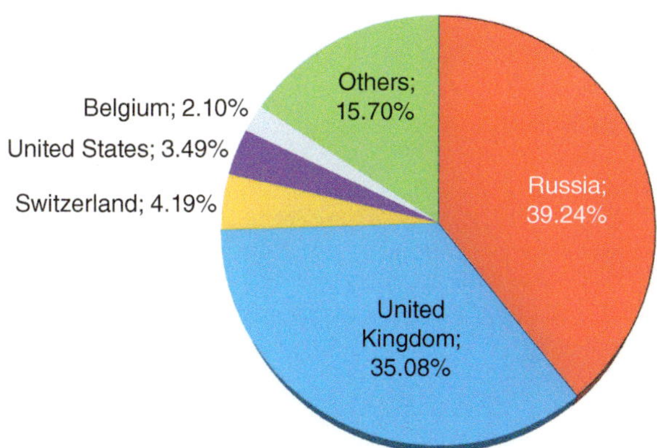

FIGURE 3.16   Inventory of radioactive material resulting from historical dumping

[16] International Atomic Energy Agency (2015). Inventory of Radioactive Material Resulting from Historical Dumping, Accidents and Losses at Sea, IAEA-TECDOC-1776. ViennaIAEA.

The Convention was amended in 1993 with the London Protocol, which specifically prohibits the dumping of radioactive waste at sea.

Since then, there have been no reported incidents of nuclear waste being dumped into the sea by any country. Instead, efforts have been focused on developing safe and sustainable disposal options for nuclear waste, such as deep geological repositories. The International Atomic Energy Agency (IAEA) and other international organisations work to promote the safe management and disposal of nuclear waste and provide guidance and support to member countries (Figure 3.16).

CHAPTER 4

Volatile Organic Compounds

4.1 Organic Compounds and Environment

Most of the registered chemical substances are organic compounds, some of them have been identified from natural sources, but the majority of organic compounds have been synthesised in laboratories or industries, which means that the majority of the organic compounds are man-made.

It is a hard task to systematise and classify organic compounds according to their interaction with the environment and living beings because of the diversity of structures, functional groups, chemical-physical properties and applications.

The first approach to systematising such interaction is to consider its volatility. The volatility of a substance is its tendency to evaporate or vaporise at a given temperature and pressure.

4.2 Volatility of Organic Compounds

Volatility refers to the tendency of a substance to vaporise or evaporate into the air. In the context of organic compounds, volatility can vary widely depending on the specific compound and its physicochemical properties.

Organic compounds are classified according to their volatility into three groups:

- Very volatile organic compounds (VVOCs)
- Volatile organic compounds (VOCs)
- Semi-volatile organic compounds (SVOCs)[1]

---

[1] Semi-volatile organic compounds will be treated in Chapters 5 and 6.

Understanding the Chemistry of the Environment, First Edition. Francisco G. Calvo-Flores.
© 2025 John Wiley & Sons Ltd. Published 2025 by John Wiley & Sons Ltd.

TABLE 4.1 Classification of organic compounds based on volatility

Acronym	BP range (°C)	Examples
VVOC	<0 to 50–100	Methane, propane, butane, ethene, isoprene and diethyl ether
VOC	50–100 to 240–250	Formaldehyde, limonene, benzene, toluene, acetone, methanol, ethanol, volatile carboxylic acids and esters
SVOC	240–260 or higher	Pesticides, plasticisers, fire retardants and some industrial by-products

Groups 1 and 2, VVOCs and organic volatile compounds (VOCs), are carbon-based chemicals that have a high vapour pressure at room temperature, so they show boiling points less than 250 °C measured at a pressure of 101.3 kPa (the atmospheric pressure standard value). They are low-molecular-weight organic compounds with about twelve or fewer carbon atoms in their structure, which easily evaporate at room temperature and are consequently emitted into the atmosphere. Both categories used to be described under the single concept of VOCs.

VOCs emitted from natural sources in terrestrial ecosystems are named biogenic volatile organic compounds (BVOCs). On the other hand, VOCs generated by human activities such as industry, agriculture, livestock, waste, etc., are named anthropogenic volatile organic compounds (AVOCs). Both groups are not mutually exclusive because there are many compounds that can be produced from both natural and anthropogenic sources.

Table 4.1 summarises the three different categories of organic compounds according to their volatility and gives some examples.

4.3 Types of Volatile Organic Compounds

According to the origin, VOCs can be divided into biogenic and anthropogenic volatile compounds.

Biogenic Volatile Organic Compounds

BVOCs are a group of organic volatile chemicals emitted by living organisms. They are released into the atmosphere as gases or vapours and play various ecological roles. BVOCs are primarily produced by plants, but they are also emitted by microbes, fungi and some animals.

Plants are the largest source of BVOCs in the natural environment. These compounds are synthesised within plant cells and then released through openings called stomata on the surface of leaves. BVOC emissions from plants are influenced by various factors, including temperature, light intensity, nutrient availability and stressors such as herbivory or pathogen attack.

The primary function of BVOCs in plants is to mediate ecological interactions. For example, they can serve as communication signals between plants and other organisms, such as pollinators or herbivores. Some BVOCs attract pollinators to aid in plant reproduction, while others repel herbivores or attract predators and parasites that feed on herbivorous insects.

The study of BVOCs is interdisciplinary, involving fields such as ecology, atmospheric science and chemistry. Researchers are interested in understanding the emission patterns of BVOCs, their ecological functions and their impacts on atmospheric chemistry and climate. It's important to note that BVOCs emissions are part of the natural carbon cycle and have occurred for millions of years. However, their contribution to climate change becomes a concern when emissions exceed the capacity of natural sinks to absorb or mitigate them.

The budget of BVOC emissions can vary depending on various factors, such as the type of vegetation, geographical location and environmental conditions.

The total global budget of BVOC emissions is challenging to estimate precisely due to the complexity and variability of emissions from different ecosystems. However, it is generally believed that BVOC emissions exceed those from anthropogenic sources. The estimates of global BVOC emissions range from hundreds of teragrams (Tg) to over a thousand teragrams per year by vegetation, predominantly forest species, with methanol and isoprenoids being the major contributors to global emissions[2].

These sources are natural, but human activities can also influence the total emissions. For example, deforestation can affect wetlands and contribute to increased methane emissions. Additionally, agricultural practices, such as rice cultivation and livestock farming, can result in significant methane release from natural sources as well.

It's important to note that biogenic methane emissions are part of the natural carbon cycle and have occurred for millions of years. However, their contribution to climate change becomes a concern when emissions exceed the capacity of natural sinks to absorb or mitigate them. Efforts are being made to understand and quantify biogenic methane emissions to better assess their impact on the climate and identify strategies for mitigation.

The specific composition and quantities of BVOC emissions can vary significantly depending on environmental factors, vegetation types and geographical locations. Detailed studies and measurements are conducted to understand the emissions from specific ecosystems and to improve our understanding of BVOCs' impact on atmospheric chemistry and climate.

Anthropogenic Volatile Organic Compounds

AVOCs are volatile organic chemicals that are emitted into the atmosphere as a result of human activities. These compounds are characterised by their ability to easily vaporise and enter the air, where they can participate in various atmospheric reactions and contribute to air pollution.

AVOCs are derived from a wide range of sources, including industrial processes, transportation, energy production and the use of consumer products.

The release of AVOCs into the atmosphere can have several environmental and health implications.

---

[2] Koppmann, R. (ed.) (2007). *Volatile Organic Compounds in the Atmosphere*. Blackwell Publishing Ltd. http://doi.org/10.1002/9780470988657. ISBN: 9780470988657.

The main issues of concern are:

1. **Harmful effects on human health:** High exposure to VOCs produces some common effects such as eye, nose and throat irritation; shortness of breath; headaches; fatigue; nausea; dizziness and skin irritation. Most of these effects disappear when exposition stops. High concentrations or long-term exposure may cause damage to the lungs, liver, kidneys or central nervous system (CNS).

2. **Undesirable effects on environment:** Contribution to tropospheric photochemical oxidant formation, increase of ground-level ozone, which can cause respiratory problems and contribute to the formation of photochemical smog, a type of air pollution that is formed by the interaction of sunlight with certain chemicals in the atmosphere. Most VOCs are net contributors to greenhouse gas emissions, stratospheric ozone depletion and global climate change.

3. **Damage to materials:** Close to their point of emission, for example, as a consequence of oxidising or corrosive properties.

In response to the environmental and health concerns associated with AVOC emissions, various regulations and measures have been implemented to control and reduce their release. These include emission standards for industrial processes and vehicles, the promotion of cleaner technologies and fuels and the development of stricter regulations on the use of certain chemicals in consumer products.

Efforts to monitor and mitigate AVOC emissions are ongoing, aiming to improve air quality and reduce the impacts of these compounds on human health and the environment.

From another point of view, VOCs are divided into two groups: outdoor and indoor volatile compounds.

The term **outdoor VOCs** refers to volatile organic compounds emitted into the atmosphere from soil, waterbodies, oceans, urban and industrial areas, dumps, etc. The term **indoor VOCs** is usually applied to volatile compounds emitted in nonindustrial indoor environments such as homes, office buildings and public buildings (schools, hospitals, theatres, cinemas, restaurants, etc.). Concentrations of many VOCs are frequently higher in indoor than outdoor spaces.

Indoor and outdoor VOCs are considered pollutants because they can lead to poor air quality and prevent the formation of photochemical smog, which is a type of air pollution characterised by a hazy appearance in the atmosphere with negative effects on human health, soil, water bodies, etc.

4.4 Representative Volatile Organic Compounds

The diversity of substances that can be considered VOCs is enormous. They are a diverse group of chemicals that have low boiling points and readily evaporate into the air at room temperature. These compounds are commonly found in various household products, industrial processes and natural sources.

In this section, we will provide a comprehensive overview of the most significant VOCs, considering two broad categories: those of natural origin and those of synthetic origin.

Such discussion is based on functional group categories, as takes place in the conventional approach for studying molecules in the field of Organic Chemistry.

4.5   Aliphatic Hydrocarbons

Aliphatic hydrocarbons are organic compounds that consist primarily of carbon and hydrogen atoms arranged in straight, branched chains and cycles. Aliphatic hydrocarbons can be further categorised into three main groups:

1. Alkanes, also known as saturated hydrocarbons, contain carbon atoms that are formed exclusively by carbon–carbon single bonds (C-C).
2. Alkenes are unsaturated hydrocarbons that contain at least one carbon–carbon double bond (C=C).
3. Alkynes are a group of unsaturated hydrocarbons that contain at least one carbon–carbon triple bond (C≡C).

With some exceptions, aliphatic volatile hydrocarbons (Table 4.3) show low toxicity at low concentrations. Exposure to high hydrocarbon concentrations may produce lung damage or affect the functions of the liver or kidney. They can easily cross the capillary alveolar membrane and cause neurological symptoms. If inhaled in high concentrations, it can cause arrhythmia, for example, from long-term exposure to gasoline.

Alkanes

Methane   Methane is the simplest hydrocarbon and the main component of natural gas. It is a colourless and odourless substance but has a sweet odour at high concentrations. In mixtures between 5 and 15% with the air, it is explosive, as it happens with fire. When emitted, it may stay in the atmosphere for ten years, and it is destroyed very gradually by photochemical reactions, worsening air quality.

While a significant portion of methane emissions come from anthropogenic sources such as agriculture and fossil fuel extraction, biogenic sources also play a role. The main sources of biogenic methane emissions are:

a. **Wetlands:** Wetlands, such as marshes and swamps, are significant natural sources of methane. Methane is produced in these areas through the breakdown of organic matter in oxygen-deprived conditions.
b. **Livestock:** Methane from livestock, primarily produced through a process called enteric fermentation. Enteric fermentation occurs in the stomachs of ruminant animals, such as cows, sheep and goats, where microorganisms help break down fibrous plant material.

   Livestock produce methane as part of their natural digestive process, and the amount generated can vary based on factors such as the type of digestive system an animal has, its diet and its age.
c. **Termites:** Termites are known to produce methane as a byproduct of their digestive processes. Termites have specialised bacteria in their guts that help break down cellulose, releasing methane in the process.

d. **Oceans:** The world's oceans also contribute to methane emissions. Methane can be generated through the decomposition of organic matter in the water column or sediments. Methane bubbles can rise to the surface and enter the atmosphere.

e. **Natural gas seeps:** Methane can naturally seep out of the ground through geologic formations such as coal seams, oil and gas reservoirs and geological faults. These seeps can release significant amounts of methane into the atmosphere.

f. **Wildfires:** Wildfires can release methane stored in vegetation and organic matter. The combustion of biomass during fires can lead to the production and release of methane into the atmosphere.

g. **Permafrost:** Permafrost refers to frozen ground in high-latitude regions. As permafrost thaws due to climate change, organic matter that was previously frozen starts to decompose, releasing methane.

Anthropogenic methane is emitted from organic waste in municipal solid waste landfills, livestock and other intensive agricultural methods, although one of the major sources of human-made methane emissions comes from the production and transport of coal, natural gas and oil.

Methane is a potent greenhouse gas, and according to recent studies from the Global Carbon Project[3], the amount of methane in Earth's atmosphere continues to rise. For the 2008–2017 decade, 576 million metric tonnes per year have been emitted. That is a 9% increase compared to the previous decade. The predominant mechanism for the removal of methane from the atmosphere is the photochemical degradation of the molecule by the hydroxyl radical (OH·)[4].

Methane is relatively non-toxic, and its health effects are associated with the fact that it displaces oxygen and may be asphyxiant.

Other Volatile Alkanes

Other common volatile alkanes are described in Table 4.2.

Propane ($CH_3CH_2CH_3$) and butane ($CH_3CH_2CH_2CH_3$) are two hydrocarbons used as combustible gases that are emitted into the atmosphere, mainly by activities related to transportation. Other hydrocarbons with a higher number of carbon atoms

TABLE 4.2   **Examples of volatile alkenes**

Name	Formula	Boiling point (°C)
Methane	CH_4	−164
Propane	$CH_3CH_2CH_3$	−42
Butane	$CH_3(CH_2)_2CH_3$	−1
Hexane	$CH_3(CH_2)_4CH_3$	69
Heptane	$CH_3(CH_2)_5CH_3$	98

[3] Jackson, R. et al. (2020). Increasing anthropogenic methane emissions arise equally from agricultural and fossil fuel sources. *Environmental Research Letters* 15: 071002.

[4] The hydroxyl radical is responsible for the breakdown and removal of most VOCs in the atmosphere.

are components of gasoline and diesel fuel. In the case of gasoline, it is typically formed by 4–12 carbon atoms with a boiling point range between 30 and 210 °C, whereas diesel fuel contains hydrocarbons with approximately 12–20 carbon atoms and a boiling point range between 170 and 360 °C. All the activities related to the production, handling or consumption of fuels are remarkable sources of emissions.

On the other hand, some simple alkane hydrocarbons such as pentane, hexane or heptane are liquids at room temperature, and they are used in industries and laboratories as solvents for reaction media, extraction solvents or in separation techniques. Such solvents show two types of risks and hazards:

1. In workspaces because they are highly flammable and toxic.
2. When emitted into the atmosphere, they participate in photochemical reactions in a similar way to methane.

A significant example is hexane ($CH_3(CH_2)_4CH_3$), which is extensively used as reaction media and for separation and extraction techniques. All over the world, the food industry uses hexane for the extraction of vegetable oils such as soja, sunflower and others from seeds or fruits. Short-term exposure to hexane vapours affects the nervous system, causing dizziness, nausea, headaches and even unconsciousness. Long-term exposure can cause severe damage to the nervous and respiratory systems, abdominal pain and damage to the skin and eyes. From the perspective of Green chemistry, hexane is strongly recommended to be substituted by heptane ($CH_3(CH_2)_5CH_3$) as a less toxic alternative.

Light Alkenes

The group of *light alkenes* is constituted by alkenes with two to five carbon atoms, as represented in Table 4.3.

Such alkenes are reactive hydrocarbons in the atmosphere, playing important roles in the photochemical production of other pollutants. C2–C4 alkenes are produced from both biogenic and anthropogenic sources, and isoprene is mainly produced by plants.

The simplest alkene is ethene (ethylene). The primary biogenic source of ethene is a consequence of microbial activity in most soil and marine environments, as well as its formation in a wide variety of plant species. It is a plant hormone and is frequently used as a ripening agent for agricultural commodities. The anthropogenic sources are related to garbage incineration, biomass burning, processes relating to

TABLE 4.3   Light alkenes

Name	Structure	Boiling point (°C)
Ethene	$CH_2{=}CH_2$	−104
Propene	$CH_2{=}CH{-}CH_3$	−47
1-Butene	$CH_2{=}CH{-}CH_2{-}CH_3$	−6.3
2-Butene (*cis* and *trans*)	$CH_3CH{=}CH{-}CH_3$	3.7 and 0.9
2-Methylpropene	$CH_2{=}CH(CH_3){-}CH_3$	−6.9

the petrochemical industry and motor vehicle exhaust. Ethene has a relatively short lifetime in the troposphere, from 14 to 32 hours, due to its reactivity with ozone and hydroxyl radical species.

These alkenes are useful synthons for the preparation of numerous organic compounds, and they are used as monomers in the polymer industry.

On the other hand, plants are major contributors to the emission of other alkenes into the atmosphere. The main examples of such alkenes (see Table 4.4) are as follows:

Isoprene (C_5H_8): Isoprene is the most abundant BVOC emitted by plants. It is primarily produced by broadleaf trees such as oaks, poplars and eucalyptus. The amount of isoprene emitted to the atmosphere by plants depends on several factors, such as the mass of the leaves, their area, the amount of light received and the temperature and light conditions. During the night, the emission of isoprene is quite reduced, but during the day, the emissions can be higher. In the middle of the day, the amount of this volatile alkene may be from 5 to 20 mg/m^2/h. Annual emission estimation is about 400–600 TgC (Tg of carbon), and about 90% of this is emitted by terrestrial plants. Isoprene is very reactive, rapidly being able to react with hydroxyl radicals in the atmosphere. It also forms hydroperoxides that can enhance ozone formation. Net aerosol formation in the atmosphere may also be influenced by isoprene.

Monoterpenes: Monoterpenes are another important class of BVOCs emitted by various plant species. Coniferous trees like pines and firs are known to emit substantial amounts of monoterpenes. Examples of monoterpenes include α-pinene, β-pinene, limonene and myrcene.

Sesquiterpenes: Sesquiterpenes are a class of BVOCs with larger and more complex chemical structures than terpenes. They are emitted by a wide range of plants, including grasses, shrubs and trees. Examples of sesquiterpenes include β-caryophyllene and farnesene.

Light alkenes are potentially toxic, mutagenic or carcinogenic due to their bioactivation by cytochrome P450s[5] to yield epoxides.

TABLE 4.4   Examples of biogenic volatile alkenes

BVOCs	Formula/structure	Relative contribution (%)	Amount emitted (Tg/y)
Isoprene		62.2	594 ± 34
Terpenes other than pinenes	$C_{10}H_{16}$	10.9	95 ± 3
Pinenes	α-Pinene β-Pinene	56.6	48.7 ± 0.8

[5] Cytochrome P450 is a set of enzymes involved in the transformation of chemical substances in organisms.

4.6   Aromatic Hydrocarbons

Aromatic compounds are a class of organic compounds that are characterised by a special type of stability and a distinct set of properties based on the cyclic structure of the aromatic ring, which provides extra stability to the molecule.

Aromatic hydrocarbons may be classified into:

1. **Monocyclic aromatic hydrocarbons (MAHs):** They are formed by a single aromatic ring, such as benzene, toluene and xylenes (see Table 4.5).

2. **Polycyclic aromatic hydrocarbons (PAHs)** are a class of compounds that consist of two or more benzene rings. When these rings share one or more common carbon atoms, they are referred to as fused rings. PAHs that are volatile or semi-volatile are known as light PAHs. These light PAHs are composed of two or three condensed rings, such as naphthalene, anthracene and phenanthrene.

Monocyclic Aromatic Hydrocarbons (MAHs)

Benzene   Benzene is the simplest aromatic hydrocarbon. It is a key molecule for the manufacture of detergents, explosives, pharmaceuticals, pesticides, cements and dyestuffs, and it has been used for decades as a solvent in paints, paint removers, adhesives, degreasing agents, denatured alcohol, rubber cements and arts and crafts supplies. In laboratories, benzene was a solvent frequently used as a reaction medium or for extraction or separation techniques.

Natural sources of benzene include volcanoes, forest fires and all kinds of spontaneous organic matter combustion. Anthropogenic benzene emissions are produced from burning coal and oil, gasoline service stations, motor vehicle exhaust and activities related to the chemical industry. Environmental tobacco smoke (ETS) is considered one of the main indoor sources of benzene.

Exposure to benzene emissions poses a very high risk to humans. Short-term exposure to benzene vapours may cause multiple effects such as drowsiness, dizziness, headaches and irritation in the respiratory tract, eyes and skin. Long-term inhalation exposure causes various disorders in the blood and reproductive system and an increased incidence of leukaemia and other cancer types. The EPA has classified benzene as a *known human carcinogen for all routes of exposure.*

TABLE 4.5   **Monocyclic aromatic hydrocarbons**

Compound	Formula	Structure	BP (°C)	Uses
Benzene	C_6H_6		80.1	Solvent, starting material
Toluene	$C_6H_5CH_3$		110.6	Solvent, starting material
Xylenes (mixture of isomers)	$C_6H_5(CH_3)_2$	*o-*, *m-* and *p*-xylene	Mixed 137	Solvents

WHO recommendations strongly promote the use of alternative solvents in industrial processes and household materials, the development of policies and legislation to remove benzene from consumer products and reducing exposure following best practices to minimise emissions from emission sources.

Toluene   Toluene is a volatile, flammable liquid that is immiscible with water and can be found naturally in petroleum crude oil but rarely exceeds one percent. Toluene is frequently used as a solvent and is often present in paints, lacquers, glues and gasoline. Despite the fact that the EPA has labelled toluene as a hazardous air pollutant and a toxic substance, it is a good substitute for benzene because their properties as solvents are very similar, and many reports have shown that toluene does not contribute to the formation of cancer in humans. For example, the Occupational Safety and Health Administration (OSHA) of the USA has labelled it as a non-carcinogenic substance. However, even though this is true, toluene must not be used carefully due to the damage it causes to the nervous system (US EPA 2005) or emitted irresponsibly to the atmosphere because, as it has been noted before, it is an air pollutant.

According to an EPA report, *it is estimated that 86% of the toluene produced is eventually released into the biosphere (primarily the troposphere), where its lifetime ranges from four days at high altitudes during the summer to several months at low altitudes during the winter. The average toluene half-life resulting from atmospheric oxidation is estimated to be 12.8 hours.*

Xylene   The name xylene refers to a commercial mixture of dimethylbenzene isomers that usually contains about 40–65% *m*-xylene and up to 20% each of *o*-xylene and *p*-xylene and other minor components (ethyl benzene [6–20%] and traces of toluene, trimethyl benzene, phenol, etc.). Xylene is used in the production of other aromatic compounds, such as ethylbenzene, as a solvent in paints, varnishes and coatings, and it is blended into gasoline. The main sources of xylene are fugitive emissions from industrial sources, vehicle motor exhaust and volatilisation when it is used as a solvent.

Xylene is emitted into the atmosphere, and its components have been detected in the air, rainwater, soils, surface water, sediments, drinking water and aquatic organisms.

Xylene single isomers or xylene mixtures produce similar effects, although specific isomers may not be equally potent in producing some effects. Short-term exposure by inhalation of mixed xylenes has been associated with dyspnea, irritation of the nose and throat, gastrointestinal effects, mild eye irritation, transient skin irritation and neurological disorders.

Long-term exposure to xylene produces neurological effects such as headaches, dizziness, fatigue, tremors, incoordination, anxiety, impaired memory and an inability to concentrate. There is no information available on the carcinogenic effects of mixed xylenes on humans.

Xylenes have short atmospheric half-lives, 0.5–1.0 days. Their removal from the atmosphere involves oxidation reactions initiated by the radical OH·.

Light Polycyclic Aromatic Hydrocarbons

Light PAHs (LPAHs) are romantic compounds that have fewer fused rings and are generally smaller in molecular size. These compounds have a wide range of uses, such as in the production of dyes, pesticides and pharmaceuticals. However, they are also known to be toxic, and prolonged exposure increases the risk of cancer and other health problems. These molecules are typically formed from incomplete combustion of organic matter and are found in various environmental sources such as coal tar, cigarette smoke and exhaust from vehicles.

Naphthalene   Naphthalene (Figure 4.1) is the simplest PAH, formed by two fused benzene rings. It sublimes very easily at room temperature, and its odour shows a characteristic that is detectable at very low concentrations.

It has been a popular substance used for decades at homes for moth-balls, although it has been replaced by other less hazardous substances such as 1,4-dichlorobenzene.

Naphthalene is obtained mainly from coal tar, and it is the starting material for numerous chemicals. Naphthalene can be detected in emissions from many chemical industries, and it is produced during biomass burning, gasoline and oil combustion and tobacco smoke.

Major exposure to naphthalene is produced mainly through inhalation, followed by ingestion, both dietary and non-dietary.

Exposure symptoms include headache, nausea, vomiting and diarrhoea, and it is associated with hemolytic anaemia, damage to the liver and neurological damage. The International Agency for Research on Cancer (IARC) classifies naphthalene as possibly carcinogenic to humans and animals.

Anthracene   Anthracene (Figure 4.2) is an aromatic polycyclic hydrocarbon consisting of three fused benzene rings.

Naphthalene is obtained from coal tar. Anthracene is mainly used as a starting material for the manufacture of 9,10-anthraquinone, a precursor for the synthesis of colourants such as alizarin (Figure 4.3). Anthracene is generated during combustion processes, and it is present in tobacco smoke.

Exposure to humans happens mainly through tobacco smoke and the ingestion of contaminated food. Anthracene may cause acute dermatitis. Prolonged dermal

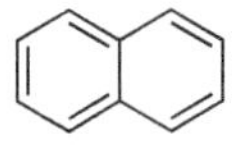

FIGURE 4.1   Naphthalene

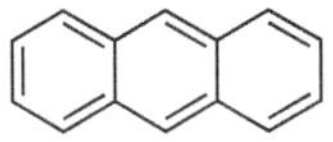

FIGURE 4.2   Anthracene

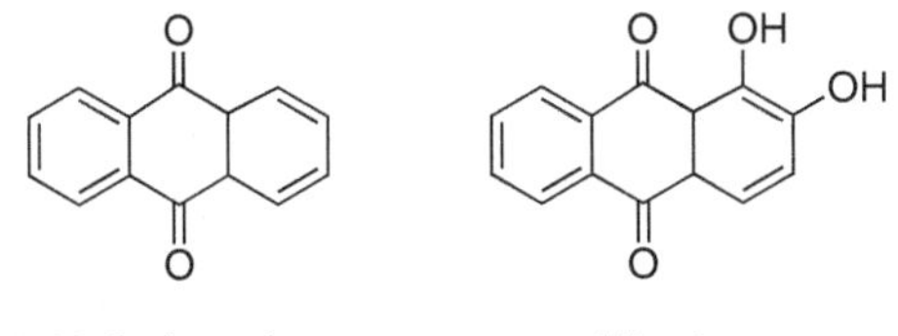

9,10-Anthraquinone   Alizarin

FIGURE 4.3   9,10-Anthraquinone and alizarin

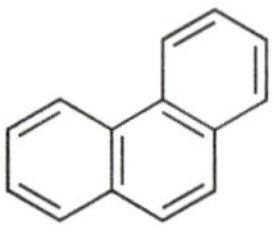

FIGURE 4.4   Phenanthrene

exposure produces pigmentation, cornification of skin surface layers and telangiecta-sis[6]. The EPA considers anthracene *not classifiable as human carcinogenicity*.

Phenanthrene   Phenanthrene (Figure 4.4) is a PAH with a formula consisting of three fused benzene rings.

Phenanthrene occurs in fossil fuels, spruce needles, tree leaves, grass and some other plants. It is present in tobacco smoke and in products of incomplete combustion of coal, wood and oil fuels.

Skin contact produces allergies and irritation and is considered phototoxic. When inhaled, it may cause respiratory tract irritation. The US Environmental Protection Agency (EPA) considers phenanthrene *not classifiable as human carcinogenic*.

Phenanthrene is widely distributed in the aquatic environment and has been detected in surface waters, wastewater and dried lake sediments.

Exposure for humans occurs primarily through inhalation of tobacco smoke and polluted air and ingestion of food or water contaminated by combustion effluents.

4.7   Halogenated Volatile Compounds

Halogenated organic compounds or *halocarbons* are molecules that have one or more halogen atoms (fluorine, chlorine, bromine or iodine) that are linked to a carbon skeleton by covalent bonds.

Halocarbons are typically classified into several groups:

1. **Haloalkanes** or alkyl halides are compounds comprised of an alkane in which one or more hydrogen atoms have been replaced by a halogen atom.

2. **Haloalkenes:** Compounds with a skeleton with one or more double bonds between carbon atoms linked to halogen atoms.

3. **Haloaromatics:** Compounds with one or more aromatic rings with halogen atoms.

Halocarbons are mostly man-made synthetic compounds. They are relatively rare in terrestrial plants and animals, but in the halogen-rich environment of the ocean, some natural products incorporate halogen atoms. The ocean is the largest known source of atmospheric methyl bromide ($BrCH_3$), methyl iodide (ICH_3) and methyl chloride ($ClCH_3$). Other significant natural contributions to methyl chloride come from biomass burning, salt marshes and wood-rotting fungi.

---

[6] Abnormal dilation of the capillaries or terminal arteries.

Chlorine and bromine derivatives linked to sp^3 carbons easily form halogen radicals by homolytic photochemical breakage, a key factor for ozone depletion in the stratosphere (Eq. 4.1).

$$C - X \rightarrow C \cdot + X \qquad (4.1)$$

4.8   Haloalkanes

Dichloromethane, Trichloromethane (Chloroform) and Carbon Tetrachloride

These three compounds have been popular solvents for decades in laboratories and industry (see Table 4.6).

Dichloromethane is a colourless liquid with a sweetish odour. It has been extensively used for the extraction of many organic molecules, including caffeine from coffee grains for decaffeinated coffee manufacture, but it has been substituted by other less hazardous solvents. Inhalation in humans produces effects on the nervous system, including decreased visual, auditory and motor functions, but these effects are reversible once exposure ceases. Common effects of chronic inhalation exposure in humans are headaches, dizziness, nausea and memory loss.

Trichloromethane (chloroform) has been used in the past as an extraction solvent for fats, oils, greases, dry cleaning spot remover, in fire extinguishers, as a fumigant and as an anaesthetic. However, chloroform is no longer used in most of these applications because of its effect on human health. In the range of 1,500 to 30,000 ppm, it produces anaesthesia and at lower concentrations (<1,500 ppm), it causes dizziness, headache and tiredness. Chronic exposure is associated with effects on the liver and kidneys.

Carbon tetrachloride has been produced in large quantities for the manufacture of refrigerants, propellants for aerosol cans, dry cleaning agents and as organic solvents in reactions, extractions, etc. Acute and chronic inhalation and oral exposures to high levels of carbon tetrachloride cause damage to the liver (swollen, tender liver, changes in enzyme levels and jaundice) and kidneys (nephritis, nephrosis and proteinuria), as well as depression of the CNS. Due to all these harmful effects, halogenated solvents tend to be eliminated.

TABLE 4.6   **Haloalkanes**

Compound	Formula	Boiling point (°C)	Uses
Dichloromethane	CH_2Cl_2	39.6	Solvent
Chloroform	$CHCl_3$	61.2	Solvent
Carbon tetrachloride	CCl_4	76.7	Solvent
Methylbromide	CH_3Br	−93.6	Biocide in agriculture

Methyl Bromide (CH_3Br)

Methyl bromide is formed naturally by *algae* or kelp in the ocean and it is synthetically obtained by the chemical industry. It has been used as a fumigant in soil to control fungi, nematodes and weeds, in food commodities (e.g., grains) and in storage facilities (such as mills, warehouses, vaults, ships and freight cars) to control insects and rodents is an irritant for the skin and eyes and acute exposure by inhalation frequently leads to neurological effects in humans and kidney and liver damage. In a human mortality study, a higher incidence of death from testicular cancer was identified in men exposed to methyl bromide. This substance is under strict regulations because of its toxicity and strong impact on ozone depletion.

4.9 Chlorofluorocarbon Compounds

Chlorofluorocarbon (CFC) compounds are a group of halogenated hydrocarbons that contain chlorine and fluorine with one or two carbon atoms. They are also commonly known by the trademark name Freons.

CFCs are very stable and unreactive under normal conditions, which makes them useful in a variety of industrial applications. They are not toxic to humans and animals at normal concentrations and they are not flammable and do not support combustion, making them safe for use in fire suppression systems.

CFCs have a high vapour pressure, which means they evaporate easily even at room temperature and can become airborne.

For decades, CFCs have been widely used as refrigerants for refrigerators, automobiles and air conditioners, propellants in aerosols and degreasing agents for the cleaning of semi-conductors and precision equipment.

Nomenclature of CFCs

In 1929, Midgley and Henne introduced a nomenclature system to classify halocarbons with carbon, hydrogen, chlorine and fluorine atoms. Subsequently, Park made further advancements to refine the system. The initial designation for the group of compounds was CFC, which later underwent subdivision into several subgroups (Table 4.7), the CFC group being those only composed of chlorine, fluorine and carbon.

TABLE 4.7 **CFCs and related compounds**

Name (acronym)	Composition
Chlorofluorocarbons (CFCs)	Chlorine, fluorine and carbon
Hydrofluorocarbons (HFCs)	Hydrogen, fluorine and carbon (no chlorine)
Hydrochlorofluorocarbons (HCFCs)	Hydrogen, chlorine, fluorine and carbon
Hydrobromofluorocarbons (HBFCs)	Hydrogen, bromine, fluorine and carbon

TABLE 4.8   **Naming system of chlorofluorocarbons and related compounds**

Number	Meaning
First	Number of double bonds, omitted if zero
Second	Number of carbon atoms −1, omitted if zero
Third	Number of hydrogen atoms +1
Fourth	Number of fluorine atoms
Fifth	Number of chlorine atoms replaced by bromine, always used with prefix '*b*' (*b*1, *b*2), omitted if zero
Suffix (additional letter)	To identify isomers, the isomer without a suffix always has the smallest mass difference on each carbon atom. If there are more isomers, the suffix is counting from *a* to *z*; it is omitted if only one isomer exists

The naming system for all those compounds consists of three components (Table 4.8): an acronym, a sequence of five numbers and an alphanumeric suffix. Each component carries specific information, as follows:

- **Body:** Represents the group type of the compound (Acronym)
- **Number:** This corresponds to the molecular formula of the compound.
- **Suffix:** Differentiates between various isomers of the compound.

In the case of cyclic structures, the number is prefixed with a 'C' for cyclic. Sometimes other bodies are found, like FC (fluorocarbon), HC (halocarbon) or R (refrigerant).

This nomenclature system provided a standardised approach to identifying and categorising halocarbon compounds based on their composition and structure.

Example:

CFC−12b1

It has no double bonds, only one carbon atom, no hydrogen atoms, two fluorine atoms, one chlorine atom replaced by bromine, and no isomers = bromochlorodifluoromethane.

Another system for naming CFCs, known as 'The Rule of 90', has been detailed in the Journal of Chemical Education[7]. According to this rule, a value of 90 is added to the CFC number, which consists of three digits (abc) (Table 4.9). The resulting number, denoted as xyz, corresponds to the Carbon-Hydrogen-Fluorine (CHF) composition of the compound.

If the carbon atoms in the CFC molecule are not saturated, meaning they are not bonded to the maximum number of other atoms, then additional chlorine atoms are required. The specific number of additional chlorine atoms needed would depend on the nature of the unsaturation in the carbon atoms.

Table 4.10 summarises some examples of chlorofluorocarbons (CFCs), their molecular formula and their common uses.

[7] Woolf, A.A. (1993). Journal of Chemical Education 70: 35–36.

TABLE 4.9 The rule of 90

Acronym	Add 90	Composition				Formula
		C	H	F	Added Cl	
CFC-11	101	1	0	1	3	CCl_3F
CFC-12	102	1	0	2	2	CCl_2F_2
HCFC-22	112	1	1	2	1	$CHClF_2$
HCFC-123	213	2	1	3	2	$CHCl_2\text{-}CF_3$
HFC-134a	224	2	2	4	0	$CH_2F\text{-}CF_3$

TABLE 4.10 Examples of CFCs

Acronym	Formula	Common uses
CFC-11	CCl_3F	Refrigeration and air conditioning
CFC-12	CCl_2F_2	Refrigeration and air conditioning
CFC-113	$C_2Cl_3F_3$	Cleaning agents and solvents
CFC-114	$C_2C_{12}F_4$	Refrigeration and air conditioning
CFC-115	C_2ClF_5	Refrigeration and air conditioning
CFC-116	C_2F_6	Refrigeration and air conditioning
CFC-218	C_2ClF_3	Solvents and blowing agents
CFC-500	$CClF_3$	Aerosol propellants and fire suppressants
CFC-502	CCl_2F	Refrigeration and air conditioning

4.10 Hydrochlorofluorocarbons

HCFCs are molecules containing hydrogen, chlorine, fluorine and carbon atoms (Table 4.11). They were introduced as a replacement for CFCs, which were found to be harmful to the ozone layer. HCFCs are less damaging to the ozone layer than CFCs, but they still have some impact on it. HCFCs have been widely used as refrigerants, solvents, blowing agents for foam insulation and in other applications. Some of them have also been used as precursors for the production of fluoropolymers. However, due to their environmental impact, their use is being phased out in many countries.

4.11 Hydrobromofluorocarbons

HBFCs are a class of chemicals that contain one, two or three carbon atoms and at least one atom each of hydrogen, bromine and fluorine. They have been used in a variety of industrial and commercial applications, including refrigeration and air conditioning systems, fire suppression and solvent cleaning. However, they are

TABLE 4.11   Examples of hydrochlorofluorocarbons

Acronym	Formula	Common name	Uses
HFC-134a	CH_2FCF_3	1,1,1,2-Tetrafluoro ethane	Refrigerant, propellant and blowing agent for foam insulation
HFC-410a	$CH_2F_2CHF_2CF_3$	R-410A	Air conditioning and refrigeration systems
HFC-23	CHF_3	Trifluoromethane	Production of refrigerants and fluoropolymers

TABLE 4.12   Examples and uses of hydrobromofluorocarbons (HBFCs)

HBFC name	Formula	Common uses
Bromofluoromethane	CH_2BrF	Fire extinguishing agent, refrigerant, propellant
Bromotrifluoromethane	$CBrF_3$	Fire extinguishing agent, refrigerant
Bromodifluoromethane	$CHBrF_2$	Fire extinguishing agent, refrigerant, propellant
Bromochlorodifluoromethane	$CBrClF_2$	Fire extinguishing agent, refrigerant, propellant
Bromochlorofluoromethane	$CHBrClF$	Fire extinguishing agent, refrigerant, propellant
Bromotrifluoroethane	C_2BrF_3	Fire extinguishing agent, refrigerant
Bromochlorotrifluoroethane	C_2BrClF_3	Fire extinguishing agent, refrigerant
Bromodichlorofluoroethane	$C_2BrCl_2F_2$	Fire extinguishing agent, refrigerant
Bromotrifluoropropene	$C_3H_2BrF_3$	Foam blowing agent, refrigerant
Bromochlorodifluoropropene	$C_3H_2BrClF_2$	Foam blowing agent, refrigerant

known to contribute to ozone depletion and are being phased out. In Table 4.12, some examples of these compounds are given.

CFCs are major contributors to the depletion of the ozone layer because of their ability to generate halogen radicals and additionally, they contribute to the greenhouse effect.

The Montreal Protocol, signed in 1994, was adopted as a framework for international cooperation regarding CFC control on the basis of the Vienna Convention for the Protection of the Ozone Layer.

4.12   Halons

Halons are defined as a group of haloalkane compounds that contain bromine and/ or chlorine, with one or two carbon atoms. Such substances are chemically stable, nonconductors of electricity and they have been extensively used as effective fire extinguishers and fire and explosion suppression agents.

TABLE 4.13 Examples of halons and their former uses

Halon	Chemical formula	Former uses
Halon 1211	$CBrClF_2$	Portable fire extinguishers, aviation and military applications
Halon 1301	$CBrF_3$	Total flooding fire suppression systems in buildings, electronic equipment and industrial machinery
Halon 2402	$CBrF_2CBrF_2$	Military applications, particularly on board ships

Halon nomenclature is relatively simple; each halon has a four-digit number *abcd*, where:

a = the number of carbon atoms

b = the number of fluorine atoms

c = the number of chlorine atoms

d = the number of bromine atoms

For example, Halon-1211 is CF_2ClBr.

However, they are now largely banned due to their ozone-depleting properties. Table 4.13 shows some examples of halons and their former uses:

Halons have an extremely high potential for ozone depletion (they are ten times more potent than CFCs) and they also act as global warming gasses. For this reason, they have been regulated. Owing to the Montreal Protocol, the phaseout of production, for example, of halons, halon 1301 ($CBrF_3$) and halon 1211 (CF_2ClBr), has been effective in the developed countries since 1994.

4.13 Haloalkenes

Vinyl Chloride (CH_2=CHCl)

Vinyl chloride is a haloalkene with the formula H_2C=CHCl. It is used as a monomer to produce the polymer polyvinyl chloride (PVC). The principal emission sources are the monomer production plants and plants where PVC and PVC products are fabricated. Acute exposure to high levels of vinyl chloride may affect the CNS. Chronic exposure to vinyl chloride through inhalation and oral exposure in humans has resulted in liver damage. Cancer is a major concern from exposure to vinyl chloride, as vinyl chloride exposure has been shown to increase the risk of a rare form of liver cancer in humans.

Tetrachloroethylene (Cl_2C=CCl_2$)

It is an excellent solvent for nonpolar molecules. It is highly stable and non-flammable. The main use of this substance is in dry cleaning and textile processing, but it is also used to degrease metal parts in the automotive and other metalworking industries.

In the past, it has been extensively used as an intermediate in the manufacture of halogenated refrigerants such as HFC-134a (CH_2FCF_3). The acute toxicity of tetrachloroethylene has been described as moderate to low. However, some international organisations, such as the European Scientific Committee on Occupational Exposure Limits (SCOEL), recommend an occupational exposure limit (8-hour time-weighted average) of 20 ppm and a short-term exposure limit (15 minutes) of 40 ppm. This recommendation is especially for workers involved in the dry cleaning branch. Tetrachloroethylene can migrate through vapours from the groundwater (or soil) up into the air of homes and buildings through vapour intrusion.

4.14 Alcohols

Alcohols are a class of organic compounds characterised by one or more hydroxyl (-OH) groups linked to a carbon skeleton. The primary source of volatile alcohols is the chemical industry, where they are produced through various synthetic processes. These processes typically involve the reaction of suitable starting materials with catalysts or other reagents to obtain the desired alcohol. In addition to synthetic production, volatile alcohols can also occur naturally in certain fruits, plants and other organic materials. Representative volatile alcohols include methanol, ethanol and 2-propanol (Table 4.14).

Methanol (CH_3OH)

The simplest alcohol is a volatile, colourless and flammable liquid, miscible with water. Historically, it was known as *wood alcohol* because it was once produced chiefly by the destructive distillation of wood in the absence of oxygen. In nature, it may be found in small amounts in normal, healthy human individuals or produced by anaerobic bacteria and phytoplankton, but most of this alcohol is produced by the chemical industry. Methanol is primarily used as an industrial solvent, but there are many other applications, such as antifreeze for automotive radiators, gasoline additive, denaturant for ethanol and as raw material for the synthesis of many chemicals, such as formaldehyde, acetic acid, methyl tert-butyl ether, methyl amines, methyl halides and biodiesel.

Methanol is also produced by plants. Some ways in which plants can emit methanol are as follows:

a. **Pectin methanolisation:** Methanol is formed through the demethylation of pectin, a structural polysaccharide found in the cell walls of plants. Pectin methanolisation occurs during the growth and development of plants, especially in tissues with high pectin content such as fruits, leaves and stems.

TABLE 4.14 **Representative volatile alcohols**

Alcohol	Formula	Boiling point (°C)
Methanol	CH_3OH	64.7
Ethanol	CH_3CH_2OH	78.4
Isopropanol	$CH_3-CHOH-CH_3$	82.5

b. **Lignin degradation:** Methanol can also be released during the degradation of lignin, a complex polymer found in the cell walls of woody plants. Lignin breakdown occurs through the action of enzymes and microorganisms, leading to the release of methanol as a by-product.

c. **Methylotrophic bacteria:** Some plants host symbiotic relationships with methylotrophic bacteria, which are capable of metabolising methanol. These bacteria can utilise methanol as a carbon and energy source, converting it into other compounds through metabolic processes. As a result, the methanol emitted by the plants is consumed by the associated bacteria.

Methanol is very toxic. Acute exposure of humans to methanol by inhalation or ingestion may result in visual disturbances, neurological damage and permanent motor dysfunction. Toxicity and death occur even after drinking a small amount, accidentally or not. For example, destroying 10 mL of pure methanol can cause permanent blindness by destroying the optic nerve and 15 mL is potentially fatal, although the median lethal dose is about 1–2 mL/kg body weight of pure methanol. Methanol can be fatal because it has a strong CNS inhibitor in a similar way as ethanol poisoning, is metabolised to give formaldehyde and later forms formic acid in a process initiated by the enzyme *alcohol dehydrogenase* in the liver.

Ethanol (CH_3CH_2OH)

Ethanol is probably the best-known alcohol. Ethanol is a volatile, flammable and colourless liquid with a slight characteristic odour. It has been produced since prehistoric times, mostly through fermentation processes. Large-scale production of ethanol is carried out either by the fermentation of cellulose, sugars or yeasts or, in the petrochemical industry, by ethylene hydration. Ethanol obtained from biomass is named bioalcohol.

On the other hand, it is an important raw material in the chemical industry for the preparation of a variety of chemicals such as ethyl halides, ethyl esters, diethyl ether, acetic acid and ethyl amines.

Ethanol is used as a multipurpose industrial solvent and is present in popular drinks such as beer, wine, spirits and other beverages. It is used as a topical disinfectant to prevent infections and is widely present in pharmaceutical preparations, cosmetics and perfumes.

Ethanol may be added to gasoline as an additive to improve its properties, and it is also used as fuel, pure or blended[8].

Nevertheless, biogenic emissions from terrestrial plants are considered to be the dominant global source of atmospheric ethanol because it is produced spontaneously in plant tissues via fermentation reactions in leaves and roots. Part of this ethanol is oxidised to acetaldehyde and acetate and metabolised by the plant, and a certain amount of the ethanol together with acetaldehyde is released to the atmosphere via leaf stomata.

---

[8] The use of ethanol as a biofuel is increasing because it is a renewable transportation fuel with a potentially lower carbon dioxide atmospheric impact compared to fossil fuels.

Some reports suggest that in recent years, the direct emission of ethanol into the atmosphere caused by anthropogenic activities has grown dramatically.

Most people are exposed to ethanol in the form of alcoholic beverages, but exposure to higher concentrations may occur in workplace settings, such as in industry. Acute inhalation can irritate the nose and throat, causing choking and coughing. At high levels, it can cause inebriation.

Isopropyl Alcohol (2-Propanol)

Isopropyl alcohol (CH_3-CHOH-CH_3) is widely used as a cleaner and disinfecting agent. Other applications of this secondary alcohol are for making cosmetics, skin and hair preparations, pharmaceuticals, perfumes, lacquer formulations, dye solutions, antifreezes and soaps.

2-Propanol causes serious eye irritation, drowsiness or dizziness, gastrointestinal irritation and gastritis. Severe intoxication produces respiratory depression, coma and finally death.

Topics of Interest: The Aroma of Rain

The aroma of rain is a distinctive and pleasant scent that is often described as fresh, earthy and clean. The smell is caused by a number of factors, including the release of oils from plants and the presence of bacteria in the soil.

One of the main reasons that rain has a distinctive smell is due to the release of VOCs from plants. These compounds, which include oils and other chemicals, are released by plants in response to humidity and other environmental factors. When it rains, the moisture in the air causes these compounds to be released into the atmosphere, creating a distinctive scent.

Another factor that contributes to the aroma of rain is the presence of bacteria in the soil. When rain falls on the ground, it stirs up the soil and releases these bacteria into the air, creating a distinct and pleasant odour.

Overall, the aroma of rain is a complex phenomenon that is influenced by a variety of factors, including the weather conditions, the types of plants and bacteria in the environment and the chemical composition of the air.

Geosmin (Figure 4.5) is a naturally occurring organic compound that is responsible for the earthy and musty smell often associated with soil and freshwater. It is produced by certain types of bacteria and blue-green algae and is also found in beets and some other plants.

The name 'geosmin' comes from the Greek words 'geo', meaning earth and 'osme', meaning smell. The compound is produced by the bacteria *Streptomyces* and several other bacterial genera, which are commonly found in soil and freshwater environments. Geosmin is also produced by blue-green algae known as cyanobacteria, which can be found in water sources such as lakes and ponds.

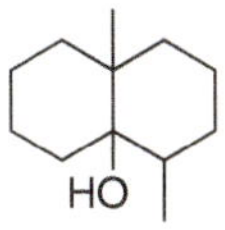

FIGURE 4.5   Geosmine structure

Geosmin is highly volatile, meaning that it easily evaporates into the air. This is why the compound is often detected as a distinct earthy smell after rainfall or when soil is disturbed. The human nose is very sensitive to geosmin and can detect it at extremely low concentrations – as low as 10 parts per trillion.

In addition to its smell, geosmin has also been found to have a taste; some people describe it as earthy, while others find it unpleasant. The compound is also used in the food and fragrance industries to add earthy notes to certain products.

4.15 Ethers (R-O-R′)

Ethers are molecules with the general formula R-O-R′ (R=R′ or R≠R′). Environmentally significant are diethyl ether and *tert*-butyl methyl ether.

Diethyl Ether [$(CH_3CH_2)_2O$]

Diethyl ether is a non-protic solvent with a moderate polarity. It is a colourless liquid, highly volatile (BP 34.6 °C), sweet-smelling and extremely flammable, tending to form explosive peroxides in the presence of oxygen in the air by a chain reaction, as shown in Figure 4.6.

To avoid this risk, it is very common to add a *stabiliser* to inhibit the formation of peroxides in ethyl ether. Adding butylated hydroxytoluene (BHT) at low ppm concentrations scavenges the free radical species responsible for peroxide formation.

Despite being a dangerous substance, it is commonly used as a solvent in laboratories for liquid–liquid extractions and as reaction media.

In the 19th century, it was used as a recreational drug to cause intoxication. For many years, it has been used as a general anaesthetic until it was found to have undesirable side effects, such as postanesthetic nausea and vomiting, so it was substituted for other anaesthetic agents that reduce these side effects.

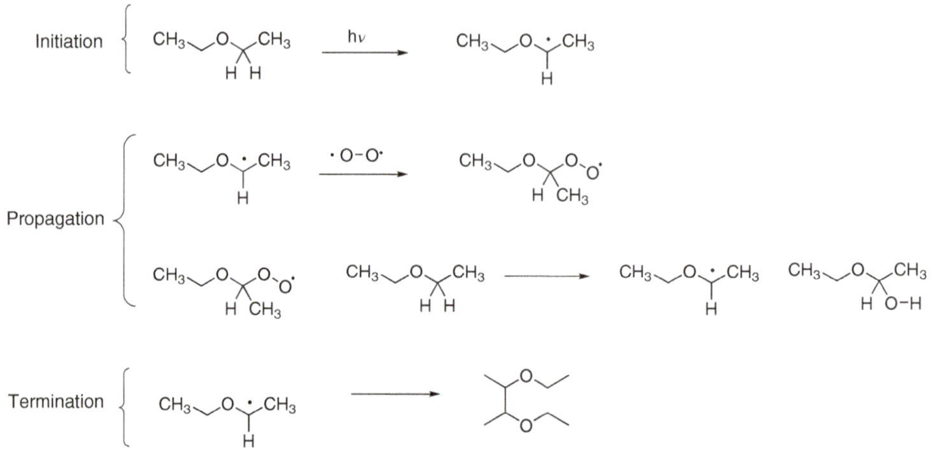

FIGURE 4.6 Peroxide formation from ethers

Methyl *tert*-Butyl Ether (CH$_3$-O-C(CH$_3$)$_3$)

Methyl *tert*-butyl ether (MTBE) is a flammable liquid (BP 55.2 °C) with a characteristic odour, used as a solvent and an antiknock additive to increase the octane number of unleaded gasoline.

Contact with the eyes and the skin causes irritation, and prolonged inhalation can cause coughing, respiratory failure, vertigo and intoxication. It is slightly water-soluble (26 g/L at 20 °C) and shows some persistence in the environment, so it travels faster and farther than many other components of gasoline when released into aquifers. Since the 1990s, the presence of MTBE in groundwater for human consumption has become noticeable and alarming.

4.16 Aldehydes and Ketones

Aldehydes and ketones are two classes of organic compounds that contain the carbonyl functional group (-C=O). Aldehydes have the carbonyl group at the terminal carbon atom of a carbon chain and ketones, on the other hand, have the carbonyl group located within the carbon chain, bonded to two other carbon atoms.

Most volatile carbonyl compounds may have a natural or anthropogenic origin (Table 4.15). Many of them are used as raw materials in the chemical industry and as solvents. They are flammable and pose certain dangers when handled. In urban areas, carbonyl compounds can be formed in explosion engines by incomplete combustion of fuel. In the troposphere, they contribute to the formation of ozone through photochemical reactions.

Low-molecular mass aldehydes and ketones are soluble in water due to hydrogen bonds formed with water molecules. Many aldehydes and ketones have characteristic odours and are used as fragrances.

Formaldehyde (HCHO)

Formaldehyde (methanal) is the simplest and most volatile (BP−19 °C) aldehyde that is produced in small amounts by most living organisms as part of normal metabolic processes. It is a colourless gas that, at low concentrations, can cause irritation to the eyes, throat and bronchial tubes.

Formaldehyde is highly soluble in water. It is used for various industrial, commercial and medical applications, including as a disinfectant, a preservative and

TABLE 4.15 **Carbonyl volatile compounds**

Carbonyl compound	Formula	BP (°C)
Formaldehyde	HCHO	19
Acetaldehyde	CH$_3$CHO	20,2
Acetone	CH$_3$COCH$_3$	56
Acrolein	CH$_2$=CH-CHO	53

in the production of resins and plastics. Formaldehyde water solutions are typically expressed as a percentage, representing the weight or volume of formaldehyde present in the solution. Common concentrations include:

1. **Formalin 37%:** This is the most commonly used concentration of formaldehyde solution. It contains 37 g of formaldehyde per 100 mL of solution.
2. **Formalin 10%:** This is a less concentrated solution, with 10 g of formaldehyde per 100 ml of solution. It is often used in histology and pathology laboratories for tissue fixation and preservation.

The choice of concentration depends on the specific application and requirements. Formaldehyde is an important precursor for the synthesis of many other chemical compounds, and it is used as a monomer for the synthesis of polymers and resins such as bachelite or urea-formaldehyde resin.

In low or moderate doses, humans metabolise formaldehyde quickly, converting it to formic acid, so it does not accumulate in the body, but at high concentrations, it can cause asthma attacks. In contact with the skin, it causes dermatitis. It is very toxic if swallowed, causing severe kidney damage, coma and even death. In 1987, the US EPA classified formaldehyde as a *probable human carcinogen under conditions of unusually high or prolonged exposure.*

Formaldehyde kills viruses, fungi and bacteria by reacting with certain groups of proteins. Aqueous solutions of formaldehyde are disinfectants and biocides for most bacteria and fungi and their spores and serve as a preservative for biological specimens in laboratories and mortuaries. Formaldehyde has a low persistence in the environment because it is quickly broken down by sunlight or by bacteria present in soil or water.

Acetaldehyde (CH_3CHO)

Acetaldehyde (ethanal) is a flammable liquid (BP 20.2 °C), it has a sweet odour and it is present in ripe fruits, produced in the normal breakdown of carbohydrates. It is an important starting material for the synthesis of many useful organic compounds, and it is present in tobacco and cannabis smoke in remarkable quantities.

Acetaldehyde is a narcotic, and in large doses, it can cause respiratory failure. Chronic exposure produces symptoms like those of alcoholism. Acetaldehyde, derived from mucosal or microbial oxidation of ethanol, tobacco smoke and diet, appears to act as a cumulative carcinogen in the upper digestive tract of humans. The IARC has listed acetaldehyde as a Group 1 carcinogen. Acetaldehyde is 'one of the most frequently found air toxins, with a cancer risk greater than one in a million'.

Acetone (CH_3COCH_3)

Acetone (propanone) is one of the most widely used organic solvents. It is a non-protic polar solvent that dissolves many organic compounds and can also be mixed with water. This highly volatile (56 °C) and flammable liquid is the solvent used in the manufacture of many varnishes, lacquers and nail polish removers.

It is produced naturally in the metabolic process, but when the metabolism of fats and carbohydrates is out of balance (for example, in starvation or diabetes mellitus), an excess of acetone is produced in the liver, a condition known as ketosis that, in severe cases, leaves the smell of acetone on the breath of a patient.

Acetone exhibits only slight toxicity for conventional uses. There is no strong evidence of chronic health effects if basic precautions are followed, and it is recognised to have low acute and chronic toxicity if ingested and/or inhaled. The majority of the acetone released into the environment is of industrial origin. Acetone evaporates rapidly, even from water and soil, and once in the atmosphere, it has a short half-life because it is degraded by UV light via methane and ethane. For all these reasons, acetone may be considered a green solvent[9].

Acrolein (CH_2=CH-CHO)

Acrolein (propenal) is an unsaturated aldehyde, a liquid at room temperature (53 °C) with an unpleasant odour and quite soluble in water. It is produced industrially from propylene and used as a building block for other chemical compounds, such as methionine and numerous other molecules. It is used as a biocide for the control of aquatic flora and fauna.

Acrolein is highly toxic. Exposure causes inflammation and irritation of the skin, respiratory tract and mucous membranes. Delayed pulmonary oedema may occur after inhalation. Systemic effects may occur after exposure by any route. Acrolein also occurs as a byproduct of the combustion of organic matter and as a metabolite of a number of compounds.

When fats and oils are heated to high temperatures, such as during frying or grilling, acrolein can be produced. This is because oils and fats may decompose into glycerol, and glycerol is thermally transformed into acrolein (Figure 4.6). The smoke that arises during high-temperature cooking contains acrolein, which can be inhaled or deposited onto the surface of food items.

However, it is important to note that the levels of acrolein formed during normal cooking processes are generally considered to be low and unlikely to pose a significant risk to human health. The World Health Organization (WHO) has established a provisional tolerable weekly intake of acrolein from all sources, including food, of 7.5 µg/kg of body weight.

To minimise acrolein formation during cooking, it is advisable to avoid excessive heating of fats and oils and to maintain proper ventilation in the cooking area to reduce exposure to smoke. Additionally, using cooking methods that involve lower temperatures, such as steaming or boiling, can help minimise acrolein production. On the other hand, acrolein is present in the smoke of tobacco. Some studies reveal connections between the acrolein gas in cigarettes and the risk of lung cancer. Other sources of acrolein are the E-cigarettes that use 1,2-propylene glycol and glycerin for the e-liquids, which are thermally degraded to form acetaldehyde, formaldehyde and acrolein when heated in electronic nicotine delivery systems (ENDS) (Figure 4.7).

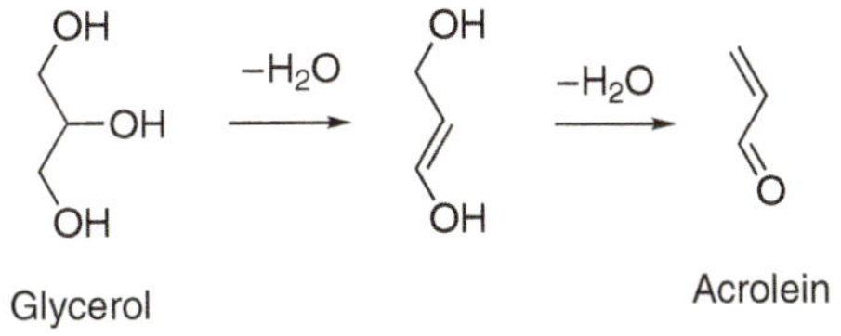

FIGURE 4.7   Acrolein from glycerol

[9] Green solvents are defined as environmentally friendly solvents because of their low toxicity and persistence.

4.17 Volatile Carboxylic Acids

Carboxylic acids are organic compounds that contain the carboxyl functional group (-COOH). The general chemical formula for a carboxylic acid is R-COOH, where R represents an organic group. Volatile organic acids (see Table 4.16) are a significant component of environmental non-methane organic compounds. They may contribute to atmospheric water acidity and participate actively in some photochemical reactions. The atmospheric sources of carboxylic acids are diverse. These sources comprise:

1. Primary anthropogenic emissions
2. Primary biogenic emissions
3. Photochemical transformations of precursors in aqueous, gaseous and aerosol phases.

Formic Acid (HCOOH)

Formic acid is the simplest carboxylic acid. It is a colourless, fuming liquid with a pungent odour. It is an irritant to the mucous membranes and blisters the skin. It naturally occurs when ant hills give off formic acid vapour (the term 'formic' comes from the Latin word formica, aunt). Formic acid can also be found in the atmosphere, primarily due to spontaneous emissions from soil or forests. Major uses of formic acid are as a preservative and antibacterial agent in livestock feed and in the production of leather, including tannin.

Acetic Acid (CH_3COOH)

Acetic acid (ethanoic acid) is the second simplest carboxylic acid. It is an important chemical starting material used primarily in the production of cellulose acetate for photographic film, polyvinyl acetate for wood glue, synthetic fibres and esters. It is produced by anaerobic fermentation of ethanol or by chemical transformation of methanol, acetaldehyde or ethylene.

TABLE 4.16   **Representative volatile carboxylic acids and their sources**

Name	IUPAC name	Formula	Sources	BP (°C)
Acetic acid	Ethanoic acid	CH_3COOH	Vinegar, fruits and grains	118
Formic acid	Methanoic acid	HCOOH	Ant venom, fruits and wood	100.8
Propionic acid	Propanoic acid	CH_3CH_2COOH	Dairy products and sweat	141.2
Butyric acid	Butanoic acid	$CH_3(CH_2)_2COOH$	Dairy products and butter	163.5
Valeric acid	Pentanoic acid	$CH_3(CH_2)_3COOH$	Valerian root and fruits	186.2

It is used as a solvent in the chemical industry, either pure or mixed with water, chloroform and hexane. The solvent and miscibility properties of acetic acid make it a useful industrial chemical solvent in the production of dimethyl terephthalate.

It is found in the household as common vinegar. Vinegar contains 4%–10% acetic acid. Diluted acetic acid is often used in descaling agents, and in the food industry, it is used as an acidity regulator.

Concentrated acetic acid is corrosive to skin, and prolonged inhalation exposure to acetic acid vapours at 10 ppm can produce some irritation of the eyes, nose and throat, as well as lung irritation.

The main anthropogenic emissions of acetic acid to the atmosphere are produced by the manufacturing of paper and paper products, the alimentary industry, textile products and other chemical industries that use this substance. Vegetation, soil and biomass burning are other sources of acetic acid emissions.

In the atmosphere, acetic acid is formed from hydrocarbons and other VOCs by photochemical oxidation. Incomplete combustion of hydrocarbons in automobile engines is an important indirect source of acetic acid in urban areas.

Other Volatile Carboxylic Acids

Carboxylic acids with 5–10 carbon atoms are volatile, and all of them have an unpleasant odour. They may be produced by skin bacteria on human sebum (skin oils), excreted by some animals and during organic matter decay, showing 'goaty' odours (explaining the odour of Limburger cheese). Common carboxylic acids and their sources are described in Table 4.16.

4.18   Volatile Carboxylic Esters

Esters are organic compounds that are derived from carboxylic acids. They are formed by the reaction between a carboxylic acid and an alcohol in a process called esterification. Esters have a general structure of R-COOR′, where R represents an organic group derived from the carboxylic acid and R′ represents an organic group derived from the alcohol.

Esters are a functional group commonly encountered in nature, for example, naturally occurring fats and oils are the fatty acid esters of glycerol.

Esters with a low molecular weight have a nice fruity smell. In fact, esters are responsible for the odours of many fruits, and they are commonly used as fragrances in the food and cosmetic industries. The odours of esters are distinctly different from those of the corresponding acids, which have unpleasant odours.

Ethyl Acetate ($CH_3COOCH_2CH_3$)

Ethyl acetate is an aprotic polar solvent that is used both in laboratories and industries as reaction media for liquid–liquid extractions, separatory chromatography, etc. It is a solvent used for the manufacture of varnishes, lacquers, dry cleaning, stains, fats and nitrocellulose. It is used as a solvent in nail polish, nail polish remover, base coats and other manicuring products.

TABLE 4.17   **Representative volatile esters**

Ester	Structure	Natural sources	BP (°C)
Ethyl acetate	$CH_3COOCH_2CH_3$	Fruits, wine and beer	77.1
Isoamyl acetate	$CH_3COOCH_2CH_2CH(CH_3)_2$	Bananas, apples	142.8
Methyl butyrate	$CH_3(CH_2)_2COOCH_3$	Apples, pineapples	102.2
Ethyl butyrate	$CH_3(CH_2)_2COOCH_2CH_3$	Pineapples, strawberries	121.2
Hexyl acetate	$CH_3COOCH_2(CH_2)_3CH_3$	Grapes, apples and pears	163.5
Ethyl hexanoate	$CH_3(CH_2)_4COOCH_2CH_3$	Apples, bananas	170.7
Isoamyl butyrate	$CH_3(CH_2)_2COOCH_2CH_2CH(CH_3)_2$	Apples, pineapples	153.6
Methyl hexanoate	$CH_3(CH_2)_4COOCH_3$	Apples, bananas	172.4

Ethyl acetate is a highly flammable liquid that shows moderate toxicity in animal studies (LD50 for rats is 5,620 mg/kg). Overexposure to ethyl acetate may cause irritation of the eyes, nose and throat. Severe overexposure may cause weakness, drowsiness and unconsciousness, causing repeated damage to internal organs.

In the environment, it has low persistence in water and soil, medium persistence in sediment and high persistence in air and is toxic to fish. The half-life of air is estimated to be 10 days. Emissions into the air can contribute to photochemical smog. Because of its low toxicity, low environmental impact and the fact that it can be obtained from renewable sources, it is considered a green solvent.

There are numerous volatile esters found in various fruits, flowers and other natural sources.

In Table 4.17, common volatile esters (boiling points less than 200 °C) and their sources are listed.

4.19 Volatile Nitrogenated Compounds: Amines

Together with ammonia, amines represent the major source of nitrogen compounds in the atmosphere. Volatile amines are emitted from both anthropogenic (animal feeding, manures, industrial combustion, composting processes, etc.) and natural (biomass oceans, oceans, vegetation, etc.) sources. They are assumed to be a significant component of the atmospheric nitrogen cycle, but there's not a lot of available data. In the atmosphere, amines act as precursors for aerosols.

Simple amines are widely used as organic bases, solvents and precursors for many organic compounds of interest. In recent years, amines have been the most widely utilised in a variety of CO_2 capture materials, such as chemical solvents, gels and porous sorbents.

Topics of Interest: Capture of CO_2 with Amines

Primary and secondary amines are able to react with CO_2 to give carbamate anions and protonated amines, where B is the Brønsted base, for example, another amine.

The most commonly used amines for carbon capture are monoethanolamine (MEA), diethanolamine (DEA) and methyldiethanolamine (MDEA). These amines are typically used in a process known as post-combustion capture, which involves capturing CO_2 from the flue gas produced by the combustion of fossil fuels.

The basic process involves passing the flue gas through a liquid solution of the amine, which reacts with the CO_2 to form a chemical compound known as a carbamate. The carbamate is then heated to release the CO_2, which can be collected and stored or used for other purposes. The amine solution is then reused in the process, and the cycle continues.

While carbon capture with amines is an effective method for reducing greenhouse gas emissions, it does have some drawbacks. One of the main issues is that the process is energy-intensive and can increase the overall energy consumption of the plant or facility. Additionally, the process can produce toxic byproducts and can be expensive to implement on a large scale.

Table 4.18 shows the list of amines with significance in the atmosphere.

TABLE 4.18   Volatile amines found in the atmosphere

Amine	Formula	Natural source	BP (°C)
Methylamine	CH_3NH_2	Putrefaction of protein matter	−6 °C
Ethyl amine	$CH_3CH_2NH_2$	Urine	16.6 °C
Dimethylamine	$(CH_3)_2NH$	Present in many foods at a level of a few mg/kg	7 °C

4.20   Volatile Organic Sulfur Compounds

Volatile organic sulfur compounds are part of the global sulfur cycle and contribute to the greenhouse gas budget in the atmosphere.

The main volatile organic sulfur compounds are organic sulfides and thiols, which are produced from the anaerobic decomposition of organic matter in considerable amounts. Sea water is a remarkable source of organosulfur compounds. So, the decomposition of plankton in the ocean produces dimethylsulfide, CH_3SCH_3 (16×10^{12} g/year).

The origin of such volatile compounds (Table 4.19) is the marine phytoplankton and seaweed metabolite dimethylsulfoniopropionate (DMSP), which is transformed into volatile dimethyl sulfide and thiols. In the sea water, organic sulfur compounds are oxidised to dimethyl sulfoxide and finally to inorganic sulfates. When emitted into the atmosphere, dimethylsulfide and thiols are transformed into oxidised sulfate aerosols.

TABLE 4.19 Sulfur-volatile compounds

Sulfur compounds	Structure	Sources
Thiols	R-SH	Anaerobic decomposition
Dimethylsulfide	CH_3-S-CH_3	Anaerobic decomposition
Dimethylsulfoxide	CH_3-SO-CH_3	Oxidation of other sulfur compounds
Dimethylsulfoniopropionate	$(CH_3)_2S^+$-CH_2CH_2-COO^-	Phytoplankton and seaweeds

Topics of Interest: Sulfur Volatile Compounds in Onions

Onions contain a variety of chemical compounds that contribute to their characteristic taste, smell and health benefits.

The primary volatile compounds responsible for the pungent odour of onions are organo-sulfur compounds, such as thiopropanal S-oxide, which are released when the onion is cut or crushed. These compounds can cause teary eyes when they react with the moisture in the air and form sulfuric acid.

Onions also contain flavonoids, which are plant pigments that have antioxidant properties. The main flavonoids in onions are quercetin and kaempferol, which have been linked to a reduced risk of certain types of cancer and heart disease.

Onions also contain fructans, which are a type of carbohydrate that can act as a prebiotic in the gut, promoting the growth of beneficial bacteria. Additionally, onions contain small amounts of vitamins and minerals, such as vitamin C and potassium.

When onions are cooked, their chemical composition changes. For example, the heat breaks down the sulfur compounds, reducing their pungency and increasing the sweetness of the onion. This is why cooked onions have a milder flavour than raw onions.

In summary, the chemistry of onions is complex and varied, with numerous compounds contributing to their taste, smell and health benefits.

4.21 Green Solvents

Solvents are substances capable of dissolving other substances to form a homogeneous mixture called a solution. They are integral to numerous chemical processes in chemistry for a wide range of purposes in research laboratories and across various industries, including pharmaceuticals, paints, coatings, adhesives, cleaning products and many others. Some common uses of solvents are:

1. **Reaction medium:** Solvents serve as reaction media by providing a medium for chemical reactions to occur. They can dissolve reactants, facilitate molecular interactions and stabilise intermediates, thereby influencing reaction rates, selectivity and yields. Many organic reactions, such as Grignard reactions or nucleophilic substitutions, are conducted in solvents like ether or acetone.

2. **Dissolution and extraction:** Solvents are primarily used to dissolve solutes and form solutions. This property is essential in processes such as extraction,

where solvents are employed to selectively dissolve and separate components from solid or liquid mixtures. For example, organic compounds can be extracted from plant materials using solvents like ethanol or hexane.

3. **Separation and other analytical techniques:** Solvents are utilised in various analytical techniques for sample preparation, separation and detection of compounds. In chromatography, solvents are used as mobile phases to elute analytes from a stationary phase. Solvents like methanol, acetonitrile and water are commonly employed in techniques such as high-performance liquid chromatography (HPLC) and gas chromatography (GC).

4. **Chemical processing** in industry, including polymer synthesis, dissolution and shaping, in electrochemical processes such as batteries, fuel cells and electroplating, pharmaceuticals and pesticides, in the formulation of paints, coatings and inks. They help dissolve and disperse pigments, resins and additives, allowing for proper application and film formation.

Many solvents are classified as VOCs due to their volatility, as has been described before. These include:

- **Hydrocarbons:** such as hexane, benzene, toluene, xylene and ethylbenzene.
- **Oxygenated solvents:** include acetone, ethanol, methanol, methyl ethyl ketone (MEK) and ethyl acetate.
- **Chlorinated solvents:** include trichloroethylene (TCE), perchloroethylene (PCE), chloroform and methylene chloride.

The environmental impact of volatile solvents stems from their propensity to evaporate into the atmosphere and contribute to air pollution in workspaces and outdoors. In recent years, the idea of using solvents that diminish their environmental impact and improves sustainability and health considerations is taking hold. They are the so-called green solvents.

A green solvent would be designed to have minimal environmental impact throughout its lifecycle. This means it would be sourced sustainably, with minimal extraction or production emissions. It would also be biodegradable, breaking down into harmless components after use. Additionally, it would have a low potential for bioaccumulation and toxicity to aquatic life, ensuring it doesn't harm ecosystems if released into the environment.

Safety is paramount, so a green solvent would be chosen or formulated to minimise risks to human health. This includes low volatility to reduce inhalation risks, non-toxicity or low toxicity to reduce health hazards and non-flammability to decrease fire risks. It would also have low odour levels, reducing potential irritations for workers and users.

Another key point is sustainability. Sustainability involves considering the long-term availability of resources and the ability to maintain ecological balance. A green solvent would be sourced from renewable materials or designed to be easily recyclable or reusable. Its production process would aim to minimise energy consumption and waste generation. Additionally, it would be designed to contribute positively to social and economic aspects, such as supporting local communities or providing fair labour conditions.

Finally, a green solvent would adhere to relevant regulations and standards, ensuring it meets or exceeds legal requirements for environmental protection, worker safety and product quality. This includes compliance with regulations such

TABLE 4.20   Some green solvents and their properties

Green solvent	Source	Properties	Applications
Ethanol	Renewable	Biodegradable, low toxicity, miscible in water	Extraction, cleaning, synthesis
Glycerol	Renewable	Biodegradable, non-toxic, high boiling point	Cosmetics, pharmaceuticals, lubricants
Limonene	Citrus fruits	Low toxicity, pleasant odour, biodegradable	Cleaning, degreasing, fragrance
Carbon dioxide	Industrial by-product	Non-toxic, non-flammable, high selectivity	Extraction, dry cleaning, synthesis
Water	Abundant	Non-toxic, polar, high dielectric constant	Cleaning, extraction, reactions
Propylene carbonate	Renewable	Low volatility, high boiling point, non-toxic	Electronics, coatings, batteries
Ionic liquids and deep eutectic solvents	Various sources	Tailorable properties, non-volatile, low toxicity	Catalysts, separations, synthesis

as REACH (Registration, Evaluation, Authorisation and Restriction of Chemicals) in the European Union or the Toxic Substances Control Act (TSCA) in the United States. Additionally, it would be labelled clearly with information on safe handling, disposal and environmental impacts to facilitate proper use and disposal by users. Table 4.20 describes some green solvents and their properties.

This table provides a snapshot of some commonly used green solvents, their sources, key properties and typical applications. It's important to note that the suitability of a green solvent for a specific application depends on factors such as solubility, compatibility with other materials and performance requirements.

CHAPTER 5

Semi-Volatile and Persistent Organic Compounds: Pesticides

5.1 Introduction

The term 'semi-volatile compounds' is a broad concept for describing organic molecules that volatilise relatively slowly at standard temperature and pressure, showing boiling points from 260 to 400 °C. Many of the semi-volatile organic compounds are persistent organic pollutants (POPs), a group of organic substances that are primarily generated by human activities such as industrial processes, pesticide and herbicide use and waste disposal.

Experts of the United Nations[1] in 1999 defined POPs as a group of chemicals that possess a combination of the following four properties:

1. **Toxicity (T):** The ability to cause adverse effects in humans or the environment
2. **Persistence (P):** The ability to resist degradation and stay in different environmental media, including air, soil, water and sediment
3. **Bioaccumulation (B):** The ability to accumulate in living tissues to levels higher than the surrounding environment
4. **Long-range transport potential (LRTP):** Evidence showing that the chemical can be transported to regions where they have never been used or produced

The toxicity of POPs is due to the fact that most of them interact with the endocrine system. The endocrine system is a complex network of glands and organs that produce and release hormones into the bloodstream. These hormones act as chemical messengers, coordinating and regulating various bodily functions and processes. The endocrine system works in close collaboration with the nervous system to maintain

---

[1] http://digitallibrary.un.org/record/277104

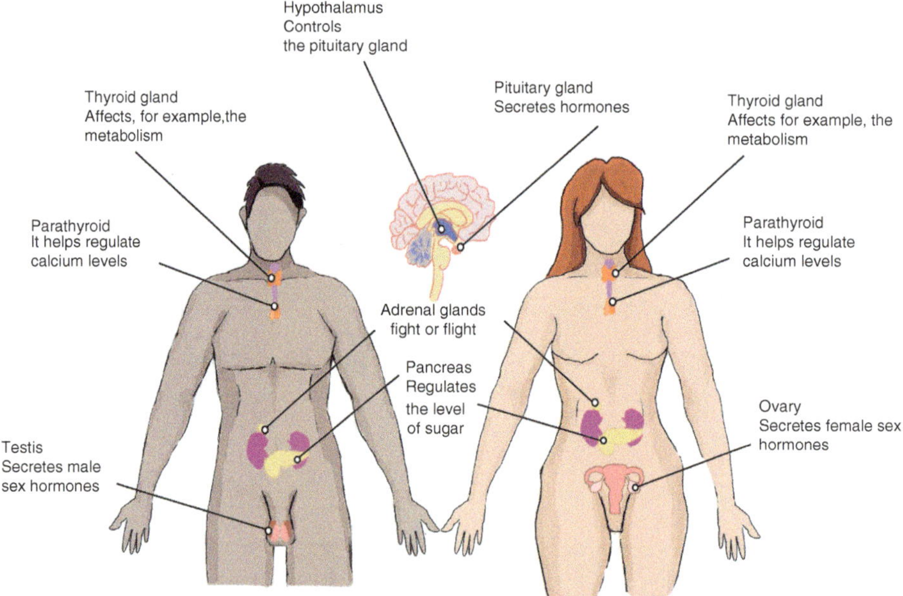

FIGURE 5.1 Human endocrine system

homeostasis, control growth and development, regulate metabolism and reproduce and perform other essential functions.

Most toxic POPs are considered endocrine disruptors, that is to say, substances that can interfere with the normal functioning of the endocrine system (Figure 5.1). These substances can mimic, block or alter the production, release, transport, metabolism or elimination of hormones, leading to distortions in the hormonal balance and potentially causing adverse health effects in humans and wildlife. Substances that behave as endocrine disruptors can be found in many products, such as pesticides, lifestyle products, flame retardants, some pharmaceuticals and many others, that can enter the body through ingestion, inhalation or skin contact. Upon entering the body, these substances have the ability to attach themselves to hormone receptors. This, in turn, leads to alterations in metabolism and interference with the regular hormonal signalling pathways, ultimately causing detrimental consequences for the development and functionality of organs and tissues. Such effects may include compromised immune function, reproductive issues, neurological disorders and an elevated susceptibility to specific types of cancer.

5.2 Common Properties of POPs

POPs are molecules that are chemically stable compounds, meaning they do not easily break down or degrade under normal environmental conditions. Many POPs show multiple halogen atoms attached to carbon atoms, aromatic moieties and other relatively inert groups that contain strong chemical bonds, which makes them highly resistant to breakdown by natural processes such as sunlight or microbial and chemical degradation. So, they can remain in the environment for many years without degradation. This means that even if the original sources of POPs are eliminated, the pollutants can persist in the environment for many years to come.

Bioaccumulation and Biomagnification

Another common property of POPs is their lipophilicity[2]. POPs are highly lipophilic, meaning that they show a preference to dissolve in fats and oils rather than in water.

This property causes them to tend to accumulate in the fatty tissues of living organisms. This is known as *bioaccumulation* (Figure 5.2). It occurs as organisms ingest or absorb these substances from their environment, such as through food, water or air. Once absorbed, these substances are often stored in the organism's tissues rather than being metabolised or excreted. As a result, the concentration of these substances can increase, progressively reaching, in some cases, concentrations that produce serious harm to living beings and even death.

Through the food chain, many POPs pass from one trophic level to the next, increasing the concentration at the next level. This process is defined as *biomagnification* (Figure 5.2). This means that organisms at higher trophic levels tend to accumulate higher concentrations of the substance than those at lower trophic levels.

The process typically starts with the accumulation of small amounts of the substance by primary producers, such as algae or plants. These primary producers are then consumed by animals at higher trophic levels. At each step of this food chain, the concentration of the substance can increase because the organism at each level consumes many individuals from the level below, thereby accumulating the substance present in each of those organisms.

To evaluate the tendency of organic compounds to move in the environment and to accumulate in tissues and organs are quite useful partition coefficient values.

Partition coefficients are a measure of the relative affinity of a compound for two immiscible phases. In the case of different environmental issues, octanol/water,

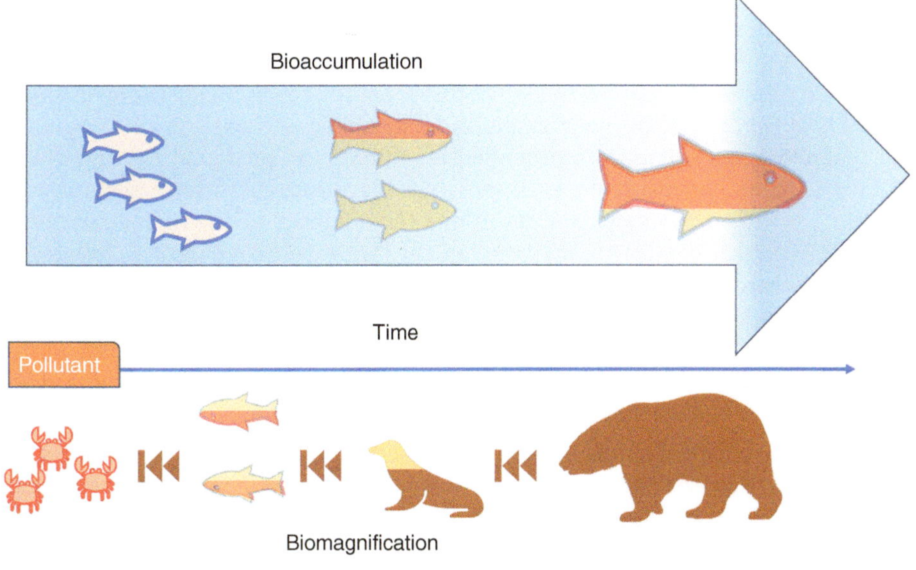

FIGURE 5.2   Bioaccumulation and biomagnification summary

---

[2]From Greek λίπος (fat) and φίλος (friendly). It refers to the ability of a molecule to dissolve in fats, oils, lipids and non-polar solvents.

octanol/air and air/water partition coefficients are useful tools to describe and predict the distribution of pollutants in air, soil, water and living beings.

a. **The octanol/water[3] partition coefficient (K_{ow})** describes the tendency for a substance to solubilise into an organic medium (lipophilicity or fat solubility) from water (hydrophilicity).

   K_{ow} = Concentration in the octanol phase/concentration in the aqueous phase. It is a good indicator to evaluate this tendency in aquatic organisms' tissues from the surrounding medium. The polarity of octanol is similar to those of tissues and, therefore, K_{ow}.

   It is very frequent to refer to K_{ow} as P (partition coefficient) or its logarithm, logP.

   There are several methods to predict the octanol/water partition coefficient, including experimental measurements and computational models:

   Experimental methods for measuring K_{ow} include the shake-flask method, reverse-phase high-performance liquid chromatography (RP-HPLC) and potentiometry. However, these methods can be time-consuming and laborious and require a significant amount of material.

   Computational models, on the other hand, offer a faster and more cost-effective way to predict K_{ow}. Some commonly used methods are based on the contribution of every group or fragment of the molecule to the final value of K_{ow}, fragment-based simulations and molecular dynamics simulations. These methods rely on the physicochemical properties and molecular structure of the compound to predict its K_{ow}.

   One popular tool for predicting K_{ow} is the ALOGPS 2.1 software, which is freely available online. ALOGPS uses a fragment-based approach and considers factors such as hydrogen bonding, steric hindrance and electronic effects to predict K_{ow}. Other software packages for predicting K_{ow} include ACD/Labs, ChemAxon and EPI Suite.

   It's important to note that while these methods can provide useful estimates of K_{ow}, they are not always accurate and may not be applicable to all compounds. Therefore, it's always recommended to validate the predicted K_{ow} values with experimental measurements if possible. Partition coefficient octanol/water (K_{ow}) is a good parameter for describing the tendency of a substance to remain in a tissue or organ.

b. **The octanol/air partition coefficient (K_{oa})** describes the distribution of a substance between environmental matrices such as soil, vegetation and aerosol particles, represented by the concentration of pollutants in octanol and air. It is a key parameter used to predict the LRTP of volatile and semivolatile organic compounds and their bioaccumulation in terrestrial biota.

c. **The air-water partition coefficient (K_{aw})** describes the distribution of a substance between air and water and therefore the tendency of a substance to escape from water to air.

---

[3] Octanol is often used as a model solvent to measure the lipophilicity or hydrophobicity, of a compound because it mimics the hydrophobic environment of biological membranes.

5.3 Global Distillation

POPs are transported over long distances through air and water currents, allowing them to spread globally and contaminate areas far away from their original source. Despite their low volatility and water solubility, they can be transported by wind and water far away from the emission source. In the air, the main mechanism for POP transportation is the so-called global distillation (Figure 5.3). Global distillation, also known as the grasshopper effect, transports semi-volatile chemicals from warmer to colder regions of the Earth. Areas near the poles and mountain tops can be affected by their presence despite the distances to the emission points.

Based on the values of octanol-water (K_{ow}) and octanol-air (K_{oa}) and air-water (K_{aw}) partition coefficients of the individual POPs, they may be classified as:

1. **Fliers:** These pollutants are carried through the air by wind or other atmospheric forces

2. **Single hoppers:** These pollutants move across the ground by hopping or jumping

3. **Multi-hoppers:** These pollutants move across the ground by hopping or jumping but have a longer range than single-hoppers

4. **Swimmers:** These pollutants move through water. Examples of swimmer pollutants include oil spills and chemical waste dumped into bodies of water

In the case of hoppers single and multi, many living beings such as fleas, grasshoppers and other animals can contribute to global transport because of their character as 'carriers'. It's worth noting that these terms are not widely used in the scientific community but are rather more commonly found in environmental education materials and outreach efforts.

Global distillation explains why relatively high concentrations of POPs have been found, for example, in the Arctic environment and in the bodies of animals and people who live there, even though most of the chemicals have not been used in the region in appreciable amounts.

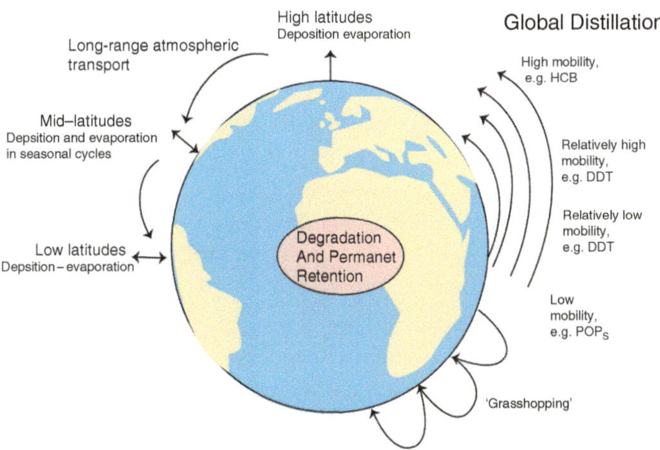

FIGURE 5.3   Global distillation

5.4 Classification of Persistent Organic Compounds

POPs include a wide range of chemicals, such as certain pesticides and many other compounds that have been used in various industrial, agricultural and household applications, but their production and use have been restricted or banned in many countries due to their adverse effects. POPs also include another group of substances that are unintentionally produced due to careless manufacturing, processing or disposal of materials.

Humans are exposed to POPS mainly through food but also through the air in the outdoors, indoors and at the workplace. As a result, POPs can be found virtually everywhere in measurable concentrations. There are two major types of POPs.

1. **Intentionally produced chemicals:** refer to substances that have been synthesised with a specific purpose, such as pest control, agriculture, manufacturing and industrial processes.

2. **Unintentionally produced chemicals:** are by-products formed in some careless or negligent industrial processes or produced during the combustion or incineration of residual organic matter and wastes.

Main groups of POPs are:

1. **Chlorinated pesticides:** Examples are dichlorodiphenyl trichloroethane (DDT) and other chlorinated pesticides that were commonly used as insecticides in the past.

2. **Polychlorinated biphenyls (PCBs):** These were widely used in electrical equipment, hydraulic fluids and other industrial applications. PCBs were banned in many countries due to their persistence and toxic effects.

3. **Polychlorinated naphthalenes (PCNs):** PCNs are a group of organic compounds that were primarily used in the past as electrical insulators and heat exchange fluids, as well as in the production of certain pesticides and flame retardants.

4. **Polybrominated diphenyl ethers (PBDEs):** These flame retardants were commonly used in electronics, furniture and textiles. PBDEs have been found to persist in the environment and can bioaccumulate in organisms.

5. **Polyfluorinated compounds (PFCs):** They are a group of man-made chemicals that contain multiple fluorine atoms attached to a carbon chain. They are known for their exceptional chemical stability, high water and oil repellency and heat resistance. PFCs have been used in a wide range of industrial and consumer applications due to their unique properties.

6. **Non-biodegradable polymers:** They are synthetic polymers that are resistant to degradation by natural biological processes. These polymers are typically derived from petrochemicals and are widely used in various industries due to their desirable properties, such as durability, flexibility and resistance to heat and chemicals.

7. **Polychlorinated dibenzo-*p*-dioxins (PCDDs) and Polychlorinated Dibenzofurans:** These chemicals are unintentional by-products of certain industrial

processes, such as waste incineration and chemical manufacturing. They are highly toxic and can cause a range of adverse health effects.

8. **Polycyclic aromatic hydrocarbons (PAHs):** PAHs are a group of organic compounds composed of multiple fused aromatic rings. They are formed through the incomplete combustion of organic materials such as fossil fuels, wood, tobacco and certain foods. PAHs can also be produced by industrial processes and released into the environment through various sources, including vehicle exhaust, industrial emissions and the burning of coal or biomass.

Groups 1–6 are or have been intentionally produced with a purpose, but groups 7 and 8 are mostly produced unintentionally as by-products in several processes and emitted to the environment.

POPs can be transported over long distances through air and water currents, leading to their global distribution. They can accumulate in the fatty tissues of animals, including humans and can be transferred from one organism to another through the food chain. This biomagnification process results in higher concentrations of POPs in organisms at the top of the food chain, such as predatory birds and mammals, including humans.

The harmful effects of POPs on human health include cancer, reproductive disorders, immune system dysfunction, neurological effects and developmental abnormalities in children. They can also have adverse effects on wildlife, such as reproductive failure and compromised immune systems.

To address the risks posed by POPs, the international community has taken action. *The Stockholm Convention on POPs*, adopted in 2001, is a global treaty that aims to eliminate or restrict the production, use and release of POPs. It currently has 182 parties committed to reducing or eliminating the production and use of POPs.

Efforts to manage and reduce POPs involve implementing regulations, promoting the use of safer alternatives, improving waste management practices and raising awareness about the risks associated with these chemicals.

Throughout this chapter, a series of substances considered semi-volatile organic compounds and POPs will be described due to their environmental impact.

5.5 Pesticides

Pesticides are a group of semi-volatile chemical substances that are used to control, repel or kill pests such as insects, weeds, fungi, rodents and other harmful organisms that damage crops, harm livestock or pets or spread diseases. They are used in agriculture, gardening, forestry, public health and other areas where pests can cause significant harm.

On the other hand, pesticides can have significant environmental impacts, which can be a cause for concern, showing unintended consequences for non-target species and contributing to soil and water pollution. When pesticides are applied to crops, they can run off into nearby streams, rivers and lakes, leading to contamination of the water supply and harm to aquatic life. Pesticides can also leach into the soil, where they can persist for years and potentially affect future crops. Another concern is the potential for pesticide residues to accumulate in the food chain, remaining on or in crops even after they are harvested and processed and ingested by animals and humans. Long-term exposure to low levels of pesticides in food can also have adverse health effects, such as cancer, developmental disorders and reproductive problems. Some groups of pesticides have been classified as persistent pollutants and have been banned globally and substituted by other compounds with lower environmental impacts.

Environmental chemistry plays an important role in understanding the behaviour and impact of pesticides in the environment. By studying their properties and the chemical reactions and interactions that occur, researchers can develop more effective and sustainable pest management strategies that minimise the environmental impacts of these chemicals.

5.6   Classification of Pesticides

Pesticides can be classified according to three criteria: Target pest, Toxicity and Chemical structure:

According to the target pest group, pesticides can be divided into several categories (Table 5.1).

In bold characters, the three most commonly used pesticides are highlighted.

1. According to toxicity, pesticides can be classified into several categories. The World Health Organization (WHO) and other organisms have their own nomenclature for classifying pesticides, which is summarised in Table 5.2.

2. Because of their great chemical diversity, an exhaustive classification of pesticides according to their chemical structure is not evident. For this reason, we will describe the more prominent families of pesticides (Table 5.3).

TABLE 5.1   Pesticide classification according to target pest

Type of pesticide	Target pest group
Acaricides	Mites
Algicides	Algae
Avicides	Birds
Bactericides	Bacteria
Fungicides	**Fungi and oomycetes**
Herbicides	**Plant**
Insecticides	**Insects**
Lampricides	Lampreys
Molluscicides	Snails
Nematicides	Nematodes
Rodenticides	Rodents
Slimicides	Algae, bacteria, fungi and slime moulds
Virucides	Viruses

TABLE 5.2   Pesticide classification according to toxicity

WHO category	Equivalence in EPA ranking system
Extremely hazardous	Category I, DANGER
Highly hazardous	Category I, DANGER
Moderately hazardous	Category II, WARNING
Slightly toxic	Category III, CAUTION
Not acutely hazardous	Category IV, CAUTION

TABLE 5.3   Classification of pesticides based on their chemical nature

Chemical group	Examples
Organochlorides	DDT and analogues, lindane, aldrin, endrin, dieldrin, chlordane, endosulfan heptachlor, toxaphene, pentachlorophenol and hexachlorobenzene
Organophosphates	Malathion, chlorpyrifos, diazinon, parathion, methamidophos and glyphosate
Carbamates and thiocarbamates	Carbaryl, carbofuran, aldicarb, ferbam, captan S-ethyldipropylcarbamothioate, molinate, vernolate, cycloate
Pyrethroids	Permethrin, cypermethrin, deltamethrin, esfenvalerate, bifenthrin, fenvalerate lambda-cyhalothrin, cyfluthrin, tetramethrin and resmethrin
Phenyl amides	Metalaxyl, mefenoxam and ametoctradin
Phenoxy alkanoates	2,4-dichlorophenoxyacetic acid (2,4-D), 2-methyl-4-chlorophenoxyacetic acid (MCPA)
Trazines	Atrazine, cyanazine, propazine, simazine
Sulfonylureas	Chlorimuron-ethyl, metsulfuron-methyl
Phtalimides	Captan, captafol and folpet
Dipyrids	Paraquat, diaquat

5.7 Organochlorine Pesticides

Organochlorine pesticides are a class of synthetic chemicals that were widely used as insecticides in agriculture and public health programmes in the mid-20th century. They are characterised by the presence of one or more chlorine atoms attached to a carbon atom in the molecular structure. Since their introduction, they turned out to be quite effective in controlling many pests in harvests or insect disease transmitters, reducing the spread of diseases like malaria and typhus. In addition, these pesticides were relatively inexpensive to produce. Their success from a short-term perspective led to their being used for decades in enormous amounts.

Organochlorine pesticides are chemically quite inert and they are not easily biodegradable, remaining in soil, water and air for long periods of time. They also bioaccumulate in the food chain meaning that higher organisms can accumulate higher levels of these pesticides in their organs and tissues as they consume lower organisms. Studies have shown that exposure to organochlorine pesticides can have negative health effects on humans and wildlife, including cancer, reproductive problems and developmental delays.

Therefore, most organochlorine pesticides are considered POPs and most countries have banned the use of these pesticides, although some of them continue to be used in certain contexts.

They can be classified into several groups according to their molecular structure:

a. **DDT and analogous**

b. **Lindane group**

c. **Cyclodienes:** aldrin, endrin, dieldrin, chlordane, endosulfan and heptaclor

d. **Chloride camphenes:** toxafen and clordecon

e. **Pentachlorophenol**

f. **Hexachlorobenzene**

Dichlorodiphenyltrichloroethane (DDT)

DDT is an organic compound that was widely used as an insecticide in the 1940s and 1950s.

DDT was first synthesised in 1874 by an Austrian chemist, Othmar Zeidler. However, it was not until 1939 that DDT's insecticidal properties were discovered by a Swiss chemist named Paul Hermann Müller, who demonstrated its effectiveness in killing insects such as mosquitoes that transmit diseases like malaria and typhus. For his discovery, Müller was awarded the Nobel Prize in Physiology or Medicine in 1948.

DDT is synthesised by a reaction between chloral (trichloroacetaldehyde) and chlorobenzene in the presence of sulfuric acid (Figure 5.4).

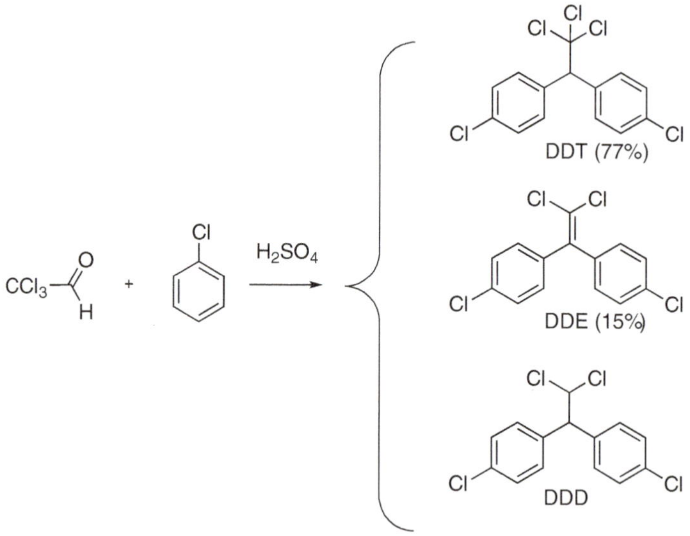

FIGURE 5.4 Synthesis of dichlorodiphenyltrichloroethane (DDT)

Nevertheless, commercial DDT is a mixture of several isomeric compounds. The major component (77%) in the mixture is the desired *p,p′*-isomer that is formed together with dichlorodiphenyldichloroethylene (DDE) (15%) and dichlorodiphenyldichloroethane (DDD), which make up the balance of impurities in commercial samples. DDE and DDD are also the major metabolites and environmental breakdown products.

It was initially seen as a breakthrough in pest control in agriculture in many countries and it was widely used in the mid-20th century as an effective insecticide for reducing malaria-carrying mosquito populations and thus reducing the spread of this disease. Between 1950 and 1993, more than 2.6 million metric tonnes were used all over the world. However, the use of DDT soon became controversial due to its environmental and health effects. DDT is a typical example of a POP that remains in the environment for a long time, accumulating in the food chain and posing a risk to wildlife and humans who consume contaminated food. It has been linked to a range of health problems, including cancer, reproductive disorders and developmental abnormalities.

DDT was banned in many countries in the 1970s due to concerns about its impact on the environment and human health. Despite its ban in many countries, DDT is still used in some parts of the world for malaria control. However, alternative methods of pest control that are less harmful to the environment and human health have been developed and are increasingly being used.

Some DDT congeners were also used as pesticides, trying to replace DDT to avoid its negative effects, such as dicofol or methoxychlor (Figure 5.5), but due to their similar persistence in the environment and potential health risks to DDT, their use has been also restricted or banned in many countries.

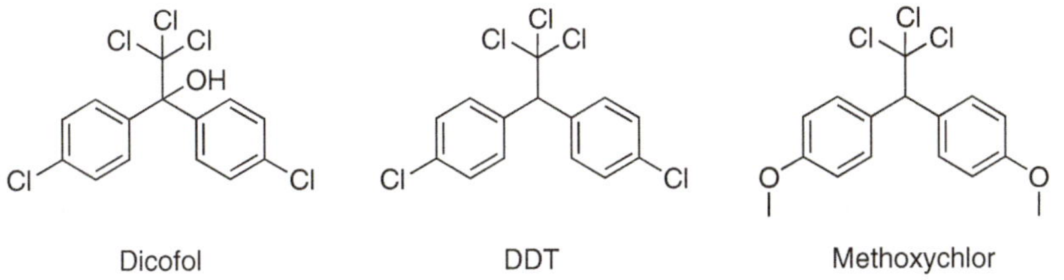

FIGURE 5.5   DDT congeners

Topics of Interest: Silent Spring

Silent Spring is a 1962 book written by American environmentalist Rachel Carson. The book is often credited with sparking the modern environmental movement in the United States. The book's title refers to the sudden absence of birdsong that Carson observed in her hometown after the spraying of the pesticide DDT. However, Carson's book drew attention to the harmful effects of DDT and other pesticides on the environment, including the potential for long-term damage to ecosystems and human health. Silent Spring had a significant impact on the public and policymakers, leading to increased awareness of environmental issues and the need for regulations to protect the environment. As a result of the book, the US eventually banned the use of DDT and other persistent pesticides.

Lindane Group Pesticides

Lindane, also known as gamma-hexachlorocyclohexane (γ-HCH) (Figure 5.6), is an organochlorine insecticide that has been used to control pests such as lice, scabies and mosquitoes. It works by interfering with the nervous system of insects.

Technical-grade hexachlorocyclohexane is produced as a mixture of isomers. The active γ-HCH (lindane) can be concentrated by treatment with methanol or acetic acid, followed by fractional crystallisation, to produce technical-grade lindane containing 99.9% of the γ-isomer. The synthesis of lindane isomers is described in Figure 5.7.

Major uses of lindane are as an insecticide for fruit and vegetable crops and in baits and seed treatments for rodent control. Commercial production of lindane in the United States began in 1945 and peaked in the 1950s, when 8,000 metric tonnes had already been manufactured.

However, lindane has been found to be persistent in the environment and highly toxic to humans and animals and its use has been banned or restricted in many countries. Prolonged or excessive exposure to lindane can cause a range of health problems, including neurological and reproductive disorders and it is classified as a possible human carcinogen by the International Agency for Research on Cancer (IARC).

Chlorinated Cyclodienes

The so-called chlorinated cyclodienes are a class of polycyclic organic compounds that were commonly used as insecticides in agriculture and public health programmes in the mid-20th century. This group includes several pesticides such as aldrin, dieldrin, endrin, chlordane and endosulfan (Figure 5.8).

Chlorinated cyclodiene pesticides are characterised by their ability to persist in the environment and accumulate in the food chain and soon it was discovered that all of them were highly toxic to a variety of non-target organisms, including birds, fish and mammals and could persist in the environment for decades. As a result, the use of chlorinated cyclodienes was banned in many countries in the 1970s and 1980s.

Despite their relatively complex structures, their preparation is quite simple based on Diels-Alder reactions from affordable starting materials such as hexachlorocyclopentadiene.

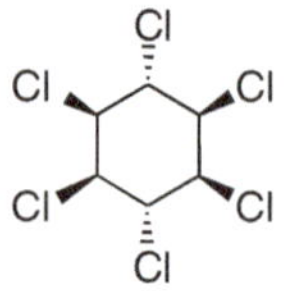

FIGURE 5.6 Lindane

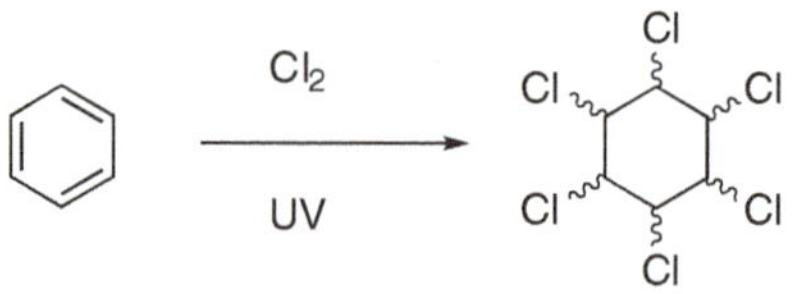

FIGURE 5.7 Lindane and isomer synthesis

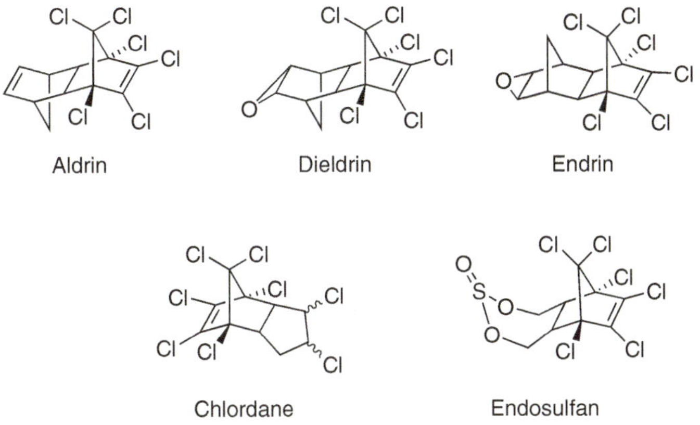

FIGURE 5.8 Chlorinated cyclodienes

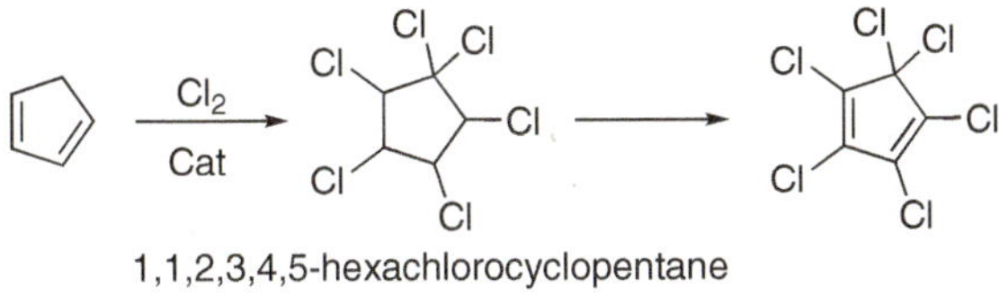

FIGURE 5.9 Synthesis of hexachlorocyclopentadiene

Hexachlorocyclopentadiene is obtained by chlorination of cyclopentadiene to give 1,1,2,3,4,5-octachlorocyclopentane, which in a second step undergoes dehydrochlorination (Figure 5.9).

Aldrin and Dieldrin

Aldrin is a colourless, crystalline solid that is highly toxic to insects, mammals and birds. Dieldrin is also formed when aldrin is metabolised in the body. Dieldrin is even more toxic than aldrin and can cause a wide range of health effects in humans and animals, including damage to the nervous system, liver and kidneys.

Aldrin is synthesised from norbornadiene[4] with hexachlorocyclopentadiene with good yields and Dieldrin is a similar chemical compound that can be synthesised from aldrin by epoxidation with a peroxy acid such as peracetic acid, as shown in Figure 5.10.

Endrin

Endrin is another chlorinated cyclodiene insecticide that was first introduced in the 1950s (Figure 5.8). It was widely used as a pesticide on crops such as cotton, corn and soybeans, as well as on fruits, vegetables and livestock. However, it was later discovered that endrin is highly toxic and can cause a range of health problems, including neurological damage, liver damage and reproductive problems. It is also known to be carcinogenic.

---

[4] Norbornadiene is a bicyclic hydrocarbon formed by a Diels-Alder reaction between cyclopentadiene and acetylene.

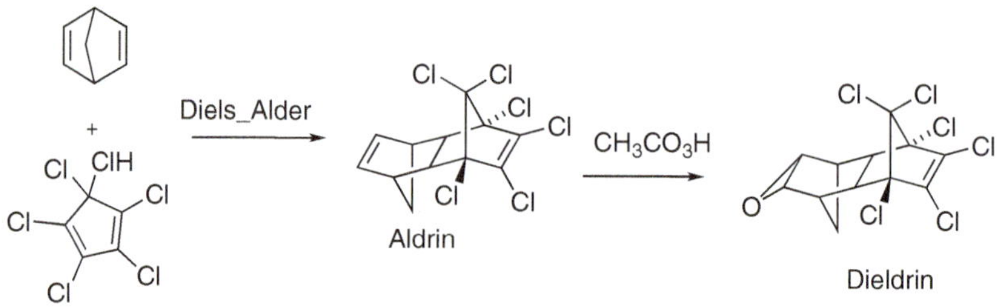

FIGURE 5.10 Synthesis of aldrin and dieldrin

Chlordane

Chlordane is an organochlorine compound that was commonly used as a pesticide in the United States from the 1940s to the 1970s. Technical chlordane development happened by chance at Velsicol Chemical Corporation during a search for possible uses of a by-product of synthetic rubber manufacturing. By chlorinating this by-product, such a potent insecticide was easily and cheaply produced as a mixture of isomers (Figure 5.11).

Endosulfan

Endosulfan is a highly toxic and persistent organochlorine insecticide. It is synthesised by a Diels-Alder reaction from hexachlorocyclopentadiene with cis-butene-1,4-diol in the presence of a catalyst such as aluminium chloride, followed by treatment with thionyl chloride ($SOCl_2$) (Figure 5.12). This reaction produces a mixture of endosulfan isomers, including alpha- and beta-endosulfan.

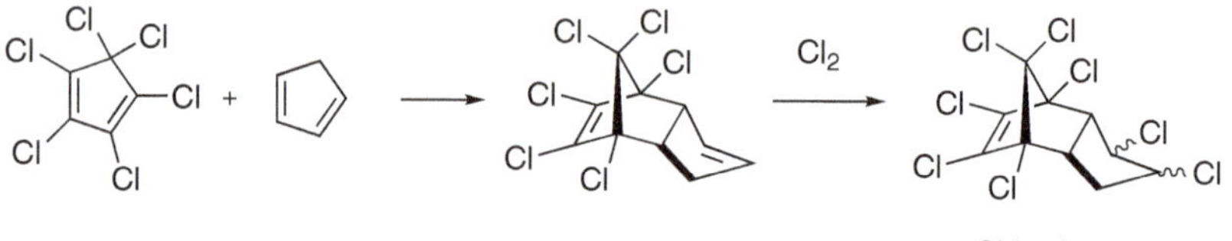

FIGURE 5.11 Synthesis of chlordane

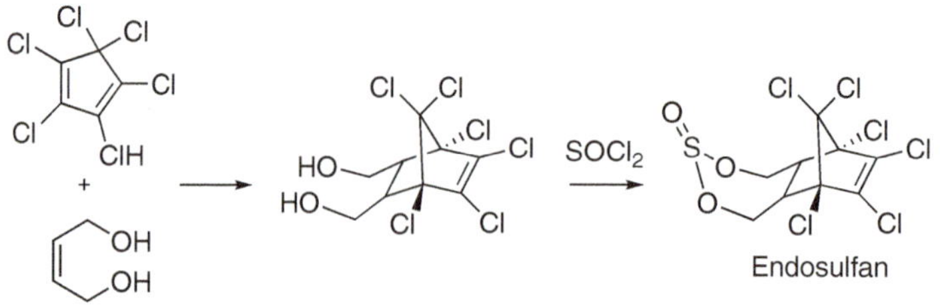

FIGURE 5.12 Synthesis of endosulfan

It has been banned in several countries, including the EU and the UE, due to its harmful effects on human health and the environment.

Chlordecone

Chlordecone, also known as Kepone, is a synthetic organochlorine insecticide that was widely used in the United States and other countries in the 1960s and 1970s.

The synthesis of chlordecone was first reported in 1952 by Gilbert and Giolito[5]. Commercial production in the United States started in 1966.

Chlordecone is prepared by dimerizing hexachlorocyclopentadiene and hydrolyzing to a ketone (Figure 5.13).

It was primarily used to control pests on crops such as tobacco, bananas and citrus fruits. However, it was later found to be highly toxic and persistent in the environment and its use was banned in many countries. Chlordecone contamination is still a significant problem in some areas, particularly in the French Caribbean, where it was used extensively in banana plantations. Efforts are ongoing to clean up contaminated areas and prevent further exposure to this toxic chemical.

Mirex

It was commonly used in the past to control fire ants, termites and other pests. It was first developed in the 1940s and was extensively used in the United States until the 1970s when its use was phased out due to concerns about its persistence in the environment and its potential health effects on humans and wildlife.

The synthesis of mirex involves several steps. The first step is the chlorination of hexachlorocyclopentadiene with chlorine gas in the presence of iron or aluminium chloride to form a mixture of hexachloro-1,3-cyclopentadiene and hexachloro-2,4-cyclopentadiene.

The second step involves the cyclization of the mixture of hexachlorocyclopentadienes to form mirex (Figure 5.14). This is typically achieved by heating the mixture with aluminium chloride in the presence of a solvent, such as benzene or chlorobenzene.

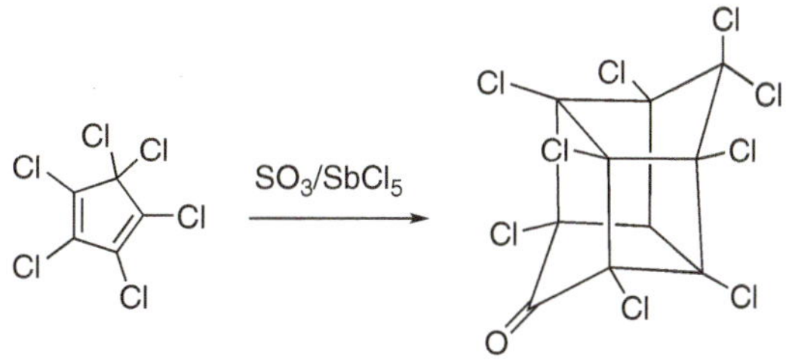

FIGURE 5.13   Synthesis of chlordecone

[5] Gilbert, E.E. and Giolito, S.L. (1952) US Patent 2.616.825 and US Patent 2.616.928, 4 November, to Allied Chemical and Dye Corporation.

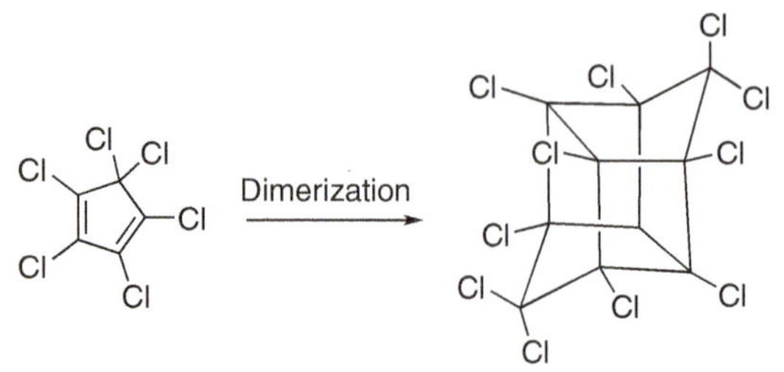

FIGURE 5.14 Mirex synthesis

Mirex is highly persistent in the environment and can accumulate in the fatty tissues of animals, including humans. It has been linked to a range of health effects, including liver damage, thyroid disruption and developmental and reproductive problems. Mirex is also a suspected carcinogen.

Both chlordecone and mirex have been banned in many countries due to their persistence and potential health effects. However, they continue to persist in the environment and can still be found in soil, water and wildlife.

Polychlorterpenes

Polychloroterpenes are substances obtained by the chlorination of natural terpenes, which are compounds commonly found in plants and trees. They have been used as pesticides in the past due to their insecticidal properties. Polychlorinated pesticides include compounds such as toxaphene. Toxaphene is actually a complex mixture of over 200 different polychlorinated camphene derivatives that were prepared by the chlorination of camphene[6] (Figure 5.15).

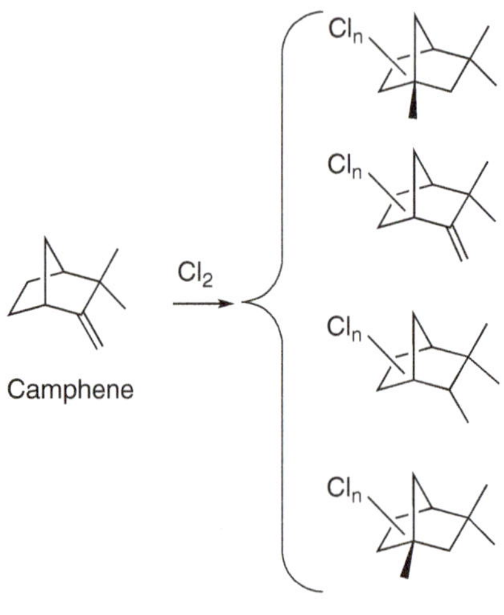

FIGURE 5.15 Toxaphene synthesis

[6] Camphene is a bicyclic monoterpene hydrocarbon that is found in many essential oils, including camphor oil, pine oil and ginger oil.

Toxaphene was primarily used as an insecticide and acaricide to control a wide range of pests, including mites, ticks, mosquitoes and other insects. It was also used as a wood preservative and in various industrial applications.

Toxaphene has been linked to a number of adverse health effects in humans and animals, including cancer, reproductive and developmental effects and neurotoxicity. As a result, its use has been banned or severely restricted in many countries, including the United States.

Chlorinated Benzenes

Chlorinated benzenes are a group of organic compounds that contain benzene rings with one or more chlorine atoms attached to them. Some of these compounds have been used as pesticides.

One of the most well-known chlorinated benzene pesticides is hexachlorobenzene (HCB), which was used extensively in the past for controlling fungal diseases in crops and as a flame retardant. HCB, a white crystalline solid, is synthesised in labs by the electrophilic aromatic substitution reaction of benzene with chlorine at 150–200 °C with a ferric chloride catalyst, but large-scale manufacture for use as a fungicide was developed by using the residue remaining after purification of the mixture of isomers of hexachlorocyclohexane in the lindane isolation The unwanted α- and β-isomers are treated with chlorine in the presence of UV light to give HCB (Figure 5.16).

Exposure to HCB can occur through inhalation, ingestion of contaminated food or water or skin contact. HCB is toxic to humans and animals and can cause a variety of adverse health effects, including liver damage, skin lesions, neurological effects and reproductive and developmental effects. It is also classified as a probable human carcinogen by the IARC.

Due to its persistence and potential for long-range transport, HCB has been found in remote areas such as the Arctic and Antarctic.

Another chlorinated benzene pesticide is pentachlorobenzene (PCBz), which was used as a wood preservative, insecticide and fungicide. It was also used as an herbicide in the past. Pentachlorophenol is prepared from phenol and chlorine in the presence of a catalyzer. In the reaction, a series of by-products such as polychlorinated phenols, PCDDs and polychlorinated dibenzofurans are even more toxic than pentachlorophenol itself (Figure 5.17).

Exposure to PCP can occur through ingestion, inhalation or skin contact. PCP is known to be toxic to humans and animals and its effects depend on the level and duration of exposure. Acute exposure can cause symptoms such as vomiting, diarrhoea, convulsions and coma. Chronic exposure to PCP can cause damage to the

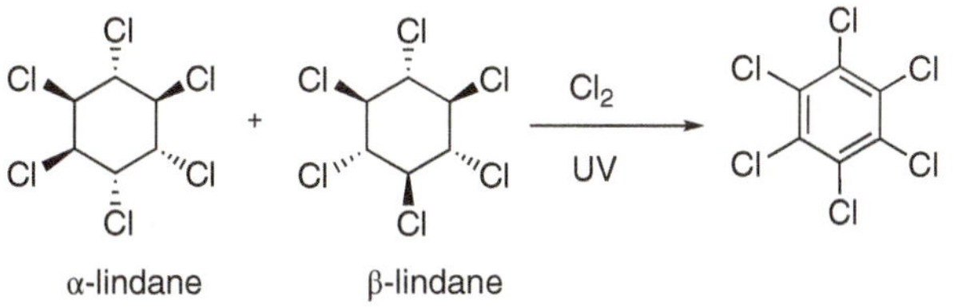

FIGURE 5.16   Synthesis of hexachlorobenezene

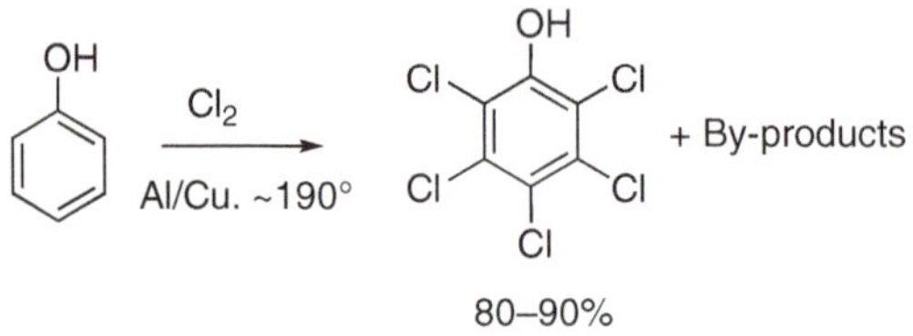

+ By-products

80–90%

FIGURE 5.17 Synthesis of pentachlorophenol

liver, kidneys and central nervous system, as well as skin rashes and respiratory problems. PCP is also classified as a probable human carcinogen. Other chlorinated benzenes that have been used as pesticides include 1,2,4-trichlorobenzene and 1,2-dichlorobenzene. These compounds are less toxic and persistent than HCB and PCBz, but their use has still been restricted or banned in many countries due to environmental concerns.

5.8 Organophosphorus Pesticides

Organophosphorus compounds are a class of chemicals that contain phosphorus in their structure and are widely used for pest treatment as insecticides, herbicides and fungicides.

They work by inhibiting an enzyme called acetylcholinesterase, which is essential for nerve function. This leads to the accumulation of the neurotransmitter acetylcholine in the nervous system, causing overstimulation and ultimately the paralysis and death of the target pests.

Unfortunately, OPs can also have harmful effects on humans and other animals. Exposure to OPs can cause a range of symptoms, including headaches, dizziness, nausea, muscle weakness and even death in severe cases. Long-term exposure to OPs has been linked to an increased risk of neurological disorders such as Parkinson's disease and Alzheimer's disease, as well as developmental and reproductive problems.

Due to these health concerns, many countries have implemented regulations to limit the use of OPs, particularly in residential and other non-agricultural settings. In addition, alternatives to OPs, such as integrated pest management practices and biological control methods, are being developed and promoted as safer and more sustainable options for pest control.

Many organophosphates are insecticides that impact the nervous system by compromising the enzyme that regulates the neurotransmitter.

Organophosphate pesticides can be classified based on their chemical structure. The structural classification of organophosphate pesticides is given in Figure 5.18.

1. **Simple esters:** Simple esters of phosphoric or thiophosphoric acid are the simplest form of organophosphate pesticides. They are formed by the reaction of phosphoric or thiophosphoric acid with an alcohol or thiol. Examples of simple esters of phosphoric acid include dimethoate, trichlorfon and monocrotophos. Examples of simple esters of thiophosphoric acid include malathion and azinphos-methyl.

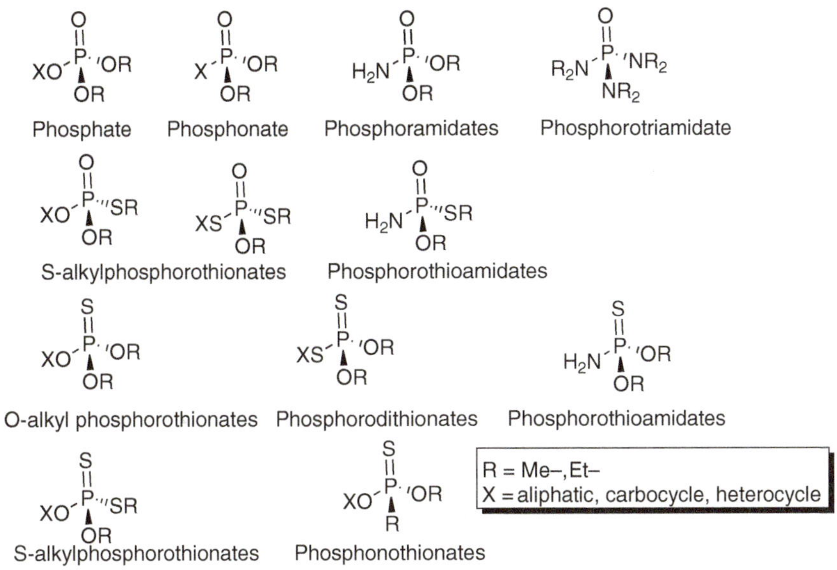

Phosphate   Phosphonate   Phosphoramidates   Phosphorotriamidate

S-alkylphosphorothionates   Phosphorothioamidates

O-alkyl phosphorothionates   Phosphorodithionates   Phosphorothioamidates

$R = Me-, Et-$
$X =$ aliphatic, carbocycle, heterocycle

S-alkylphosphorothionates   Phosphonothionates

FIGURE 5.18   Structural classification of organophosphate pesticides

2. **Dialkyl phosphates:** Dialkyl phosphates are formed by the reaction of phosphorous acid with two molecules of alcohol. Examples of dialkyl phosphates include chlorpyrifos, dichlorvos and diazinon.

3. **Diethyl phosphates:** Diethyl phosphates are a subclass of dialkyl phosphates that have a specific structure with two ethyl groups. Examples of diethyl phosphates include malathion, coumaphos and fenthion.

4. **Trialkyl phosphates:** Trialkyl phosphates are formed by the reaction of phosphorus oxychloride with three molecules of alcohol. Examples of trialkyl phosphates include parathion, methyl parathion and ethion.

5. **Phosphonates:** Phosphonates are structurally similar to phosphates but have a C-P bond instead of a P-O bond. Examples of phosphonates include glyphosate and glufosinate.

6. **Phosphoramidates:** Phosphoramidates are formed by the reaction of phosphorus trichloride with an amine. Examples of phosphoramidates include tetrachlorvinphos and phosmet.

5.9   Examples of Organophosphate and Organothiophosphate Pesticides

Methamidophos

It is a phosphorothioamidate used to control a variety of pests on fruits, vegetables and grains (Figure 5.19). Methamidophos is highly toxic to humans and can cause a range of health effects, including nausea, vomiting, abdominal pain, diarrhoea,

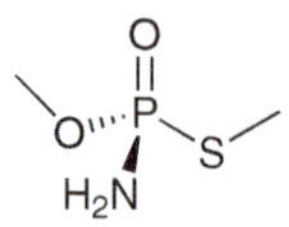

FIGURE 5.19 Methamidophos

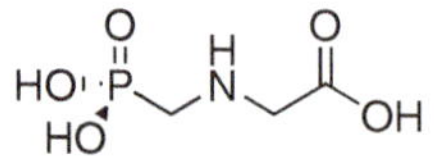

FIGURE 5.20 Glyphosate

dizziness, blurred vision, difficulty breathing, seizures and even death in severe cases. Long-term exposure to methamidophos has been associated with an increased risk of neurological disorders such as Parkinson's disease. Due to its high toxicity, many countries have banned or severely restricted the use of methamidophos. However, it is still used in some parts of the world, particularly in developing countries where regulations and safety standards may be less stringent.

Glyphosate

Glyphosate is a widely used organophosphonate herbicide that was first introduced by the agricultural company Monsanto in the 1970s under the brand name Roundup (Figure 5.20). Glyphosate works by inhibiting an enzyme that is essential for plant growth, effectively killing weeds and other unwanted plants.

Topics of Interest: Glyphosate Controversy

The safety of glyphosate has been a contentious issue. While glyphosate has been approved for use by regulatory agencies, including the US Environmental Protection Agency (EPA), the European Food Safety Authority (EFSA) and the WHO, some studies have raised concerns about its potential health risks, including the possibility of being a human carcinogen. In 2015, the IARC, a specialised agency of the WHO, classified glyphosate as 'probably carcinogenic to humans' based on evidence from animal studies and limited evidence from human studies.

Glyphosate has also been detected in soil, water and air samples in various parts of the world, raising concerns about its potential for environmental contamination and impacts on non-target organisms. It has also been the subject of regulatory and legal battles. Regulatory agencies in different countries have adopted different approaches to glyphosate regulation, with some countries imposing bans or restrictions on its use, while others continue to approve it for use. In the United States, for example, glyphosate has been approved for use by the EPA, but some states and municipalities have imposed restrictions or bans on its use. Glyphosate has also been the subject of numerous lawsuits, with some individuals claiming that their exposure to glyphosate caused their cancers, resulting in significant legal settlements.

Industry Perspectives: The use of glyphosate is supported by the agricultural and chemical industries, which argue that it is an effective tool for controlling weeds and improving crop yields. These industries argue that the regulatory agencies' evaluations and approvals of glyphosate demonstrate its safety when used as directed. They also argue that glyphosate has been extensively studied and that the evidence supporting its safety outweighs any potential risks.

Public Concerns and Advocacy: Glyphosate has become a highly debated and controversial topic among the general public and advocacy groups. Some individuals and organisations express concerns about the potential health and environmental risks associated with glyphosate and call for more stringent regulations, bans or restrictions on its use. They also advocate for alternative methods of weed control, such as organic farming and agroecology. On the other hand, there are also individuals and organisations that support the continued use of glyphosate, citing its importance in modern agriculture and the lack of robust evidence linking it to significant health or environmental risks.

Malathion

Malathion (Figure 5.21) is a phosphorodithioate that has been used for many years to control a variety of pests, including mosquitoes, flies and agricultural pests such as aphids, mites and thrips. It is generally considered to be effective and relatively safe when used according to label instructions, although there have been concerns about potential health effects associated with exposure to the chemical. Due to its potential toxicity, malathion is regulated by government agencies such as the US EPA and the European Union's European Chemicals Agency (ECHA). These agencies set limits on the amount of malathion that can be used and establish guidelines for safe use to minimise risks to human health and the environment.

Chlorpyrifos

Chlorpyrifos (Figure 5.22) is a phosphorothioate that has been extensively used since the 1960s to control a variety of pests, including insects and worms. It is used to control a wide range of pests on crops, fruits, vegetables and nuts. Chlorpyrifos can cause a range of health effects, including nausea, dizziness, headaches, muscle twitching and, in severe cases, convulsions and respiratory paralysis. In recent years, there has been growing concern about the potential health risks associated with exposure to chlorpyrifos, particularly for children and pregnant women. Studies have linked exposure to chlorpyrifos to developmental delays, neurobehavioral effects and other

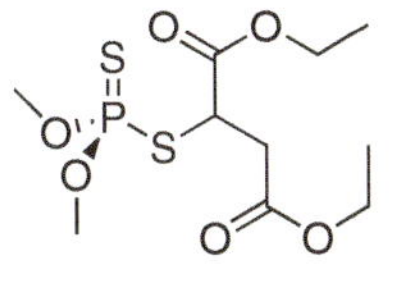

FIGURE 5.21   Malathion

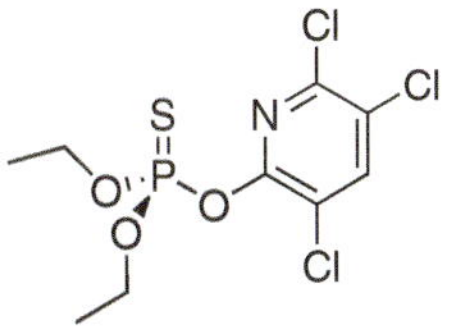

FIGURE 5.22   Chlorpyrifos

adverse health outcomes. In response to these concerns, several countries, including the United States, have restricted or banned the use of chlorpyrifos in certain applications.

Diazinon

Diazinon (Figure 5.23) is a phosphorothioate insecticide that is commonly used in agriculture and residential settings to control insects. Due to its toxicity and potential health risks, Diazinon has been banned for residential use in several countries, including the United States. However, it is still used in some agricultural applications in certain parts of the world. Exposure to Diazinon can cause a range of symptoms, including headaches, dizziness, nausea, vomiting, difficulty breathing, muscle weakness and cancer. In severe cases, seizures and coma. Long-term exposure to Diazinon has also been linked to neurological and developmental disorders, as well as certain types of cancer. It is used to control pests in crops, soil and livestock.

Parathion

Parathion (Figure 5.24) phosphorothioate is used to control insects on a variety of crops, including cotton, fruit and vegetables. Exposure to parathion can occur through inhalation, ingestion or skin contact and symptoms of poisoning may include headache, nausea, vomiting, diarrhoea, abdominal pain, blurred vision and difficulty breathing. In severe cases, parathion poisoning can lead to convulsions, coma and even death.

To prevent exposure to parathion, it is important to follow safety guidelines and regulations when handling or using pesticides, including wearing protective clothing and equipment, using pesticides in accordance with label instructions and storing pesticides securely. However, due to its extreme toxicity and potential to cause harm to humans and the environment, its use has been restricted or banned in many countries.

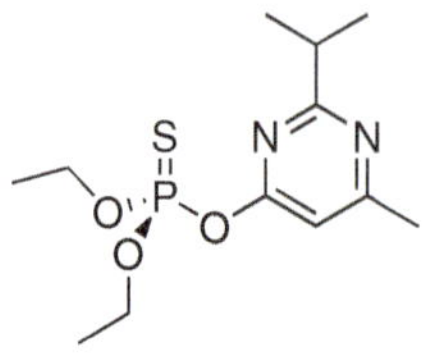

FIGURE 5.23   Diazinon

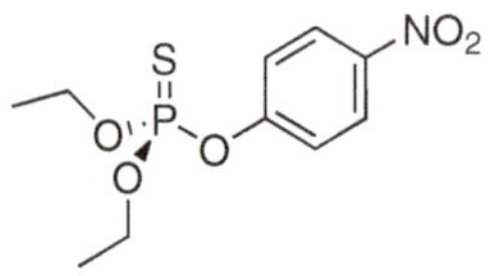

FIGURE 5.24   Parathion

5.10 Carbamates and Thiocarbamates

Carbamate pesticides are esters of carbamic acid (Figure 5.25) that are commonly used in agriculture to control a variety of pests, including insects, mites and nematodes. They are relatively polar, highly soluble in water and chemically reactive.

They inhibit the activity of the enzyme acetylcholinesterase in insects, leading to an accumulation of acetylcholine in their nervous system and ultimately causing paralysis and death.

The first carbamate insecticide to be commercialised was Carbaryl (Table 5.4), which was introduced by Union Carbide in 1956 under the trade name Sevin. Carbaryl quickly became a popular insecticide due to its broad-spectrum activity and relatively low toxicity to humans and other mammals. However, concerns began to emerge in the 1960s and 1970s about the environmental impact of carbamates, particularly their effects on non-target organisms and their persistence in the environment.

Despite these concerns, carbamates continued to be widely used throughout the second half of the 20th century and new compounds were developed and introduced. One notable example is carbofuran, which was introduced by FMC Corporation in 1969 and quickly became popular for its effectiveness against a wide range of pests.

Together with Carbaryl, other commonly used carbamate pesticides include carbofuran and aldicarb (Table 5.4).

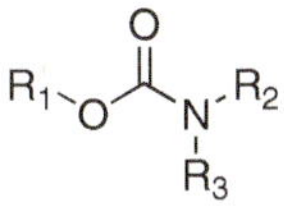

FIGURE 5.25 General structure of carbamates

TABLE 5.4 **Carbamate pesticides**

Carbamate	Uses
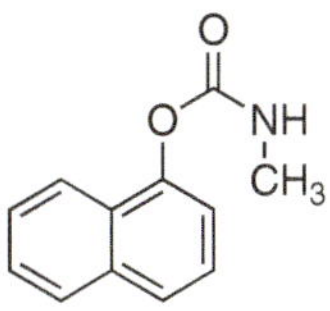 Carbaryl	It is used on a broad range of insect pests on crops, ornamental plants, lawns and in homes and gardens
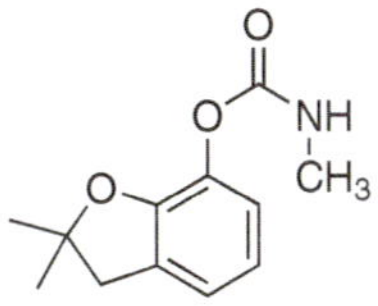 Carbofuran	It is used to control a wide range of insects in crops such as vegetables, fruit and tobacco, as well as in turf, ornamental plants and forestry
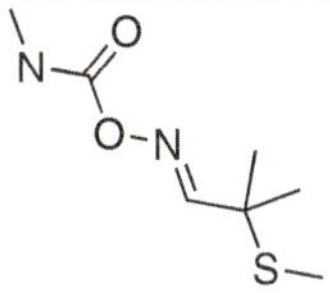 Aldicarb	It is used to control a wide range of insects in crops such as cotton, soybeans, peanuts and potatoes

Currently, carbamates continue to be used in agriculture and public health programmes, although their use has declined somewhat in recent years due to concerns about their environmental and health impacts. Some carbamate insecticides, such as aldicarb, have been phased out or restricted in many countries due to their high toxicity and persistence in the environment.

While carbamate pesticides are generally considered to be less toxic than other types of insecticides, such as organophosphates, they can still pose risks to human health and the environment if used improperly. Exposure to high levels of carbamate pesticides can cause a range of acute health effects in humans, including headaches, dizziness, nausea, vomiting and respiratory distress. Long-term exposure to low levels of carbamate pesticides may also increase the risk of certain types of cancer and neurological disorders.

Topics of Interest: Physostigmine

Calabar bean paste is a traditional African ingredient that is made from the seeds of the Calabar bean (*Physostigma venenosum*). The Calabar bean is native to West Africa and has been used for centuries in traditional medicine and as a poison. In some communities in West Africa, such paste was used to *reveal* the guilt or innocence of people accused of witchcraft. If the alleged 'witches' died after being forced to eat calabar bean paste, then they were indeed witches; if not, then they were declared innocent. The active carbamate in the calabar bean was physostigmine (Figure 5.26).

Thiocarbamates (thiourethanes) are a family of sulfur analogues of carbamates. There are two isomeric forms of thiocarbamates: O-thiocarbamates, ROC(=S)NR2 (esters) and S-thiocarbamates, RSC(=O)NR$_2$ (thioester) (Figure 5.27).

Thiocarbamates are widely used as herbicides to control weeds in agricultural and horticultural crops. They work by inhibiting the growth of plants and interfering with their ability to photosynthesise. Some common examples of thiocarbamate herbicides include:

- **EPTC (S-ethyl dipropyl carbothioate):** This herbicide is commonly used to control annual grasses and broadleaf weeds in corn, soybeans and other crops (Figure 5.28).

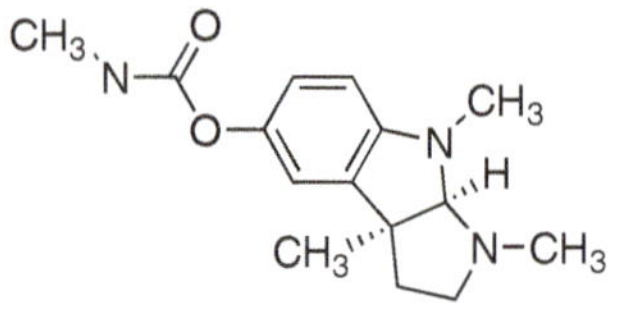

FIGURE 5.26   Physostigmine

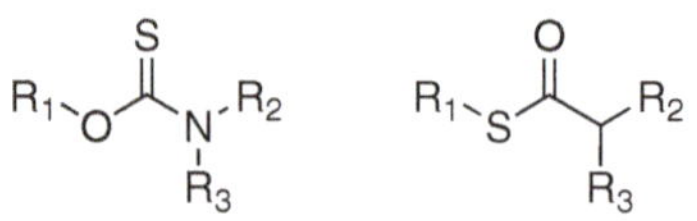

FIGURE 5.27   Isomeric forms of thiocarbamates

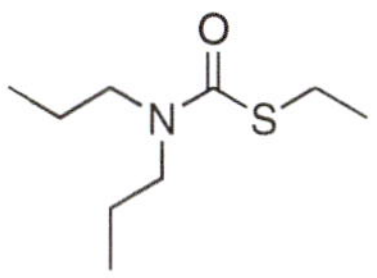

FIGURE 5.28   EPTC (S-ethyl dipropyl carbothioate)

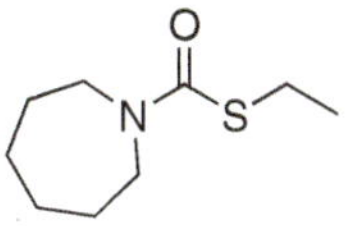

FIGURE 5.29   Molinate

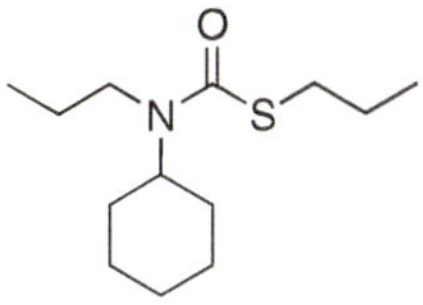

FIGURE 5.30   Cycloate

- **Molinate:** This herbicide is used to control annual grasses and broadleaf weeds in rice paddies and other flooded crops (Figure 5.29).
- **Cycloate:** This herbicide is used to control annual grasses and broadleaf weeds in various crops, including sugar beets and potatoes (Figure 5.30).

Although thiocarbamates are generally considered safe for humans and the environment, they can be toxic if ingested or absorbed through the skin. Symptoms of toxicity may include nausea, vomiting, diarrhoea, abdominal pain, headaches, dizziness and muscle weakness. Long-term exposure to high levels of thiocarbamates may also increase the risk of certain types of cancer.

5.11   Pyrethrins and Pyrethroids

Pyrethrum refers to a naturally occurring mixture of chemicals and natural insecticides derived from the dried flowers of certain species of chrysanthemum plants, particularly *Chrysanthemum cinerariaefolium* and *Chrysanthemum coccineum*. The active compounds in pyrethrum are known as pyrethrins, which are neurotoxins that target the nervous systems of insects, causing paralysis and death. Pyrethrins are actually a mixture of six ester-related compounds (Figure 5.31).

Pyrethroids are synthetic compounds that are typically derived from natural pyrethrins. Examples of pyrethroids are given in Table 5.5.

Pyrethrins and pyrethroids are widely used in commercial insecticides and are considered to be relatively safe for humans and other mammals because they have a low toxicity and are quickly degraded by sunlight and air. They are often used in household insecticides, pet flea and tick treatments and agricultural pesticides, but

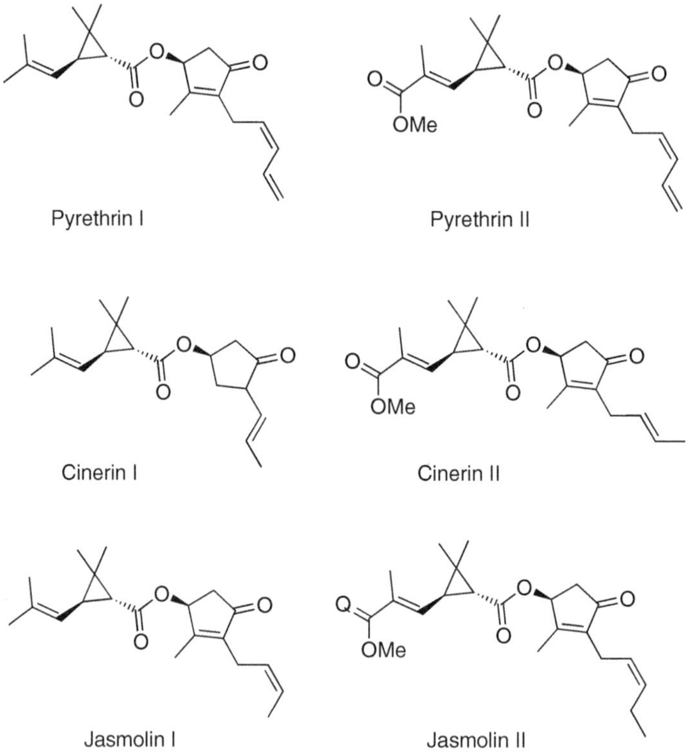

Pyrethrin I

Pyrethrin II

Cinerin I

Cinerin II

Jasmolin I

Jasmolin II

FIGURE 5.31 Pyrethrins

TABLE 5.5 **Examples of pyrethroids**

Pyrethroid	Structure	Uses
Permethrin	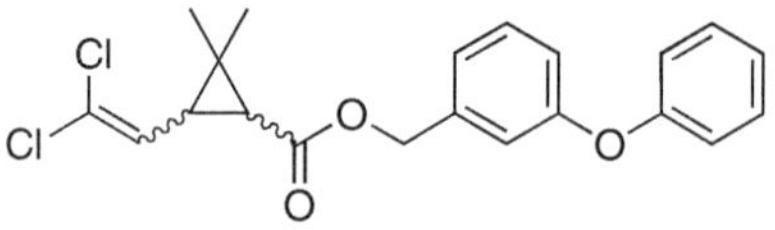	Insecticide for control of mosquitoes, ticks and lice
Cypermethrin	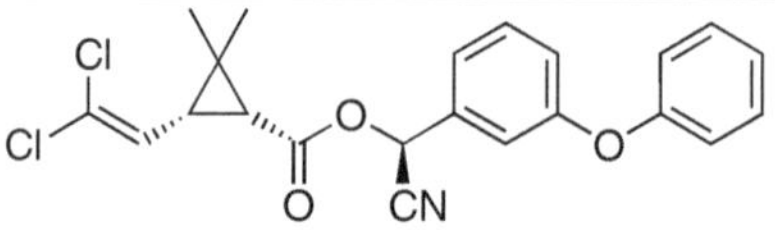	Agricultural insecticides and mosquito control
Deltamethrin	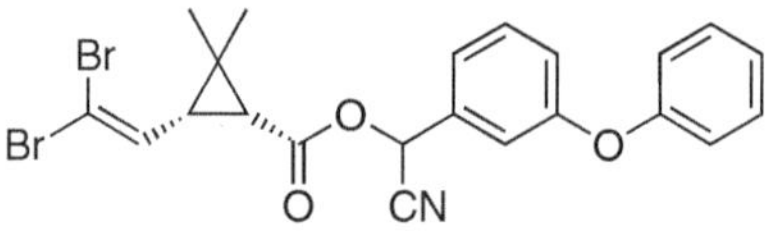	Pest control in agriculture, public health and households
Lambda-cyhalothrin	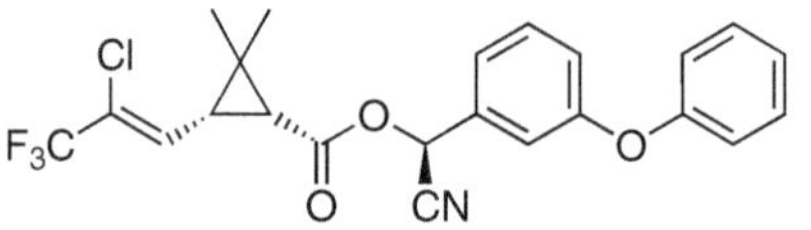	Crop protection, public health, mosquito control
Esfenvalerate	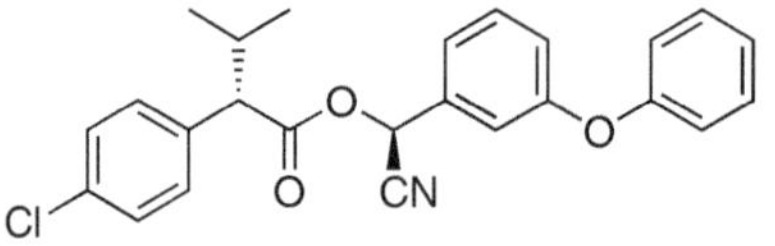	Insecticide for agriculture, turf and ornamental plants
Bifenthrin	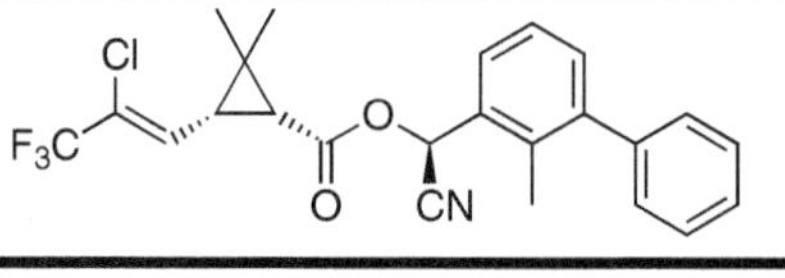	Lawn and garden insecticide, termite control

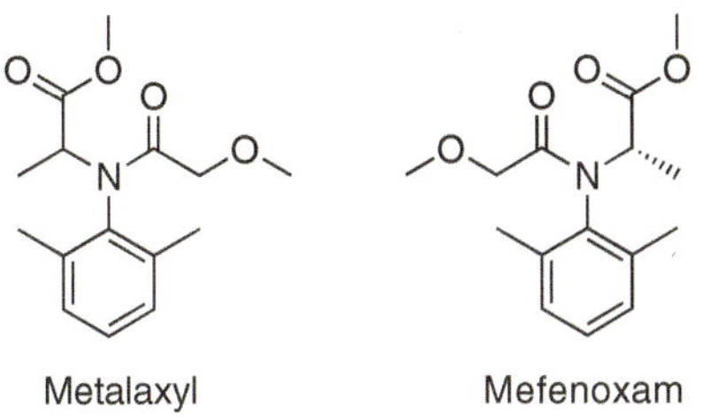

Metalaxyl Mefenoxam

FIGURE 5.32 Metalaxyl and mefenoxam

they can still cause skin irritation and other adverse effects if used improperly or in high doses. On the other hand, they can be highly toxic to non-target organisms, such as fish and other aquatic organisms, so care should be taken to avoid contaminating water sources.

5.12 Phenylamide Pesticides

Phenylamide pesticides are a group of pesticides that contain the phenyl amide functional group (-C(O)NPh$_2$) in their molecules. They are commonly used to control fungal diseases in crops such as potatoes, grapes and soybeans.

Some commonly used phenylamide pesticides include metalaxyl and mefenoxam (Figure 5.32). These pesticides work by inhibiting the growth of fungi by interfering with the biosynthesis of cell wall components or affecting the respiration of the pathogen.

Phenylamide pesticides have been shown to have toxic effects on aquatic organisms and may also have negative impacts on human health. As a result, their use is regulated by various government agencies to ensure that they are used safely and appropriately.

5.13 Phenoxy Herbicides

Phenoxy herbicides are a class of herbicides that mimic the natural plant hormone auxin, which controls the growth and development of plants. Phenoxy herbicides work by selectively killing broadleaf weeds while leaving grasses relatively unharmed. They are commonly used to control weeds in crops such as cereals, corn, soybeans and wheat.

Some of the most commonly used phenoxy herbicides include 2,4-dichlorophenoxyacetic acid (2,4-D), 2-methyl-4-chlorophenoxyacetic acid (MCPA) and mecoprop (Figure 5.33). An equal mixture of 2,4-dichloro phenoxyacetic

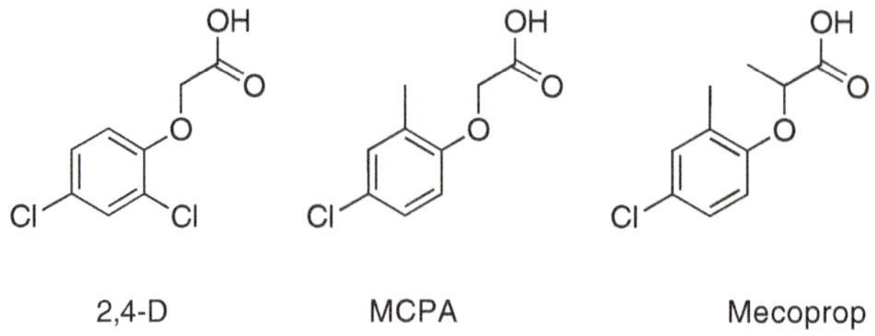

FIGURE 5.33   2,4-D, MCPA and mecoprop

acid (2,4-D) and MCPA was extensively used by the US military during the Vietnam War as a defoliant, known as Agent Orange because of its colour[7].

The use of phenoxy herbicides has been controversial due to concerns over their potential health and environmental impacts. They have been linked to health issues such as cancer, reproductive problems and developmental delays, as well as environmental issues such as soil and water pollution. It's important to note that the use of Agent Orange has been banned in many countries since the 1970s due to its toxicity and health effects.

5.14   Triazines

Triazines are a class of pesticides that are widely used in agriculture to control weeds. The most common triazine pesticides are atrazine, simazine, prometryn, terbuthylazine and cyanazine (Table 5.6).

These pesticides inhibit the photosynthesis process in plants, which leads to their death. Triazines are highly effective at controlling a wide range of weeds and are often used as pre-emergent herbicides, meaning they are applied to the soil before the weed seeds germinate.

However, triazines are also highly persistent in the environment and can leach into groundwater and surface water, where they can remain for long periods of time. Atrazine, in particular, has been found to be a contaminant in drinking water and has been linked to various health problems, including reproductive issues and cancer.

Due to their potential environmental and health impacts, the use of triazines is heavily regulated in many countries. Some countries have even banned the use of atrazine entirely.

5.15   Sulfonylureas

Sulfonylureas are a class of pesticides that are used in agriculture to control weeds. These pesticides work by inhibiting the activity of an enzyme in plants that is necessary for the production of amino acids. Without amino acids, the plant cannot survive and eventually dies. A large number of sulfonylureas have been used as

[7]The use of Agent Orange in the Vietnam War has been highly controversial and has been the subject of lawsuits and compensation claims by US veterans and Vietnamese civilians who

TABLE 5.6   Examples of triazines

Triazine pesticide	Structure	Uses
Atrazine	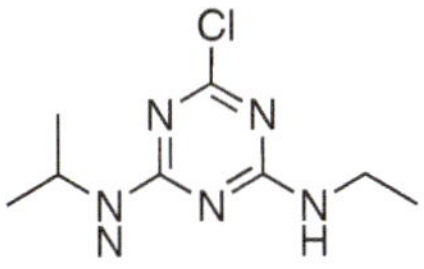	Broad-spectrum herbicide primarily used on corn, sorghum, sugarcane and other crops to control broadleaf and grassy weeds. It is also used on golf courses and residential lawns
Simazine	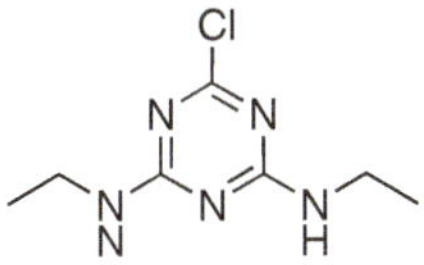	Selective herbicides are used to control broadleaf weeds and grasses in agricultural crops, such as corn, sorghum and sugarcane. It is also used in non-crop areas like industrial sites and roadsides
Prometryn	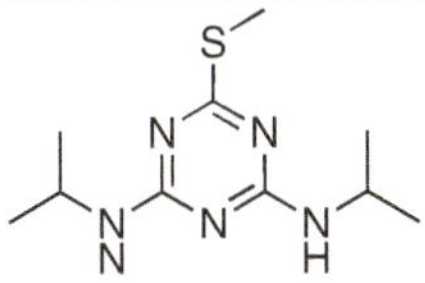	Herbicides are used to control annual grasses and broadleaf weeds in crops such as corn, soybean, cotton and vegetables. It is also used in non-crop areas like industrial sites, railways and roadways
Terbuthylazine	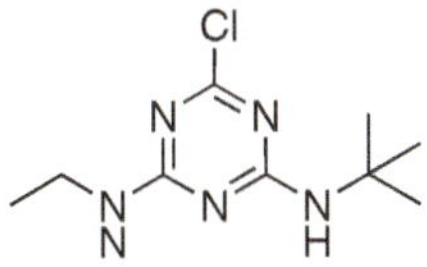	It is a selective herbicide used for pre- and post-emergence control of annual grasses and broadleaf weeds in crops like corn, sorghum, sugar cane and fruits
Cyanazine	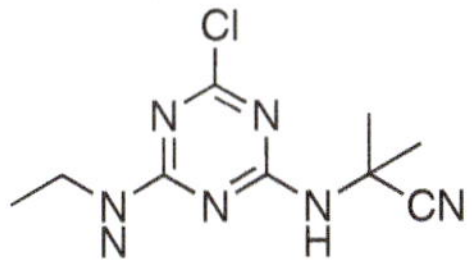	Broad-spectrum herbicide used to control grasses and broadleaf weeds in crops like corn, sorghum and sugarcane. It is also used on non-crop areas like industrial sites and roadsides

herbicides since the 1980s and they are generally considered to be effective and safe when used according to label instructions.

Some examples of sulfonylurea pesticides are included in Table 5.7.

However, concerns have been raised about the potential for these pesticides to persist in the environment and accumulate in the food chain.

Some studies have suggested that sulfonylurea pesticides may have negative impacts on non-target organisms such as bees, aquatic organisms and beneficial insects. Additionally, there is some evidence to suggest that these pesticides may pose health risks to humans if they are consumed in high concentrations over long periods of time.

5.16 Phthalimides

Phthalimides are a class of organic derivatives of phthalic acid with a cyclic structure containing a nitrogen atom. Phthalimide pesticides are known for their broad-spectrum activity against various pests, including insects, mites, fungi and weeds (Table 5.8).

TABLE 5.7 Examples of sulfonylureas

Name	Structure	Uses
Chlorimuron-ethyl		Controls broadleaf weeds in various crops, including soybeans, peanuts, cotton and corn. It is also used in turf management
Metsulfuron-methyl		Effective against a wide range of broadleaf weeds and some grasses in crops like wheat, barley, corn, rice and sugarcane. It is also used for weed control in non-crop areas, such as roadsides and industrial sites
Thifensulfuron-methyl		Used to control broadleaf weeds in crops like wheat, barley, corn, rice and sugarcane. It is also employed for weed management in non-crop areas and turfgrass
Tribenuron-methyl		Primarily used for controlling broadleaf weeds in wheat and barley crops. It is effective against many common weeds, including wild radish, wild mustard and wild turnip
Chlorsulfuron		Used to control broadleaf weeds and some grasses in various crops, such as wheat, barley, corn, rice and sugarcane. It is also applied in non-crop areas like rights-of-way and industrial sites
Sulfometuron-methyl		Used to control various broadleaf weeds and grasses in crops like wheat, barley, corn, soybeans and sugarcane. It is also applied in non-crop areas and forestry settings

TABLE 5.8   **Examples of phthalimides**

Name	Structure	Primary use
Captan	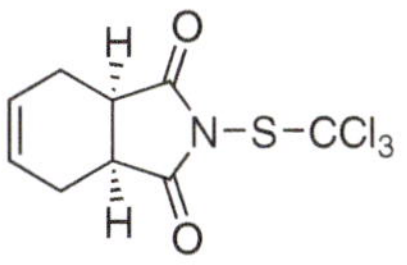	Broad-spectrum fungicide
Captafol	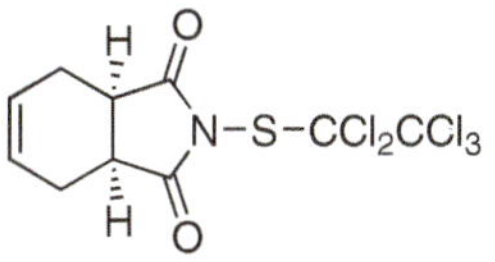	Protective fungicide and mildewcide

One of the most widely used phthalimide pesticides is Captan, which is used in agriculture to control various fungal diseases in crops such as fruits, vegetables and ornamental plants. Other examples of phthalimide pesticides include Captafol and Folpet.

Captafol was first introduced in the 1960s and was widely used until the 1980s when concerns over its potential health and environmental effects led to its discontinuation. Captafol has been associated with several health hazards, including skin and eye irritation, respiratory problems and carcinogenicity. The IARC has classified captafol as a possible human carcinogen and the US EPA has banned its use since 1999 for ornamentals plants and other crops. It belongs to the class of phthalimides and works by inhibiting the growth of fungi by disrupting their metabolic processes.

Folpet has been used since the 1950s and it is considered to be relatively safe for humans and the environment. However, it can cause skin and eye irritation and prolonged exposure may result in respiratory problems. The US EPA has classified folpet as a Group E chemical, which means that there is no evidence of human carcinogenicity.

Some concerns have been raised about the potential for residues to accumulate in the environment and food products. The European Union has established maximum residue levels for folpet in food products and some countries have banned its use altogether. The use of folpet is also subject to regulatory oversight in many countries.

5.17   Bipyridyl Pesticides

Bipyridyl pesticides are commonly used to control weeds and other unwanted plants, particularly in agriculture. This group includes paraquat and diquat (Figure 5.34).

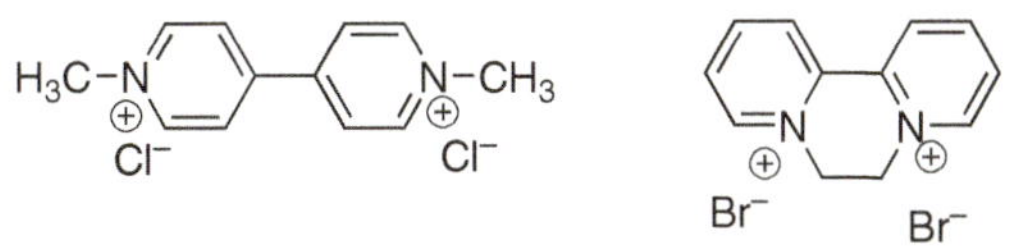

FIGURE 5.34   Paraquat and Diquat structures

Paraquat (1,1'-dimethyl-4,4'-bipyridinium) is a highly toxic herbicide that acts as a contact poison, meaning it kills plants on contact by damaging the plant's cell membranes. It is commonly used to control weeds in crops such as cotton, soybeans and corn. However, paraquat is also highly toxic to humans and animals and can cause serious health problems if ingested or absorbed through the skin or eyes. In fact, paraquat poisoning can be fatal and there is no known antidote.

Diquat (6,7-dihydrodipyridol [1,2-a:2',1'-c]pyrazinediium) is another herbicide that is used to control weeds in a variety of settings, including agriculture, aquatic environments and non-crop areas such as roadsides and industrial sites. Diquat works by disrupting the photosynthesis process in plants, leading to their death. Diquat is less toxic than paraquat, but it can still cause skin and eye irritation as well as respiratory problems if inhaled.

Due to their high toxicity, many countries have banned or restricted the use of paraquat and diquat. In some cases, they are only available to licenced professionals who have specialised training in their safe use and handling. Both paraquat and diquat are classified as restricted-use pesticides by the US EPA.

5.18   Pesticide Exposure

Humans are exposed to pesticides when they are manufactured, transported or used in agriculture and other branches. This is a minority group, but exposure to residual amounts of pesticides is a global health problem because they are present in consumer products, food, drinking water, air and soil.

The main routes for pesticide exposure in workspaces such as factories, farms, professional gardening and transportation are as follows:

1. Ingestion is caused by incidental ingestion (e.g., from hand-to-mouth contact) or when eating or drinking contaminated food or water. This can happen when pesticides are sprayed on crops, food commodities or packaging. The residues remain on the fruits, vegetables and grains.

2. Inhalation when breathing in contaminated air. This can happen when pesticides are sprayed or applied as dust, fog or mist.

3. Dermal contact occurs when they come into contact with the skin. This can happen when handling contaminated plants, soil or equipment or when coming into contact with surfaces that have been treated with pesticides.

4. Eye contact when they come in contact with the eyes. This can happen when handling pesticides without wearing proper eye protection or when they are in areas where pesticides are being sprayed or applied.

In the case of household products such as sprays, baits, pet collars and mosquito repellents for personal protection, these four routes are unlikely. Nevertheless, it is always important to take precautions when using or coming into contact with pesticides to minimise exposure and potential harm to health. This includes following label directions carefully, wearing protective clothing and equipment and avoiding contact with pesticides whenever possible.

5.19 Residual Pesticides in Air, Soil, Drinking Water and Food

Residual pesticides in the air may occur in the presence of small amounts of pesticides that have been released into the atmosphere but have not yet settled to the ground, degraded or dissipated. Pesticides can enter the air in several ways, including spray drift during application, volatilization from soil or plants and evaporation from stored pesticide products. Once in the air, pesticides can be transported long distances by wind and can potentially impact human health and the environment. The effects on human health of pesticides in the air depend on how toxic the pesticides are, how much pesticide is in the air and how much a person breathes in or gets exposed to. Exposure to residual pesticides in the air can occur through inhalation or skin contact with contaminated air or dust.

To reduce the risk of exposure to residual pesticides in the air, it is important to follow proper application and handling procedures for pesticides and to minimise the use of pesticides whenever possible. Regular monitoring and testing of air quality can also help identify and address any potential pesticide contamination issues.

The presence of residual pesticides in soil can result in soil contamination and have negative impacts on soil health, soil microbial activity and the overall ecosystem. Residual pesticides can also enter the food chain through the consumption of contaminated plants and animals. To mitigate the effects of residual pesticides in the soil, it is important to use pesticides judiciously and follow proper application procedures. Farmers and other pesticide users should also use environmentally friendly alternatives and engage in crop rotation to reduce pesticide buildup in the soil.

Furthermore, regular soil testing can help identify the presence of residual pesticides in the soil and guide the development of appropriate remediation strategies. Soil remediation techniques such as bioremediation, phytoremediation and chemical treatments can help remove or reduce the levels of residual pesticides in the soil.

Residual amounts of pesticides may remain in food after it has been treated with pesticides. While the use of pesticides in agriculture can be beneficial for preventing crop damage and increasing yields, it can also lead to the presence of residual pesticides in food. Regulations regarding pesticide use and residual limits vary by country.

The potential health effects of residual pesticides in food are a matter of ongoing research and debate. Some studies suggest that exposure to residual pesticides may increase the risk of cancer, neurological problems and other health issues. However, the actual risk posed by residual pesticides in food depends on a number of factors, including the type of pesticide, the amount present and the frequency of exposure. To minimise exposure to residual pesticides in food, it is recommended to:

1. Wash fruits and vegetables thoroughly before eating them.
2. Buy organic produce whenever possible, as organic farming practices typically use fewer pesticides.
3. Choose foods that are in season and grown locally, as they may be less likely to have been treated with pesticides.
4. Be aware of which foods are most likely to contain residual pesticides.

The Environmental Working Group publishes an annual 'Dirty Dozen' list of the most pesticide-contaminated fruits and vegetables.

When pesticides are used on crops, they can find their way into nearby water sources such as rivers, lakes and groundwater and may be incorporated into drinking water supply networks. Water treatment facilities can also help reduce the levels of residual pesticides in drinking water[8].

Even at low concentrations, residual pesticides in drinking water can have adverse effects on human health. As in the case of contaminated food, in drinking water, pesticides have been linked to various health problems, including cancer, reproductive problems, neurological disorders and immune system damage.

To reduce the risk of residual pesticides in food and drinking water, it is essential to use pesticides in a responsible manner. Farmers and other agricultural workers should follow best practices for pesticide use, including using the minimum amount of pesticide necessary, applying them at the right time and avoiding spraying near water sources.

Individuals can also take steps to reduce their exposure to residual pesticides in drinking water. They can choose to drink filtered or bottled water or use it at home.

---

[8] Treatment methods such as activated carbon filtration, reverse osmosis and ozone treatment can effectively remove pesticides from water sources.

CHAPTER 6

Semi-Volatile and Persistent Organic Compounds: Industrial POPs

6.1 Introduction to Industrial Persistent Organic Pollutants

As mentioned in Chapter 5, persistent organic pollutants (POPs) are a class of organic chemicals that remain in the environment, bioaccumulate in food chains and pose a risk to human health and the environment. In this chapter, it will be treated first as POPs that have been produced with specific purposes mainly for industrial uses, followed by other POPs unintentionally produced as by-products of industrial processes, combustion and incineration.

6.2 Polychlorinated Biphenyls (PCBs)

Polychlorinated biphenyls (PCBs) are a group of synthetic organic chemicals that were widely used as electrical insulators and heat transfer agents in a variety of industrial applications. PCBs are composed of two biphenyl rings connected by chlorine atoms (Figure 6.1). The number and arrangement of chlorine atoms on the biphenyl rings can vary, resulting in different PCB congeners with varying toxicological properties. There are a total of 209 possible congeners. PCBs were highly valued for their chemical stability, electrical insulating properties and resistance to heat and fire.

Understanding the Chemistry of the Environment, First Edition. Francisco G. Calvo-Flores.
© 2025 John Wiley & Sons Ltd. Published 2025 by John Wiley & Sons Ltd.

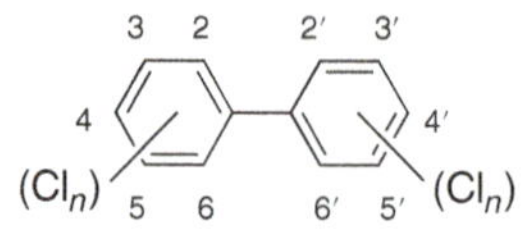

FIGURE 6.1 Polychlorinated biphenyls (PCBs)

They were also used in consumer products such as adhesives, caulking and paints. PCBs were manufactured from the 1930s until they were banned in the United States in 1979 due to concerns about their toxicity and persistence in the environment.

However, PCBs can occur as a result of industrial processes that involve the use of certain chemicals containing chlorine and benzene, which can react with each other to form PCBs as by-products. For example, the production of certain types of dyes, pigments and plastics.

Additionally, PCBs can also be formed through combustion processes, such as waste incineration or fires, in the presence of chlorine and benzene compounds. Another source of PCBs is when old electrical equipment or building materials are burned. The heat can cause the PCBs to break down and reform into different types of PCBs or other toxic chemicals.

6.3 Polychlorinated Naphthalenes

Polychlorinated naphthalenes (PCNs) are a group of organic compounds that consist of naphthalene moieties with one or more chlorine atoms attached to them (Figure 6.2). They are chemically similar to PCBs, which have similar properties and are also widely used in industrial applications. PCNs were primarily used as industrial chemicals in electrical equipment, lubricants and other applications where their high-temperature stability and electrical insulating properties were desirable. However, they are very inert, with a high persistence in the environment and can accumulate in the food chain, leading to potential health risks for humans and wildlife.

Due to their toxicity and persistence, the production and use of PCNs have been banned in many countries. The major sources of PCNs in the environment have been the historic production and release of technical PCN formulations, impurities in commercial PCB mixtures and other chemicals, but nowadays the major source is the unintentional formation from industrial thermal processes.

6.4 Polybrominated Diphenyl Ethers (PBDEs)

Polybrominated diphenyl ethers (PBDEs) are a class of chemicals used as flame retardants in a variety of consumer products, including electronics, textiles and plastics (Figure 6.3). They are composed of two phenyl rings linked by an ether group, with

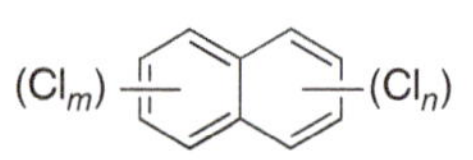

FIGURE 6.2 Polychlorinated naphthalenes

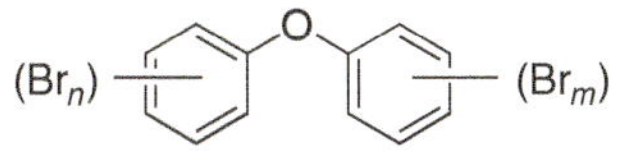

FIGURE 6.3   Polybrominated diphenyl ethers (PBDEs)

varying degrees of bromination. They have been found in several environmental media, including air, water, sediment and biota, as well as in human breast milk, blood and adipose tissue. There are three main commercial formulations of PBDEs: pentaBDE, octaBDE and decaBDE, with 5, 8 and 10 bromine atoms, respectively. PentaBDE and octaBDE have been banned or phased out in many countries due to their potential adverse effects on human health and the environment. DecaBDE, which is less bioaccumulative and less toxic than the other two formulations, is still in use in some countries. Exposure to PBDEs has been associated with a variety of adverse health effects in humans and animals, including developmental and reproductive toxicity, neurotoxicity and endocrine disruption. The most vulnerable populations to PBDE exposure are pregnant women, infants and young children, as well as wildlife living in contaminated areas.

6.5   Polyfluorinated Aliphatic Compounds

Perfluorinated aliphatic compounds (PFCs) are a group of man-made chemicals that are highly resistant to heat, water and oil. They are used in a variety of industrial and consumer products, including non-stick cookware, waterproof clothing and firefighting foam. PFCs are considered POPs because they do not break down easily in the environment and can accumulate in the food chain. They have been detected in the air, water, soil and wildlife around the world, as well as in human blood and breast milk.

Some common types of PFCs include perfluorocarboxylic acids and perfluorosulfonic acids.

Perfluorocarboxylic Acids

Perfluorocarboxylic acids (PFCAs) are a class of synthetic organic compounds that contain a chain of carbon atoms with all the hydrogen atoms replaced by fluorine atoms and a carboxylic acid functional group (-COOH) at one end.

PFCAs are highly stable and resistant to degradation. They have strong carbon-fluorine (C-F) bonds, which make them resistant to chemical, thermal and biological breakdown. PFCAs exhibit excellent water and oil-repellency properties. This makes them useful in a variety of applications, such as water-resistant coatings for fabrics, firefighting foams and stain-resistant treatments for carpets and upholstery.

PFCAs have a tendency to bioaccumulate due to their stability and resistance to degradation. They can persist in the environment for extended periods, bioaccumulate in aquatic organisms such as fish and shellfish and disrupt hormonal systems in wildlife. PFCAs have also been detected in drinking water sources, raising concerns about human exposure.

PFCAs are highly persistent in the environment, meaning they do not easily break down over time. Once released into the environment, they can remain for years or even decades. This persistence leads to their widespread distribution in the air,

TABLE 6.1   **Relevant perfluorocarboxylic acids and their uses**

Name	Formula	Uses
Perfluorohexanoic acid (PFHxA)	$CF_3(CF_2)_4COOH$	Surfactant in various industrial applications, including in the production of textiles, paper and coatings.
Perfluorooctanoic acid (PFOA)	$CF_3(CF_2)_6COOH$	Production of non-stick coatings (e.g., teflon) and stain-resistant found in some food packaging materials. Surfactant in certain industrial applications.
Perfluorononanoic acid (PFNA)	$CF_3(CF_2)_7COOH$	Surfactant in the production of coatings, including for textiles and paper. Found in certain food packaging materials.
Perfluorodecanoic acid (PFDA)	$CF_3(CF_2)_8COOH$	Surfactant in the production of coatings, including for textiles and paper. Found in some food packaging materials.

water, soil and biota. Some PFCAs, such as perfluorooctanoic acid (PFOA), have been found to be toxic to humans and wildlife. Studies have linked these compounds to various adverse health effects, including developmental and reproductive problems, liver toxicity, immune system disorders and potential carcinogenicity.

PFCAs have been detected in various parts of the world, including remote regions far from industrialised areas. They can be transported over long distances through air and water currents, resulting in their presence in ecosystems where they were never directly used or released.

Due to their persistence, toxicity and potential for long-range transport, PFCAs have garnered significant attention from regulatory bodies and international agreements. Many countries have implemented restrictions on the production, use and release of PFCAs and efforts are ongoing to find safer alternatives and remediate contaminated sites. Table 6.1 summarises the more relevant PFCAs and their uses.

Perfluorosulfonates

Perfluorosulfonates (PFSAs) are a class of synthetic chemicals that belong to a larger group known as per- and polyfluoroalkyl substances (PFAS). PFSAs are characterised by a carbon chain with fluorine atoms attached to a sulfonic acid ($-SO_3H$) or sulfonate ($-SO_3O^-$) group. They are known for their exceptional resistance to heat, chemicals and water, which makes them useful in various industrial and consumer applications.

These compounds, as happens with perfluorocarboxylic acids, are highly resistant to degradation, meaning they can persist in soil, water and living organisms for long periods of time. PFSAs have been detected in various environmental compartments worldwide, including surface water, groundwater, soil and even remote regions like the Arctic.

The presence of PFSAs in the environment has raised concerns due to their potential toxicity and bioaccumulation. Studies have suggested that PFSAs can have adverse effects on human health and wildlife. In animal studies, they have been linked

TABLE 6.2   **Relevant perfluorosulfonate compounds and their uses**

Compound	Formula	Uses
Perfluoromethanesulfonic acid (PFMS)	$CF_3\text{-}SO_3H$	Fuel cell membranes, catalysts and electroplating
Perfluoromethanesulfonate (PFMS) salts	$CF_3\text{-}SO_3M$	Electroplating, catalysts, surfactants and fuel cell membranes
Perfluorobutanesulfonic acid (PFBS)	$CF_3\text{-}(CF_2)_3SO_3H$	Stain and water repellents, coatings, metal plating and electroplating
Perfluorobutane Sulfonate (PFBS) salts	$CF_3\text{-}(CF_2)_3SO_3M$	Stain and water repellents, surfactants and cleaning agents
Perfluorohexanesulfonic acid (PFHxS)	$CF_3\text{-}(CF_2)_5SO_3H$	Stain and water repellents, coatings, metal plating, electroplating and firefighting foam
Perfluorohexane Sulfonate (PFHxS) salts	$CF_3\text{-}(CF_2)_5SO_3M$	Stain and water repellents, surfactants and cleaning agents
Perfluorooctanesulfonic acid (PFOS)	$CF_3\text{-}(CF_2)_7SO_3H$	Firefighting foam, stain and water repellents, industrial applications, electroplating and photoresist production
Perfluorooctanesulfonyl fluoride (POSF)	$CF_3\text{-}(CF_2)_7SO_3M$	Intermediate for the synthesis of other perfluorosulfonate compounds
Perfluorodecanesulfonic acid (PFDS)	$CF_3\text{-}(CF_2)_9SO_3H$	Electroplating, metal finishing and anti-corrosion coatings
Perfluorodecanesulfonate (PFDS) salts	$CF_3\text{-}(CF_2)_9SO_3M$	Electroplating, surfactants, anti-corrosion coatings, cleaning agents

to developmental issues, liver toxicity, reproductive disorders and immune system suppression in studies on animals. Additionally, some studies have associated PFAS exposure with certain cancers, although more research is needed to establish a definitive link.

Due to their environmental persistence and potential health risks, regulatory actions have been taken to reduce the use and release of PFSAs. For example, the production and use of PFOS and PFOS-related compounds have been phased out or restricted in many countries, which have called for their elimination. Relevant PFSA compounds and their uses are summarised in Table 6.2.

Topics of Interest: The Carbon-Fluorine Bond

C-F bonds possess unique properties that make them highly desirable in various applications. Carbon fluor bonds are one of the strongest single bonds known. The C-F bond is exceptionally stable and requires a significant amount of energy to break. This high bond strength makes carbon fluor compounds highly resistant to chemical reactions and degradation.

The strong C−F bond imparts exceptional chemical inertness to carbon floor compounds. They are resistant to attack by most acids, bases, oxidising agents and reducing agents. This inertness makes carbon fluor compounds suitable for applications where chemical stability is crucial, such as solvents, lubricants and high-temperature environments.

Carbon–fluorine compounds are highly hydrophobic, meaning they repel water. The electronegativity difference between carbon and fluorine atoms leads to a polarised C-F bond, with fluorine being more electronegative. This results in a significant charge separation, creating a strong dipole moment. This hydrophobic nature finds applications in water-repellent coatings, waterproofing materials and non-stick surfaces.

Carbon–fluorine compounds exhibit excellent thermal stability. They can withstand high temperatures without decomposing or losing their desirable properties. This thermal stability makes them valuable in applications requiring heat resistance, such as fire retardants, electrical insulation and aerospace materials.

The presence of fluorine atoms in these compounds contributes to low surface energy. This property reduces friction and adhesion, making them ideal for applications involving low friction, anti-fouling and anti-stick surfaces. Carbon–fluorine compounds are commonly used as lubricants, release agents and coatings to reduce friction and stickiness.

Carbon–fluorine compounds possess excellent electrical insulating properties. They have a high dielectric strength and a low dielectric constant, making them suitable for insulating materials in electrical and electronic devices.

In certain forms, such as polytetrafluoroethylene (PTFE), carbon–fluorine compounds exhibit biocompatibility. PTFE is used in various medical and pharmaceutical applications due to its non-toxic nature and resistance to biological fluids.

Carbon–fluorine bonds are known for their exceptional strength and stability. The C-F bond is highly resistant to chemical and biological degradation, making it challenging to break down naturally in the environment. As a result, fluorinated compounds, including those containing carbon–fluorine bonds, often persist for extended periods.

Biodegradation refers to the breakdown of organic substances by microorganisms, such as bacteria or fungi, into simpler compounds through natural biological processes. However, the microbial enzymes responsible for biodegradation typically do not possess the necessary mechanisms to cleave carbon–fluorine bonds effectively. This limited ability of microorganisms to degrade C-F bonds has led to concerns about the environmental impact of certain fluorinated compounds.

Fluoropolymers, which are synthetic materials containing carbon–fluorine bonds, are particularly resistant to biodegradation. Examples of commonly known fluoropolymers include PTFE, used in non-stick coatings and polyvinylidene fluoride (PVDF), used in various applications due to its excellent chemical resistance.

While some studies have reported the potential for partial degradation of certain fluorinated compounds under specific conditions, the overall biodegradation of carbon–fluorine bonds remains slow and inefficient in natural environments. Environmental persistence can lead to the accumulation of these compounds over time, potentially causing long-term ecological concerns.

It is important to note that researchers are actively investigating strategies to enhance the biodegradation of fluorinated compounds. These include the development of engineered enzymes or microbial systems capable of breaking down carbon–fluorine bonds. However, such approaches are still in the early stages of development and have not yet reached widespread implementation.

6.6 Environmental Impact of Polymers: Non-Biodegradable Polymers

Polymers are macromolecules composed of repeating units known as monomers. These monomers are chemically bonded together to form long chains or networks.

Polymers can be found in various natural and synthetic materials and they play a crucial role in many aspects of our daily lives.

Synthetic polymers are created through a process called polymerisation, where monomers undergo a chemical reaction to form those macromolecules. When two or more monomers are used in the polymerisation process, the macromolecule is named copolymer. The polymerisation of several monomers into copolymers is called copolymerisation. The majority of polymers are commonly known as plastics because of their ability to be moulded and shaped.

Synthetic polymers are widely used in various applications due to their versatility, durability and cost-effectiveness, including packaging, automotive, construction, electronics, textiles, healthcare and consumer goods. Because of its multiple applications, the market for polymers is a vast and dynamic industry that plays a crucial role in numerous sectors of the global economy. The global polymer market has been experiencing steady growth over the years. According to industry reports, the market was valued at around \$600 billion in 2020 and is projected to reach over \$900 billion by 2027. The demand for polymers is driven by factors such as population growth, urbanisation, rising disposable income, technological advancements and increasing industrialisation in emerging economies. The demand for polymers varies across different regions. Asia-Pacific is the largest consumer and producer of polymers, primarily driven by countries like China and India. North America and Europe are also significant markets, where the United States and Germany play key roles. Developing regions in Latin America, the Middle East and Africa are also witnessing substantial growth in polymer consumption.

A key aspect related to the environmental impact of synthetic polymers is their biodegradability. So, they can be classified into two major groups:

1. **Non-biodegradable polymers**, also known as persistent polymers, are materials that do not readily degrade through natural biological processes. They typically have a backbone composed of long chains of carbon atoms that the enzymes produced by microorganisms in the environment cannot break down their complex chemical structures.

2. **Biodegradable polymers** are designed to break down and degrade naturally over time through biological processes. These materials can be broken down by microorganisms into simpler compounds, such as water, carbon dioxide and biomass. The backbone structure of biodegradable polymers often consists of repeating units linked by covalent bonds with ether, ester or amide linkages that are more susceptible to enzymatic or chemical hydrolysis, facilitating the breakdown of the polymer.

While polymers have numerous benefits and applications, they also raise several environmental concerns. Some of the key environmental issues associated with polymers are:

1. **Non-biodegradability:** Many polymers, such as polyethylene (used in plastic bags) and polyethylene terephthalate (used in plastic bottles), are non-biodegradable. This means they do not break down naturally over time, leading to the accumulation of plastic waste in landfills, oceans and other natural environments. This poses a significant threat to wildlife, marine ecosystems and the overall balance of ecosystems.

2. **Plastic pollution:** Improper disposal and inefficient recycling systems have resulted in widespread plastic pollution. Plastics often find their way into water bodies, where they can harm marine life through ingestion and entanglement. Microplastics, small plastic particles, are of particular concern as they can enter the food chain and potentially impact human health.

3. **Chemical pollution:** Polymers can contain additives, such as plasticisers, flame retardants and colourants, which may be toxic and can leach into the environment. These chemicals can contaminate soil, water sources and ecosystems, posing risks to both wildlife and human health.

4. **Recycling challenges:** While recycling is often promoted as a solution to mitigate plastic waste, it faces numerous challenges. Certain polymers are more difficult to recycle due to their composition or contamination, limiting the effectiveness of recycling processes. Inadequate waste management systems, lack of infrastructure and low recycling rates also hinder the recycling potential of polymers.

As a result, they persist in the environment for a long time, leading to pollution and waste management challenges. Non-biodegradable polymers are commonly derived from petrochemicals[1].

Non-biodegradable polymers show a strong environmental impact caused by some characteristics such as:

1. **Slow decomposition:** Non-biodegradable polymers undergo decomposition processes at an extremely slow rate, taking decades or even centuries to break down into smaller fragments. This slow decomposition rate leads to their persistence in landfills, oceans and other natural environments. This is because non-biodegradable polymers exhibit high resistance to the actions of microorganisms, enzymes and other biological agents that typically break down organic matter. On the other hand, polymers are chemically very stable showing strong molecular structures, which contribute to their resistance to degradation. They often have tightly bonded polymer chains, making them highly durable and resistant to environmental factors such as sunlight, heat and moisture. Reactions such as abiotic photolysis, hydrolysis and oxidations are not capable of degrading polymers quickly.

2. **Accumulation:** Due to their resistance to degradation, non-biodegradable polymers tend to accumulate in the environment, leading to pollution and adverse ecological effects. They can persist in various habitats, including soil, water bodies and the atmosphere.

3. **Potential toxicity:** Some non-biodegradable polymers contain chemical additives, such as plasticisers, flame retardants and colourants, which can be toxic or have harmful effects on living organisms. As these polymers degrade or break down into smaller particles, these additives may be released, posing potential risks to ecosystems and human health.

---

[1] It's important to note that the distinction between biodegradable and non-biodegradable polymers is not always black and white. Some polymers may degrade under specific conditions or require industrial composting facilities to break down effectively. Additionally, the rate of degradation can vary depending on factors such as temperature, moisture and exposure to microorganisms.

4. **Microplastics formation:** Non-biodegradable polymers, when exposed to environmental factors such as weathering and mechanical forces, can fragment into smaller pieces called microplastics. Microplastics are tiny plastic particles that can persist in the environment for extended periods and have been found in various ecosystems, including oceans, rivers and even the air.

In this section, two groups of non-biodegradable polymers will be the subject of attention: those that do (fluorinated polymers), which are studied separately due to their special characteristics and those that do not have fluorine atoms in their structure.

6.7 Fluorinated Polymers

Fluorinated polymers offer a unique combination of properties, including high thermal stability, chemical resistance, low surface energy and excellent electrical insulation. Their versatility has led to their use in a wide range of applications across industries such as automotive, electronics, chemical processing and medical devices. Fluoropolymers, including perfluorinated polymers, have raised concerns about their environmental impact and persistence, because they show strong resistance to degradation by natural processes, leading to increased regulation and research on the safety of these materials.

Some key points regarding their environmental considerations are as follows:

1. **Persistence:** Carbon-fluor bonds are highly resistant to chemical and biological degradation, meaning these polymers persist in the environment for long periods of time. This property can lead to their accumulation in various environmental compartments, such as water, soil and biota.

2. **Potential for bioaccumulation:** Some fluoropolymers, such as PFOA and perfluorooctane sulfonate (PFOS), have been identified as POPs that can bioaccumulate in organisms. They have been found in wildlife, including fish, birds and mammals, as well as in humans.

3. **Toxicity:** There is evidence that certain perfluorinated compounds can have adverse effects on human health and the environment. Studies have linked them to various health issues, including developmental and reproductive problems, liver toxicity and disruption of hormonal systems in both animals and humans.

4. **Manufacturing process:** The production of fluoropolymers involves the use of perfluorinated compounds and other chemicals that can be hazardous to human health and the environment. Additionally, the manufacturing process can result in the emission of greenhouse gasses and other air pollutants.

5. **Waste management:** Disposal of fluoropolymer-containing products can pose challenges. Incineration of fluoropolymers may release toxic gasses and compounds, including perfluorinated by-products. Landfilling can result in the leaching of these compounds into soil and groundwater.

Regulatory actions and industry initiatives have been implemented to address the environmental concerns associated with fluoropolymers. For example, PFOA and PFOS have been phased out or restricted in many countries. Manufacturers are also

TABLE 6.3   **Structure of perfluorinated polymers and their starting monomers**

Polymer	Monomer
Polytetrafluoroethylene $-(CF_2-CF_2)_n-$	Tetrafluoroethylene (TFE) $CF_2=CF_2$
Polyvinylidene fluoride $-(CF_2CH_2)_n-$	Vinylidene fluoride (VDF) $CF_2=CH_2$
Perfluoroalkoxy alkane $-(CF_2-CF_2)_n-(CF_2-CFO-CF_2-R)_m-$	Tetrafluoroethylene (TFE) and perfluoroalkyl vinyl ether $CF_2=CF_2$ / $CF_2=CF-O-CF_2-R$
Ethylene tetrafluoroethylene $(CH_2-CH_2-CF_2-CF_2)_n-$	Ethylene and tetrafluoroethylene (TFE) $CH_2=CH_2$ / $CF_2=CF_2$

working on developing more environmentally friendly alternatives and improving the recyclability of fluoropolymer materials.

Examples of perfluorinated polymers are:

1. **Polytetrafluoroethylene (PTFE):** PTFE is perhaps the most well-known fluoropolymer. It is highly non-reactive and has a low coefficient of friction, making it an excellent material for non-stick coatings, such as in cookware (e.g., Teflon). PTFE is also used in various industrial applications, including electrical insulation, gaskets, seals and bearings.

2. **Polyvinylidene fluoride (PVDF):** PVDF combines excellent chemical resistance with high thermal stability. It is often used in applications that require resistance to harsh chemicals, such as piping systems, chemical tanks and semi-conductor manufacturing. PVDF is also used in the production of films, membranes and coatings.

3. **Perfluoroalkoxy alkane (PFA):** PFA is a copolymer of tetrafluoroethylene and perfluoroalkyl vinyl ether. It shares many properties with PTFE, but it has improved processability at lower temperatures. PFA is commonly used in the chemical industry for lining tanks, pipes and fittings, as well as in the production of wiring insulation and tubing.

4. **Ethylene tetrafluoroethylene (ETFE):** ETFE is a fluoropolymer known for its exceptional transparency and mechanical strength. It is often used as a light-weight and durable alternative to glass in architectural structures, such as roofs and facades. ETFE is also utilised in the aerospace industry, electrical insulation and in the production of flexible printed circuit boards. Table 6.3 describes the structure and starting monomers of these examples.

Topics of Interest: New EU Rules for Banning the Use of 'Forever Chemicals'

The European Union has established a strategy in order to ban the use of the so-called 'forever chemicals' to mitigate adverse health effects of per- and polyfluorinated alkyl substances or PFASs in food contact packaging.

Negotiators have set a target for reusable packaging for alcoholic and non-alcoholic beverages and other food contact packaging (excluding products such as milk, wine, aromatised wine and spirits) to reach at least 10% by 2030. Member states may grant a five-year extension for compliance under specific conditions. Final distributors of beverages and takeaway food in the food service sector are required to allow consumers to bring their own containers. Additionally, they must strive to offer 10% of their products in reusable packaging by 2030. Furthermore, in response to Parliament's request, member states are to incentivise restaurants, canteens, bars, cafés and catering services to provide tap water (where available) in reusable or refillable formats, either for free or for a low service fee.

6.8 Persistent Non-Fluorinated Polymers

Non-fluorinated polymers or fluorine-free polymers, are a class of polymers consisting of long carbon chains bonded to hydrogen and other atoms or fragments different from fluorine in their molecular structure. Nevertheless, these polymers exhibit exceptional resistance to chemicals. They are widely used in various industrial applications where high-performance materials with unique properties are required. Examples of these non-biodegradable polymers are included in Table 6.4.

TABLE 6.4 **Examples of non-biodegradable polymers**

Polymer	Monomer	Examples of use
Polyethylene (PE) $-(CH_2\text{-}CH_2)_n-$	Ethylene (ethene) $CH_2{=}CH_2$	Plastic bags, milk jugs, water bottles, pipes
Polypropylene (PP) $-(CH_2\text{-}CH[CH_3])_n-$	Propylene (propene) $CH_2{=}CH\text{-}CH_3$	Food containers, automotive parts, packaging
Polystyrene (PS) $-(CH_2\text{-}CH[C_6H_5])_n-$	Styrene $C_6H_5\text{-}CH{=}CH_2$	Foam cups, packaging, disposable cutlery
Polyvinyl chloride (PC) $-(CH_2\text{-}CHCl)_n-$	Vinyl chloride $CH_2{=}CH\text{-}Cl$	Pipes, window frames, flooring, medical tubing

Topics of Interest: Biodegradable Polymers

As an alternative to non-biodegradable polymers, biodegradable polymers are considered more environmentally friendly because they have the potential to reduce waste accumulation and minimise long-term environmental impact. Such polymers show oxygen or nitrogen atoms along the chains, such as polyesters, polyethers, polyamides and polyurethanes. This characteristic makes them accessible to microorganisms and therefore more easily biodegradable. Some examples of biodegradable polymers are described in Table 6.5.

TABLE 6.5   Examples of biodegradable polymers

Biodegradable polymer	Starting monomer	Common uses
Poly(lactic acid) (PLA)	Lactic acid	Packaging materials, disposable cutlery, medical implants
Polyhydroxyalkanoates (PHA)	Various hydroxyalkanoic acids	Packaging films, agricultural mulch and drug delivery systems
Poly(butylene succinate) (PBS)	Butanediol, succinic acid	Packaging materials, disposable products, agricultural films
Poly(caprolactone) (PCL)	ε-Caprolactone	Drug delivery systems, tissue engineering scaffolds, 3D printing
Poly(ethylene oxide) (PEO)	Ethylene oxide	Personal care products, pharmaceuticals, water-soluble films
Poly(butylene adipate-*co*-terephthalate) (PBAT)	Butanediol, adipic acid, terephthalic acid	Packaging films, bags, agricultural mulch
Poly(vinyl alcohol) (PVA)	Vinyl acetate, hydrolysis	Packaging films, textile coatings, adhesives
Polycaprolactone carbonate (PCLC)	ε-Caprolactone, carbon dioxide	Drug delivery systems, tissue engineering, coatings
Starch-based polymers	Starch, various monomers	Packaging materials, disposable products, agricultural applications
Poly(hydroxybutyrate-*co*-hydroxy valerate) (PHBV)	Various hydroxybutyric acids and hydroxyvalerianic acid	Packaging films, disposable cutlery and biomedical applications

Unintentionally Produced POPs: Polychlorinated Dibenzo-*p*-Dioxins, Dibenzofurans and Polycyclic Aromatic Hydrocarbons

Unintentionally Produced Organic Pollutants are a subset of Persistent Organic Pollutants (POPs) that are not deliberately manufactured but are instead created as by-products of industrial processes, combustion, or chemical reactions.

These pollutants are characterized by their persistence in the environment, potential for long-range transport, bioaccumulation in human and animal tissue, and significant adverse effects on human health and the environment. These include substances like Polychlorinated Dibenzo-*p*-Dioxins (PCDDs), Polychlorinated Dibenzofurans, Polycyclic Aromatic Hydrocarbons (PAHs), and Polycyclic Aromatic Hydrocarbons (PAHs) whose characteristics, properties, and formation mechanism are described below

6.9   Polychlorinated Dibenzo-p-Dioxins (PCDDs)

PCDDs (Figure 6.4) are a group of chemical compounds that belong to the class of dioxins. The basic structure of dioxins consists of two benzene rings connected by two oxygen atoms in a central six-membered ring called a dioxin skeleton.

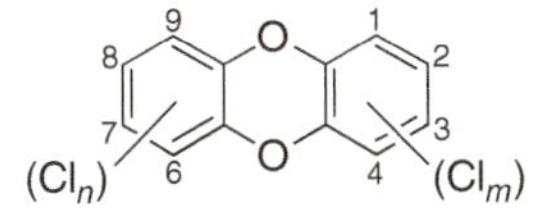

FIGURE 6.4 Polychlorinated dibenzo-*p*-dioxins

Common sources and processes that can lead to the formation of dioxins:

1. **Waste incineration:** When municipal solid waste, medical waste or hazardous waste is burned in incinerators, dioxins can be formed as by-products. The presence of chlorine-containing substances in the waste, such as PVC plastics or chlorinated organic compounds, can contribute to dioxin formation.

2. **Industrial processes:** Certain industrial activities, such as the manufacturing of chlorine-based chemicals, bleaching of pulp and paper, production of certain herbicides and pesticides and metal smelting, can release dioxins into the environment. Again, these processes involve the combustion or chemical reactions of chlorine-containing compounds.

3. **Forest fires:** During large-scale wildfires or forest fires, dioxins can be formed due to the combustion of vegetation and organic matter that contains chlorine. Dioxins released during these fires can enter the atmosphere and settle in surrounding areas.

4. **Diesel engines:** Dioxins can be emitted from diesel engines, particularly older models or those running on poor-quality fuel. The combustion of diesel fuel, which often contains trace amounts of chlorine, can lead to the formation of dioxins in the exhaust gases.

Such dioxins are a family of chemicals that includes 75 different congeners with a different number and placement of chlorine atoms on the dibenzo-*p*-dioxin structure, showing different toxicity levels.

The most toxic congener is 2,3,7,8-tetrachlorodibenzo-*p*-dioxin (TCDD) which is taken as a reference (Table 6.6).

TEF values indicate the relative toxicity of a particular PCDD congener compared to 2,3,7,8-TCDD, which is considered the most toxic and assigned a TEF of 1. The TEF values are determined based on the ability of the congeners to bind to and activate the

TABLE 6.6 Dioxin toxic equivalency factor

PCDD congener	TEF (toxic equivalency factor)
2,3,7,8-TCDD	1 (reference compound)
1,2,3,7,8-PeCDD	0.5
1,2,3,4,7,8-HxCDD	0.1
1,2,3,6,7,8-HxCDD	0.1
1,2,3,7,8,9-HxCDD	0.1
2,3,4,6,7,8-HxCDD	0.1
1,2,3,4,6,7,8-HpCDD	0.01
OCDD	0.0003

aryl hydrocarbon receptor (AhR), which is the primary mechanism of their toxicity. However, the toxicity of each isomer depends on the number and position of the chlorine atoms on the dibenzo-*p*-dioxin structure, as well as the species exposed, duration and route of exposure and other factors. Some isomers, such as TCDD, are highly toxic and can cause a range of adverse health effects in humans and animals, including cancer, immune system dysfunction and reproductive and developmental problems. People can be exposed to dioxins by eating contaminated food, inhaling air containing dioxins or through skin contact with contaminated soil or water.

Dioxins can accumulate in the food chain, particularly in fatty tissues of animals and can be found in foods such as meat, dairy products and fish. The concentrations of dioxins in food are typically measured in picograms per gram (pg/g) or parts per trillion (ppt). Detectable concentrations of dioxins in various food items can vary depending on factors such as the specific dioxin compound, the source of contamination and the sampling and analysis methods used. Table 6.7 provides an approximate range of detectable concentrations of dioxins in certain food items:

It is important to keep in mind that regulatory standards for acceptable dioxin levels in food may differ between countries or regions. Regular monitoring and adherence to food safety regulations help ensure that the dioxin concentrations in food remain within safe limits.

To reduce the formation of dioxins, it is important to minimise the use of chlorine in industrial processes and to properly manage waste incineration. This can be achieved through the use of cleaner production technologies. Some ways to reduce dioxin emissions are:

1. **Reduce waste incineration:** Dioxins are produced when organic matter is burned and waste incineration is a major source of dioxin emissions. Reducing waste incineration by promoting waste reduction, recycling and composting can help reduce dioxin emissions.

TABLE 6.7 **Approximate range of detectable concentrations of dioxins in certain food items**

Food item	Dioxin concentration (pg/g)
Chicken (meat)	0.1–10
Beef (meat)	0.2–20
Pork (meat)	0.1–10
Fish (oily)	0.1–10
Fish (non-oily)	0.05–5
Milk (whole)	0.05–5
Eggs	0.05–5
Butter	0.1–10
Cheese (hard)	0.1–10
Vegetables (various)	0.05–5
Fruits (various)	0.05–5
Grains (cereal, rice, wheat)	0.05–5

2. **Implement pollution control technologies:** Pollution control technologies such as scrubbers, baghouses and electrostatic precipitators can help capture dioxin emissions from industrial processes before they are released into the environment.

3. **Use alternative fuels:** Using alternative fuels to coal and oil, such as renewable energy sources like solar and wind power, can help reduce dioxin emissions from energy production.

4. **Improve waste management practices:** Proper waste management practices can help reduce dioxin emissions. This includes ensuring that hazardous waste is properly stored, transported and disposed of in accordance with regulations.

5. **Promote sustainable agriculture:** Dioxins can also be produced when certain chemicals are used in agriculture. Promoting sustainable agriculture practices, such as using organic farming methods and reducing the use of pesticides and herbicides, can help reduce dioxin emissions from this source.

By implementing these strategies, it is possible to reduce dioxin emissions and protect human health and the environment.

Topics of Interest: Agent Orange

Agent Orange is a herbicide that was used by the US military during the Vietnam War to defoliate forests and remove vegetation in order to deny the enemy cover and concealment. The herbicide was named after the orange band that was used to mark the containers in which it was stored.

Agent Orange was a mixture of two herbicides, 2,4-dichlorophenoxyacetic acid (2,4-D) and 2,4,5-trichlorophenoxyacetic acid (2,4,5-T), which were combined to increase their effectiveness. Unfortunately, the manufacturing process for 2,4,5-T produced a byproduct called 2,3,7,8-TCDD which is a highly toxic and persistent compound.

The use of Agent Orange during the Vietnam War resulted in the widespread contamination of soil, waterways and vegetation with TCDD, which has been linked to a wide range of health problems in humans and animals. Many Vietnam War veterans who were exposed to Agent Orange, as well as civilians in Vietnam and their descendants, have suffered from serious health issues as a result of this exposure.

The use of Agent Orange during the Vietnam War remains a controversial and contentious issue, with many people still advocating for compensation for those affected by its use. The long-term environmental and health impacts of this herbicide are still being studied and understood, but it is clear that the use of this chemical had a significant and lasting impact on both human health and the environment.

Polychlorinated Dibenzofurans

PCDFs consist of two benzene rings connected by a furan ring, with chlorine atoms substituting hydrogen atoms at various positions. The chlorine substitution pattern can vary, leading to different PCDF congeners (Figure 6.5).

As dioxins, the production of PCDFs occurs through combustion or thermal processes involving organic materials containing chlorine, such as certain types of plastics, wood preservatives and pesticides. These processes involve high temperatures and chlorine-containing precursors, which can lead to the formation of PCDFs

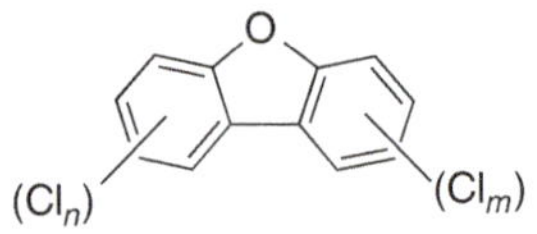

FIGURE 6.5   Polychlorinated dibenzofurans (PCDFs)

as unintended by-products. The specific production mechanisms and conditions can vary depending on the industrial processes involved.

PCDFs are considered highly toxic and can have detrimental effects on human health and the environment. They are classified as POPs due to their resistance to degradation and their ability to bioaccumulate in the food chain. The toxicity of PCDFs depends on factors such as the degree of chlorination, the specific congeners present and their interactions with biological systems.

The main toxic effects of PCDFs include:

1. **Carcinogenicity:** Some PCDF congeners are known to be carcinogenic and have been associated with an increased risk of various cancers, including liver, lung and skin cancer.

2. **Developmental and reproductive effects:** PCDF exposure has been linked to adverse effects on reproduction and development. It can lead to reduced fertility, impaired foetal growth and developmental abnormalities.

3. **Immunotoxicity:** PCDFs can suppress the immune system, making individuals more susceptible to infections and diseases.

4. **Endocrine disruption:** PCDFs can interfere with the normal functioning of the endocrine system, which regulates hormones in the body. This disruption can lead to hormonal imbalances and various health effects.

5. **Neurotoxicity:** Exposure to PCDFs has been associated with neurological effects, including cognitive impairments, learning disabilities and behavioural changes.

Due to their toxic nature, PCDFs are subject to strict regulations and control measures in many countries that aim to reduce or eliminate the production and release of PCDFs into the environment.

These regulations include measures to reduce the formation of PCDFs during industrial processes, the proper handling and disposal of waste materials containing chlorine and the monitoring of environmental levels to ensure compliance with safety standards.

Efforts are also made to remediate contaminated sites and reduce human exposure to PCDFs through measures such as air and water pollution control, waste management practices and the promotion of clean technologies that minimise or eliminate PCDF formation.

Polycyclic Aromatic Hydrocarbons (PAHs)

PAHs are a class of organic compounds composed of fused aromatic rings. Fused aromatic rings refer to a molecular structure composed of multiple aromatic rings that share one or more common carbon atoms, resulting in a connected ring system.

They are produced through incomplete combustion of organic matter such as coal, oil, gasoline and wood. PAHs are also found in tobacco smoke, charbroiled foods and other sources.

PAHs are known to be toxic and have the potential to cause adverse effects on human health and the environment. The toxicity of PAHs can vary depending on factors such as the specific compound, concentration, duration of exposure and route of exposure (inhalation, ingestion or dermal contact). It is important to note that the toxicity of PAHs can vary depending on their chemical structure and the specific compound in question. The main effects of PAH exposure are:

1. **Carcinogenicity:** Many PAHs have been identified as potent human carcinogens. Long-term exposure to high levels of PAHs has been associated with an increased risk of developing various types of cancer, including lung, skin, bladder, liver and gastrointestinal cancers.

2. **Mutagenicity:** PAHs have the ability to induce mutations in DNA, which can lead to genetic damage. They can interfere with the normal functioning of genes and increase the risk of heritable genetic disorders.

3. **Developmental and reproductive effects:** PAH exposure during pregnancy has been linked to adverse effects on foetal development, including low birth weight, developmental abnormalities and impaired neurological development. Some PAHs can also interfere with reproductive function in both males and females.

4. **Respiratory effects:** Inhalation of PAHs can irritate the respiratory system and cause respiratory symptoms such as coughing, wheezing and shortness of breath. Prolonged exposure to high levels of PAHs may contribute to the development or exacerbation of respiratory conditions such as asthma.

5. **Skin effects:** Direct contact with PAHs can cause skin irritation, redness and dermatitis. Certain PAHs are also known to be photoallergic, meaning they can cause an allergic reaction when exposed to sunlight.

PAHs are also known to be persistent in the environment and can accumulate in soil, water and sediment, leading to long-term exposure. Regulatory agencies such as the US Environmental Protection Agency (EPA) have established guidelines for acceptable levels of PAHs in various environmental media and efforts are ongoing to reduce their emissions and control their impact on human health and the environment. Some PAHs are known to be carcinogenic and are considered harmful to human health and the environment.

EPA has developed a list of 16 priority PAHs that are considered the most harmful to human health and the environment. This list is commonly referred to as the 'EPA blacklist'. The 16 PAHs on the EPA blacklist are shown in Figure 6.6. They are divided into two groups according to their molecular weight: low molecular weight (LMW) PAHs and high molecular weight (HMW) PHAs.

The EPA blacklist is used as a tool to regulate and monitor the presence of these harmful PAHs in the environment, particularly in air, water and soil.

These unintentionally produced POPs are a significant concern because they can persist in the environment for long periods of time and accumulate in the food chain, posing a risk to human health and the environment. Therefore, efforts are being made to reduce their production and release into the environment, as well as to develop effective remediation strategies to address the legacy of past contamination.

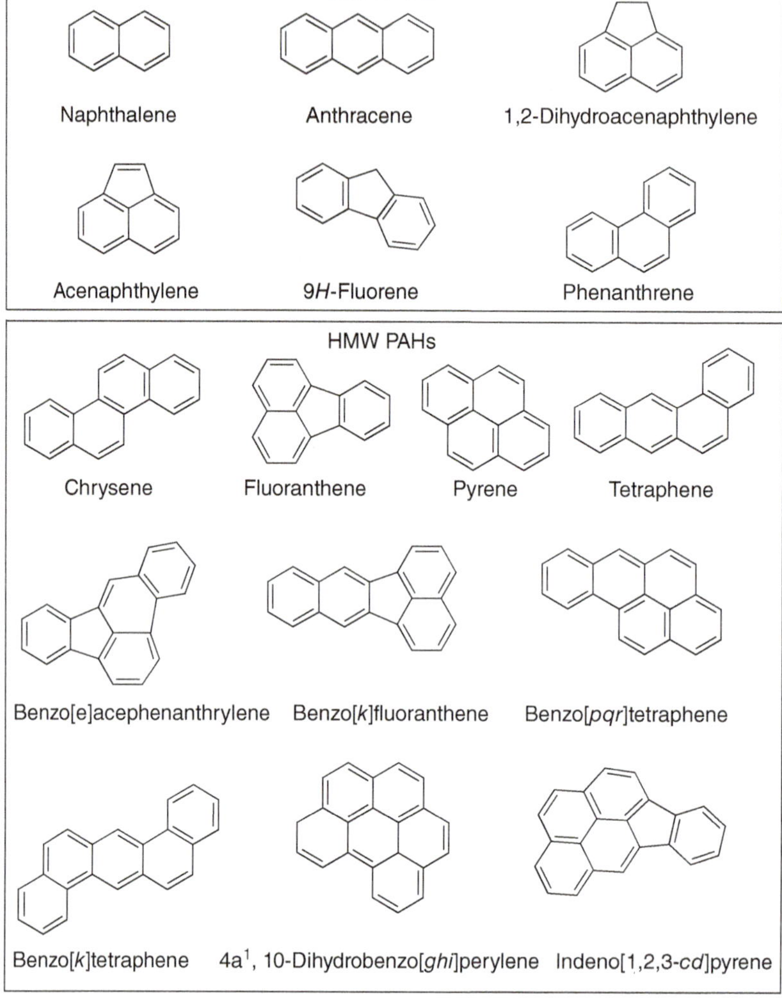

FIGURE 6.6 The 16 PAHs' EPA blacklist

Topics of Interest: PAHs and Cooking

One common source of PAH exposure is through the consumption of grilled, charred or smoked foods. When bread, meat, poultry or fish are cooked at high temperatures, such as on a barbecue grill or through other methods like pan-frying or broiling, PAHs can be formed. The fat and juices from the meat can drip onto the heat source, causing flare-ups and the production of PAHs as the flames come into contact with the food.

PAHs can also be formed when foods are smoked or charred directly over an open flame. This includes processes like smoking meats or charring vegetables. The smoke generated during these cooking methods contains PAHs, which can be deposited on the surface of the food.

Consuming foods that contain PAHs can potentially increase the risk of certain types of cancer. However, the actual risk depends on various factors, including the level and duration of exposure, individual susceptibility and overall diet and lifestyle.

To reduce PAH exposure from food cooking, you can take several precautions:

1. **Use gentler cooking methods:** Options for cooking methods that involve lower temperatures and less direct exposure to open flames. Examples include baking, steaming, boiling or stewing.

2. **Reduce charring and avoid flare-ups:** Avoid excessive charring or blackening of foods. Trim excess fat from meat to minimise drippings and flare-ups. You can also partially pre-cook meat or fish before grilling to reduce the overall cooking time and the formation of PAHs.

3. **Use marinades:** Marinating meats before cooking has been found to help reduce the formation of PAHs. Ingredients like vinegar, lemon juice or herbs can be added to the marinade to further minimise PAH formation.

4. **Avoid prolonged cooking times:** Limit the cooking time to avoid overcooking or burning the food. PAH formation increases with longer cooking times, especially at high temperatures.

5. **Opt for leaner cuts of meat:** PAHs tend to form in fat and drippings. Choosing leaner cuts of meat can reduce the overall fat content and decrease the potential for PAH formation.

6. **Maintain a clean grill:** Clean the grill grates regularly to remove any residue or charred debris that may contain PAHs.

7. **Diversify your diet:** Incorporate a variety of foods into your diet, including fruits, vegetables, whole grains and legumes. These foods provide important nutrients and antioxidants that can help mitigate the potential harmful effects of PAHs.

It's important to note that while these precautions can help reduce PAH exposure from food, complete elimination is unlikely. Therefore, maintaining a balanced and varied diet, along with a healthy lifestyle, is essential for overall well-being.

6.10 Mechanisms of Dioxins, Polychlorinated Dibenzofurans and Polycyclic Aromatic Hydrocarbon Formation in Combustion Processes

During the combustion process, the majority of the fuel-carbon, wood, waste, etc., is transformed into carbon dioxide and water vapour. However, a fraction produces materials from incomplete combustion, such as:

- Carbon monoxide
- Small-sized aromatic hydrocarbons
- Polycyclic aromatic hydrocarbons
- Soot
- Other substances.

All these substances can be formed during incomplete combustion of organic materials, when carbon-containing molecules are not fully burned and undergo pyrolysis (thermal decomposition) instead, caused by the high temperatures and limited oxygen availability during combustion, to give highly reactive intermediates, such as free radicals, which can undergo different reactions.

These materials from incomplete combustion come into contact with a series of substances containing halogen atoms, especially chlorine, which act as precursors for the formation of dioxins and PCDFs.

The most common precursors are:

- Short-chain aliphatic halogenated compounds
- 2,4-dichlorophenol (DCP), 2,4,5-trichlorophenol (TCP) and 2,4,6-trichlorophenol (TCP)
- Polychlorinated biphenyls (PCBs)
- Chlorinated herbicides like 2,4-dichlorophenoxyacetic acid (2,4-D) and 2,4,5-trichlorophenoxyacetic acid (2,4,5-T)

Soot particles can act as a surface for the adsorption of chlorine-containing compounds, allowing them to come into close proximity and enhancing the likelihood of chemical reactions leading to dioxin and PCDF formation. However, it is important to note that soot itself does not possess catalytic properties in the strict sense of the term.

The formation of dioxins and PCDDs is complex and depends on various factors, including temperature, oxygen availability, composition of the fuel or waste being burned, proportion of chlorine derivatives and the efficiency of combustion processes. In Figure 6.7, all these transformations are summarised.

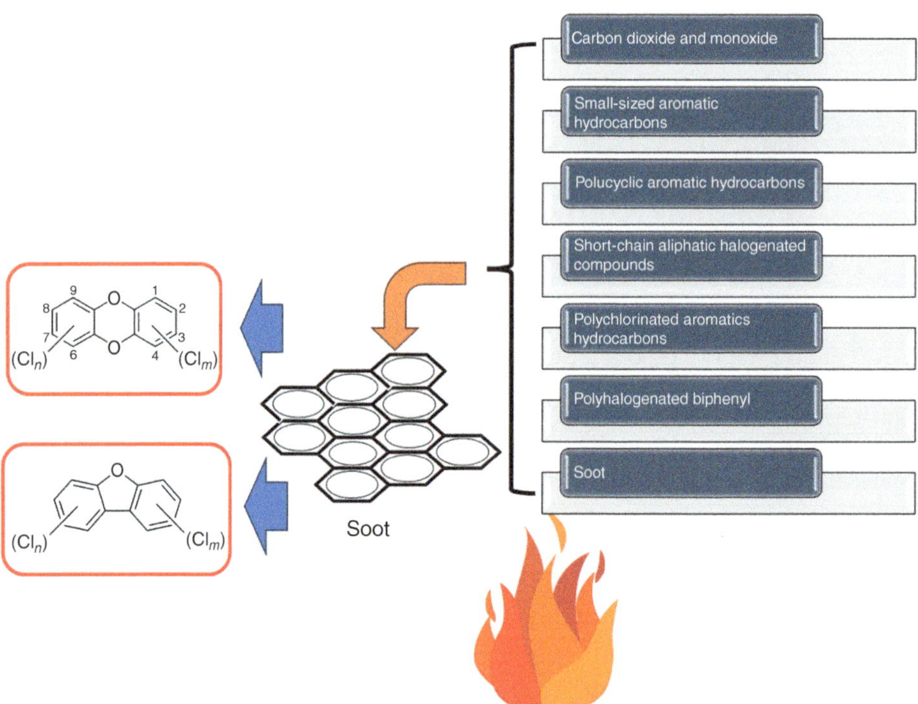

FIGURE 6.7   Mechanism of dioxins, PCDFs and PAH formation

6.11   Stockholm Convention

The United Nations Environment Programme (UNEP) has been actively involved in researching and investigating POPs for many years because of their serious health and environmental problems working with governments, organisations and communities around the world to reduce and eliminate the use and release of POPs[2].

The Protocol of Stockholm, also known as the Stockholm Convention on Persistent Organic Pollutants, is an international treaty signed in 2001 to address the global issue of POPs. The Stockholm Convention seeks to eliminate or restrict the production and use of these twelve POPs. The treaty requires countries to take measures to reduce and eliminate the release of these chemicals into the environment and to work towards developing safer alternatives.

Under the treaty, countries are required to develop national implementation plans and report on their progress in implementing the treaty. The Convention also establishes a fund to provide financial and technical assistance to developing countries to help them comply with the treaty's provision.

The initial focus was on what became known as the 'Dirty Dozen'. These were a group of 12 highly persistent and toxic chemicals: **aldrin, chlordane, DDT, dieldrin, endrin, heptachlor, hexachlorobenzene, mirex, polychlorinated biphenyls, polychlorinated dibenzo-*p*-dioxins, polychlorinated dibenzofurans and toxaphene** (Table 6.8).

TABLE 6.8   The dirty dozen

Substance	Use
Aldrin	Insecticide used to control soil-dwelling insects
Chlordane	Insecticide used for termite and ant control
DDT	Insecticide used to control malaria and other diseases
Dieldrin	Insecticide used to control termites, ants and soil pests
Endrin	Insecticide used to control rodents
HCB	Industrial chemical used in the production of pesticides
Heptachlor	Insecticide used to control termites and other pests
Mirex	Insecticide used to control fire ants and other pests
Polychlorinated biphenyls (PCBs)	Used in electrical equipment, lubricants and paints
Polychlorinated dibenzo-*p*-dioxins (PCDDs)	By-product of industrial processes
Polychlorinated dibenzofurans (PCDFs)	Byproduct of industrial processes
Toxaphene	Insecticide used to control a variety of pests

---

[2] https://www.pops.int/default.aspx.

Many of the pesticides in this group are no longer used for agricultural purposes but a few continue to be used in developing countries.

The Stockholm Convention has been ratified by more than 180 countries, making it one of the most widely supported environmental agreements in the world. It has played an important role in reducing the production and use of POPs and in protecting human health and the environment from the harmful effects of these chemicals.

It's important to note that the list of POPs is regularly updated and revised based on new scientific information and the availability of safer alternatives.

Between 2017 and 2019 the following chemicals were added (Table 6.9).

TABLE 6.9   Implementation of first list of POPs

Compound	Structure	Use
Pentabromodiphenyl diphenyleter	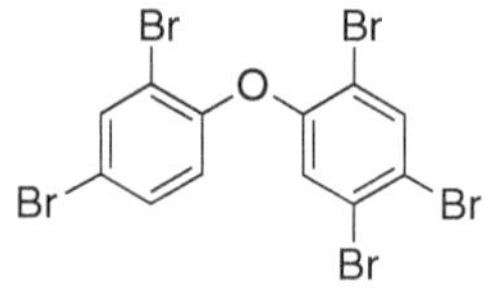	Flame retardant used in foam cushions and plastics
Commercial octaBDE mixtures	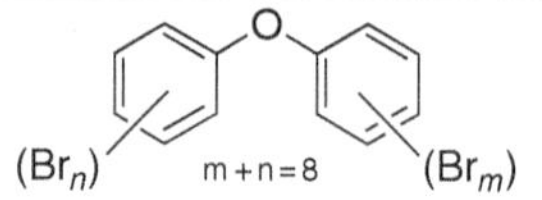	Flame retardants used in electronic equipment, furniture and other products
Hexachlorobutadiene (HCBD)	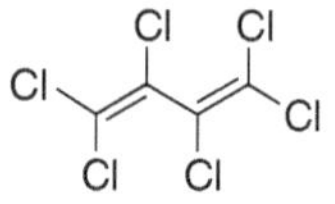	Intermediate chemical for the production of rubber compounds and resins
Endosulfan	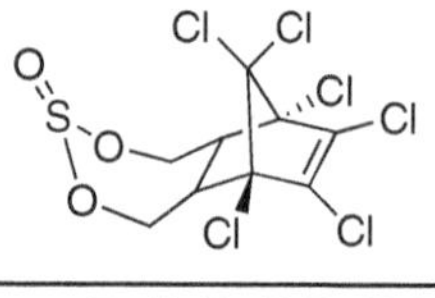	Insecticide used on fruits, vegetables and other crops
Perfluorooctane sulfonic acid (PFOS), its salts	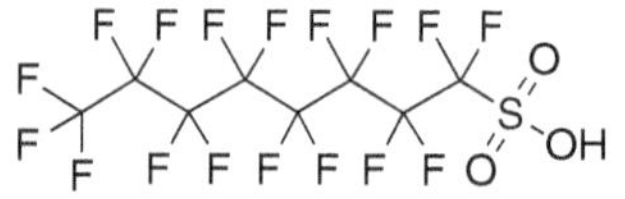	Surfactant used in various industrial processes and consumer products
Perfluorooctane sulfonyl fluoride (PFOSF)	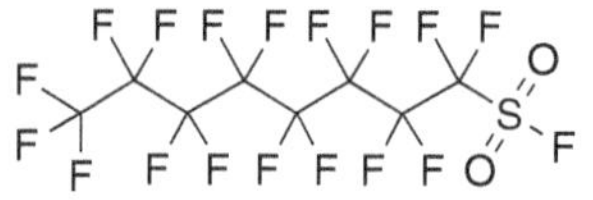	Coatings, textiles, paper and firefighting foams. Reagent in organic synthesis, for production of sulfonamides and sulfonate esters.
Alpha hexachloro-cyclohexane (α-HCH)	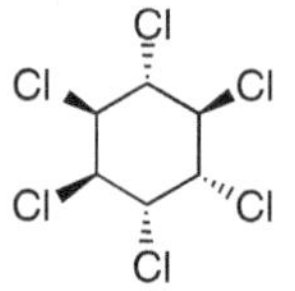	Insecticide and acaricide used on crops, animals and in public health

TABLE 6.9   **(Continued)**

Compound	Structure	Use
Decabromodiphenyl ether (DecaBDE)	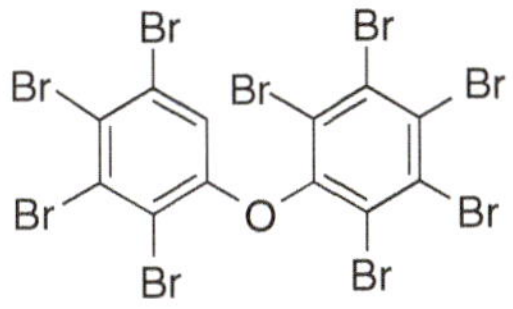	Flame retardant used in plastics, textiles and other products
Short-chain chlorinated paraffins (SCCPs): mixtures of chlorinated hydrocarbons with a chain length of 10–13 carbon atoms and a chlorine content of 40%–70%.	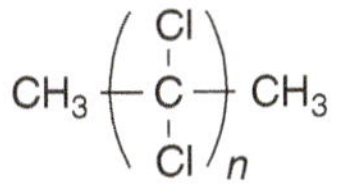	Plasticisers and flame retardants used in rubber, textiles and other products
Dicofol	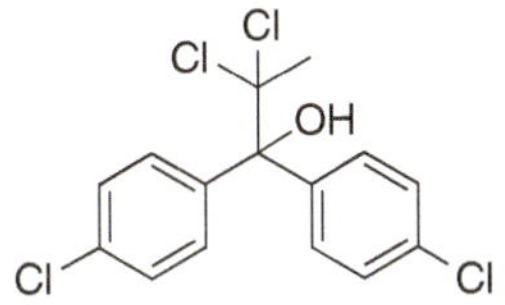	Miticide used on fruits, vegetables and ornamental plants
Perfluorooctanoic acid (PFOA)	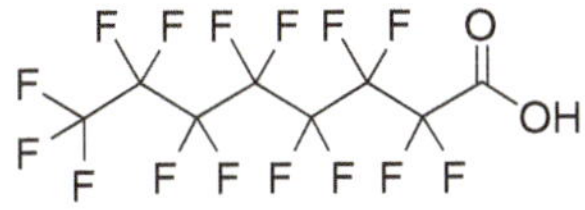	Surfactant used in the production of fluoropolymers, such as teflon

1. PentaBDE (pentabromodiphenyl ether) and commercial octaBDE mixtures
2. Hexachlorobutadiene (HCBD)
3. Endosulfan
4. Perfluorooctane sulfonic acid (PFOS), its salts and perfluorooctane sulfonyl fluoride (PFOSF)
5. Alpha hexachlorocyclohexane (α-HCH)
6. Decabromodiphenyl ether (DecaBDE)
7. Short-chain chlorinated paraffins (SCCPs)

Finally, in 2021, a new set of POPs were added to the Stockholm Convention through a revision process. The new POPs include:

1. **Dicofol:** A pesticide that can break down into DDT, a well-known POP.
2. **Perfluorooctanoic acid (PFOA), its salts and PFOA-related compounds:** These are used in a variety of industrial applications, such as in the production of non-stick coatings, water-resistant fabrics and firefighting foam.
3. **Hexachlorobutadiene (HCBD):** This is used in the production of rubber compounds and as a solvent for oils and waxes.

4. **Pentachlorophenol (PCP) and its salts and esters:** These are used as wood preservatives, as well as in the production of pesticides and herbicides.

The addition of these new POPs to the Stockholm Convention means that their production, use and trade will be restricted or eliminated, depending on the specific chemical and its intended use. This is an important step towards protecting human health and the environment from these harmful chemicals.

CHAPTER 7

Emerging Pollutants

7.1 Introduction

Emerging pollutants are chemical substances that are not yet regulated by agencies and organisations responsible for environmental issues, but they are under surveillance due to the concern that they generate in the population and in the scientific community for their possible effect on human health and their impact on the environment.

The concept of emerging pollutants is focused on the following groups of substances:

1. Substances related to technologies or industrial processes that have not been previously regulated or monitored and are suspected to produce harmful effects

2. Substances associated with Kwon technologies or industrial processes that have remained undetectable until advances in analytical techniques have provided their identification in the environment and have been linked to hazards for humans or the environment

3. Substances that were previously considered safe but whose role on living beings has been reviewed and are suspected to produce undesirable effects on humans or the environment

4. Substances that have become more prevalent in the environment due to changes in land use, climate change or other factors.

The main groups of emerging pollutants are pharmaceuticals and personal care products (PPCPs), illegal drugs, flame retardants, per- and polyfluoroalkyl substances (PFAS), surfactants, microplastics and nanoparticles.

7.2 Pharmaceuticals as Emerging Pollutants

Pharmaceuticals are an essential part of modern medicine and have significantly improved the quality of life and lifespan of many individuals. They play a critical role in managing chronic diseases, preventing and treating infections, reducing

Understanding the Chemistry of the Environment, First Edition. Francisco G. Calvo-Flores.
© 2025 John Wiley & Sons Ltd. Published 2025 by John Wiley & Sons Ltd.

pain and inflammation and providing vital support to the healthcare industry. Pharmaceuticals are typically administered in various forms, including pills, capsules, tablets, injections, creams, ointments or inhalers, and they work by interacting with the body's chemical processes to alleviate or manage medical conditions. There are many different types of pharmaceuticals that are commonly prescribed for humans, depending on the medical condition being treated.Another group of pharmaceuticals are veterinary pharmaceuticals that are widely used to prevent, treat and manage diseases in livestock and pets. Common pharmaceuticals used in veterinary medicine include antibiotics, anti-parasitics, anti-inflammatories, vaccines and anaesthetics. These medications may be administered in a similar way as in humans: orally, topically or through injection.These substances are increasingly detected in the environment, including surface water, groundwater, soil, sediment and even in our food.

One of the primary reasons for concern is that pharmaceuticals are designed to be biologically active even at low concentrations and many of them are able to persist in the environment for extended periods without breaking down. As a result, many of these substances remain in the environment and require careful management to mitigate their impact on the environment and human health.

The most common way that pharmaceuticals can enter the environment in an uncontrolled way is through the following four mechanisms:

a. **Through excretion by humans and animals**. The primary route of drug excretion is through the kidneys, which filter the blood and eliminate waste products, including drugs and their metabolites, in the urine. Other routes of excretion include faecal elimination, which occurs primarily through the liver and exhalation of volatile drugs through the lungs. By this process, residual pharmaceuticals and their metabolites can be incorporated into the environment, especially in wastewater due to their widespread use, dispersing through water bodies and soil. The concentration of drugs found in water bodies and soils is thought to be directly proportional to population densities and the yearly sale of prescription as well as non-prescription drugs in every country.

b. **By the disposal of unused or expired medications**, which are usually flushed down the toilet or thrown in the trash. They usually end up in wastewater or landfills, where they can eventually leach into the environment.

c. **Through the effluents from hospitals and clinics**. Liquid wastes from hospitals may contain fluids such as blood and other bodily fluids and wastewater from sinks, showers and toilets with appreciable concentrations of pharmaceuticals. Hospitals typically have specialised waste management systems in place to handle some waste safely and effectively, but in many cases, some of these liquid wastes can pass into the urban sanitation network.

d. **Via agricultural runoffs**. They can contain a variety of substances, such as fertilisers, pesticides and pharmaceuticals used in farming, that are excreted by the body and end up in the environment. When agricultural land is irrigated, runoff containing veterinary pharmaceuticals flows into nearby water sources, which can create a potential hazard.

As commented before, one remarkable source of residual pharmaceuticals is municipal wastewater. Wastewater treatment plants are designed to remove a variety of pollutants, however, not all pharmaceuticals can be effectively removed by conventional wastewater treatment processes. Advanced treatment technologies, such as ozonation and activated carbon adsorption, are required to remove them from wastewater. Usually, these technologies can be expensive and may not be feasible for all wastewater treatment plants, leaving detectable amounts of pharmaceuticals in treated waters and distributed in the environment.

Most pharmaceuticals produce adverse effects on fauna and flora, although in higher concentrations than usual in those found in different environmental matrices. The long-term effects of such interactions between living beings and pharmaceuticals and the combined and often synergistic impact of multiple emerging pollutants on human health and the environment remain unknown and need further study.

7.3 Antibiotics

Antibiotics are a class of pharmaceuticals used to treat bacterial infections. Antibiotics are a group of emerging pollutants present in wastewater as a result of human and animal excretion, as well as from pharmaceutical manufacturing and hospital discharges. The main concern about this is that they can potentially contribute to the development of antibiotic-resistant bacteria, which is a growing concern in modern medicine as it can make bacterial infections much harder to treat and can result in prolonged illness, increased healthcare costs and even death.

The two main sources of residual antibiotics are:

1. **Animal-derived food:** Antibiotics are used in the treatment of bacterial infections in animals to prevent and control the spread of infectious diseases and promote growth. Residual amounts of antibiotics may be found in animal-derived food products, such as meat, milk and eggs, that are consumed by humans. Such practices can promote the development of antibiotic-resistant bacteria when people consume food that contains detectable amounts of antibiotics. Regulatory agencies in many countries have established maximum residue limits (MRLs) for antibiotics in food products to minimise the risk of antibiotic resistance. These MRLs represent the maximum amount of residual antibiotics that are allowed to be present in food products and food producers are required to comply with these limits to ensure the safety of their products.

2. **Municipal wastewater:** Wastewater treatment plants are designed to remove many pollutants from wastewater. However, antibiotics may not be completely removed during the treatment process. Antibiotics should undergo various specialised processes such as adsorption, biodegradation and oxidation, but this is not the case. As a consequence, small amounts may still be present in the treated wastewater that is discharged into rivers, lakes or oceans.

Once in the environment, residual antibiotics can have a range of effects on ecosystems and human health. The main concerns about antibiotics as emerging pollutants in the environment are:

a. Exposure to low levels of antibiotics in the environment can select bacteria that are resistant to these drugs, making it more difficult to treat such infections in both humans and animals. This occurs when bacteria evolve mechanisms to protect themselves from antibiotics or when mutations in their genetic material allow them to survive in the presence of such drugs. With the conventional use of antibiotics, resistance is a growing concern because it can lead to the failure of antibiotic treatments, making infections more difficult to treat and potentially more dangerous. This can result in longer hospital stays, higher healthcare costs and an increased risk of death. The presence of antibiotics in the environment as micropollutants can increase this unwanted effect.

b. Effects on non-target organisms, such as fish, birds and aquatic invertebrates. For example, exposure to the antibiotic erythromycin[1] has been shown to reduce the feeding and swimming behaviour of fish and exposure to the antibiotic ciprofloxacin has been found to impair the growth and development of frog tadpoles.

c. Affects the microbial communities in soils and aquatic environments, reducing the diversity of bacterial communities and disrupting ecosystem processes such as nutrient cycling and decomposition.

Topics of Interest: Antibiotic Resistance

Antibiotic resistance is the ability of bacteria or other microorganisms to resist the effects of antibiotics, rendering them ineffective in treating bacterial infections. This occurs when bacteria evolve and develop mechanisms to counteract the actions of antibiotics, making the drugs less potent or completely ineffective.

This phenomenon is a significant global health concern because it can lead to more severe and prolonged infections, increased healthcare costs and higher mortality rates. There are several factors contributing to the development and spread of antibiotic resistance:

1. **Misuse and overuse of antibiotics:** The inappropriate use of antibiotics, such as taking them for viral infections or not completing the full course prescribed by a healthcare professional, can contribute to antibiotic resistance. When antibiotics are used unnecessarily or incorrectly, bacteria have more opportunities to develop resistance. A significant case is the use of antibiotics in agricultural practices to promote growth and prevent diseases in livestock, which can lead to the development of antibiotic-resistant bacteria that can then be transmitted to humans through food consumption or environmental exposure.

2. **Poor infection control:** Inadequate infection control practices in healthcare settings can facilitate the spread of antibiotic-resistant bacteria. For example, improper hand hygiene, inadequate sterilisation of medical equipment and improper waste disposal can contribute to the transmission of resistant bacteria between patients.

3. **Lack of new antibiotics:** The discovery and development of new antibiotics have significantly decreased in recent decades. The limited availability of new drugs means that healthcare providers have fewer treatment options for resistant infections, increasing the likelihood of further antibiotic resistance.

4. **Global travel and trade:** International travel and trade can contribute to the spread of antibiotic resistance across borders. Resistant bacteria can be carried by individuals or

---

[1] Erythromycin is an antibiotic that belongs to the class of macrolide antibiotics and is commonly used to treat infections such as respiratory tract infections (including pneumonia), skin and soft tissue infections, urinary tract infections and certain sexually transmitted infections such as chlamydia.

through contaminated food, water or other goods, resulting in the global dissemination of resistant strains.

Antibiotic resistance is a complex issue requiring global collaboration and concerted efforts from healthcare professionals, policymakers, researchers and the general public to ensure the effectiveness of antibiotics for future generations. It has already been detected at the same time as the discovery of different types of antibiotics.

Here's a brief timeline of the discovery of antibiotics and the evolution of antibiotic resistance. The key facts about the history of antibiotics are:

- **1928s:** Alexander Fleming discovers penicillin, the first antibiotic and notes its potential for treating bacterial infections.
- **1939s:** Howard Florey and Ernst Chain developed a method for mass-producing penicillin, leading to widespread use during World War II.
- **1940s:** Streptomycin, chloramphenicol and tetracycline are discovered and become widely used antibiotics.
- **1940s–1960s:** Antibiotics are prescribed liberally for a range of infections, leading to widespread use and overuse.
- **1950s–1970s:** Resistance to antibiotics becomes more common as bacteria develop mechanisms to defend against them.
- **1960s–1980s:** New antibiotics are discovered, but resistance continues to emerge and spread.
- **1990s–present:** Antibiotic resistance becomes a major public health concern as more and more bacteria become resistant to multiple drugs. Efforts to develop new antibiotics and reduce overuse are ongoing.
- **2020s:** With the emergence of new bacterial strains that are resistant to multiple antibiotics, the World Health Organization (WHO) has declared antibiotic resistance one of the top 10 global public health threats facing humanity.

Figure 7.1 summarises when the discovery of the most common antibiotics took place, the years of their first use and when the first episodes of resistance were detected.

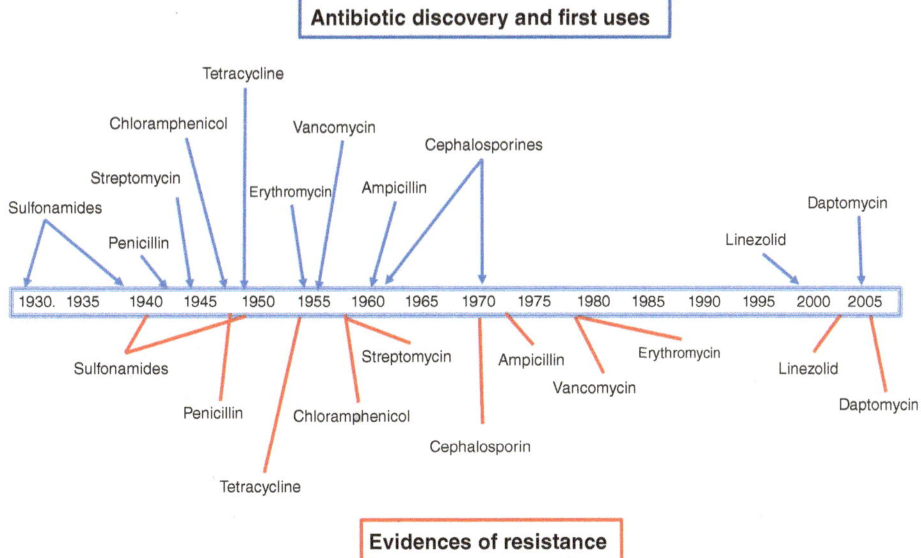

FIGURE 7.1   Timeline of antibiotic discovery and resistance

7.4 Antibiotics Found in Environmental Issues

1. **Sulfonamides:** This class of antibiotics is used in both human and veterinary medicine. Sulfonamides have been in use since the 1930s, and they have been instrumental in treating a wide range of bacterial infections such as urinary tract, middle ear, certain types of pneumonia, bronchitis, traveler's diarrhoea and meningitis. The general formula of sulfonamides is described in Figure 7.2.

   The overuse and misuse of sulfonamides have contributed to the development of antibiotic-resistant bacteria, which are a growing global health concern. They are often found not only in wastewater but also in surface waters, with median concentrations of around 50 ng/L and peak concentrations of up to 4–6 µg/L. Such values found in the environment are susceptible to altering microbial community structure and diversity.

2. **β-lactams:** This class of antibiotics includes penicillins, cephalosporins, carbapenems and monobactams, which are commonly used in human medicine (Figure 7.3).

   They can also be used in veterinary medicine. β-lactam antibiotics can enter the environment through a variety of routes, including wastewater treatment plants, agricultural runoff and improper disposal of pharmaceuticals. Once in the environment, they can persist for long periods and potentially affect aquatic organisms, soil microorganisms and human health. The concentrations of β-lactams in residual waters can vary widely depending on factors such as the antibiotic type, the type of wastewater treatment used and the location of the wastewater treatment plant. The concentrations can vary from a few ng/L to several hundred µg/L, depending on the location and the antibiotic being measured.

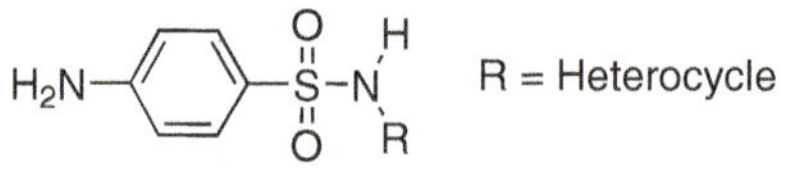

FIGURE 7.2 General formula of sulfonamides

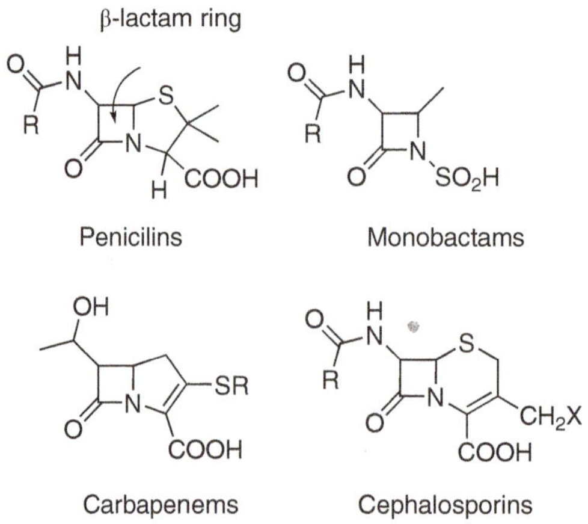

FIGURE 7.3 β-lactam antibiotics

Studies have shown that β-lactam antibiotics can have negative impacts on aquatic organisms, such as fish and amphibians by disrupting their growth and development. In addition, exposure to β-lactam antibiotics can lead, as with the other families of antibiotics, to the development of antibiotic-resistant bacteria, which can pose a threat to human health. β-lactams can be found in residual waters and can contribute to the development of antibiotic resistance.

3. **Tetracyclines:** Tetracycline molecules comprise a group of antibiotics formed by a linear fused polycyclic core with rings designated A, B, C and D, to which a variety of functional groups are linked to this structure derived from the tetracycline antibiotic (Figure 7.4).

    This class of antibiotics is commonly used in both human and veterinary medicine. Some commonly used tetracycline antibiotics are included in Table 7.1.

    They are often found in wastewater and they can persist in the environment for a long time. In domestic wastewater, detected concentrations of tetracyclines are low ($1\,\mu g/L$), whereas hospital wastewater has a relatively higher concentration, about $100\,\mu g/L$.

4. **Fluoroquinolones:** These antibiotics are commonly used to treat infections in humans. They work by inhibiting bacterial DNA synthesis, which ultimately prevents the bacteria from reproducing and spreading. Fluoroquinolones are effective against a wide range of bacterial infections, including respiratory, urinary, gastrointestinal and skin infections. They are also used in veterinary medicine. The generic structure of fluoroquinolones is described in Figure 7.5.

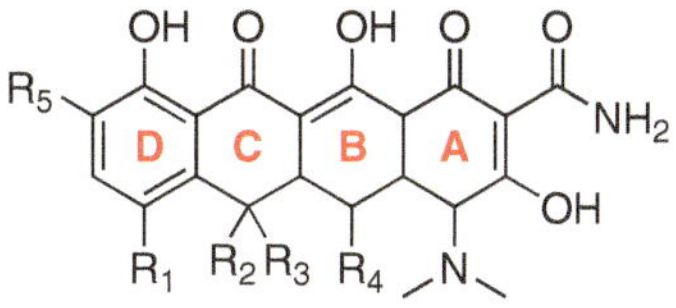

FIGURE 7.4   General structure of tetracycline

TABLE 7.1   **Tetracycline antibiotics**

Antibiotic	Structure	Uses
Tetracycline	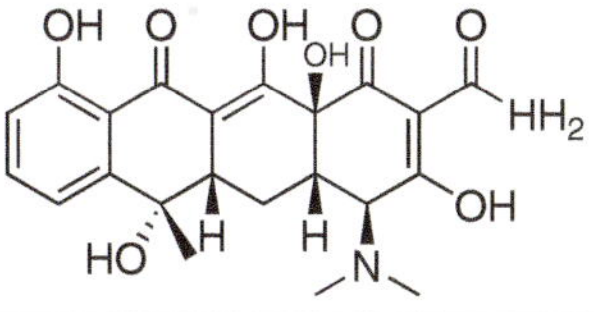	It is used to treat various bacterial infections, including respiratory tract infections, urinary tract infections, etc.
Doxycycline	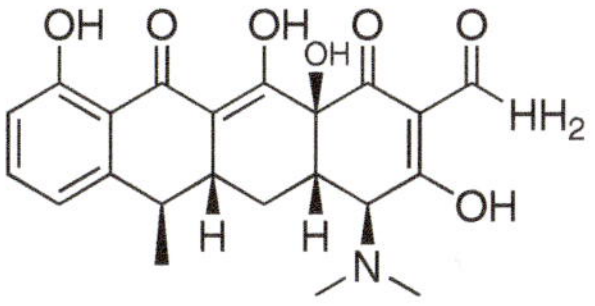	It is used to treat respiratory tract infections, urinary tract infections, sexually transmitted infections, etc.
Minocycline	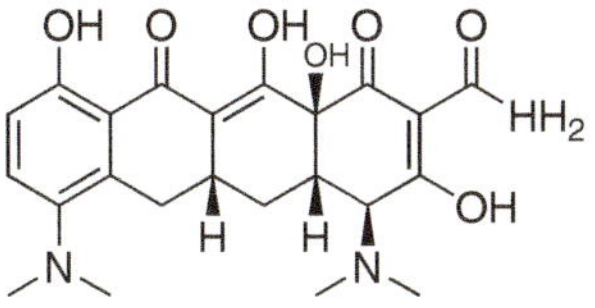	It is used to treat various infections, including respiratory tract infections, acne, urinary tract infections, etc.

TABLE 7.1 (Continued)

Antibiotic	Structure	Uses
Demeclocycline	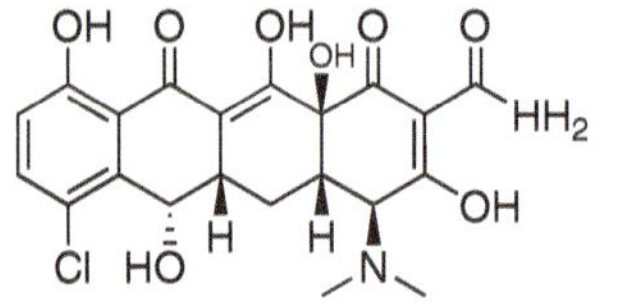	It is used to treat certain hormonal imbalances and conditions like SIADH (syndrome of inappropriate anti-diuretic hormone).
Oxytetracycline	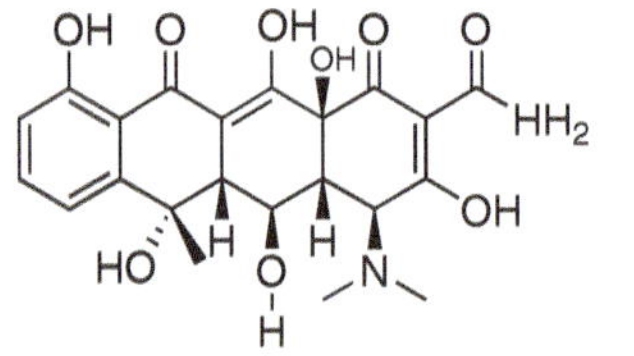	It is used topically to treat skin infections, eye infections and other localised infections.

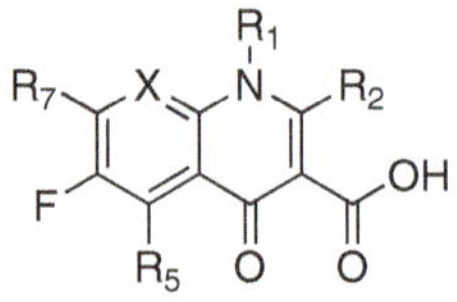

FIGURE 7.5 Fluoroquinolone generic structure

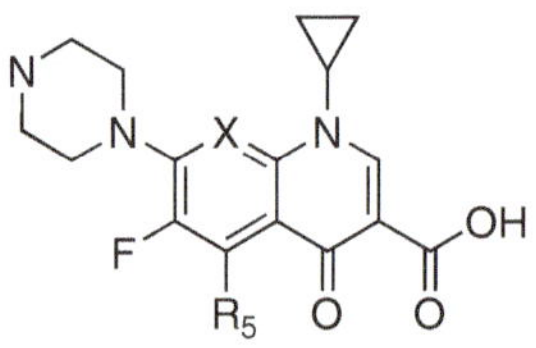

FIGURE 7.6 Ciprofloxacin

Ciprofloxacin (Figure 7.6) is one of the most used fluoroquinolones. It has a broad spectrum of activity against many bacteria, making it effective in treating a wide range of infections. Ciprofloxacin is widely prescribed for urinary tract infections, respiratory tract infections, gastrointestinal infections, skin and soft tissue infections and certain sexually transmitted infections. It is also frequently used as a prophylactic treatment for individuals at high risk of anthrax exposure.

Fluoroquinolones are often found in residual water. It has been described that concentrations of different fluoroquinolones range from 0.3 ng/L up to 31 µg/L.

5. **Macrolides:** These antibiotics are commonly used to treat respiratory and skin infections in humans and are also used in veterinary medicine. Macrolide antibiotics are a class of antibiotics that have a complex macrocyclic lactone ring structure. The general structure of macrolide antibiotics consists of a lactone ring with one or more sugars attached to it. The lactone ring is composed of 12- to 16-membered macrocyclic rings. The size of the ring and the number and position of the functional groups attached to it vary depending on the specific macrolide antibiotic (Figure 7.7).

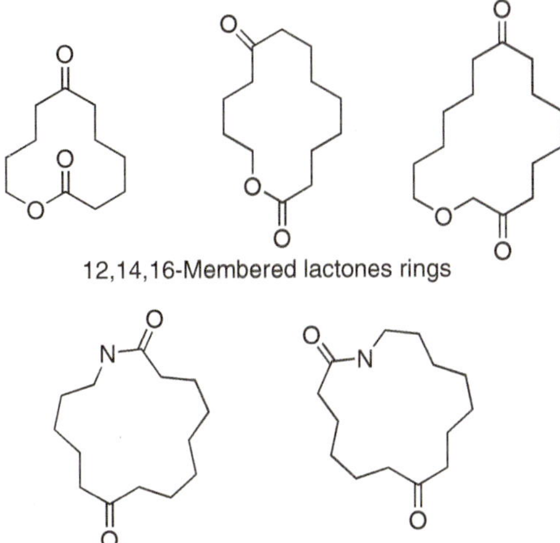

FIGURE 7.7   Core structure of macrolide antibiotics

Macrolides have been primarily used to treat respiratory tract infections such as pneumonia, bronchitis and sinusitis. They are also effective against other bacterial infections, including skin and soft tissue infections, as well as some sexually transmitted infections. Some commonly used macrolide antibiotics include Azithromycin, Clarithromycin and Erythromycin.

Due to their extensive use, macrolides have been detected in various environmental matrices, such as surface waters, sediments and soils. Studies have found that macrolide concentrations in wastewater can range from trace amounts to several micrograms per litre ($\mu g/L$) or even higher in certain cases. However, it's important to note that concentrations can fluctuate over time and be influenced by changes in drug consumption patterns and treatment practices.

7.5   Analgesic and Anti-Inflammatory Pharmaceuticals

Analgesics and anti-inflammatory drugs are pharmaceuticals used to relieve pain and inflammation in the body. Common analgesics include opioids such as oxycodone and hydrocodone, as well as non-steroidal anti-inflammatory drugs (NSAIDs) such as aspirin, paracetamol and ibuprofen or naproxen. A large fraction of pharmaceutical pollution in water is composed of anti-inflammatory and analgesic drugs, which are rapidly excreted in urine.

Some of the most used analgesic and anti-inflammatory pharmaceuticals include:

1. **Non-Steroidal Anti-Inflammatory Drugs (NSAIDs):** NSAIDs are a group of molecules that block the production of prostaglandins, which are responsible for causing pain and inflammation. Examples of NSAIDs include aspirin, paracetamol, ibuprofen and naproxen (Table 7.2).

TABLE 7.2 Analgesic and anti-inflammatory pharmaceuticals

Name	Structure
Aspirin: Pain reliever, reducing inflammation and fever, used for preventing blood clots, which can reduce the risk of heart attacks and strokes.	
Paracetamol: Used to alleviate mild to moderate pain associated with headaches, toothaches, muscle aches, menstrual cramps and the common cold.	
Ibuprofen: Used to relieve pain, reduce inflammation and lower fever by inhibiting the production of prostaglandins.	
Naproxen: Is used to relieve pain, inflammation and stiffness caused by conditions such as osteoarthritis, rheumatoid arthritis, gout, menstrual cramps and minor injuries.	
Prednisone: Is prescribed to treat a variety of conditions, including autoimmune disorders, allergic reactions, asthma and certain types of cancers. It works by suppressing the immune system's response to inflammation and reducing swelling, redness, itching and allergic reactions.	
Dexamethasone: Is used to treat a variety of conditions, primarily those related to inflammation and immune system dysregulation. It's often prescribed for conditions like allergic disorders, skin conditions, asthma, arthritis, certain types of cancer and autoimmune diseases.	
Celecoxib: Is used to relieve pain, inflammation and swelling caused by various conditions such as arthritis, acute pain, menstrual cramps and certain types of musculoskeletal disorders.	
Morphine: It is a potent opioid analgesic medication used to relieve severe pain. It is derived from the opium poppy plant and is one of the oldest known pain-relieving substances. Morphine works by binding to opioid receptors in the central nervous system, including those in the brain and spinal cord, thereby altering the perception of pain.	

TABLE 7.2   (Continued)

Name	Structure
Codeine: It is used to treat mild to moderate severe pain and to relieve coughing. It is derived from the opium poppy plant, like morphine and works by binding to opioid receptors in the brain, spinal cord and gastrointestinal tract, thereby altering the perception of pain and suppressing the cough reflex.	
Fentanyl is a powerful synthetic opioid medication primarily used to treat severe pain, particularly associated with surgery, cancer or other chronic conditions. It is one of the most potent opioids available for medical use, estimated to be approximately 50–100 times more potent than morphine.	

2. **Corticosteroids:** Corticosteroids are a group of molecules that are used to reduce inflammation. They work by inhibiting the production of cytokines and other inflammatory molecules. Examples of corticosteroids include prednisone and dexamethasone.

3. **COX-2 inhibitors:** COX-2 inhibitors are a group of molecules that specifically target the COX-2 enzyme, which is responsible for producing prostaglandins that cause inflammation. Examples of COX-2 inhibitors include celecoxib.

4. **Opioids:** Opioids are a group of molecules that work by binding to specific receptors in the brain and spinal cord to reduce pain. Examples of opioids include morphine, codeine and fentanyl.

Commonly used painkillers such as acetaminophen, ibuprofen and naproxen have been detected in wastewater at concentrations ranging from ng/L to µg/L. and they can persist for long periods of time. The presence of these drugs and their metabolites in wastewater can have potential implications for the environment, as they can affect aquatic organisms and ecosystems. For example, a study conducted on zebrafish found that exposure to paracetamol, one of the most consumed analgesics in the world, at environmentally relevant concentrations led to reduced hatching rates and impaired swimming behaviour.

Moreover, there is also growing concern over the potential for these compounds to accumulate in the food chain and impact human health. Although the levels of analgesics found in drinking water sources are generally low, long-term exposure to low levels of these compounds could have health implications for humans, particularly for vulnerable populations such as pregnant women and children.

Topics of Interest: Diclofenac Crisis in India

Diclofenac (Figure 7.8) is an NSAID that has been widely used in India for both human and veterinary purposes. In the 1990s, the drug was approved for veterinary use in India to treat livestock, particularly for the treatment of pain and fever in cattle.

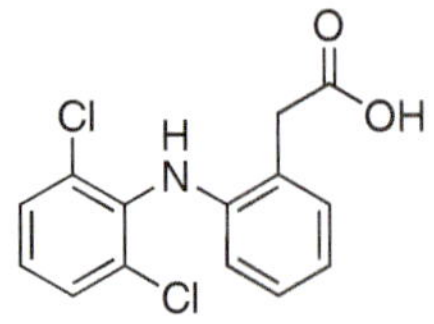

FIGURE 7.8 Diclofenac

The Indian vulture crisis, also known as the South Asian vulture crisis, was a major ecological disaster that occurred in the late 1990s and early 2000s. The crisis was caused by the massive use of diclofenac. It was discovered that the drug was responsible for the deaths of millions of vultures in India, Pakistan and Nepal. Vultures are essential for the ecosystem in South Asia, as they help dispose of animal carcasses and prevent the spread of disease.

The vultures were dying because they were feeding on the carcasses of cattle that had been treated with diclofenac. The drug caused renal failure in the vultures, leading to their rapid decline in numbers. The decline of vultures had a cascading effect on the ecosystem, as it led to an increase in the number of feral dogs and rats, which spread disease and caused other ecological problems.

The crisis was recognised in the early 2000s and the Indian government banned the use of diclofenac in veterinary medicine in 2006. The drug was replaced with an alternative, meloxicam, which is safer for vultures.

The crisis highlights the importance of considering the wider ecological impact of drugs, particularly those used in agriculture and veterinary medicine. It also emphasises the need for careful regulation and monitoring of drugs to ensure their safety for both humans and the environment.

However, diclofenac is still available for human use in India and it is a popular and widely used painkiller. There have been concerns about the safety of diclofenac for human use, particularly its link to cardiovascular events such as heart attacks and strokes. In 2013, the Indian government imposed restrictions on the sale of diclofenac, limiting its availability to prescription-only.

7.6 Anti-Depressants

Anti-depressants are drugs used to treat depression and other mood disorders. Some commonly prescribed anti-depressants include selective serotonin reuptake inhibitors (SSRIs), such as Prozac and Zoloft, as well as tricyclic anti-depressants such as Elavil (Table 7.3).

Anti-depressants are emerging pollutants due to the fact that a significant percentage of the anti-depressants consumed by humans are excreted from the body in their original form or as metabolites and then enter the sewage system through wastewater. Studies have shown that exposure to anti-depressants can have negative effects on aquatic organisms, such as altering their behaviour, growth and reproductive capabilities. For example, some studies have found that exposure to anti-depressants can disrupt the reproduction of fish, causing them to produce fewer eggs or change their sex. There are also concerns about the potential long-term effects of chronic exposure to these compounds on aquatic ecosystems, such as changes in the composition of microbial communities and the potential for the emergence of new antibiotic-resistant bacteria.

TABLE 7.3   Anti-depressants

Name	Chemical structure	Uses
Prozac		Treats depression, obsessive-compulsive disorder, bulimia nervosa and panic disorder
Zoloft		Treats depression, obsessive-compulsive disorder, panic disorder, social anxiety disorder and post-traumatic stress disorder
Elavil		Treats depression, nerve pain and certain types of headaches

7.7   Antipsychotics

These are drugs that are used to treat psychotic disorders such as schizophrenia and bipolar disorder. Examples of antipsychotic medications and their primary uses are given in Table 7.4.

They have also been identified as emerging pollutant antipsychotics that can enter the environment through various pathways, including wastewater effluent from pharmaceutical manufacturing facilities, hospital wastewater and residential wastewater.

TABLE 7.4   Examples of antipsychotic medications and their primary uses

Antipsychotic	Structure	Primary use
Risperidone		Schizophrenia, bipolar disorder and irritability associated with autism
Olanzapine		Schizophrenia, bipolar disorder and treatment-resistant depression

Studies on the adverse effects of antipsychotic drugs polluting the environment on the health of aquatic organisms are very scarce. For example, some antipsychotics can disrupt the growth and reproduction of fish, alter the behaviour of invertebrates and inhibit photosynthesis in algae, but at higher concentrations than usual in those found in different environmental matrices. In addition to their impact on the environment, antipsychotics can also pose a risk to human health. Workers in pharmaceutical manufacturing facilities and healthcare settings may be exposed to these compounds and people who consume contaminated water or seafood may also be at risk.

7.8 Anxiolytics

Anxiolytics, also known as anti-anxiety medications, are a class of drugs that are used to treat anxiety disorders. They work by reducing the symptoms of anxiety, such as nervousness, tension and fear. Anxiolytics can be classified into different groups based on their chemical structure, mode of action and potential for addiction.

Benzodiazepines are one of the most common types of anxiolytics. They work by enhancing the activity of a neurotransmitter called GABA, which has a calming effect on the brain. Examples of benzodiazepines include alprazolam (Xanax), diazepam (Valium) and lorazepam (Ativan) (Table 7.5). Benzodiazepines can be habit-forming and have a high potential for abuse, so they are usually only prescribed for short-term use.

Once anxiolytics enter the wastewater system, they are difficult to remove through conventional wastewater treatment processes, which can result in their presence in treated wastewater and potentially in surface water and groundwater.

TABLE 7.5   **Examples of anxiolytics**

Drug	Chemical structure	Uses
Alprazolam (Xanax)	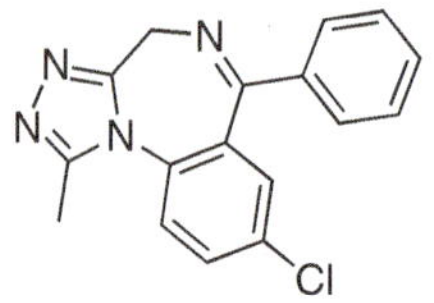	Anxiety disorders, panic disorders and anxiety are associated with depression.
Diazepam (Valium)	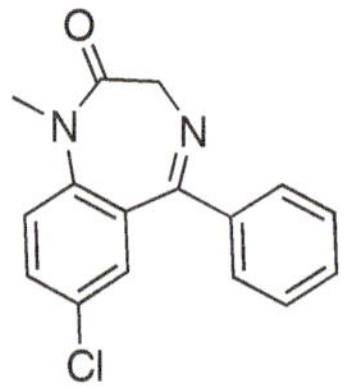	Anxiety disorders, alcohol withdrawal symptoms, muscle spasms and seizures.
Lorazepam (Ativan)	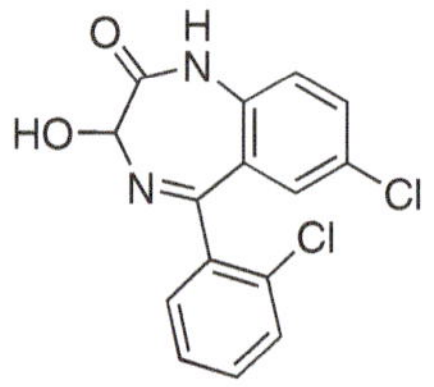	Anxiety disorders, insomnia and seizure disorders.

Studies have found detectable levels of anxiolytics in wastewater and receiving bodies of water all around the world. For example, a study conducted in Spain found trace levels of several anxiolytics, including alprazolam and lorazepam, in treated wastewater and in the receiving river. Another study conducted in the United States detected the presence of several anxiolytics, including diazepam and oxazepam, in surface water and groundwater.

The potential environmental impacts of anxiolytics in wastewater are not yet fully understood, but research suggests that they may have adverse effects on aquatic organisms, such as disrupting their behaviour and physiology. Additionally, there is concern that prolonged exposure to low levels of anxiolytics in the environment may contribute to the development of antibiotic resistance in bacteria.

Efforts are being made to develop more effective methods for removing anxiolytics and other pharmaceuticals from wastewater, such as advanced oxidation processes and membrane filtration. However, reducing the occurrence of anxiolytics in wastewater in the first place through proper disposal of unused medications and other measures may be the most effective approach to minimising their environmental impact.

Several studies have detected the presence of anxiolytics in wastewater and receiving bodies of water. For instance, a study conducted in Spain analysed treated wastewater and the receiving river and found trace levels of anxiolytics, including alprazolam and lorazepam. Similarly, another study conducted in the United States identified the presence of anxiolytics, such as diazepam and oxazepam, in surface water and groundwater. These findings indicate that anxiolytics can enter the environment through wastewater and may have the potential to accumulate in aquatic ecosystems.

7.9 Anti-Hypertensives

These are drugs that are used to lower high blood pressure. Some commonly prescribed anti-hypertensives include beta-blockers such as propranolol, atenolol and metoprolol; angiotensin-converting enzyme (ACE) inhibitors such as lisinopril and calcium channel blockers such as amlodipine (Table 7.6).

One common anti-hypertensive drug that has been detected in the environment is amlodipine. It has been found in surface water, sediment and soil and has been shown to have adverse effects on fish and other aquatic organisms.

TABLE 7.6 **Examples of anti-hypertensives**

Medication	Structure	Uses
Propranolol		Beta-blocker: Used to treat hypertension (high blood pressure). Anti-arrhythmic: Helps control irregular heart rhythms. Tremor control: Used for essential tremors and tremors associated with certain conditions.

TABLE 7.6 (Continued)

Medication	Structure	Uses
Atenolol	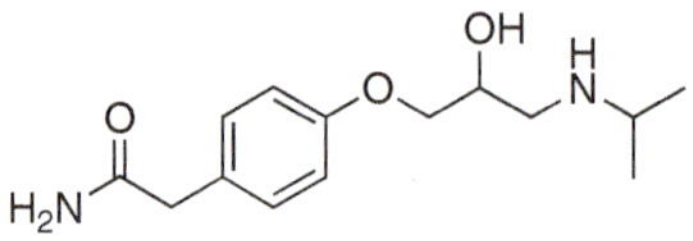 (chemical structure)	Beta-blockers are used. It helps treat hypertension. Angina: It helps reduce the frequency and severity of angina (chest pain). Some cardiovascular conditions. Beta-blockers: Used to treat hypertension.
Metoprolol	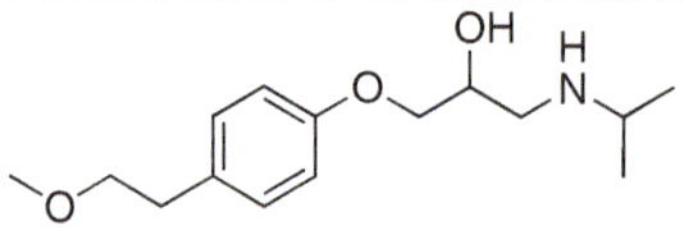 (chemical structure)	Angina: It helps reduce the frequency and severity of angina. Heart failure: Can improve survival and reduce the risk of hospitalisation.
Lisinopril	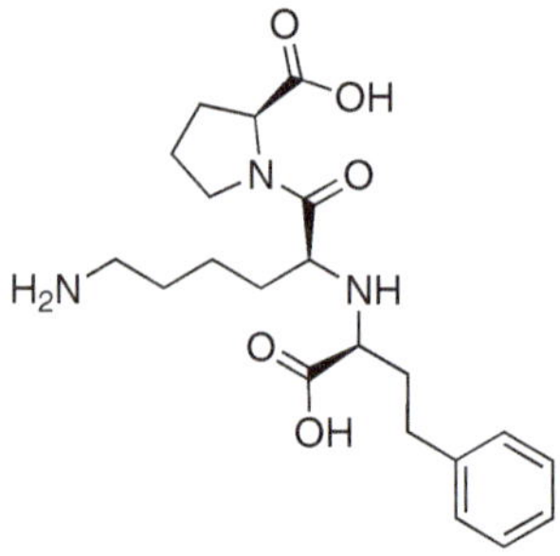 (chemical structure)	ACE inhibitors are used to treat hypertension. Heart failure: It helps improve the heart's ability to pump blood. Kidney protection: Can slow the progression of kidney disease in some cases.
Amlodipine	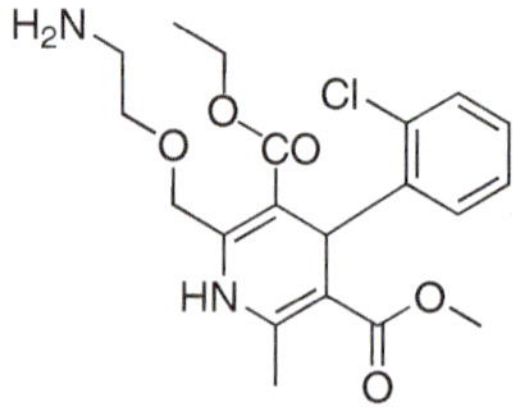 (chemical structure)	Calcium channel blockers are used to treat hypertension. Angina: It helps reduce the frequency and severity of angina.

Other anti-hypertensive drugs that have been detected in the environment include beta-blockers such as propranolol, atenolol and metoprolol.

7.10 Imagine Testing Substances

1. **Iodine-based contrast agents (IBCAs):** Imaging tests such as CT and MRI scans of the kidneys, ureters and bladder during an IVP allow for the detection of abnormalities or obstructions in the urinary tract. Arteriography and venography involve the injection of IBCA into arteries or veins to study their anatomy and diagnose vascular conditions. Some examples of IBCAs are commonly used in medical imaging (Figure 7.9).

   a. **Iohexol:** This is a water-soluble IBCA used in CT scans and angiography to enhance blood vessels and highlight specific structures for better visualisation.

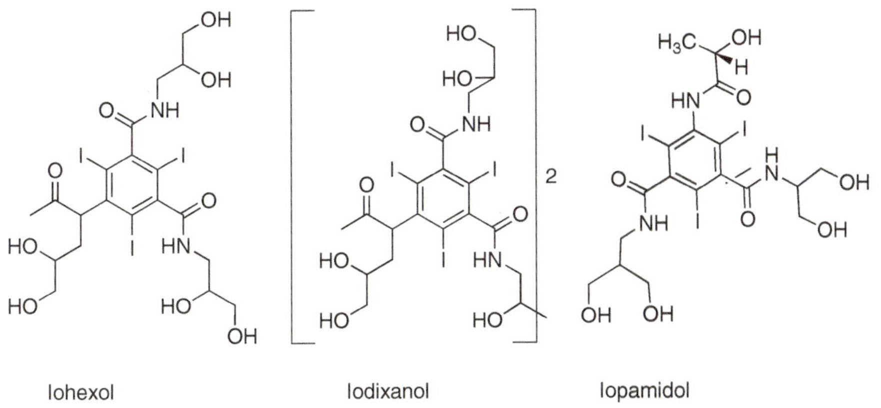

Iohexol   Iodixanol   Iopamidol

FIGURE 7.9 Examples of iodine-based contrast agents (IBCA)

   b. Iodixanol: This is another water-soluble IBCA used in CT scans and angiography. It is commonly used due to its lower toxicity compared to older contrast agents.

   c. Iopamidol: This is a non-ionic IBCA that is used in various radiographic procedures, including CT scans and intravenous urography.

They have been designed to resist biodegradation for the period necessary to perform the test. After administration, IBCAs are eliminated from the body primarily through the kidneys and excreted in urine or faeces and they may persist in wastewater treatment systems. Traditional wastewater treatment processes may not effectively remove these compounds.

As a result, IBCAs can potentially enter the environment through the effluent discharged from wastewater treatment plants, and they can often remain unchanged after entering the environment or be transformed into many different by-products in complex physical, chemical and biological processes.

In certain areas, large amounts of IBCA and their by-products can be found in natural waters, groundwater, drinking water (up to 100 µg/L) and even soil, where they can pose a potential threat to the inhabitants of these environments. The release of IBCAs into the environment, even in trace amounts, can have potential ecological impacts. Iodine is an essential element and increased levels in the environment can disturb natural balances and potentially affect aquatic organisms and ecosystems.

2. **Barium-based contrast agents:** These are used in gastrointestinal imaging, such as barium swallow and barium enema procedures. Barium is opaque to X-rays, meaning that it blocks X-rays from passing through it. When the contrast agent is introduced into the body, it outlines the internal structures and organs, making them more visible in the resulting images. Barium sulfate ($BaSO_4$) is the primary active ingredient in the contrast agent. When excreted, $BaSO_4$ does not dissolve well in water, and it can persist for a long time in the environment. $BaSO_4$ and barium carbonate ($BaCO_3$) are the barium compounds most commonly found in soil and water. If $BaSO_4$ and $BaCO_3$ are released onto land, they will combine with particles of soil. The amount of barium found in food and water is usually not high enough to be a health concern. However, information is still being collected to determine if long-term exposure to low levels of barium causes any health problems.

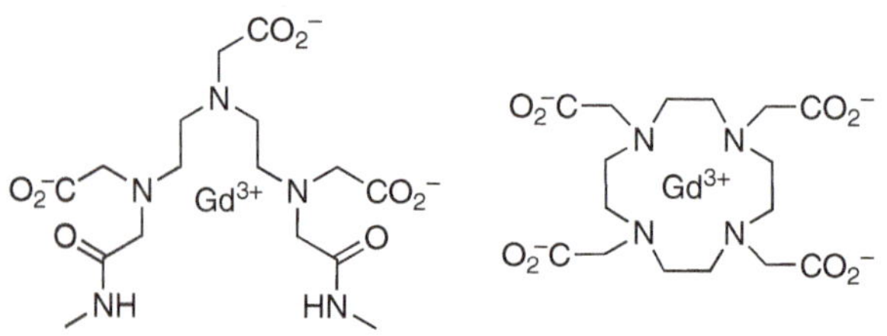

FIGURE 7.10   Gadolinium-based contrast agents (GBCAs)

3. **Gadolinium-based contrast agents (GBCAs):** These are used in medical imaging to enhance the quality of magnetic resonance imaging (MRI) scans. These contrast agents contain gadolinium, a rare earth metal that has magnetic properties and can be detected by MRI machines. Generally, GBCAs consist of a gadolinium salt and a chelating ligand that binds to the metal ion to stabilise it and reduce toxicity (Figure 7.10). The chelating ligand is usually a linear or cyclic polyamino carboxylate compound.

When injected into a patient's bloodstream, GBCAs enhance the contrast between different tissues and organs in the body, making it easier for doctors to identify abnormalities and diagnose medical conditions. GBCAs are often used to detect tumours, inflammation and blood vessel abnormalities. However, some studies have suggested that gadolinium may accumulate in the brain and other organs of patients who receive repeated doses of GBCAs, particularly in those with impaired kidney function. This can lead to a condition called gadolinium deposition disease, which can cause symptoms including pain, skin changes and cognitive impairment. As a result, the use of GBCAs has been the subject of increased scrutiny in recent years and some regulatory agencies have issued warnings and restrictions on their use. Their occurrence was confirmed in surface waters (up to 1100 ng/L), sediments (up to 90.5 µg/g) and living organisms. Based on a literature review, there is a need to investigate the contamination of different ecosystems and ascertain the environmental fate of Gd.

7.11   Illegal Drugs: Cocaine, Heroin, Methamphetamine, and Cannabis

Illegal drugs are substances that are prohibited by law for use, possession, sale and distribution. These drugs are considered harmful and pose a threat to public health and safety. The use of illegal drugs can lead to addiction, physical and mental health problems, social and economic problems and, in extreme cases, death.

Examples of illegal drugs include heroin, cocaine, methamphetamine, LSD, ecstasy and marijuana (in some countries).

Illegal drugs, such as cocaine, heroin and methamphetamine, can be considered emerging pollutants due to their impact on the environment. When these drugs are used, metabolised and excreted by humans, they can enter the environment through various pathways, including wastewater treatment plants, surface water runoff and soil contamination.

The presence of illegal drugs in the environment can have negative impacts on aquatic life and soil organisms. For example, cocaine has been found to cause behavioural changes in fish, while methamphetamine has been found to have toxic effects on aquatic invertebrates. Additionally, the presence of these drugs in the environment can also affect human health, particularly for those who live or work in areas where drug use and production are prevalent.

Efforts are being made to reduce the environmental impact of illegal drugs. For example, some wastewater treatment plants are being upgraded to remove pharmaceuticals and other emerging pollutants, including illegal drugs, from the wastewater before it is discharged into waterways. Additionally, there are campaigns aimed at reducing drug use and production to prevent the release of these substances into the environment.

Cocaine

Cocaine is a stimulant drug that is rapidly metabolised by the body.

When cocaine is consumed, it is broken down into various metabolites, which are then excreted from the body through urine or faeces. The most commonly tested cocaine metabolites in drug testing are (Figure 7.11):

1. **Benzoylecgonine (BE):** This is the primary metabolite of cocaine and the most commonly used marker for cocaine use in drug testing. BE has a longer half-life than cocaine, which means it can be detected in urine for a longer period of time.
2. **Ecgonine methyl ester (EME):** This is a minor metabolite of cocaine and is less commonly used in drug testing. EME has a shorter half-life than BE and is usually detectable in urine for up to 48 hours after cocaine use.
3. **Cocaethylene:** This is a metabolite that is produced when cocaine is used in combination with alcohol. Cocaethylene has a longer half-life than cocaine or BE and can be detected in urine for up to five days after use.

Cocaine is rapidly metabolised in the body to BE, which is excreted in urine, making it a reliable marker for cocaine use. BE is stable and can be detected in wastewater for up to several days, providing a time-weighted average of drug use in a community.

It's important to note that the detection time of cocaine metabolites in drug testing depends on various factors, including the dose and frequency of cocaine use, the individual's metabolism and the type of drug test being used.

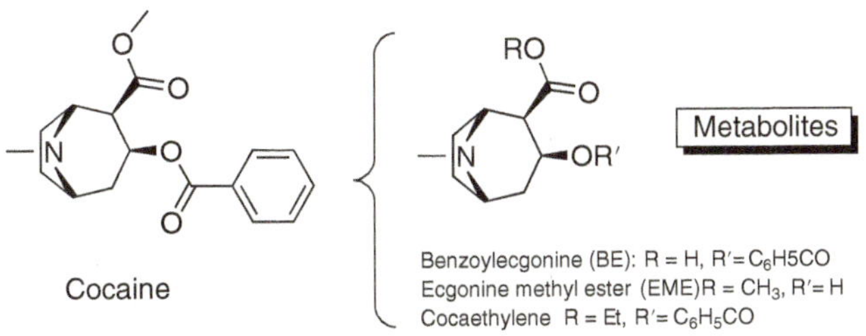

FIGURE 7.11 Cocaine and its metabolites

The presence of cocaine and its metabolites in wastewaters reflects the level of drug use in the population. Wastewater-based epidemiology (WBE) is a relatively new approach that can provide a snapshot of drug use in a community by analysing wastewater for drug metabolites.

A study conducted in Europe reported that the mean concentration of cocaine in wastewater was around 13.5 mg/1,000 people per day. However, the concentration can range from 1 to 100 mg/1,000 people per day, depending on the location and time of day.

Several studies have shown that wastewater analysis can be used to estimate the prevalence and patterns of drug use in a population. For example, a study conducted in Italy found that cocaine use was highest in urban areas and on weekends. Another study in the Netherlands found that cocaine use increased during festivals and other large events.

Wastewater analysis can also be used to monitor changes in drug use over time and evaluate the effectiveness of drug control measures. However, there are challenges in interpreting wastewater data, such as the difficulty in distinguishing between drug use and environmental contamination and the variability in drug excretion rates among individuals.

In recent years, there has been growing concern about cocaine's impact as an emerging pollutant and its effects on wildlife. Once in the environment, cocaine can have a range of ecological and health impacts. For example, studies have shown that cocaine can harm fish and other aquatic organisms by altering their behaviour, metabolism and reproductive systems. Cocaine has also been found to accumulate in the tissues of fish, birds and mammals, potentially leading to chronic exposure and health problems.

Overall, the emergence of cocaine as a pollutant highlights the need for greater awareness and action to address the environmental and health impacts of illicit drug use and production.

Heroin

Heroin is a highly addictive opioid drug that is synthesised from morphine, which is derived from the opium poppy (Figure 7.12).

It is commonly sold as a white or brownish powder or as a black, sticky substance known as black tar heroin.

When someone uses heroin, it binds to the opioid receptors in their brain, which can cause a rush of pleasure and euphoria as well as a feeling of warmth and relaxation. However, heroin use also carries many risks, including addiction, overdose and the transmission of infectious diseases such as HIV and hepatitis. While the use and abuse of heroin have long been recognised as serious public health issues, its emergence as a pollutant is a relatively new phenomenon.

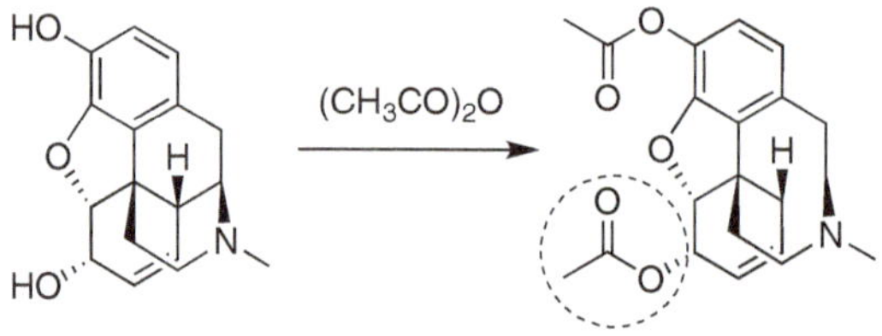

FIGURE 7.12 Heroin synthesis

The main way that heroin enters the environment is through its excretion by users. Heroin and its metabolites can be detected in wastewater and surface waters and recent studies have shown that they are present in high concentrations in some urban and suburban areas.

The detection of heroin in wastewater can provide valuable information about drug use in a particular area. WBE is a technique that uses the analysis of wastewater to estimate drug use within a population. Heroin and its metabolites can be detected in wastewater, which can then be used to estimate the total consumption of heroin in a community.

This is of concern because heroin is a potent and persistent drug that can have harmful effects on aquatic life and ecosystems.

Methamphetamine

Methamphetamine, also known as crystal meth, is a highly addictive and potent central nervous system stimulant (Figure 7.13). It is a synthetic drug that stimulates the release of dopamine, a neurotransmitter in the brain that plays a role in pleasure, reward and motivation. Methamphetamine is typically smoked, snorted or injected and its effects can last for several hours.

Methamphetamine is actually a chiral compound. The (R)-enantiomer of methamphetamine is also known as levomethamphetamine, while the (S)-enantiomer is known as dextromethamphetamine. It is important to note that the two enantiomers of methamphetamine can have different effects on the body. While both forms have stimulant properties, dextromethamphetamine is more potent and has a stronger effect on the central nervous system. Additionally, the two forms can have different pharmacokinetic properties, meaning they can be metabolised and eliminated from the body at different rates.

The term methamphetamine drug properly refers to racemic and/or dextromethamphetamine in their pure forms.

Both racemic and dextromethamphetamine are illicitly trafficked and sold.

In addition to its harmful effects on human health, methamphetamine can also pose a threat to the environment as an emerging pollutant. Methamphetamine and its metabolites can be excreted by users into wastewater, which can eventually end up in the environment through sewage treatment plants or septic systems.

A study published in 2019 found that the concentration of methamphetamine in wastewater samples collected from New York City ranged from 2.1 to 62.5 ng/L, with higher concentrations observed during the summer months.

It's worth noting that these concentrations are typically very low, measured in nanograms per litre and are not an indication of the number of people using the drug. However, they can provide insight into drug use trends and patterns in a community over time.

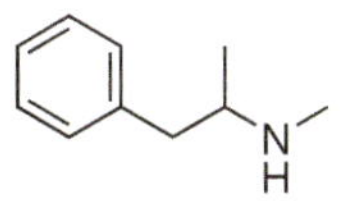

FIGURE 7.13   Methamphetamine

Studies have shown that methamphetamine and its metabolites can persist in the environment for extended periods of time and may accumulate in soils and water sources, posing a potential threat to aquatic and terrestrial ecosystems. Methamphetamine has also been detected in surface water, groundwater and soil in areas where it is commonly abused.

Cannabis

Cannabis is a psychoactive drug that comes from the Cannabis sativa plant. It contains compounds called cannabinoids, the most well-known of which are δ-9-tetrahydrocannabinol (THC) and CBD, which are the most well-known cannabinoids in cannabis. THC is the psychoactive compound responsible for the 'high' experienced by users, while CBD is non-psychoactive but has potential therapeutic effects. Both THC and CBD undergo metabolism in the liver (Figure 7.14). THC is the component responsible for the drug's psychoactive effects, including altered perception, mood and cognition.

Cannabis can be consumed in various ways, including smoking, vaporising and ingesting it in food or drink for recreational purposes, but it also has some medicinal uses, such as pain relief and appetite stimulation.

The use of this drug has been legalised in some countries for both medical and recreational purposes. It is important to note that the use of cannabis can have potential negative health effects, especially when used regularly or in high doses. These effects can include impaired memory and concentration, lung damage from smoking and an increased risk of mental health issues such as anxiety and depression. THC and its metabolites have been detected in wastewater, surface water and drinking water. In particular, 11-nor-9-carboxy-Δ^9-tetrahydrocannabinol (THC–COOH) (Figure 7.15).

As the body metabolises cannabinoids, the breakdown products, such as THC–COOH, are eventually excreted through urine. This is why drug tests often look for THC metabolites in urine as evidence of cannabis use. Cannabis compounds and their metabolites were not commonly classified as 'emerging pollutants' in the same

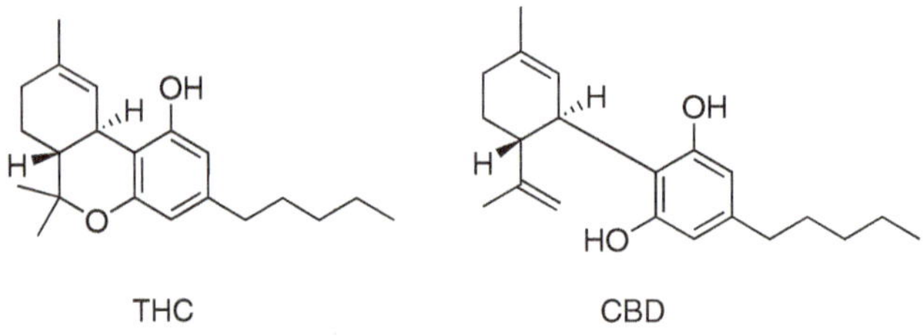

FIGURE 7.14 Cannabinoids

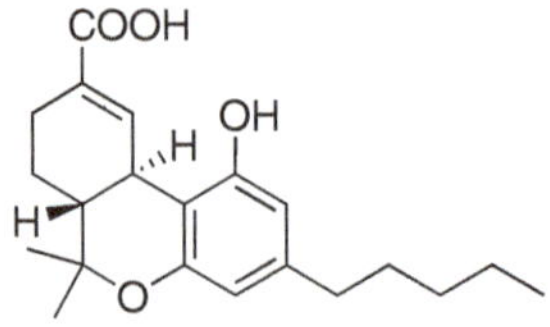

FIGURE 7.15 11-nor-9-carboxy-Δ^9-tetrahydrocannabinol

sense as other chemical contaminants like pharmaceuticals, PCPs or industrial chemicals, but there is a growing concern due to their presence in urban wastewater and potential adverse effects on ecosystems in the long term.

Topics of Interest: Wastewater-Based Epidemiology

WBE is a scientific approach that involves analysing sewage or wastewater to identify the presence of specific markers or biomolecules, such as viruses, bacteria and drugs, that can provide information about the health and habits of a population. This approach can provide early warning of disease outbreaks and help identify populations that are at risk.

WBE has gained attention during the COVID-19 pandemic as a tool for monitoring the prevalence of the virus in communities. By analysing wastewater for the presence of SARS-CoV-2 RNA, researchers can estimate the number of people infected with the virus in a particular area. This can help public health officials identify hotspots and respond quickly to prevent the spread of the virus.

WBE can also be used to monitor the use of drugs in a population, as traces of drugs and their metabolites can be detected in wastewater. This information can help public health officials understand patterns of drug use, identify emerging drug trends, and allocate resources for prevention and treatment.

Wastewater chemical analysis has been increasingly used as a tool to estimate drug consumption within a population, both legal and illegal, by analysing the concentration of certain drugs in wastewater.

Pharmaceuticals such as painkillers, anti-depressants and antibiotics have been found in wastewater and their concentration levels can provide insight into the overall use of these drugs within the population. Similarly, illegal drugs such as cocaine, heroin and marijuana can also be detected in wastewater, providing an indication of their prevalence within the community.

The study of the concentration of illegal drugs in wastewater is an emerging field that provides important insights into drug use patterns and trends in a given population.

Samples are collected from sewage treatment plants, which can reveal information about drug use in a particular community with varying results depending on the location and time of sampling. Some general conclusions that have been drawn from these studies are as follows:

1. WBE can provide accurate estimates of drug use patterns and trends in a population and may be more reliable than traditional self-reported surveys.

2. The concentration of illegal drugs in wastewater varies widely depending on the location and time of sampling. Factors such as population density, drug use habits and wastewater treatment practices can all affect the concentration of drugs in wastewater.

3. The concentration of drugs in wastewater tends to be higher on weekends and during major events, such as festivals and holidays, when drug use is more likely to occur.

4. The use of cocaine and amphetamines is widespread in many parts of the world, while the use of opioids is more localised and varies by region.

5. The United States and Canada have been particularly affected by the opioid epidemic in recent years, with high rates of prescription opioid and heroin use.

6. Wastewater analysis can also provide information on emerging drug trends, such as the use of new psychoactive substances (NPS) or designer drugs.

It is important to note that wastewater analysis can only provide an estimate of drug consumption and cannot provide specific information on individual drug use. However, the data can be useful for public health officials and law enforcement agencies to identify trends and patterns of drug consumption within a population and to develop appropriate interventions and policies.

7.12 Life-Style Substances: Caffeine and Nicotine

Caffeine

Caffeine (Figure 7.16) is a stimulant that is commonly found in coffee, tea and other beverages. Between 2016/2017 and 2019/2020, Europe, Asia & Oceania and North America saw increases of 1.8%, 1.9% and 2.1%, respectively, in their coffee consumption, according to data from the International Coffee Organisation (ICO) (http://www.ico.org/).

Caffeine has been identified as an emerging contaminant (EC) of concern due to its widespread occurrence in the aquatic environment and potential to be biologically active, pointing out that caffeine is more diluted in marine waters, resulting in lower concentrations when compared to freshwater and wastewater. Some reports indicate that long-term exposure to contaminated environments has led to the bioaccumulation of caffeine in the tissues of various coastal and marine organisms, including macroalgae, coral reefs, bivalves and fish.

Despite the longstanding belief that caffeine poses minimal risk to aquatic ecosystems, the existing data suggest that concerning levels of caffeine detected in both abiotic and biotic samples necessitate that effective measures be taken to eliminate or significantly reduce the presence of caffeine and its derivatives in water systems.

Nicotine

When tobacco products are used, nicotine (Figure 7.17) is released into the environment in various ways, such as through cigarette smoke and discarded cigarette butts. Cigarette butts are one of the most common forms of litter found worldwide and are often not properly disposed of, leading to nicotine and other harmful chemicals leaching into soil and water bodies.

Though nicotine itself is not considered an 'emerging' pollutant, the broader issue of tobacco-related pollution and its impact on the environment, particularly in terms of litter and water pollution, continues to be a concern. Cigarette litter can harm wildlife, and the chemicals from tobacco products can contaminate water sources, affecting aquatic ecosystems.

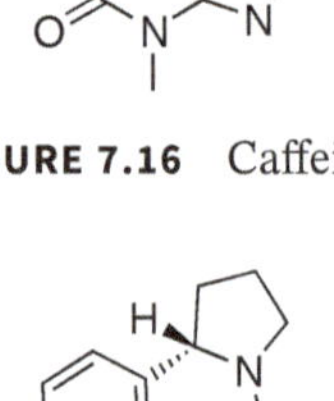

FIGURE 7.16 Caffeine

FIGURE 7.17 Nicotine

7.13 Personal Care Products (PCPs): Parabens, Phthalates, Formaldehyde, and Triclosan

PCPs are products that people use for personal hygiene, health and beautification purposes. They include a wide range of items such as soaps, shampoos, deodorants, toothpaste, makeup, lotions and perfumes.

There are some concerns about certain components of PCPs that will be treated in this section, especially those with biocidal properties used as preservatives, but it is important to note that not all PPCPs contain these chemicals and that many manufacturers have started to remove or reduce the use of potentially harmful ingredients in their products.

PCPs are used by millions of people worldwide and are an important part of daily life. However, they can also have negative impacts on the environment and human health. PCPs can enter the environment through various routes. Once in the environment, some PCP components can persist for a long time and even accumulate in the food chain.

These products contain a variety of chemicals, some of which may have potential health and environmental risks. Here are some common chemicals found in PCPs.

Parabens

Parabens are a type of preservative commonly used in cosmetic and personal care products to prevent the growth of bacteria, fungi and other microbes since the 1930s. They are a group of chemical compounds derived from p-hydroxybenzoic acid (Figure 7.18).

Table 7.7 summarises examples of parabens, their uses and the potential health concerns about them.

Parabens, now recognised as ECs, are a significant global concern due to their continuous discharge from human activities, their ability to persist in the environment and the ecological threats they pose. Their physicochemical properties contribute to their environmental persistence and their possible harmful effects on organisms.

Some studies have suggested that parabens may have oestrogenic effects, meaning they can mimic the hormone oestrogen in the body and potentially disrupt the endocrine system. However, the evidence for this is still inconclusive and regulatory agencies such as the US Food and Drug Administration (FDA) consider parabens to be safe for use in cosmetics at low levels.

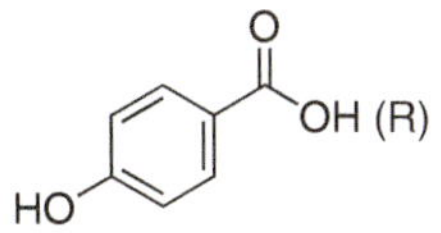

FIGURE 7.18   General structure of parabens

TABLE 7.7 Examples of parabens, uses and concerns

Paraben	R	Common uses	Potential health concerns
Benzylparaben	$C_6H_5\text{-}CH_2\text{-}$	Cosmetics and personal care products	Skin irritation
Butylparaben	$CH_3CH_2CH_2CH_2\text{-}$	Cosmetics, personal care products and food	Endocrine disruption, skin irritation
Ethylparaben	$CH_3CH_2\text{-}$	Cosmetics, personal care products and food	Endocrine disruption, skin irritation
Isobutylparaben	$(CH_3)_2CH_2CH_2\text{-}$	Cosmetics, personal care products and food	Endocrine disruption, skin irritation
Methylparaben	$CH_3\text{-}$	Cosmetics, personal care products and food	Endocrine disruption, skin irritation
Propylparaben	$CH_3CH_2CH_2\text{-}$	Cosmetics, personal care products and recognised food	Endocrine disruption, skin irritation

Nevertheless, many consumers choose to avoid products containing parabens due to these concerns, and there are many products available in the market that are labelled as 'paraben-free'.

Parabens can also have an impact on the environment. Products containing parabens can enter the environment through wastewater, and they may not be effectively removed by conventional wastewater treatment methods. This can lead to the accumulation of parabens in rivers, lakes and other bodies of water, potentially causing harm to aquatic organisms and ecosystems. Additionally, some studies have suggested that parabens can react with chlorine in water to form toxic byproducts known as chlorinated parabens. These byproducts may have harmful effects on aquatic life and the environment. As a result of these concerns, some countries have taken steps to restrict the use of certain types of parabens in personal care products. For example, in the European Union, five types of parabens have been banned from use in cosmetic products due to their potential environmental and health risks. These are isopropylparaben, isobutylparaben, phenyl paraben, benzylparaben and pentylparaben.

The ban came into effect in July 2015 and was introduced as part of the EU's ongoing efforts to regulate the use of chemicals in consumer products. The European Commission, which is responsible for implementing EU legislation, cited concerns that these parabens could disrupt the body's hormonal balance and potentially cause cancer. It's worth noting that other types of parabens, such as methylparaben, ethylparaben and propylparaben, are still allowed for use in cosmetic products in the EU, but their use is regulated and restricted to specific concentrations.

The decision to ban these parabens in the EU was based on scientific evidence and reflects a precautionary approach to protecting public health and the environment. It highlights the importance of ongoing research into the safety of chemicals used in consumer products and the need for robust regulatory frameworks to ensure their responsible use.

While the evidence for the environmental impact of parabens is still emerging, it is clear that they have the potential to cause harm to the environment. As such, it is important to use personal care products containing parabens responsibly and to dispose of them properly to minimise their impact on the environment.

Phthalates in PCPs

Phthalates (Figure 7.19) are a class of chemicals derived from phthalic acid.

Commonly used in a variety of consumer products, including personal care products, such as lotions, shampoos and perfumes (Table 7.8).

They are primarily used as plasticisers, which are added to materials to make them more flexible and durable. However, phthalates have been linked to a range of potential health risks, including endocrine disruption, developmental and reproductive toxicity and cancer. In recent years, there has been growing concern about the use of phthalates in personal care products, particularly among consumers who are seeking safer, more natural alternatives. Some companies have begun to

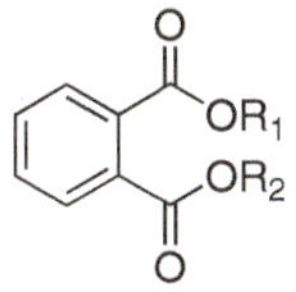

FIGURE 7.19 General structure of phthalates

TABLE 7.8 **Examples of phthalates and their uses**

Phthalate	R_1, R_2	Common uses
Diethyl phthalate (DEP)	CH_3CH_2- CH_3CH_2-	Fragrances, cosmetics, personal care products and solvents
Dibutyl phthalate (DBP)	$CH_3CH_2CH_2CH_2$- $CH_3CH_2CH_2CH_2$-	Nail polish, adhesives, sealants and plastics
Diisobutyl phthalate	$(CH_3)_2CH_2CH_2$- $(CH_3)_2CH_2CH_2$-	Adhesives, sealants and plastics
Butyl benzyl phthalate	$CH_3CH_2CH_2CH_2$- C_6H_5-CH_2-	Vinyl flooring, automotive parts, adhesives and sealants
Di(2-ethylhexyl) phthalate (DEHP)	$CH_3(CH_2)_4CH_2(CH_3CH_2)CH_2$-	PVC plastics, medical tubing, wire and cable coatings
Diisononyl phthalate (DINP)	$(CH_3)_2(CH_2)_6CH_2$-	Vinyl flooring, wire and cable coatings, automotive parts
Diisodecyl phthalate (DIDP)	$(CH_3)_2(CH_2)_7CH_2$- $(CH_3)_2(CH_2)_7CH_2$-	Vinyl flooring, wire and cable coatings, automotive parts
Di-*n*-octyl phthalate (DnOP)	$CH_3(CH_2)_6CH_2$- $CH_3(CH_2)_6CH_2$-	PVC plastics, automotive parts, wire and cable coatings

voluntarily remove phthalates from their products, while others have been required to do so by regulations such as the EU's REACH regulations and the US FDA's ban on certain phthalates in children's toys. Consumers can take steps to avoid phthalates in personal care products by reading ingredient labels and choosing products that are labelled as phthalate-free or that contain only natural ingredients. Some common phthalates to look out for include dibutyl phthalate (DBP), diethyl phthalate (DEP) and dimethyl phthalate (DMP).

Formaldehyde

Formaldehyde (Figure 7.20) is a colourless gas with a strong, pungent odour.

Exposure to formaldehyde can cause a variety of health problems, including skin irritation, respiratory problems and even cancer. Despite this fact, it is a common ingredient in many personal care products, particularly hair straightening treatments, nail polish and some cosmetic preservatives. Formaldehyde is used in these products because of its ability to prevent the growth of bacteria and other microorganisms that can spoil or contaminate them.

The use of formaldehyde in personal care products is regulated by government agencies such as the US FDA and the European Union's Cosmetics Regulation. In the United States, for example, the FDA limits the concentration of formaldehyde in personal care products to no more than 0.2%.

Consumers concerned about the use of formaldehyde in personal care products can look for products that are labelled 'formaldehyde-free' or that do not contain formaldehyde-releasing preservatives, such as quaternium-15, DMDM hydantoin, imidazolidinyl urea and diazolidinyl urea. They can also check the ingredient list on product labels to see if formaldehyde or formaldehyde-releasing preservatives are listed.

Triclosan

Triclosan is a synthetic anti-microbial agent that has been used in a wide range of consumer products (Figure 7.21), including soaps, toothpaste and other personal care items. It is often added to these products to help prevent bacterial growth and reduce odours. Triclosan has been used in these products for several decades, but concerns have been raised about its safety and potential impacts on human health and the environment.

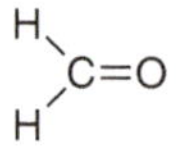

FIGURE 7.20   Formaldehyde

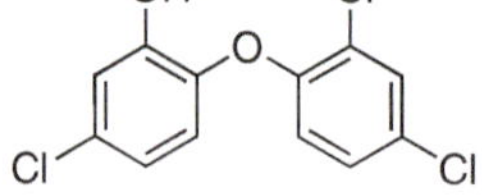

FIGURE 7.21   Triclosan

There is some evidence to suggest that triclosan may have negative effects on human health. For example, studies have found that triclosan can disrupt hormone function in animals, and may have similar effects in humans. It has also been shown to promote the growth of antibiotic-resistant bacteria.

Triclosan has also been detected in the environment, including in waterways and in the tissues of aquatic animals. Some studies have suggested that it may be toxic to certain types of aquatic organisms, although the long-term environmental effects of triclosan are still not fully understood.

In recent years, there has been a growing trend among manufacturers to remove triclosan from their products, and some countries have banned or restricted its use. However, triclosan is still used in some products, and consumers should be aware of its potential risks and consult with healthcare professionals if they have concerns about exposure.

7.14 Sunscreens

Sunscreens, also known as sun blockers or sun creams, are lotions, sprays, gels or other topical products that absorb or reflect some of the sun's ultraviolet (UV) radiation and help protect the skin from sunburn and other damage caused by UV rays. Using sunscreen can help prevent skin cancer, which is itself an important public health issue with environmental implications, as UV radiation can damage DNA and cause mutations in living organisms. Sunscreens are labelled with a sun protection factor (SPF) number, which refers to the product's ability to block UV rays.

Sunscreens can contain a variety of ingredients, but some common components include substances that act as filters for UV radiation. There are two main types of UV filters:

a. Physical UV filters, such as zinc oxide and titanium dioxide, reflect UV rays away from the skin.

b. Chemical UV filters that provide broad-spectrum protection against both UVA and UVB rays absorb such radiation. Common chemical UV filters include:

   i. **Octinoxate** (Figure 7.22) is a chemical sunscreen filter that protects against UVB rays but not UVA rays. It is often combined with other UV filters to provide broad-spectrum protection.

   ii. **Homosalate** (Figure 7.23) is a chemical sunscreen filter that helps enhance the UVB protection provided by other UV filters.

   iii. **Octisalate** (Figure 7.24) is a chemical sunscreen filter that provides additional UVB protection and is often combined with other UV filters to achieve broad-spectrum protection.

   iv. **Oxybenzone** (Figure 7.25) is a chemical sunscreen filter that is effective against both UVA and UVB rays but has been associated with potential health risks and is not allowed in certain countries.

FIGURE 7.22 Octinoxate

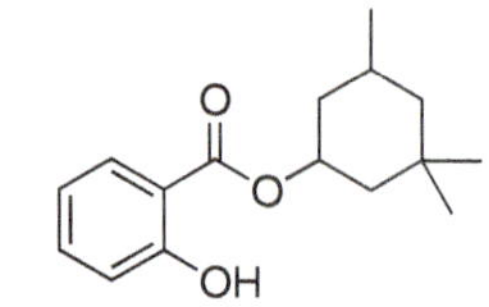

FIGURE 7.23 Homosalate

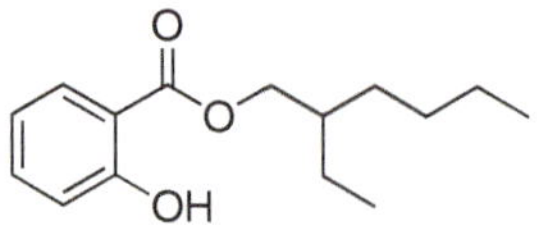

FIGURE 7.24 Octisalate

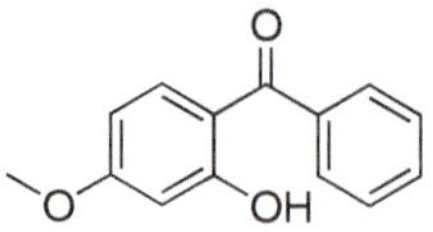

FIGURE 7.25 Oxybenzone

There are some concerns associated with their use:

a. **Skin irritation:** Some people may be sensitive to certain chemical UV filters and experience skin irritation, redness or other allergic reactions.

b. **Hormone disruption:** Some studies suggest that certain chemical UV filters, such as oxybenzone and octinoxate, may disrupt hormones in the body, particularly oestrogen. This could potentially lead to adverse effects, such as reproductive issues or developmental delays.

c. **Environmental impact:** Chemical UV filters can enter the environment through wastewater and cause damage to aquatic organisms, particularly coral reefs. This has led some countries to ban the use of certain chemical UV filters in sunscreen products.

d. **Photodegradation:** Some chemical UV filters can break down in the presence of sunlight, which reduces their effectiveness in protecting the skin from UV radiation.

On the other hand, one of the main concerns with sunscreens for the environment is their potential to harm coral reefs and other marine life. This is because some of the chemicals commonly used in sunscreen products, such as oxybenzone and octinoxate, have been shown to be toxic to coral and other marine organisms. When people swim in the ocean while wearing sunscreen, these chemicals can be released into the water and have negative effects on the marine ecosystem.

Because of the negative effects on human health and the environment, regulations have been put in place to control the use of chemical UV filters. In the United States, the FDA regulates the use of chemical UV filters in sunscreens. The FDA has identified 16 active ingredients that are commonly used in sunscreens and has set guidelines for their use. These guidelines include maximum concentrations, labelling requirements and testing requirements.

In the European Union, chemical UV filters are regulated under the Cosmetics Regulation. The regulation includes a list of approved UV filters and sets limits on their concentrations in cosmetic products. The EU also requires that cosmetic products containing UV filters undergo safety testing before they can be sold.

In addition to these regulations, some countries have banned certain chemical UV filters altogether. For example, Hawaii has banned the use of oxybenzone and octinoxate, two commonly used chemical UV filters, due to concerns about their negative impact on coral reefs.

7.15   Insect Repellents

Insect repellents work by disrupting an insect's ability to locate and feed on a host. The chemicals in insect repellents either block an insect's olfactory receptors or interfere with its ability to locate a host by mimicking the scent of a predator or confusing the insect's sense of direction.

The active ingredients in many insect repellents are compounds that are often derived from natural sources, such as plants or animals. Some common active ingredients in insect repellents include:

1. **DEET (*N,N*-diethyl-*m*-toluamide**) (Figure 7.26) is one of the most widely used insect repellents and works by blocking the insect's olfactory receptors.

    DEET is effective against a wide range of insects, including mosquitoes, ticks and biting flies. It was first synthesised in 1953 by the United States Department of Agriculture and has been used in insect repellents since the 1950s. It can have negative environmental impacts if not disposed of properly.

    DEET is known to be a persistent organic pollutant, which means it does not easily break down in the environment and can accumulate in organisms over time. DEET has been detected in surface water, groundwater, soil and sediment in various parts of the world, including in remote areas such as the Arctic. Studies have shown that DEET can be toxic to aquatic organisms, including fish, algae and crustaceans, even at low concentrations. DEET has also been linked to developmental and reproductive effects in some aquatic species.

2. **Picaridin or Icardin** (2-(2-hydroxyethyl)-1-piperidinecarboxylic acid), also known as KBR 3023 (Figure 7.27), is an insect repellent commonly used in personal care products such as lotions, sprays and wipes.

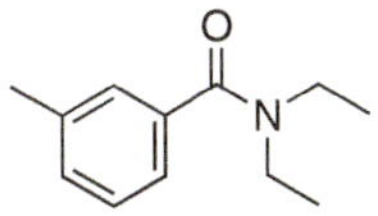

FIGURE 7.26   DEET (*N,N*-diethyl-*m*-toluamide)

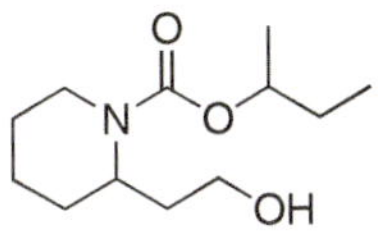

FIGURE 7.27   Picaridin

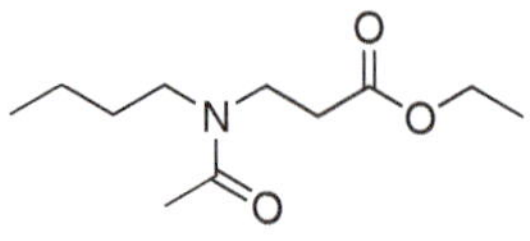

FIGURE 7.28   Ethyl butylacetylaminopropionate

KBR 3023 was developed in the 1990s. It is structurally similar to piperine, which is found in black pepper. It has gained popularity as an alternative to DEET due to its similar effectiveness against insects, lack of odour, and non-greasy feel on the skin.

However, picaridin has been identified as an emerging pollutant due to its increasing presence in surface waters and aquatic organisms. It has been found to have toxic effects on aquatic organisms, including fish and zooplankton, at concentrations found in some surface waters.

Picaridin is believed to enter the environment through direct discharge from personal care products and wastewater treatment plants. Its persistence and potential for bioaccumulation make it a concern for environmental and human health.

3. **Ethyl butylacetylaminopropionate**, also known as IR3535 (Figure 7.28), is a synthetic compound that was developed in the 1970s.

It works by interfering with an insect's ability to locate a host by confusing its sense of direction. IR3535 is effective against a wide range of insects, including mosquitoes, ticks and biting flies. While R3535 has not been extensively studied as a pollutant, it is considered an emerging pollutant due to its potential to accumulate in the environment and potentially impact aquatic organisms. Some studies have shown that R3535 can be toxic to algae and may interfere with the growth and reproduction of certain aquatic species.

Additionally, R3535 has been detected in wastewater treatment plant effluent and surface water samples, indicating that it can persist in the environment and potentially impact aquatic ecosystems. As with any emerging pollutant, further research is needed to fully understand the potential impacts of R3535 on the environment and human health.

7.16   Surfactants: Sodium Lauryl Sulfate and Others

Surfactants are compounds that are commonly used in a variety of products, including detergents, shampoos, soaps and cleaning agents, among others. They are added to these products to reduce the surface tension of water and help break up and remove dirt, oil and other impurities.

While surfactants are generally considered safe for human use and the environment, some types of surfactants can pose a risk to aquatic life and the environment if they are released into waterways or the environment in large quantities. In this group of surfactants with special concern are included compounds like nonylphenol ethoxylates (NPEs), octylphenol ethoxylates (OPEs) and perfluoroalkyl acids and perfluoro sulfonyl acids Figure 7.30.

Sodium Lauryl Sulfate (SLS)

Sodium lauryl sulfate (SLS) (Figure 7.29) is a synthetic detergent and surfactant commonly used in a variety of personal care and cleaning products. It is anionic in nature and is derived from lauryl alcohol, which is obtained from coconut or palm kernel oil. SLS is known for its excellent foaming and emulsifying properties, making it widely used in products such as shampoos, toothpastes and other personal care products. SLS has been the subject of some controversy. Some people may experience skin irritation or dryness from products containing SLS, especially those with sensitive skin. In such cases, individuals might prefer products labelled 'SLS-free' or 'sulfate-free'. SLS can cause skin irritation. Also, it has been linked to the development of certain cancers.

Other Surfactants with Special Environmental Concerns

a. **Nonylphenol and NPEs** (Figure 7.30) are a group of synthetic surfactants that are commonly used in industrial and commercial applications such as detergents, emulsifiers and wetting agents.

   NPEs are formed by ethoxylating nonylphenol, which is a type of alkylphenol, with ethylene oxide to create a variety of different ethoxylate chain lengths.

   NPEs have been found to be persistent in the environment and can accumulate in living organisms. They are also known to be endocrine disruptors, meaning they can interfere with the hormonal systems of animals and humans, which can lead to adverse effects on reproduction, development and other physiological processes.

   Due to their potential environmental and health impacts, many countries have placed restrictions on the use of NPEs. For example, the European Union has banned the use of NPEs in many consumer products and several US states have also restricted their use in certain applications.

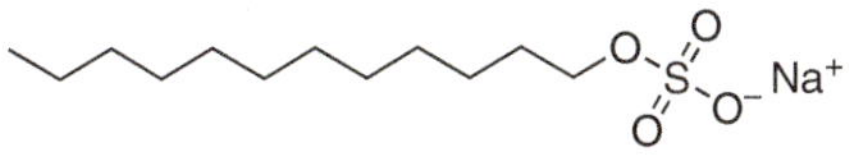

FIGURE 7.29 Sodium lauryl sulfate (SLS)

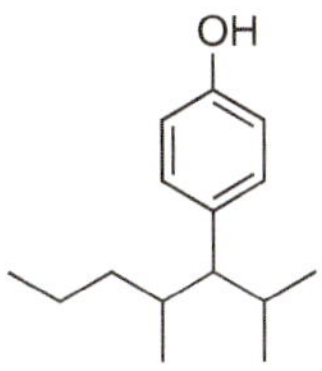

Nonylphenol

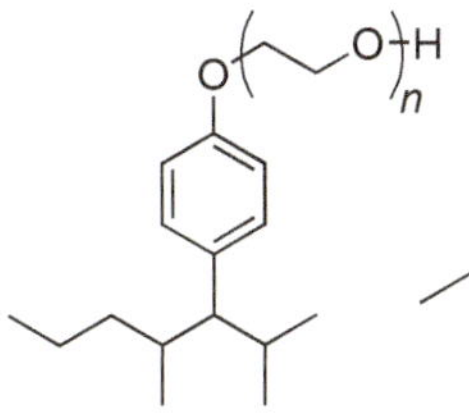

Nonylethoxylates

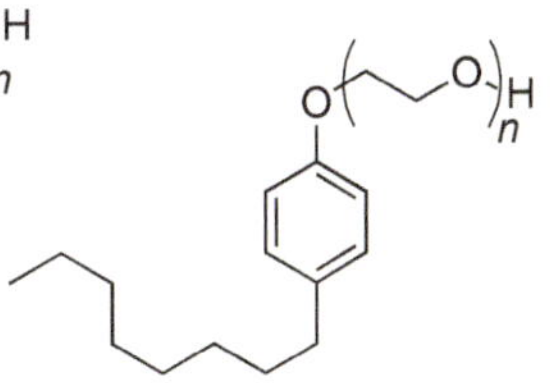

Octylphenol ethoxylates

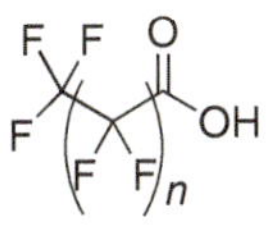

Perfluoroalkyl acids

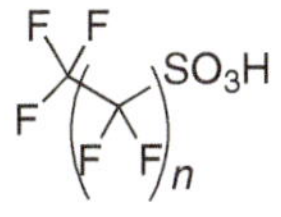

Perfluorosulphonic acids

FIGURE 7.30 Surfactants with special environmental concerns

OPEs are another class of nonionic surfactants that are widely used in a variety of industrial and household applications, such as detergents, emulsifiers, wetting agents and dispersants. OPEs are synthesised by reacting octylphenol with ethylene oxide, which results in a range of different ethoxylate chain lengths.

As NPEs, OPEs have been found to be persistent and potentially harmful to the environment. They can accumulate in aquatic environments and be toxic to aquatic organisms. OPEs are also known to disrupt the endocrine system of wildlife and may have similar effects on human health.

Due to these concerns, OPEs have been regulated or banned in some countries, such as the European Union and Canada. In the United States, OPEs are still commonly used, but there is increasing pressure to find safer alternatives.

b. **PFAS,** also known as fluorosurfactants, are a subgroup of PFAS[2] that are used as surfactants in a variety of applications and are commonly used in cleaning and coating products.

Fluorosurfactants are valued for their ability to reduce surface tension and their resistance to heat, chemicals and degradation. They are used in a range of industrial and consumer products, such as cleaning agents, coatings, adhesives and personal care products.

c. **Perfluorooctanoic acid (PFOA) and perfluorooctanesulfonic acid (PFOS)** are two of the most well-known fluorosurfactants, though they're being phased out due to environmental and health concerns. PFOA has been used in the production of nonstick coatings (like Teflon) and is found in various consumer products.

Like other PFAS, fluorosurfactants have raised concerns about their persistence in the environment and potential health effects. They have been found in water sources and wildlife, and some studies have linked their exposure to health problems such as developmental effects, immune system dysfunction and cancer.

As with other PFAS, efforts are being made to reduce or eliminate the use of fluorosurfactants in products and to clean up areas that have been contaminated with these chemicals. In some countries, regulations have been implemented to restrict or ban the use of certain fluorosurfactants.

7.17   Flame Retardants

Flame retardants are chemicals that are added to various materials, such as plastics, textiles and electronics, to reduce their flammability and prevent them from catching fire or slowing down the spread of fire. They work by interrupting the chemical reactions that occur during combustion, making it more difficult for a fire to start or continue burning.

There are many different types of flame retardants, but they can generally be grouped into two categories: halogenated and non-halogenated. Halogenated flame

---

[2] PFAS are a group of synthetic chemicals that are used in a wide range of products, including non-stick cookware, waterproof clothing and firefighting foam. They are persistent in the environment, meaning that they do not break down easily and can accumulate in the tissues of animals and humans. PFAS have been linked to a range of negative health effects, including cancer, liver damage and developmental issues.

retardants contain elements such as chlorine or bromine, while non-halogenated flame retardants do not. Both types have been extensively used in a wide range of products, including furniture, building materials and electronics.

Some examples of halogenated flame retardants include:

1. **Polybrominated diphenyl ethers (PBDEs), mentioned above:** PBDEs are a group of flame retardants that have also been used in various products, including electronics, textiles and furniture (Figure 7.31). They are persistent organic pollutants and have been found to accumulate in the environment and in human and animal tissues.

2. **Tetrabromobisphenol A (TBBPA) (Figure 7.32):** TBBPA is a widely used flame retardant in electronics, such as printed circuit boards. It has been detected in the environment and in human tissues.

3. **Chlorinated paraffins:** Chlorinated paraffins (Figure 7.33) are a group of flame retardants that are used in plastics, rubber and textiles. They have been found to be persistent in the environment and have been detected in human tissues.

4. **Hexabromocyclododecane (HBCD) (Figure 7.34):** HBCD is a flame retardant used in polystyrene foam insulation, textiles and electronic equipment. It has been detected in the environment and in human tissues.

5. **Decabromodiphenyl ether (DecaBDE) (Figure 7.35):** DecaBDE is a flame retardant used in electronics, textiles and plastics. It is persistent in the environment and has been detected in human tissues.

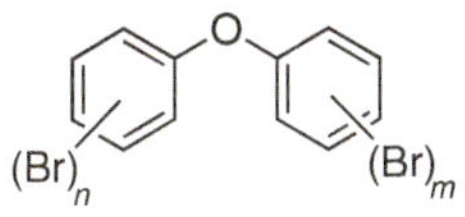

FIGURE 7.31   Polybrominated diphenyl ethers

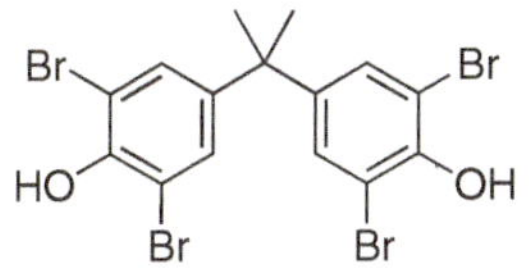

FIGURE 7.32   Tetrabromobisphenol A (TBBPA)

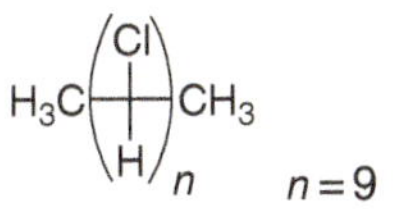

FIGURE 7.33   Chlorinated paraffins

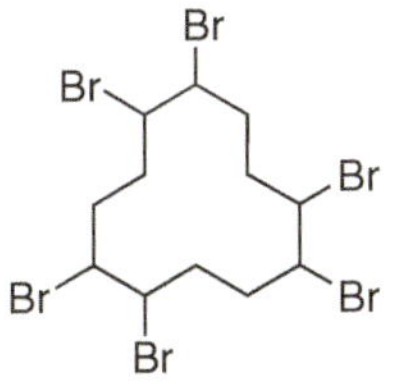

FIGURE 7.34   Hexabromocyclododecane (HBCD)

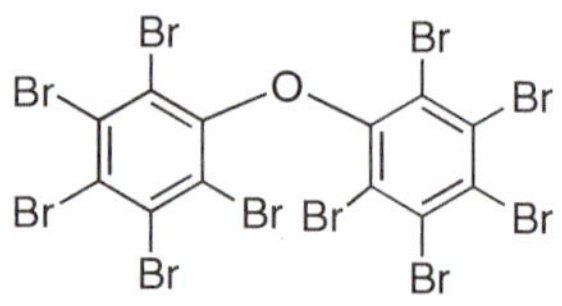

FIGURE 7.35 Decabromodiphenyl ether

While flame retardants can provide important fire safety benefits, there are concerns about their potential health and environmental impacts due to their persistence. They can enter the environment through various pathways, including the manufacturing and disposal of products containing such compounds.

Studies have shown that halogenated flame retardants can accumulate in the environment and in living organisms, including humans. They are associated with a range of health effects, including developmental and reproductive problems, neurotoxicity and endocrine disruption.

7.18 Plasticisers

Plasticisers are additives that are added to polymers to improve their flexibility, durability and workability. They are typically low-molecular-weight compounds that can be easily mixed with the polymer and reduce the glass transition temperature of the material. This makes the polymer more malleable and easier to shape.

Plasticisers are commonly used in the production of plastics, synthetic rubbers and other polymer-based materials. They are also used in a wide range of applications, such as adhesives, sealants, paints and coatings.

Some plasticisers are classified as emerging pollutants because they can be harmful to the environment and human health. Here are some examples of plasticisers that are considered emerging pollutants:

1. **Phthalates:** Phthalates are a class of chemicals that are commonly used as plasticisers (Figure 7.19). They are added to a wide variety of products, including vinyl flooring, shower curtains and food packaging. Phthalates are known to be endocrine disruptors, which means they can interfere with hormone systems in the body. Some phthalates have also been linked to developmental and reproductive problems in humans and animals.

2. **Bisphenol A (BPA) (Figure 7.36):** Bisphenol A (BPA) is a chemical compound that has been widely used in the production of plastics and resins since the 1960s. It is commonly found in many everyday products, including water bottles, food containers, baby bottles, dental sealants, and the lining of metal cans used for food and beverages. BPA is used to make polycarbonate plastics and epoxy resins, which are known for their toughness and durability, and as a plasticiser in the polymer industry.

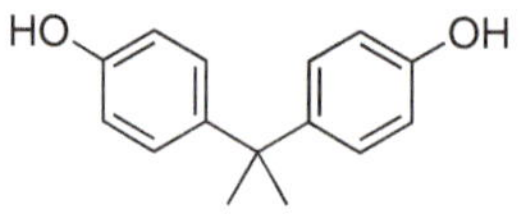

FIGURE 7.36 Bisphenol A

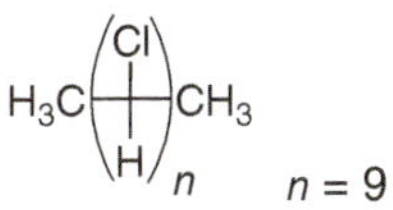

$$H_3C \overset{Cl}{\underset{H}{\overline{\underline{}}}} CH_3 \Big)_n \qquad n = 9$$

FIGURE 7.37   Chlorinated paraffins

However, over the years, concerns have been raised about the potential health effects of BPA exposure. Research has shown that BPA can leach from the containers or products it is used in and enter the human body, mainly through ingestion and skin contact. One of the primary ways people are exposed to BPA is through the consumption of food and beverages stored in BPA-containing containers.

Some studies have suggested that BPA can act as an endocrine disruptor, meaning it may interfere with the body's hormonal systems. The compound has been associated with a range of health concerns, including:

a. **Hormonal disruption:** BPA has been shown to mimic oestrogen, a hormone that plays a crucial role in various bodily functions. Excessive exposure to BPA may lead to hormonal imbalances, potentially affecting reproductive and developmental processes.

b. **Reproductive effects:** There is evidence suggesting that BPA exposure may impact fertility in both men and women. It has been linked to changes in sperm quality, ovarian dysfunction and menstrual cycle irregularities.

c. **Developmental issues:** Prenatal exposure to BPA has raised concerns about its potential impact on foetal development. Some studies have linked BPA exposure to developmental issues in children, although more research is needed to establish conclusive links.

d. **Cardiovascular effects:** Some research has indicated a possible association between BPA exposure and an increased risk of cardiovascular diseases.

   Due to these concerns, many countries and regions have taken steps to regulate the use of BPA in certain products, particularly those intended for use by infants and children. Some manufacturers have also voluntarily shifted away from using BPA in their products, opting for alternative materials considered safer.

3. **Chlorinated paraffins (Figure 7.37):** Chlorinated paraffins are a group of chemicals that are used as plasticisers in a wide range of products, including rubber, textiles and paints. They have been found to be persistent in the environment and can accumulate in living organisms. Chlorinated paraffins have been linked to a range of health problems, including liver and kidney damage.

Topics of Interest: Alternatives to High-Impact Plasticisers

There are several alternatives to phthalates, bisphenol A (BPA) and other plasticisers that are being used by manufacturers today. Some of these alternatives include:

1. **Non-phthalate plasticisers:** Non-phthalate plasticisers are a group of chemicals that are used to soften plastics and make them more flexible. Examples include citrates, adipates and sebacates. These alternatives have been found to be safer for human health and the environment.

2. **Bio-based plasticisers:** Bio-based plasticisers are derived from natural sources, such as vegetable oils or starches. These plasticisers are renewable and biodegradable, making them more sustainable and environmentally friendly.

3. **Epoxidised soybean oil (ESBO):** ESBO is a vegetable oil-based plasticiser that is used in PVC applications. It has been found to be a safe and effective alternative to phthalates and other traditional plasticisers.

4. **Polymeric plasticisers:** Polymeric plasticisers are polymers that are used to improve the flexibility and durability of plastics. Examples include polyethylene glycol, polyesters and polyacrylates.

5. **Liquid polymers:** Liquid polymers are used as plasticisers in various applications, such as vinyl flooring and artificial leather. Examples include polyurethanes and acrylics.

7.19 Microplastics

Microplastics are small plastic particles, less than 5 mm in size, that can be found in various environmental settings such as oceans, rivers, soil and air. They can be generated from the breakdown of larger plastic items as well as from products such as personal care items, synthetic textiles, packaging materials and granular infill material used on artificial sport surfaces.

Also, they can be formed through a variety of processes, including:

1. **Mechanical breakdown:** This occurs when larger plastic items, such as bottles, bags and fishing nets, are exposed to the elements and break down into smaller fragments over time.

2. **Degradation:** Plastic materials can also degrade due to exposure to sunlight, heat and chemicals. This process can cause plastic to become brittle and break down into smaller pieces.

3. **Manufacturing:** Some microplastics are intentionally manufactured for specific purposes, such as microbeads in personal care products or microfibers in clothing.

4. **Abrasion:** Microplastics can also be created through the abrasion of larger plastic items, such as car tyres, during use.

5. Machine washing of clothes made of synthetic fibres such as polyester, nylon and acrylic. They shed tiny fibres that are too small to be filtered out by waste-water treatment plans.

The presence of microplastics in the environment is a growing environmental concern for several reasons.

a. First, microplastics can be ingested by wildlife, such as fish and birds, causing harm to their digestive systems and potentially disrupting their behaviour and reproductive cycles.

b. Second, microplastics can absorb and concentrate toxic chemicals, such as pesticides and persistent organic pollutants, which can then be transferred up the food chain.

c. Third, microplastics can also have an impact on human health, as they can enter the human food chain through the consumption of seafood and drinking water, although the long-term effects of this are still unclear.

It's important to note that the field of microplastics research is still evolving and further studies are needed to gain a comprehensive understanding of their impacts on living beings. Exposure to microplastics can occur through various routes (Figure 7.38), including ingestion, inhalation and dermal contact.

1. **Ingestion:** Microplastics can enter the food chain when aquatic organisms, such as fish and shellfish, consume them along with their natural food sources. If humans consume these contaminated organisms, they can be exposed to microplastics indirectly. Additionally, microplastics have been detected in drinking water, both bottled and tap, which can also contribute to ingestion exposure.

2. **Inhalation:** Microplastics can become airborne, especially in environments where plastic particles are present, such as during the production or degradation of plastic materials. People working in industries associated with plastic manufacturers or recycling may be at higher risk of inhalation exposure to microplastics. However, the extent of inhalation exposure and its potential health effects are still areas of ongoing research.

3. **Dermal contact:** Microplastics can also come into contact with the skin through the use of personal care products containing microbeads or by handling plastic materials directly. While the potential health effects of dermal exposure to microplastics are not well understood, they are generally believed to be lower compared to ingestion or inhalation routes.

The long-term health impacts of exposure to microplastics are still being studied, and the current scientific understanding is limited. Some studies suggest that microplastics may accumulate in tissues and organs, potentially causing inflammation and disrupting normal cellular processes. However, more research is needed to fully understand the potential health risks and the levels of exposure required to trigger adverse effects in living beings.

To minimise exposure to microplastics, individuals can take the following steps:

1. **Reduce single-use plastics:** Minimise the use of disposable plastic items such as bags, bottles and straws to limit the overall production and release of microplastics into the environment.

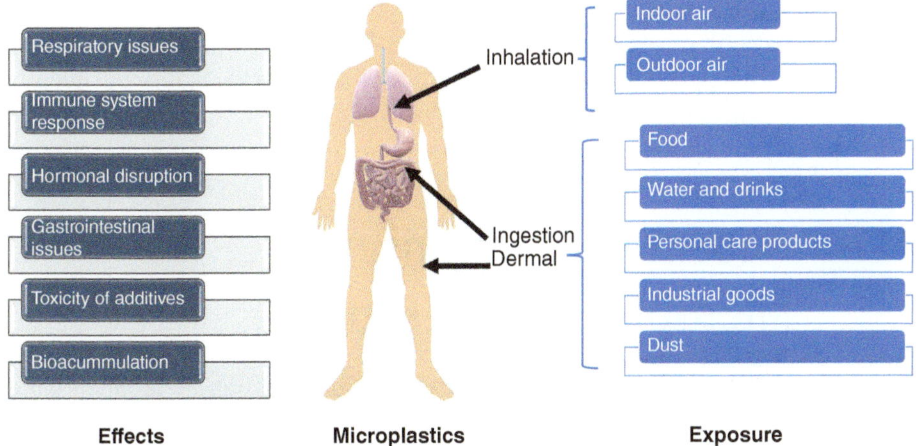

FIGURE 7.38   Route of exposure to microplastics

2. **Properly dispose of plastic waste:** Ensure that plastic waste is disposed of correctly, recycling wherever possible. This helps reduce the amount of plastic that ends up in natural ecosystems.

3. **Choose natural fibres:** When purchasing clothing and textiles, opt for natural fibres like cotton or wool over synthetic materials, as synthetic fibres shed microplastics during washing.

4. **Be informed:** Stay updated on research and developments regarding microplastics to understand the potential risks and contribute to informed decision-making and advocacy for better plastic waste management.

The detection and evaluation of microplastics in water can be achieved through various methods, including both laboratory-based techniques and field sampling. Here are some common approaches:

1. **Filtration:** Water samples are passed through a filter with a specific pore size (typically 0.45–1 μm) to collect suspended particles, including microplastics. The filter is then examined under a microscope to identify and count the particles.

2. **Density separation:** This method involves adjusting the density of the water sample using a solution with a known density, such as sodium chloride. Microplastics, being less dense, will float to the surface or sink to the bottom, where they can be collected and analysed.

3. **Spectroscopic analysis:** Techniques like Fourier-transform infrared spectroscopy (FTIR) and Raman spectroscopy can be used to identify and characterise microplastic particles. These methods analyse the unique molecular vibrations of the plastic polymers, allowing for identification.

4. **Microscopy:** Microscopes, such as optical microscopes and electron microscopes, are commonly used to observe and measure microplastics. Optical microscopes are suitable for larger particles, while electron microscopes provide higher magnification and can analyse smaller particles in greater detail.

5. **Fluorescence imaging:** Some microplastics, particularly those made of fluorescent polymers or containing additives, can be detected using fluorescence microscopy. Specific dyes or fluorescent tags are applied to the sample, causing the microplastics to emit light under specific wavelengths.

6. **Automated imaging systems:** Advanced imaging systems equipped with image recognition algorithms can automate the detection and analysis of microplastics. These systems use machine learning algorithms to identify and quantify microplastics based on their visual characteristics.

7.20 Recent Regulations on Microplastics in Europe

The European Commission has been one of the first organisations to adopt measures that restrict microplastics intentionally added to products under the EU chemical legislation REACH in 2023. The new rules will prevent the release into the environment of about half a million tonnes of microplastics. They will prohibit the sale of microplastics as such, and of products to which microplastics have been added on purpose and that release those microplastics when used. When duly justified, derogations and transition periods for the affected parties to adjust to the new rules apply.

The adopted restriction uses a broad definition of microplastics. It covers all synthetic polymer particles below five millimetres that are organic, insoluble and resist degradation. The purpose is to reduce emissions of intentional microplastics from as many products as possible. Some examples of common products in the scope of the restriction are:

- The granular infill material used on artificial sport surfaces – the largest source of intentional microplastics in the environment;
- Cosmetics, where microplastics are used for multiple purposes, such as exfoliation (microbeads) or obtaining a specific texture, fragrance or colour;
- Detergents, fabric softeners, glitter, fertilisers, plant protection products, toys, medicines and medical devices, just to name a few.

The initial measures, such as the prohibition of loose glitter and microbeads, will take effect in 20 days when the restriction comes into force. In certain instances, the sales ban will be implemented after a more extended period to allow affected stakeholders sufficient time to explore and transition to alternative solutions.

7.21 Nanoparticles

Nanoparticles are particles with sizes ranging from 1 to 100 nm in diameter. They can be composed of various materials, including metals, ceramics and polymers. Due to their small size, they have unique physical, chemical and biological properties that differ from those of their bulk counterparts.

Nanoparticles have a wide range of applications in various fields, including medicine, electronics, energy and environmental remediation. In medicine, nanoparticles are used as drug delivery systems, diagnostic tools and imaging agents. In electronics, they are used to improve the performance and efficiency of electronic devices. In energy, nanoparticles are used to improve the efficiency of solar cells and fuel cells. In environmental remediation, nanoparticles are used to remove pollutants from the air and water.

One of the main concerns with nanoparticle pollution is their small size, which allows them to penetrate deeply into the body and cause toxic effects.

The effects of nanoparticles in the body remain uncertain, but there are many concerns related to their composition, size, shape and surface properties, and how they are introduced into the body. Some potential effects of nanoparticles in the body are as follows:

1. **Biodistribution:** Nanoparticles can distribute differently throughout the body based on their properties, leading to varying effects in different organs.

2. **Organ and tissue accumulation:** Nanoparticles can accumulate in specific organs or tissues, potentially causing localised effects. For instance, nanoparticles used in drug delivery may accumulate in specific target tissues.

3. **Blood circulation:** Nanoparticles introduced into the bloodstream may interact with blood components and affect clotting mechanisms.

4. **Barrier permeability:** Nanoparticles can affect the permeability of barriers in the body, such as the blood-brain barrier. This can have implications for drug delivery and potential neurological effects.

5. **Cellular uptake:** Nanoparticles can be taken up by cells through various mechanisms, such as endocytosis. This can impact cell function and behaviour.

6. **Inflammation:** Certain nanoparticles may trigger an inflammatory response when they interact with immune cells. Chronic inflammation can contribute to various health issues.

7. **Oxidative stress:** Some nanoparticles can generate reactive oxygen species (ROS), leading to oxidative stress. This can damage cells and biomolecules, potentially causing various diseases.

8. **Genotoxicity:** Nanoparticles may interact with DNA, potentially leading to DNA damage and genetic mutations.

9. **Immunomodulation:** Nanoparticles can interact with the immune system, potentially altering immune responses and affecting overall immune function.

10. **Allergic reactions:** Some individuals may develop allergic reactions to certain types of nanoparticles.

11. **Cytotoxicity:** Nanoparticles can directly damage cells, affecting their viability and function.

12. **Nanoparticle aggregation:** Nanoparticles can aggregate in the body, potentially leading to blockages in blood vessels or other unwanted effects.

13. **Long-term accumulation:** Some nanoparticles might accumulate over time due to slow clearance mechanisms, potentially leading to chronic exposure effects.

14. **Endocrine disruption:** Certain nanoparticles may interfere with hormonal systems, potentially causing endocrine-related disorders.

15. **Nanoparticle interactions with drugs:** Nanoparticles used in drug delivery can affect the pharmacokinetics and pharmacodynamics of the encapsulated drug.

It's important to note that not all nanoparticles will exhibit these effects, and the extent of these effects can vary widely based on the specific properties of the nanoparticles and the exposure levels. The field of nanotoxicology aims to understand these potential effects and develop strategies to minimise risks associated with nanoparticle exposure.

In addition to their impact on human health, nanoparticles can also have adverse effects on the environment. They can accumulate in soil, water and air and may have toxic effects on plants and animals. For example, nanoparticles can disrupt ecosystems by affecting the growth and development of plants and can also accumulate in the food chain, leading to bioaccumulation and biomagnification.

However, the effects of nanoparticles depend on a range of factors, such as their size, shape, surface charge and coating, as well as the route of exposure, dose and duration of exposure. More research is needed to fully understand the potential risks associated with different types of nanoparticles and to develop effective strategies for minimising exposure and mitigating their effects.

Therefore, it is important to monitor and regulate the release of nanoparticles into the environment and to develop effective methods for their removal and remediation. This can include measures such as improving industrial processes to minimise the release of nanoparticles, using alternative materials that are less toxic, and developing effective methods for the detection and removal of nanoparticles from the environment.

CHAPTER 8

Soil Pollution

8.1 Introduction

The Food and Agriculture Organization (FAO) defines soil pollution as the *presence of substances in soil, either naturally occurring or human-made, that exceed levels that are normally found and that may have negative effects on human health, the environment or soil functions.*

Some common sources of soil pollution include industrial activities such as mining and manufacturing, improper disposal of hazardous waste, agricultural practices such as the use of pesticides and fertilisers, urbanisation and construction.

The effects of soil pollution can be wide-ranging and include reduced soil fertility, loss of biodiversity and contamination of groundwater and surface water. Additionally, soil pollution can lead to health problems in humans and animals, such as respiratory problems, cancer and neurological disorders.

Preventing soil pollution requires a combination of measures, including better waste management practices, the use of environmentally friendly agricultural practices and the regulation and enforcement of laws related to pollution prevention. Additionally, remediation techniques, such as soil testing and cleanup, can be used to address existing soil pollution.

8.2 Sources of Soil Pollution

Soil pollution is caused by a variety of factors, including:

1. **Industrial activities:** Industries that produce chemicals, oil and gas and mining activities can release toxic substances into the soil.

2. **Agricultural activities:** The use of pesticides, herbicides and fertilisers can also lead to soil pollution.

3. **Improper waste disposal:** The disposal of household waste, plastics, batteries and other hazardous waste can cause soil pollution.

4. **Landfills:** The accumulation of waste in landfills can cause the leaching of harmful chemicals into the soil.

Understanding the Chemistry of the Environment, First Edition. Francisco G. Calvo-Flores.
© 2025 John Wiley & Sons Ltd. Published 2025 by John Wiley & Sons Ltd.

5. **Construction activities:** Construction activities, including the use of heavy machinery, excavation and demolition, can damage the soil structure and contaminate the soil with chemicals.

6. **Transportation activities:** Emissions from vehicles, including exhaust fumes and oil leaks, can contribute to soil pollution.

7. **Nuclear activities:** Accidents in nuclear power plants, nuclear testing and nuclear waste disposal can lead to radioactive contamination of the soil.

8. **Natural causes:** Natural causes such as erosion, landslides and volcanic eruptions can also lead to soil pollution.

8.3 Types of Soil Pollution

Soil pollution can be classified from three points of view if it is caused by physical, chemical and biological factors in the affected area.

According to the first criteria, we have:

1. **Physical soil pollution:** It refers to the presence of physical objects or substances in the soil that can be harmful to plant growth and other organisms living in the soil. This can include things like plastic debris, glass shards, construction waste and other forms of litter or materials that are not biodegradable or slow to degrade. The soil is compacted or eroded, which can lead to a loss of soil fertility and make it difficult for plants to grow. Another cause of physical pollution is if the temperature of the soil is altered due to human activities such as the construction of buildings or paving over natural surfaces.

2. **Chemical pollution:** Chemical soil pollution occurs when toxic chemicals are released into the soil, either through human activities such as municipal activities, industrial processes, agricultural practices or any other activity that leads to soil contamination with chemicals such as pesticides, fertilisers, heavy metals, radioactive substances or any other industrial or municipal waste containing harmful chemical substances.

3. **Biological pollution:** This occurs when the soil is contaminated with harmful microorganisms such as bacteria, viruses and parasites, which can pose a threat to human and livestock health or affect crops or vegetation.

According to the area affected, soil pollution may be classified into three main categories:

1. **Point source pollution:** This type of pollution occurs from a single, identifiable source, such as a chemical spill or a leaking underground storage tank. The contaminated area is typically limited to the immediate vicinity of the source.

2. **Area source pollution:** This type of pollution results from activities that occur over a large area, such as landfills or mining sites. The contaminated area can be extensive and can have a significant impact on nearby ecosystems and human health.

3. **Non-point-source pollution or diffuse pollution:** Pollutants are spread out over a large area and come from multiple sources more unspecific and arise from

multiple sources, such as agricultural runoff, atmospheric deposition or urban stormwater runoff. The contaminated area can be more extensive and difficult to define than in point-source pollution.

8.4 Symptoms and Effects of Soil Pollution

Soil pollution can have a significant impact on the environment, agriculture and human health. When soil is polluted, a range of symptoms can be detected depending on the type and extent of pollution. Some common symptoms of soil pollution include:

a. **Changes in soil colour and texture:** Polluted soil can have a different colour and texture than healthy soil. It may appear darker, have a different consistency or contain visible pollutants.

b. **Reduced crop yields:** Polluted soil can negatively impact the growth and health of plants, leading to reduced crop yields.

c. **Contaminated surface waters and groundwater:** Polluted soil can release contaminants into groundwater or nearby bodies of water, leading to contamination of drinking water and harm to aquatic life.

d. **Health problems in animals:** Livestock and other animals that consume contaminated soil can experience health problems, including illness or death.

e. **Increased risk of disease:** Polluted soil can contain harmful bacteria and other pathogens that can increase the risk of disease in humans and animals.

f. **Bad odour:** Soil pollution can cause bad odour due to the presence of organic pollutants.

g. **Presence of hazardous substances:** Soil pollution can contain hazardous substances, such as heavy metals and chemicals, which can harm human health and the environment.

On the other hand, oil pollution can have a range of negative effects on both the environment and human health. Some of the main effects are:

a. **Reduced soil fertility:** Soil pollution can reduce the soil's ability to support plant life, which can lead to reduced crop yields and lower food quality.

b. **Pollution of water bodies:** Pollutants from contaminated soil can run off or leach into rivers, lakes and other surface water bodies and also in the groundwater, contaminating water supplies and potentially causing serious problems in humans livestock and agriculture.

c. **Soil erosion:** Polluted soil may be more susceptible to erosion, leading to loss of topsoil and reduced soil quality.

d. **Damage to ecosystems:** Soil pollution can harm the delicate balance of ecosystems, leading to loss of biodiversity and damage to natural habitats.

e. **Human health risks:** Exposure to contaminated soil can pose serious health risks, including respiratory problems, skin irritations and even cancer.

f. **Food contamination:** Polluted soil can lead to the accumulation of harmful substances in crops, which can then be consumed by humans and animals, leading to health problems.

8.5 Soil Chemical Pollutants

There are various types of soil pollutants that can affect the quality of soil and its ability to support plant and animal life, both inorganic and organic. A few examples of the many types of soil pollutants that can impact soil quality and environmental health include:

a. **Heavy metals:** Heavy metals such as lead, arsenic, cadmium and mercury are toxic and can accumulate in soil through industrial waste, mining and agricultural practices.

b. **Fertilisers:** When fertilisers are overused or applied improperly, they can contribute to soil pollution by causing nutrient imbalances, increasing soil acidity and promoting the growth of harmful bacteria and algae.

c. **Radioactive materials:** Nuclear testing and accidents can release radioactive materials into the soil, which can remain in the environment for many years and cause long-term health effects.

d. **Atmospheric pollutants:** Atmospheric pollutants can have a significant impact on soil pollution. Pollutants such as nitrogen oxides, sulfur dioxide and particulate matter can deposit onto soil surfaces and have a range of effects on soil chemistry, biology and ecology. Nitrogen oxides and sulfur dioxide can react with rainwater to form acid rain, which can have a corrosive effect on soil, lowering pH and causing aluminium and other toxic metals to leach from the soil. This can make the soil less hospitable to plants and reduce biodiversity.

e. **Organic chemicals:** Chemicals such as benzene, toluene and xylene can be released into the soil from industrial and manufacturing processes and can persist for long periods of time.

f. **Pesticides:** These chemicals are used to control pests and weeds in agriculture, but they can also harm non-target organisms and persist in the soil for long periods of time.

g. **Petroleum products:** Oil spills and leaks from underground storage tanks can contaminate soil with petroleum products, which can be toxic to plants and animals.

8.6 Interaction of Chemical Pollutants with Soil

Chemical pollutants can interact with soil in a variety of ways, depending on their properties and the characteristics of the soil. Some possible interactions between pollutants and soil are as follows (Figure 8.1):

a. **Lixiviation:** It is the movement of pollutants through the soil via water. When water containing pollutants percolates through soil, it can dissolve or suspend the pollutants, carrying them downward into the soil profile. The extent of lixiviation depends on the solubility and mobility of the pollutant, the texture and structure of the soil, the amount and intensity of rainfall or irrigation and the slope of the land.

b. **Runoff:** It refers to the movement of water across the surface of the soil, carrying with it any pollutants that may be present on the surface. The extent of runoff depends on

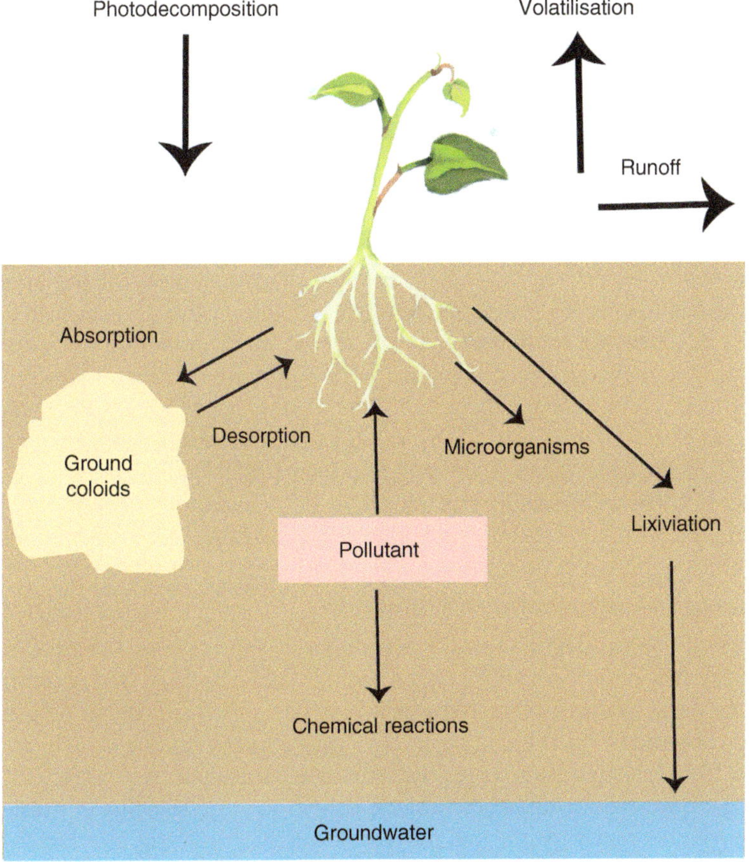

FIGURE 8.1   Interaction between pollutants and soil

the amount and intensity of rainfall or irrigation, the slope and texture of the land, the vegetation cover and the presence of impervious surfaces such as pavement.

c. **Adsorption:** Absorption occurs when pollutants are retained in the soil through processes that involve the adhesion of pollutants onto the surface of soil particles, or absorption, which involves the uptake of pollutants into the soil particles by various mechanisms such as electrostatic forces, hydrogen bonding and van der Waals forces. In general, pollutants that are highly polar or charged tend to be absorbed more strongly by soils than those that are nonpolar. This can reduce the concentration of the pollutants in the soil solution and prevent them from leaching into groundwater or being taken up by plants.

d. **Desorption:** It occurs when pollutants that are initially absorbed by the soil are released back into the environment. This can occur through processes such as desorption, which involves the release of pollutants from the soil particles back into the surrounding water or air.

e. **Volatilisation:** Pollutants in soil are converted into a gaseous state and released into the atmosphere. Some common pollutants that can undergo volatilisation from soil include volatile organic compounds (VOCs) and some pesticides. The extent of volatilisation depends on the physical and chemical properties of the pollutant as well as the soil characteristics, such as texture, moisture content and pH.

f. **Degradation:** Some chemical pollutants can be broken down or transformed in the soil through chemical or biological processes:

i. **Chemical degradation of pollutants**

This process can occur naturally in soil through various mechanisms, such as oxidation, hydrolysis and photolysis.

Oxidation is a common mechanism for the degradation of pollutants in soil, where oxygen molecules react with pollutants to form less harmful substances. This process can occur through various pathways, such as biological oxidation by microorganisms, chemical oxidation by oxidising agents like hydrogen peroxide and photochemical oxidation through exposure to light.

Hydrolysis is another mechanism for the degradation of pollutants in soil, where water molecules react with pollutants to break them down into simpler, less harmful substances. This process is commonly observed in the degradation of organic pollutants like pesticides and herbicides.

Photolysis refers to the degradation of pollutants through exposure to light. Ultraviolet radiation in sunlight can cause the breakdown of many organic pollutants in soil, such as polycyclic aromatic hydrocarbons (PAHs) and chlorinated hydrocarbons.

ii. **Biological degradation of pollutants**

The biological degradation of pollutants in soil involves the breakdown of contaminants by microorganisms such as bacteria, fungi and actinomycetes. These microorganisms use the pollutants as a food source, converting them into simpler substances.

The biodegradation of pollutants in soil depends on several factors, including the type and concentration of the pollutant, the availability of nutrients and oxygen, temperature, pH and moisture content. The rate of biodegradation can also be affected by the presence of other contaminants or toxic substances.

8.7 Recuperation of Polluted Soils: Remediation

The process of removing or reducing contaminants from soil to restore its quality and reduce the risks associated with the pollutants is known as remediation. There are various techniques that can be used to remediate polluted soils, including physical, chemical and biological methods. The choice of remediation technique depends on several factors, including the type and extent of contamination, the site location, the severity of pollution and the cost of the remediation process. Some remediation techniques can be expensive and time-consuming, while others are more affordable and faster.

8.8 Physical Remediation Methods

Physical remediation of soil refers to the use of physical techniques to restore soil that has been contaminated with pollutants or other harmful substances. There are several physical remediation methods available, including:

1. **Excavation:** This involves the physical removal of contaminated soil from a site, which is then transported to a designated disposal facility (Figure 8.2).

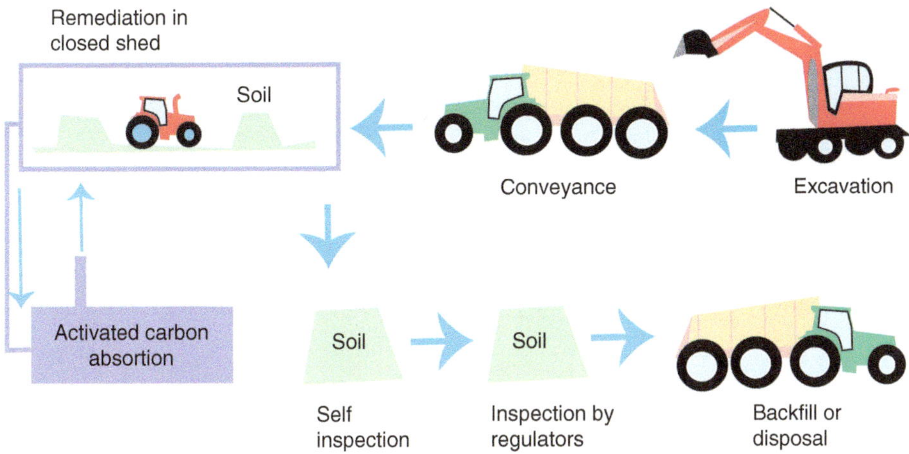

FIGURE 8.2   Schematic representation of remediation by excavation

This technique is typically used when the contamination is deep in the soil and cannot be easily treated in place. The process of soil excavation typically involves the following steps:

a. **Site assessment:** A site assessment is conducted to determine the extent and nature of the contamination.

b. **Excavation plan:** A plan is developed for excavation, including the depth and area to be excavated, the equipment and personnel required and the disposal method for the contaminated soil.

c. **Excavation:** The contaminated soil is excavated using heavy equipment, such as backhoes or excavators. The soil is loaded onto trucks and transported to a disposal facility.

d. **Soil disposal:** The contaminated soil is disposed of at a licensed landfill or other approved facility.

e. **Backfilling and restoration:** Once the contaminated soil has been removed, the site is backfilled with clean soil and restored to its original condition.

2. **Washing:** Soil washing is a type of soil remediation process that involves separating contaminated soil from its contaminants through a washing process. This technique is typically used for soils that have been contaminated with hazardous chemicals such as heavy metals or toxic organic substances. During the soil washing process, the contaminated soil is excavated and then treated with various washing agents such as water, surfactants and solvents. The soil is then separated from the contaminants by either physical or chemical means (Figure 8.3).

3. **Aeration:** Aeration involves the use of mechanical or natural methods to increase the oxygen supply to the soil, which can accelerate the natural degradation of contaminants by microorganisms (Figure 8.4).

4. **Solidification/stabilisation:** This method involves adding materials such as cement, lime or other chemicals to contaminated soil to create a solid block or stabilise the contaminants in place, preventing their migration and exposure (Figure 8.5).

5. **Thermal treatment** of soil refers to the process of using heat to treat contaminated soil in order to reduce or eliminate the concentration of pollutants. This is typically done by heating the soil to a high temperature for a specified amount

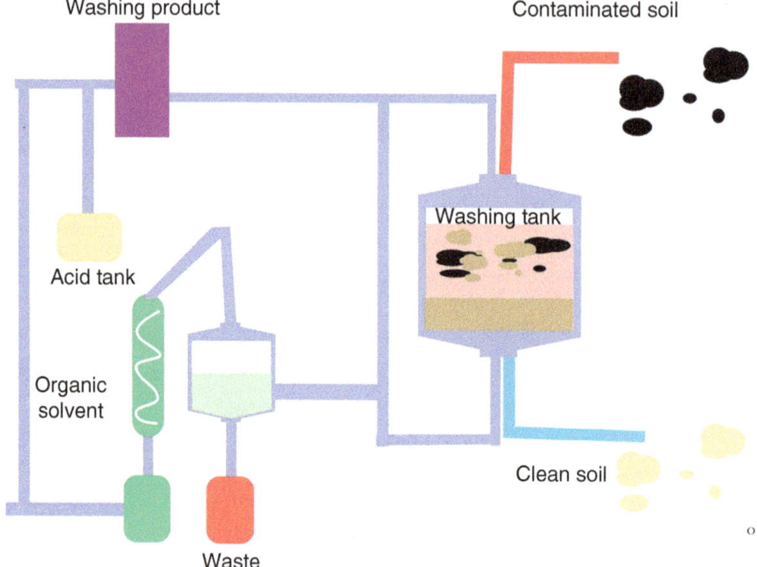

FIGURE 8.3 Scheme of washing of polluted soils

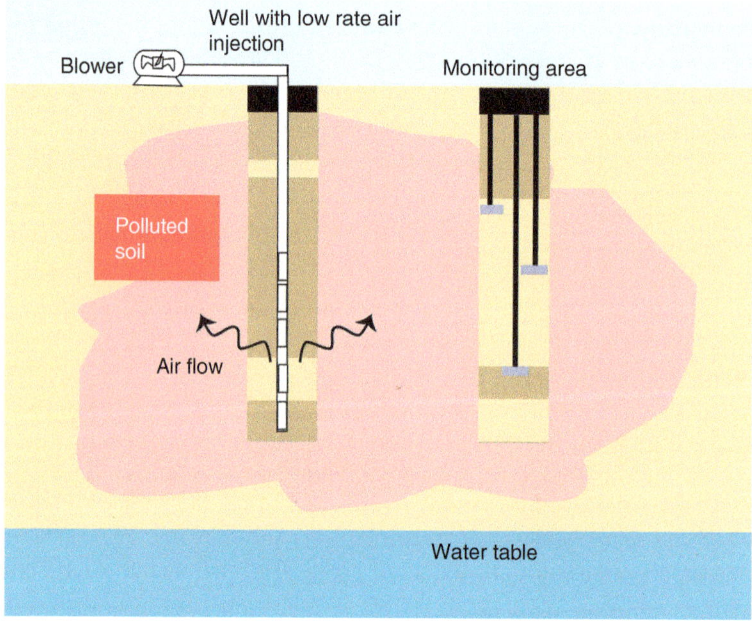

FIGURE 8.4 Scheme of aeration of polluted soils

of time, which causes the contaminants to break down or evaporate. There are several types of thermal treatment methods that can be used, including:

Incineration: This involves heating the soil to temperatures of 800–1,200 °C in an enclosed furnace, which causes the organic contaminants to break down into ash and gases (Figure 8.6).

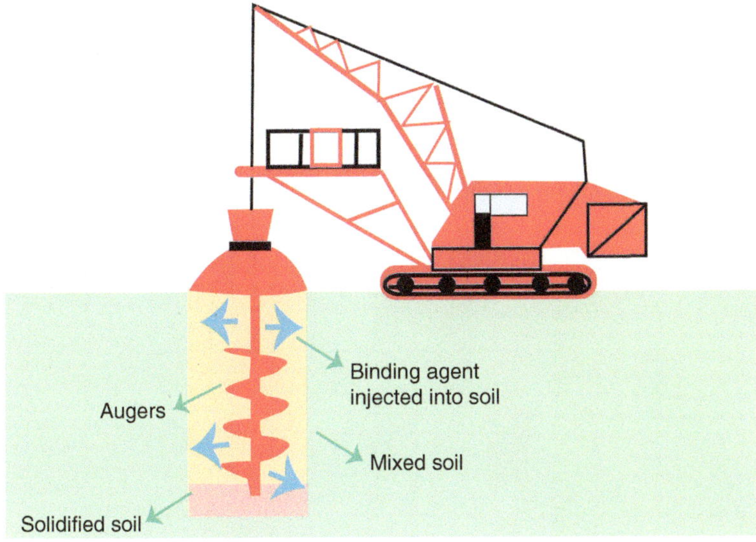

FIGURE 8.5 Solidification and stabilisation of polluted soil

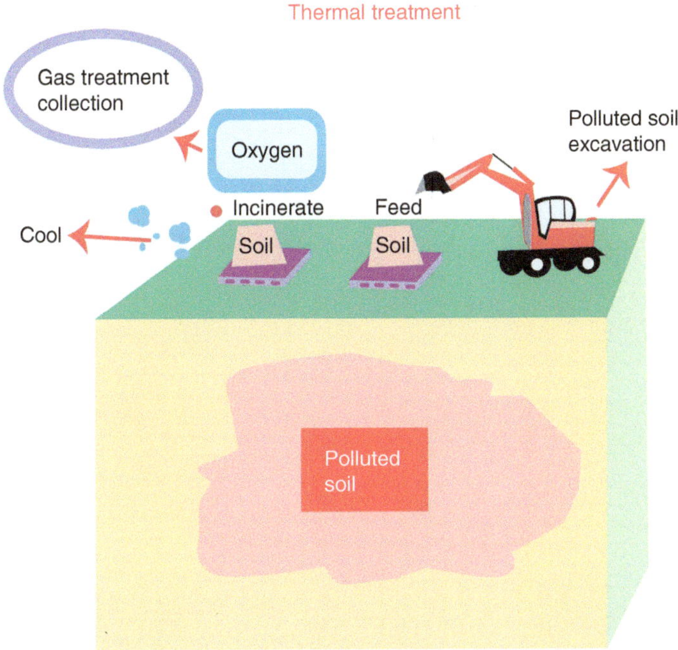

FIGURE 8.6 Incineration of pollutants in polluted soil

Thermal desorption: This process involves heating the soil to temperatures of 250–500 °C in a rotating kiln or other enclosed chamber, which causes the organic contaminants to evaporate and be captured in a gas treatment system (Figure 8.7).

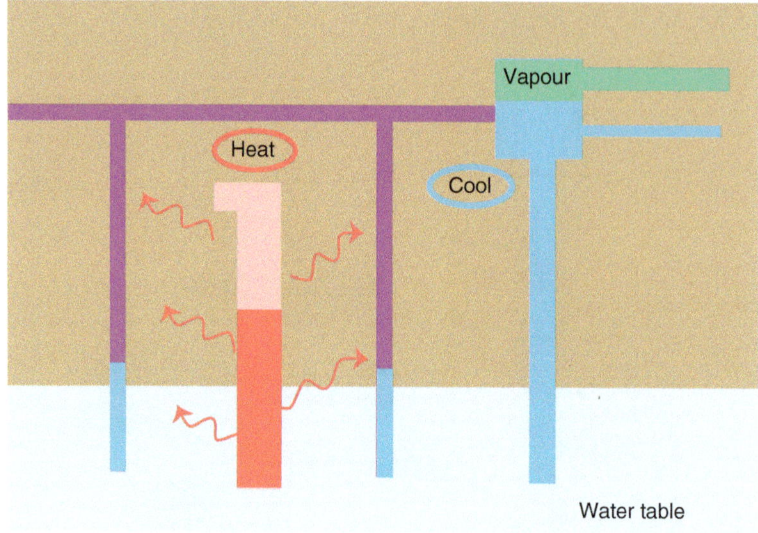

FIGURE 8.7 Thermal desorption of pollutants in soil

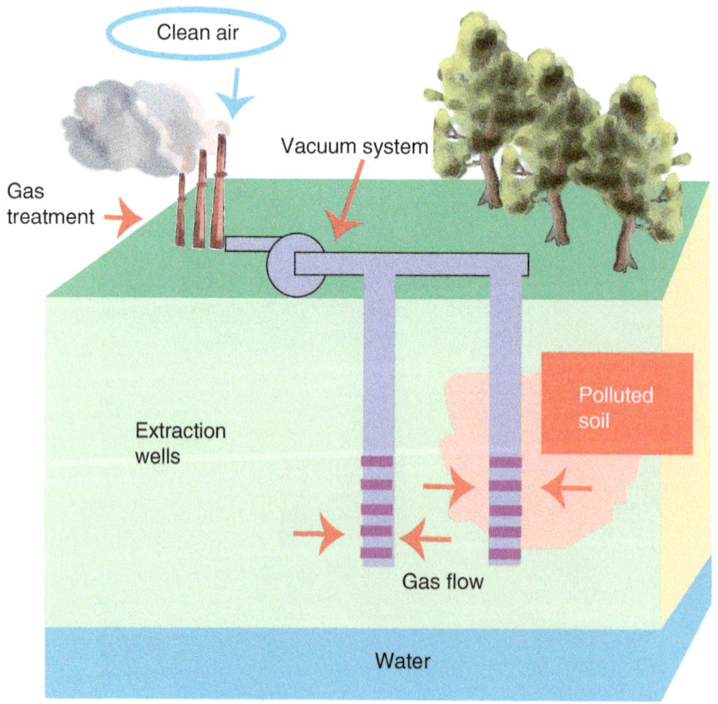

FIGURE 8.8 Soil vapour extraction

6. **Soil vapour extraction:** This involves heating the soil to temperatures of 50–100 °C and then extracting the resulting vapours, which contain VOCs, using a vacuum system (Figure 8.8).

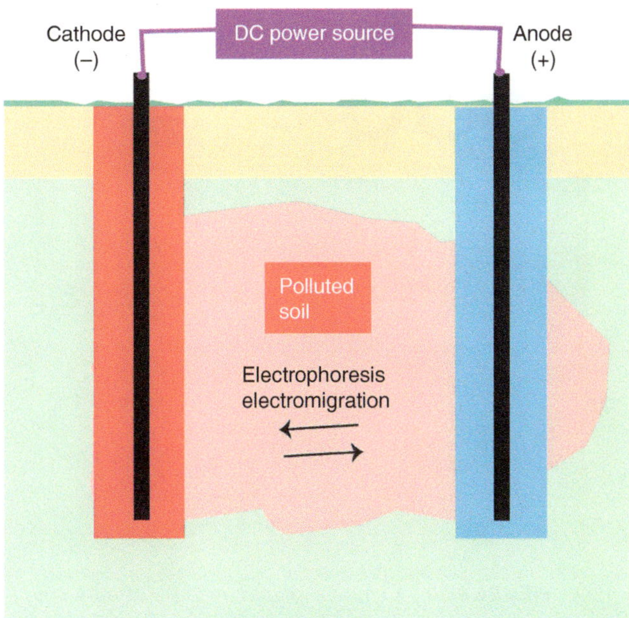

FIGURE 8.9   Electrokinetic remediation

7. **Pyrolysis:** This process involves heating the soil in the absence of oxygen to temperatures of 300–700 °C, which causes the organic contaminants to break down into gases and a solid residue.

8. **Electrokinetic remediation** is a technique used to clean up contaminated soil and groundwater by applying an electric field to the affected area. This process involves the use of electrodes, which are placed in the ground or in a groundwater well, and a power source that supplies an electric current to the electrodes. The electric field created by the electrodes causes charged particles, such as ions and colloids, to move through the soil and groundwater. This movement can help mobilise contaminants, such as heavy metals and organic compounds, which can then be captured and removed by various methods, such as pumping or adsorption onto an electrode (Figure 8.9). Electrokinetic remediation can be effective in treating a wide range of contaminants, including both organic and inorganic pollutants. It can also be used in a variety of soil types, including fine-grained soils such as clay.

8.9 Chemical Remediation

Chemical treatments can be used for soil remediation in a variety of situations, such as when the soil is contaminated with hazardous chemicals, pollutants or heavy metals. Here are some common chemical treatments for soil remediation:

Chemical oxidation: This involves using chemicals to break down or oxidise the contaminants in the soil. The chemical oxidants commonly used in soil remediation include hydrogen peroxide, permanganate, ozone and persulfate. These oxidants react with the contaminants in the soil and convert them into less toxic or non-toxic

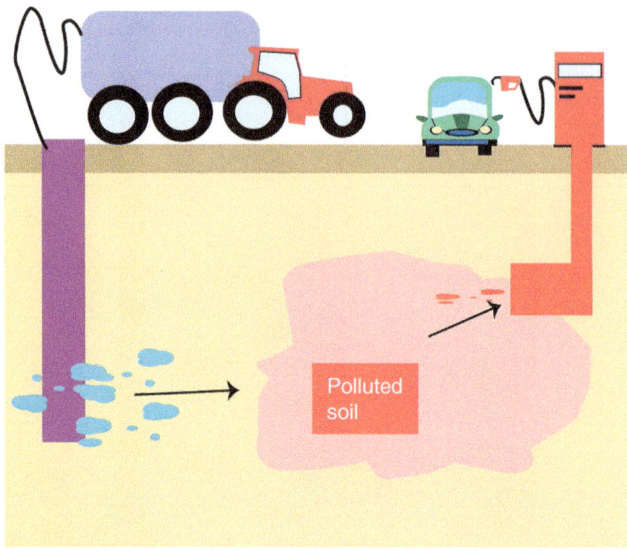

FIGURE 8.10 Chemical oxidation of pollutants

substances. Chemical oxidation can be used to treat a wide range of organic contaminants, including petroleum hydrocarbons, chlorinated solvents and pesticides. It is often used in combination with other soil remediation techniques, such as excavation, to remove contaminated soil. The effectiveness of the oxidation process depends on the type and concentration of the contaminants, the type of oxidant used and the soil conditions (Figure 8.10).

Chemical reduction is a commonly used technique for soil remediation, particularly for the treatment of soils contaminated with heavy metals, metalloids and other toxic compounds. The process involves adding a reducing agent to the soil, which chemically converts the contaminants into less harmful or non-toxic forms. One common reducing agent used for soil remediation is zero-valent iron (ZVI). ZVI can react with a wide range of contaminants, including chlorinated solvents, heavy metals and nitrates, to convert them into less harmful forms. ZVI works by donating electrons to the contaminants, which reduces them to their elemental or lower oxidation states. Another reducing agent that can be used for soil remediation is sodium dithionite. This chemical is a strong reducing agent that can reduce a wide range of contaminants, including heavy metals, chlorinated solvents and other organic compounds, to less toxic forms. Sodium dithionite works by breaking the bonds between the contaminants and the soil, allowing the contaminants to be removed or immobilised. Chemical reduction is often used in conjunction with other soil remediation techniques, such as soil vapour extraction, bioremediation and soil washing. The effectiveness of chemical reduction depends on the nature and extent of the contamination as well as the type and amount of reducing agent used.

Chemical stabilisation: Chemical stabilisation is a process used to reduce the mobility and availability of contaminants in soil. It involves the use of various chemical amendments to transform the contaminants into less toxic or less mobile forms. The aim of chemical stabilisation is to immobilise the contaminants in the soil, making them less likely to leach into groundwater or become airborne. It can be applied to a wide range of contaminants, including heavy metals, organic compounds and radioactive materials. Some common chemical amendments used for soil stabilisation include:

1. **Lime:** Lime is a commonly used chemical for stabilising soil contaminated with heavy metals. It raises the soil pH, which can cause the heavy metals to precipitate out of solution and become less mobile.

2. **Portland cement:** Portland cement is a common chemical stabiliser for contaminated soils. It reacts with the soil to form a hard, durable material that can immobilise contaminants.

3. **Fly ash:** Fly ash is a byproduct of coal-fired power plants and can be used as a chemical stabiliser for soils contaminated with heavy metals. It contains high levels of calcium and other elements that can react with the contaminants in the soil to form stable compounds.

4. **Phosphates:** Phosphates are commonly used to immobilise heavy metals in contaminated soils. They react with the metals to form stable compounds that are less likely to leach into groundwater.

5. **Organic matter:** Organic matter, such as lignin, compost or manure, can be used to stabilise soils contaminated with organic compounds. Organic matter can stimulate the growth of microorganisms that can break down contaminants. Chemical stabilisation is often used in combination with other remediation techniques, such as soil washing or excavation, to achieve more effective cleanup of contaminated sites.

8.10 Bioremediation

Bioremediation is the process of using living organisms, such as bacteria, fungi or plants, to detoxify and remove or degrade contaminants from contaminated soil. Bioremediation can be used to clean up a wide range of contaminants, including petroleum products, solvents, pesticides and other organic chemicals.

In bioremediation, microorganisms or plants break down the contaminants into harmless byproducts such as water and carbon dioxide. The process of bioremediation can occur naturally or be enhanced by adding nutrients or other materials to stimulate microbial activity. Bioremediation is often a cost-effective and environmentally friendly alternative to traditional methods of cleaning up contaminated sites, such as excavation and disposal.

There are two main types of bioremediation: in situ and ex situ.

1. **In situ bioremediation** involves treating the contaminated site directly. There are several different types of in-situ bioremediation, including:

   a. **Aerobic bioremediation:** This process uses microorganisms that require oxygen to break down pollutants.

   b. **Anaerobic bioremediation:** This process uses microorganisms that can function in environments with little or no oxygen.

   c. **Biostimulation:** This involves adding nutrients, such as nitrogen or phosphorus, to the contaminated area to promote the growth of naturally occurring microorganisms that can break down pollutants.

   d. **Bioaugmentation:** This involves adding specially selected microorganisms to the contaminated area to enhance the bioremediation process.

2. **Ex situ bioremediation** involves removing contaminated soil from the site and treating it in a separate location in a controlled environment. In this process, contaminated soil is excavated and brought to a treatment facility, where the pollutants are biodegraded by microorganisms under carefully controlled conditions. The contaminated soil is typically placed in a lined bioreactor, where it is mixed with water, oxygen and nutrients to create an environment that is conducive to the growth of microorganisms. The microorganisms then break down the pollutants into harmless byproducts such as water and carbon dioxide.

   Ex situ bioremediation can be used to treat a wide range of contaminants, including petroleum hydrocarbons, solvents, pesticides and heavy metals. It is often used in conjunction with other remediation techniques, such as excavation or soil vapour extraction, to provide a more complete cleanup solution.

3. **Phytoremediation** is a process in which plants are used to remove, detoxify or immobilise environmental pollutants from contaminated soils. This is an environmentally friendly and cost-effective method of remediation that has gained a lot of attention in recent years (Figure 8.11). There are several ways in which plants can be used for phytoremediation of polluted soils:

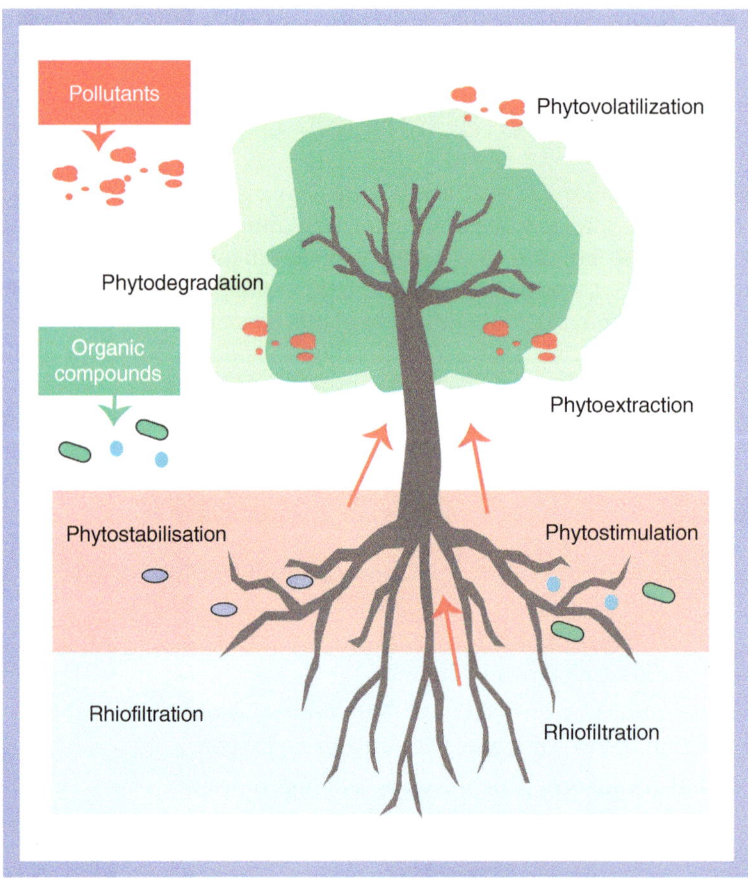

FIGURE 8.11 Phytoremediation

a. **Phytoextraction:** Plants can accumulate and concentrate pollutants in their tissues, which can then be harvested and disposed of properly. This method is particularly effective for heavy metals.

b. **Rhizofiltration:** This method involves using plants with deep root systems to absorb and filter contaminants from the soil. This is particularly effective for hydrophobic organic compounds.

c. **Phytodegradation:** Certain plants can break down organic pollutants through a process called phytodegradation. This is particularly effective for petroleum hydrocarbons.

d. **Phytostabilisation:** This method involves using plants to immobilise pollutants in the soil, reducing their bioavailability and preventing them from spreading.

Phytoremediation has several advantages over traditional remediation methods such as excavation and incineration. It is cost-effective, environmentally friendly and does not disturb the natural ecosystem. However, it is not suitable for all types of contaminants or soil types, and it can be a slow process. Additionally, the harvested plants may need to be disposed of properly to prevent further pollution.

Phytoremediation techniques have been used in various settings to remove or detoxify pollutants in soil to treat contaminated industrial sites such as oil refineries, chemical factories and mine tailings; to detoxify pollutants in landfill sites, such as heavy metals and organic compounds; to remove excess nutrients and contaminants from agricultural land, improving soil quality and reducing runoff into nearby water bodies; and to mitigate pollution in urban environments, such as along roadsides, in parks and on green roofs.

8.11 Treatment of Polluted Soil: Barriers

The displacement of pollutants in soil can occur due to various factors, including natural processes such as erosion and leaching as well as human activities such as construction and industrial activities. When pollutants are displaced, they can contaminate groundwater, surface water and even air, posing significant risks to human health and the environment. To prevent the displacement of pollutants in soil, barriers can be constructed to contain them. Barrier systems can be designed to prevent the movement of pollutants through soil and groundwater by creating a physical or chemical barrier. Physical barriers can include liners or impermeable membranes, while chemical barriers can include reactive materials that can immobilise or degrade pollutants.

There are various types of barrier systems that can be used in soil construction, including:

1. **Sheet pile walls:** Sheet pile walls are commonly used in soil remediation projects to contain contaminated soil and prevent its spread to surrounding areas. Sheet piles are made of steel or other materials and are driven vertically into the ground to create a barrier that prevents the contaminated soil from spreading (Figure 8.12).

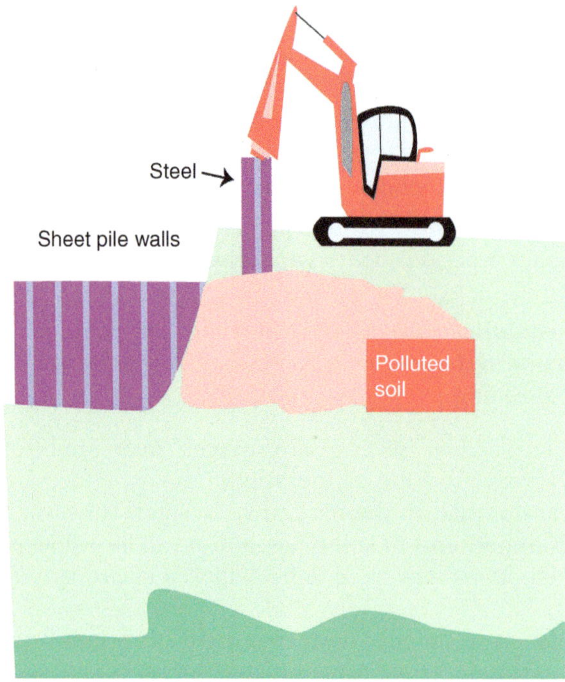

FIGURE 8.12   Sheet pile walls

2. **Impermeable membranes:** These are synthetic membranes that are used to create a physical barrier that prevents the migration of pollutants through the soil. The use of impermeable membranes can help prevent the spread of contaminants and protect surrounding areas from further contamination.

   There are different types of impermeable membranes that can be used for soil remediation, including synthetic liners, geotextiles and concrete barriers. These membranes can be used in various ways, such as creating a barrier around the contaminated area, lining the bottom of a pond or lagoon to prevent leakage or capping contaminated soil to prevent the release of pollutants into the air or water. One of the benefits of using impermeable membranes for soil remediation is that they can be effective in reducing the risk of exposure to hazardous chemicals or pollutants. They can also help prevent the spread of contamination to other areas, which can reduce the overall cost of remediation. However, it's important to note that impermeable membranes are not a one-size-fits-all solution for soil remediation. The effectiveness of a membrane depends on factors such as the type of contaminant, the concentration of the contaminant and the site-specific conditions. In addition, impermeable membranes can be costly to install and maintain, and they may not be feasible in all situations (Figure 8.13).

3. **Grout curtains:** Grout curtains involve injecting grout into the soil to create a barrier that immobilises pollutants and prevents their migration. This technique involves the injection of a cementitious grout into the soil to create a barrier that prevents the flow of water and contaminants. To create a grout curtain, a series of holes are drilled into the soil to a predetermined depth and spacing. Then, a mixture of water and cement is pumped under high pressure into the

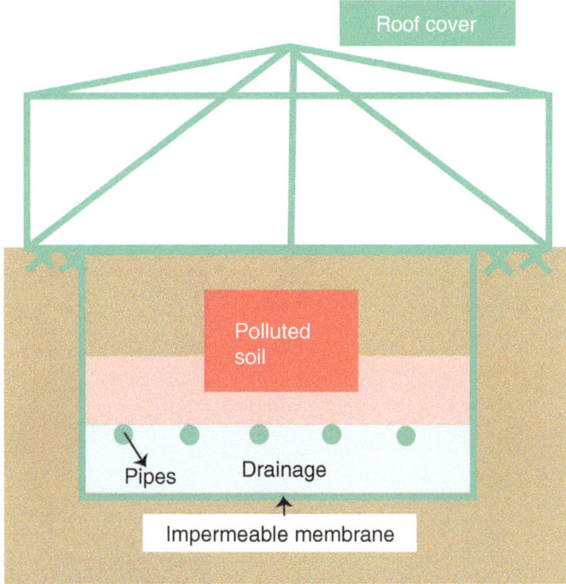

FIGURE 8.13   Impermeable membranes

holes, filling the voids and creating a solid barrier. Grout curtains can be used in a variety of soil types, including clay, silt and sand. They are commonly used in areas with contaminated groundwater, such as industrial sites, landfills and oil and gas facilities.

4. **Reactive barriers:** Reactive barriers use reactive materials to immobilise or degrade pollutants as they flow through the barrier (Figure 8.14). The choice of barrier system depends on the type of pollutant, the soil type and other factors specific to the site. It is important to design and implement barrier systems carefully to ensure their effectiveness in preventing contamination. Active barriers used for soil remediation can be broadly categorised into two types: chemical and biological.

   a. **Chemical barriers:** These barriers use chemical reactions to remediate contaminants in the soil. Some examples include:

      i. Permeable reactive barriers (PRBs) that use reactive materials, such as iron, to treat contaminants as they pass through the barrier.

      ii. Solidification and stabilisation (S/S) techniques that use additives to bind and immobilise contaminants in the soil.

      iii. Chemical oxidation or reduction techniques that use oxidising or reducing agents to break down contaminants in the soil.

   b. **Biological barriers:** These barriers use microorganisms to degrade contaminants in the soil. The barrier is typically composed of a layer of living or dead biological material, such as plant roots, microbial biomass or a combination of both. This layer is placed in the soil between the contaminated area and the surrounding environment.

The semipermeable nature of the barrier is achieved through the use of natural or synthetic materials that allow for the passage of water and certain nutrients while

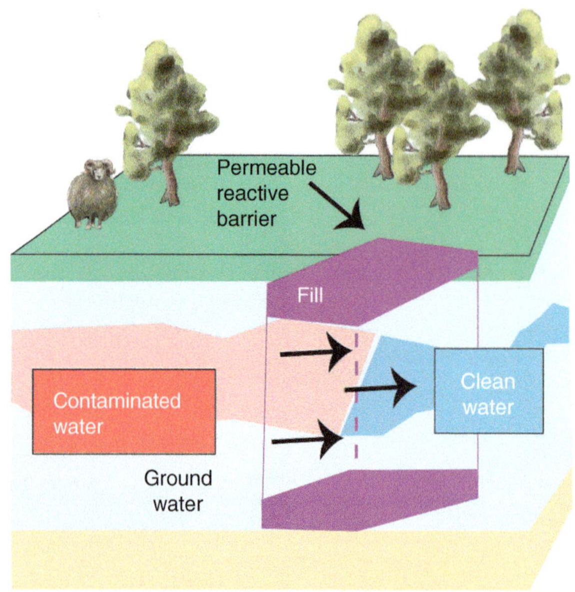

FIGURE 8.14   Reactive barriers

blocking the movement of contaminants. The barrier can be designed to target specific contaminants based on their size, charge or other physical or chemical properties.

Both chemical and biological barriers can be effective for soil remediation, and the selection of the appropriate barrier depends on the specific contaminants present, site characteristics and other factors. In some cases, a combination of chemical and biological barriers may be used for optimal remediation results.

CHAPTER 9

Water Pollution

9.1 Water Pollution

Water pollution refers to the contamination of bodies of water, such as rivers, lakes, oceans and groundwater, with harmful substances that make the water unfit for human use or consumption or harm ecosystems. Water pollution can be caused by a variety of sources, both natural and anthropogenic, including industrial waste, sewage, agricultural runoff, oil spills etc.

Water pollution can be divided into three major types.

1. **Physical pollution:** This type of pollution involves the presence of physical objects, particles of sediments in the water or anomalous values of temperature, light or noise.

2. **Chemical pollution:** This is caused by the discharge of harmful chemicals such as pesticides, fertilisers, radionuclides or industrial waste into water bodies. These chemicals can cause a range of health problems for humans and animals that come into contact with the water.

3. **Biological pollution:** This is caused by the presence of harmful bacteria, viruses and other microorganisms in water bodies. This can lead to the spread of water-borne diseases such as cholera, typhoid and hepatitis.

Water quality can be assessed based on various parameters that measure different aspects of its physical, chemical and biological characteristics.

9.2 Organoleptic Parameters: Taste, Odour and Colour

Taste and Odour

Polluted waters can have a variety of tastes and odours, depending on the type and level of contaminants present. The odour of polluted water can be caused by various factors, including:

Understanding the Chemistry of the Environment, First Edition. Francisco G. Calvo-Flores.
© 2025 John Wiley & Sons Ltd. Published 2025 by John Wiley & Sons Ltd.

1. **Decomposition of aquatic life and residual organic matter:** The decomposition of dead fish, algae or other aquatic organisms can produce a strong odour and displeasure taste.

2. Chemicals, such as pesticides, fertilisers and industrial waste, can cause the water to have a strong chemical taste and odour.

3. **Sewage:** Sewage contamination can cause a foul odour as it contains organic matter that produces gases as it decomposes (H_2S, NH_3, organic acids, etc.).

It's important to note that the odour of polluted water can also indicate the presence of harmful substances, such as bacteria, viruses or chemicals, which can pose a health risk to humans and animals. Therefore, it's important to avoid contact with water that has a foul odour and to seek advice from local authorities on how to properly dispose of contaminated water.

Polluted waters can have a variety of tastes, depending on the type and level of contaminants present. Here are some examples:

1. **Metallic or bitter taste:** Heavy metals like lead, copper and zinc can give water a metallic or bitter taste. This taste can also be caused by industrial waste, such as from mining or metal processing.

2. **Chlorine, chemical or medicinal taste or odour:** Chlorine is often added to water as a disinfectant, but if the levels are too high, it can give water a strong chlorine taste and odour.

3. **Earthy or musty odour:** This odour is often caused by algae or other aquatic plants that produce geosmin and other compounds that give water an earthy or musty smell. This can be particularly common in stagnant water or bodies of water with high levels of nutrients.

4. **Fishy or rotten egg odour:** These odours can be caused by organic matter such as decaying plants or animals or by sulfur bacteria that produce hydrogen sulfide.

5. **Petroleum or gasoline taste or odour:** If there is an oil spill or leak, water can take on a strong petroleum or gasoline smell.

It's important to note that some contaminants in water may not have a taste or odour, making it difficult to detect their presence. Therefore, it's important to test water regularly for contaminants and to follow local guidelines for safe drinking water.

Colour

The colour of water can vary depending on various factors, such as the amount of soluble organic and inorganic matter and suspended particles present in it. Generally, natural water appears colourless or slightly blue due to the absorption and scattering of light by water molecules, but sometimes, water shows colour which is a clue about the presence of pollutants. Some common natural water colours include:

1. **Brown:** Brown water is often caused by tannins leaching from decaying plant material, such as leaves or bark. Tannins are a group of compounds found in many plants that can give water a brown or tea-like colour.

2. **Red:** Red water can be caused by high levels of iron in the water. Iron can come from natural sources like soil and rocks or from human activities like mining and industrial processes.

3. **Yellow:** Yellow water is often caused by dissolved organic matter, which can come from plant material or soil. Humic acid and fulvic acid are two types of organic matter that can give water a yellow colour.

4. **Green:** Green water can be caused by algae blooms when there is an excess of nutrients in the water or dissolved minerals such as calcareous materials.

5. **Blue:** Blue water can be caused by the scattering of sunlight in the water as well as by the reflection of the sky. Some bodies of water, like the ocean, can appear blue because of their depth and the way they absorb and scatter light.

True and Apparent Water Colour

- **True colour:** The so-called **true** colour of natural waters refers to the colour of the water that is caused by the presence of dissolved organic and inorganic substances, such as tannins, humic acids and iron. These substances can give water a yellow, brown or reddish hue.

- **Apparent colour** is the colour of a body of water as reflected from the surface of the water and it is caused by both dissolved and suspended components in the water. Apparent colour may also be changed by variations in sky colour or the reflection of nearby vegetation. After filtering suspended materials, the true colour of the water remains.

9.3 Physical Parameters: Temperature, Turbidity, and Conductivity

Temperature

Thermal pollution is the increase in temperature of natural bodies of water, such as lakes, rivers and oceans, due to human activity. This increase in temperature can be caused by the discharge of heated water from industrial processes, power plants and other sources.

When water is heated, its physical and chemical properties change. As the temperature of water increases, its ability to hold dissolved oxygen (DO) decreases, which can harm aquatic organisms. Fish and other aquatic animals have adapted to survive within a certain range of water temperatures and a sudden increase in temperature can cause stress or even death. Temperature increases are also able to modify the concentration of dissolved species, both organic and inorganic.

Thermal pollution can also affect the growth and reproduction of aquatic plants, algae and other organisms that form the base of the food chain. When the temperature of the water rises above a certain threshold, it can cause blooms of harmful algae that can lead to oxygen depletion and fish kills.

In addition to its effects on aquatic life, thermal pollution can also impact water quality and availability for human use. It can alter the chemical composition of water, making it less suitable for drinking, irrigation and industrial processes.

Therefore, it is important to monitor and regulate the discharge of heated water into natural bodies of water to prevent thermal pollution and protect aquatic ecosystems.

Turbidity

Turbidity refers to the measure of the degree to which water loses its transparency due to the presence of suspended particles. Natural waters such as rivers, lakes and oceans can have varying levels of turbidity depending on several factors, including weather conditions, geological features and human activities. In polluted waters, high levels of turbidity can indicate the presence of various contaminants, including organic matter, sediment, nutrients, bacteria and chemicals. These contaminants can be harmful to aquatic life and can also affect water quality, making it unsafe for human consumption.

Some natural factors that can increase the turbidity of water include erosion of rocks and soil, organic matter decay and algal blooms. Human activities such as land development, agriculture and mining can also increase turbidity by increasing the amount of sediment and pollutants in the water.

Measuring turbidity is important because it can affect water quality and aquatic ecosystems. High levels of turbidity can reduce the amount of light that penetrates the water, which can affect photosynthesis and the growth of aquatic plants. It can also interfere with fish feeding and breeding habits.

There are several methods for measuring turbidity, including using a turbidimeter, which measures the amount of light scattered by suspended particles in the water. The results are usually reported in NTUs (nephelometric turbidity units) or FTUs (formazin turbidity units). The World Health Organization (WHO) recommends a maximum turbidity level of 5 NTUs for drinking water, although this can vary depending on local regulations and standards.

Conductivity

The conductivity of water is a measure of its ability to conduct electrical current. It is a useful indicator of the total dissolved solids (TDS) or dissolved salts in water. The higher the TDS or salinity of the water, the greater its conductivity.

Natural waters such as rivers, lakes and groundwater can have a wide range of conductivity values depending on the types and amounts of dissolved ions present. Typically, freshwater has a conductivity range of 50–1,500 to 30,000–60,000 (μS/cm), while seawater has a much higher conductivity range of 30,000–60,000 μS/cm.

Polluted waters may contain a variety of pollutants, such as chemicals, heavy metals, organic compounds and microorganisms, all of which can affect conductivity in different ways. For example, some pollutants may increase conductivity by adding ions to the water, while others may decrease conductivity by interfering with the movement of charged particles.

TABLE 9.1   Conductivity of several types of water

Water type	Conductivity range (µS/cm)
Deionized water	0.05–0.1
Drinking water	50–800
Freshwater rivers	50–1,500
Seawater	50,000–60,000
Moderately polluted water	800–10,000
Heavily polluted water	10,000+

In general, highly polluted waters tend to have higher conductivity levels than clean waters, as the presence of pollutants can increase the concentration of ions in the water. However, the relationship between conductivity and pollution can be complex and other factors such as temperature, pH and salinity can also affect conductivity levels in water.

Therefore, it is important to measure conductivity along with other water quality parameters to fully understand the nature and extent of water pollution.

Conductivity can be measured using a handheld conductivity metre or a more sophisticated device like a multi-parameter water quality metre. It is an important parameter to monitor in natural waters because changes in conductivity can indicate changes in water quality, such as the presence of pollutants or changes in salinity due to natural or human-caused factors (Table 9.1).

9.4   Chemical Parameters: Salinity, Hardness, Total Dissolved Solids, Alkalinity Acidity and pH, Radioactivity, and Dissolved Oxygen

Chemical parameters of water refer to various characteristics and properties that are used to assess the quality and composition of water. These parameters are essential for understanding the health of aquatic ecosystems and ensuring the safety of water for human use. Some key chemical parameters commonly measured in water bodies are described below.

Salinity

Salinity is defined as the total amount of salts dissolved in water. Water bodies can have varying levels of salinity depending on factors such as location, climate, geology and human activities. Salinity is usually expressed in terms of conductivity in Siemens per metre (S/m) or units of concentration or as

- **Grams per litre (g/L) or parts per million (ppm):** These units are commonly used in freshwater systems or for measuring the salinity of specific solutions or samples. They represent the number of grams of dissolved salts per litre of water or the number of parts of dissolved salts per million parts of water, respectively.

- **Parts per thousand (ppt) or ‰:** This is a commonly used unit for expressing salinity in oceanography and marine biology. It represents the number of grams of dissolved salts in 1,000 g of seawater. For example, seawater with a salinity of 35 ppt contains 35 g of dissolved salts per 1,000 g of seawater.

The salinity of natural waters can have important implications for ecosystems and human activities (see Table 9.2).

For example, changes in salinity can affect the distribution and abundance of aquatic plants and animals and can also affect the quality of water for drinking and irrigation purposes.

Plants have evolved various mechanisms to cope with salinity, such as osmotic adjustment, ion exclusion and antioxidant defence. However, these mechanisms are not always sufficient to counteract the negative effects of salinity, especially in plants that are not well adapted to salt-rich environments.

An excess of salinity can have a negative impact on plant growth and health. Here are some of the effects of salinity on plants:

1. **Water stress:** Salinity can cause water stress in plants by reducing the availability of water in the soil. This is because the presence of high levels of salt in the soil can reduce the ability of plants to take up water from the soil.

2. **Ion toxicity:** Salt ions can accumulate in plant tissues, leading to ion toxicity. This can disrupt various cellular processes and cause damage to plant cells.

3. **Nutrient imbalance:** High levels of salt in the soil can also disrupt the balance of nutrients in plants. For example, excess sodium can interfere with the uptake of other essential nutrients like potassium, calcium and magnesium.

TABLE 9.2 **Salinity of different types of natural waters**

Water type	Salinity range (ppm)
Freshwater	0–500
Brackish water	500–30,000
Estuaries	1,000–35,000
Coastal waters	5,000–35,000
Ocean water	35,000–50,000
Salt pans	50,000–300,000
Salt lakes	100,000–350,000
Dead sea	290,000–340,000
Great salt lake	150,000–280,000
Salt-concentrated ponds	250,000–350,000

4. **Reduced growth and yield:** Salinity can significantly reduce plant growth and yield. This is because the stress caused by high salt levels can limit photosynthesis, reduce the efficiency of nutrient uptake and cause damage to plant tissues.

Polluted waters can have higher or lower salinity levels, depending on the source and type of pollutants. Pollutants such as heavy metals and organic compounds can affect the salinity levels of water by altering the balance of ions in the water. For example, heavy metals such as lead and cadmium can bind to chloride ions in saltwater, reducing the overall concentration of chloride ions and thus reducing the salinity of the water.

Hardness

Hardness refers to the concentration of dissolved calcium and magnesium ions in water, usually measured in terms of calcium carbonate equivalents (CCE) or in units of ppm or milligrams per litre (mg/L). Table 9.3 describes the main salts responsible for this parameter.

These salts are commonly found in the earth's crust and can dissolve in water. When water comes into contact with rocks and minerals containing calcium and magnesium compounds, such as limestone or dolomite, it begins to dissolve these compounds, releasing the respective ions into the water.

Hardness is not strictly a pollution phenomenon, but the presence of these calcium and magnesium salts can cause a variety of undesirable properties in water, such as:

1. **Soap scum:** Hard water reacts with soap to form a sticky residue called soap scum, which can build up on sinks, bathtubs and other surfaces.

2. **Clogged pipes and appliances:** The minerals in hard water can accumulate inside pipes and appliances, causing blockages and reducing their efficiency.

TABLE 9.3   Salts responsible for water hardness

Salt	Chemical formula
Aluminium sulfate	$Al_2(SO_4)_3$
Calcium carbonate	$CaCO_3$
Calcium bicarbonate	$Ca(HCO_3)_2$
Calcium sulfate	$CaSO_4$
Calcium chloride	$CaCl_2$
Iron sulfate	$FeSO_4$
Magnesium carbonate	$MgCO_3$
Magnesium bicarbonate	$Mg(HCO_3)_2$
Magnesium sulfate	$MgSO_4$
Magnesium chloride	$MgCl_2$

3. **Reduced water flow:** The buildup of minerals in pipes can also reduce the flow of water through them.

4. **Stains:** Hard water can leave stains on clothes, dishes and other surfaces.

5. **Dry skin and hair:** The minerals in hard water can strip natural oils from skin and hair, leaving them dry and itchy.

6. **Increased energy consumption:** Appliances that use hard water, such as water heaters and washing machines, may use more energy to heat and run, leading to higher energy bills.

7. **Reduced lifespan of appliances:** The minerals in hard water can also reduce the lifespan of appliances such as dishwashers and washing machines, leading to more frequent repairs or replacements.

8. **Unpleasant taste:** Some people find that hard water has an unpleasant taste, which can affect the taste of drinks and food made with it. Hard water can contribute to the formation of both kidney and biliary stones. Hard water contains high levels of minerals such as calcium and magnesium. When these minerals accumulate in the kidneys or gallbladder, they can form solid masses known as stones.

9. **Kidney and biliary diseases:** Calcium salts from hard water diseases can form small, hard deposits in kidneys that can cause significant pain when they pass through the urinary tract. In addition to hard water, other factors that can contribute to the formation of kidney stones include dehydration, a high-protein diet and certain medical conditions. Biliary stones, also known as gallstones, can also form as a result of hard water. The minerals in hard water can combine with cholesterol and other substances in the bile to form stones. These stones can cause pain and other symptoms and may require treatment such as surgery or medication to dissolve or remove them. To reduce the risk of developing kidney or biliary stones, it is important to drink plenty of water to stay hydrated, avoid a diet high in protein and sodium and limit consumption of alcohol and caffeine. If you live in an area with hard water, using a water softener or filtration system may also help reduce the mineral content of your water.

There are two types of hardness:

1. **Temporary hardness:** It is caused by the presence of dissolved bicarbonate minerals in water, which can be removed by boiling the water. Bicarbonates decompose to form insoluble carbonates and CO_2 (Scheme (9.1).

$$Ca(HCO_3)(aq) \rightarrow CO_2(g) + H_2O + CaCO_3(s)$$
$$Mg(HCO_3)(aq) \rightarrow CO_2(g) + H_2O + MgCO_3(s)$$

(Scheme 9.1)

2. **Permanent hardness:** It is caused by the presence of other salts, such as $CaCl_2$, $MgCl_2$, $CaSO_4$, $MgSO_4$, $FeSO_4$ and $Al_2(SO_4)_3$, which cannot be removed by boiling.

The hardness of water is typically expressed in units of mg/L or ppm of calcium carbonate ($CaCO_3$) equivalents. Water with a hardness of less than 60 mg/L (or ppm) is considered soft, while water with a hardness greater than 120 mg/L (or ppm) is

TABLE 9.4   **Classification of waters based on hardness**

Classification	Hardness range (mg/L or ppm)
Soft	0–60
Slightly hard	61–120
Moderately hard	121–180
Hard	181–250
Very hard	Above 250

considered hard. Water with a hardness between these levels is considered moderately hard. The units used in the table are mg/L. The classification of waters according to their salinity can vary slightly depending on the source or organisation, but this is a commonly used range (Table 9.4).

The hardness of water can be measured using a variety of methods, including titration with a standardised solution, colorimetric tests and electronic instruments. Water softening techniques, such as ion exchange, reverse osmosis and lime softening, can be used to remove minerals and reduce the hardness of water.

Total Dissolved Solids (TDS)

TDS refer to the total amount of inorganic and organic substances that are dissolved in water. TDS can include minerals, salts, metals and other compounds that are present in water. High levels of TDS can affect the taste, odour and appearance of water and can also have health implications if certain substances are present in excessive amounts. TDS is typically measured in the following units:

- **Parts per million (ppm):** This is the most common unit for measuring TDS. One ppm is equivalent to one milligram of dissolved solids per litre of water.

- **Milligrams per litre (mg/L):** This unit is also used to measure TDS, with 1 mg/L being equivalent to 1 ppm.

- **Parts per billion (ppb):** This unit is used for TDS levels that are very low. One ppb is equivalent to one microgram of dissolved solids per litre of water.

The acceptable level of TDS in drinking water varies depending on the source and the intended use of the water. Generally, TDS levels below 500 ppm are considered acceptable for drinking water, while levels above 1,000 ppm may affect the taste and have health implications. However, the specific TDS levels that are considered safe or acceptable may vary depending on the specific substances present in the water and local regulations.

In polluted waters, the TDS levels may be higher than in clean water. This is because pollutants can dissolve in water and contribute to the TDS levels. Elevated levels of TDS can have negative impacts on aquatic life and can also make the water unsuitable for drinking or other uses. The specific TDS levels that are considered harmful depend on the type of pollutant and the intended use of the water. For

TABLE 9.5	Common values of total dissolved solids (TDS) for different types of waters
Water type	**TDS range (mg/L)**
Freshwater lakes	50–500
Drinking water	50–1,000
Rivers	100–1,000
Groundwater	100–10,000
Seawater	Around 35,000
Brackish water	1,000–35,000
Mineral water	Varies greatly
Tap water	Varies depending on source
Distilled water	Close to 0
Spring water	Varies depending on source

example, drinking water standards typically set a maximum allowable TDS level of 500 mg/L, while the EPA recommends a TDS level of less than 1,000 mg/L for aquatic life protection.

Monitoring TDS levels in polluted waters is important for identifying and mitigating water pollution. Treatment methods, such as reverse osmosis, distillation and ion exchange, can be used to remove TDS from water and improve its quality. Common values of TDS for different types of waters are described in Table 9.5.

Alkalinity Acidity and pH

Alkalinity is defined as the quantitative capacity of water to buffer or neutralise acids. It is dependent on the presence of certain chemicals in the water, such as bicarbonates, carbonates and hydroxides. Acidity is defined as the quantitative ability to buffer or neutralise bases.

In the context of water bodies, acidity is often a result of dissolved acids or hydrolyzed cationic salts, which can enter water systems through various processes like industrial discharges, atmospheric deposition and runoff from polluted areas. Natural processes, such as the decomposition of organic matter, can also contribute to the acidity of water.

The acidity of water bodies is typically measured by the pH scale. Some factors that can influence the pH level of natural water sources include:

1. **Dissolved minerals:** Certain minerals, such as limestone, can increase the alkalinity of water.

2. **Organic matter:** Decaying organic matter can produce acidic compounds that can lower the pH of water.

3. **Acid rain:** Rainwater with a pH below 5.6 is considered acidic and can lower the pH of natural water sources.

TABLE 9.6   pH values for different natural waters

Water type	pH range
Pure water	7
Rainwater	5.0–5.5
Surface water	6.0–8.5
Drinking water	6.5–8.5
Groundwater	6.0–8.5
Seawater	7.5–8.4
Freshwater lakes	6.0–8.5
Acid rain	<5.0
Polluted water	Varies widely

4. **Soil type:** The type of soil in the area can affect the pH level of nearby water sources.

5. **Volcanic activity:** Volcanic ash and gas can release acidic compounds into the air and water.

6. **Human activities:** Mining, industrial and municipal waste, accidental spills, etc.

Generally, freshwater tends to have a pH of around 6.5–8.5, with 7 being neutral. However, it's important to note that human activities such as industrialization, mining and agricultural practices can affect the pH levels of freshwater. Seawater, on the other hand, typically has a pH of around 7.5–8.4, with an average of about 8.1. This slightly alkaline pH is maintained by the presence of dissolved carbon dioxide and bicarbonate ions, which are in turn influenced by factors such as temperature and the amount of sunlight penetrating the water (Table 9.6).

It is important to monitor the pH level of natural water sources, as extreme levels of acidity or alkalinity can have negative effects on aquatic life and human health. For example, water with a pH below 6 can be harmful to fish and other aquatic organisms, while water with a pH above 8.5 can cause skin irritation and other health issues in humans.

However, when water is contaminated with pollutants, pH levels can fall outside this range. For instance, if water is contaminated with acid rain or industrial chemicals such as sulfuric acid, nitric acid or hydrochloric acid, the pH can drop significantly and become highly acidic, with values below 4.0.

On the other hand, if water is contaminated with substances that are alkaline in nature, such as ammonia (NH_3) or certain detergents, the pH can increase, becoming more basic, with values above 8.5.

Radioactivity

Radioactivity is present in natural waters due to the presence of naturally occurring radioactive isotopes such as uranium, thorium and potassium-40 from geological materials or human activities such as nuclear power plants, nuclear waste disposal sites and nuclear accidents.

The most common radioactive isotopes found in water bodies are radium, uranium and radon. These isotopes decay over time and emit ionising radiation in water, which can be measured by several parameters, including:

- **Activity concentration:** It is the amount of radioactive material present in the water, expressed in Becquerels per litre (Bq/L). The activity concentrations of different isotopes can be measured separately using gamma spectrometry or liquid scintillation counting.

- **Total alpha and beta activity:** This parameter measures the total amount of alpha and beta particles emitted by all radioactive isotopes present in the water sample. It is expressed in Bq/L and can be measured using a liquid scintillation counter.

- **Specific radionuclides:** It is the concentration of a specific radionuclide in water, usually expressed in Bq/L. The most common radionuclides found in water are radium-226, radon-222, tritium and uranium.

- **Gross alpha and beta activity:** This parameter measures the total amount of alpha and beta particles emitted by all radioactive isotopes present in the water sample, except for radon-222. It is expressed in Bq/L and can be measured using a liquid scintillation counter.

- **Radon concentration:** Radon is a colourless, odourless and tasteless gas that is produced by the decay of radium-226. It is a major contributor to natural background radiation exposure. Radon concentration in water is usually expressed in Bq/L or picocuries per litre (pCi/L). It can be measured using a liquid.

Radioactivity in water can be measured using a variety of techniques, including:

1. **Geiger counter:** A handheld Geiger counter can be used to measure the level to reduce the risk of exposure to radioactive substances in water bodies, it is important to regularly monitor water quality and take appropriate measures to ensure that radioactive contaminants are kept at safe levels. This may involve fungi, the treatment of water sources or the implementation of protective measures such as the installation of barriers or shields of radioactivity in water. This device detects ionising radiation and produces an audible or visual signal indicating the level of radiation present.

2. **Liquid scintillation counter:** This is a laboratory instrument used to measure the radioactivity in liquid samples. It works by adding a scintillation cocktail to the water sample, which absorbs the radiation and produces light. The light is then measured by the instrument, which calculates the radioactivity level.

3. **Gamma spectroscopy:** This technique involves using a gamma ray spectrometer to detect the energy levels of gamma rays emitted by radioactive isotopes in the water sample. The spectrum obtained can be used to identify the specific isotopes present and their concentration.

4. **Alpha spectroscopy:** Similar to gamma spectroscopy, alpha spectroscopy measures the energy levels of alpha particles emitted by radioactive isotopes in the water sample. This technique is used to detect isotopes such as radium and uranium.

It is important to note that different types of radiation have different levels of danger and the measurement of radioactivity in water is an important safety consideration, especially in relation to public health. The risk of health effects depends on the amount and type of radioactive substances present in the water, as well as the duration and frequency of exposure.

Dissolved Oxygen (DO)

DO refers to the amount of oxygen that is dissolved in water and it is typically measured in mg/L or ppm. DO is an important parameter in natural waters, essential for the survival of aquatic organisms. In general, natural waters with high levels of DO are healthier for aquatic life than those with low levels of DO. If the DO level drops below a certain threshold, Fish and other aquatic organisms may become stressed or die off, leading to declines in biodiversity and disruptions to food webs.

Low oxygen concentrations are called hypoxia. Hypoxia occurs when the DO levels in water bodies drop below the level necessary to support most aquatic life. This can occur naturally or as a result of human activities, such as pollution and nutrient runoff. The term anoxia refers to the condition in which there is an absence or depletion of DO in the water. This can occur in both freshwater and marine environments and can have significant impacts on aquatic ecosystems, including being deadly for aquatic organisms.

There are several factors that can affect the level of DO in natural waters.

1. **Temperature** is one of the most important factors affecting the concentration of DO in water. In general, the solubility of oxygen decreases as the temperature of water increases. This is because as the temperature of water increases, the kinetic energy of water molecules increases, which reduces their ability to hold onto dissolved gases like oxygen. This leads to a decrease in the solubility of oxygen in water. For example, at a temperature of $0\,°C$ the solubility of oxygen in water is approximately $14.6\,mg/L$, while at a temperature of $25\,°C$, the solubility of oxygen in water is only $7.6\,mg/L$. This decrease in the solubility of oxygen with increasing temperature can have important implications for aquatic organisms that rely on DO to survive, as warmer water temperatures can lead to lower DO concentrations and potentially even hypoxic (low-oxygen) conditions.

2. **Atmospheric pressure:** At standard atmospheric pressure (1 atm), the solubility of oxygen in freshwater is approximately $8\,mg/L$ at $0\,°C$ and decreases to $6\,mg/L$ at $30\,°C$. In seawater, the solubility of oxygen is lower due to the presence of other dissolved ions, such as chloride and sulfate and is approximately $6.6\,mg/L$ at $0\,°C$ and $4.6\,mg/L$ at $30\,°C$.

3. **Salinity** is another important factor that can affect the concentration of DO in natural waters. In general, the solubility of oxygen in water decreases as salinity increases. This is because dissolved salts in water (such as sodium chloride) can compete with oxygen molecules for the available space between water molecules, thereby reducing the amount of oxygen that can dissolve in the water. This means that water with a higher salinity will have a lower solubility of oxygen than freshwater. However, the effect of salinity on DO concentration can vary depending on other factors such as temperature, pressure and the mixing of water layers. For example, in some areas of the ocean, colder and saltier water can actually have higher levels of DO than warmer and less salty water due to the greater solubility of oxygen in colder water.

4. **Presence of aquatic living beings:** Biological processes associated with living beings, such as photosynthesis and respiration, can also affect DO concentrations in natural waters. Photosynthesis, which occurs in aquatic plants and algae, can produce oxygen, while respiration, which occurs in all aquatic organisms, can consume oxygen. These processes can be influenced by a variety of factors, including temperature, light availability, nutrient availability and the presence of pollutants.

5. **Pollution** can have a significant impact on DO levels in water. When pollutants, such as organic matter and nutrients, enter water bodies, they stimulate the growth of algae and other aquatic plants. As these organisms grow and decompose, they consume oxygen from the water, leading to a reduction in DO levels. Pollution can also introduce chemicals and other substances into water that can directly reduce DO levels. For example, industrial discharges and untreated sewage can introduce high levels of organic matter and nutrients into water, leading to oxygen depletion.

9.5 Biological Parameters in Polluted Water

The biological parameters of polluted water can vary depending on the type and degree of pollution. Here are some general biological parameters that can be affected by pollutants in water, including bacteria, viruses and other microorganisms:

1. **Bacterial indicators:** Bacterial indicators are commonly used to assess water quality and determine the presence of pollution. Examples include total coliform bacteria, faecal coliform bacteria and *Escherichia coli* (*E. coli*). Elevated levels of these bacteria indicate the potential presence of faecal contamination and indicate a higher risk of waterborne diseases. Other pathogenic bacteria, in polluted water, which are capable of causing diseases, include *Salmonella*, *Vibrio cholerae* (causing cholera) and *Campylobacter jejuni* (causing campylobacteriosis). These bacteria can cause severe gastrointestinal infections and other health issues.

2. **Viruses:** Viruses are microscopic infectious agents that can be present in polluted water. Common waterborne viruses include enteroviruses, noroviruses and hepatitis A virus. These viruses can cause gastrointestinal illnesses and other health problems.

3. **Algae:** Water pollution can lead to the growth of excessive algae, a phenomenon called algal bloom. Some algae can produce toxins, such as cyanobacteria (blue-green algae), which can contaminate the water and pose a threat to aquatic life and human health.

4. **Protozoa:** Protozoa are single-celled organisms that can be found in polluted water. They include parasites like *Giardia lamblia* and *Cryptosporidium parvum*, which can cause gastrointestinal illnesses when ingested.

5. **Fungi:** Fungal organisms can also be present in polluted water. Although they are less commonly associated with waterborne diseases compared to bacteria and viruses, certain fungi, such as *Aspergillus* and *Fusarium*, can cause infections in individuals with compromised immune systems.

9.6   Consumption of Oxygen by Organic Matter in Water: Biochemical Oxygen Demand and Chemical Oxygen Demand

When organic matter, such as sewage, agricultural runoff or industrial waste, enters a water body, bacteria and other microorganisms decompose it. During this decomposition process, oxygen from the water is consumed to support their metabolic activities. As a result, the DO levels in the water decrease, leading to a decrease in the overall oxygen concentration.

The amount of oxygen consumption can be established using two parameters.

Biochemical Oxygen Demand (BOD)

BOD is a measure of the amount of oxygen required by aerobic microorganisms to decompose organic matter in water. It is a commonly used parameter in wastewater treatment and water quality assessments. The BOD test involves measuring the amount of DO consumed by microorganisms over a specified period of time (usually five days) in a water sample kept at a constant temperature. The difference in DO concentration between the initial and final measurements is used to calculate the BOD value. A high BOD value indicates that there is a large amount of organic matter present in the water, which can deplete oxygen levels and harm aquatic life. BOD is also an indicator of the effectiveness of wastewater treatment processes, as the goal is to reduce the BOD of the effluent to acceptable levels before it is discharged into the environment. The BOD values of several types of water are summarised in Table 9.7.

BOD Test   The estimation of BOD is typically done using the BOD bottle test, which involves the following apparatus:

a. **BOD bottles:** These are specially designed glass bottles with a stopper and a side arm (Figure 9.1). BOD tests are usually performed in multiple bottles to allow for replication and control.

TABLE 9.7   Biochemical oxygen demand values of several types of waters

Water type	BOD (mg/L)
Drinking water	1–3
Surface water	1–10
Untreated wastewater	200–600
Treated wastewater	10–30
Industrial wastewater	Varies widely depending on the industry

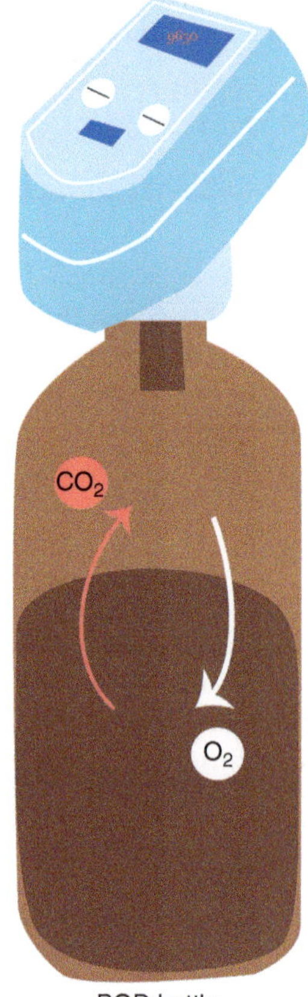

FIGURE 9.1 BOD bottle

b. **Sample containers:** Containers are used to collect representative water samples from the water body being tested. The samples should be handled carefully to minimise the introduction of air bubbles.

c. **DO metre:** A DO metre is used to measure the initial DO concentration in the water sample. This provides a baseline value for comparison with the oxygen levels at the end of the BOD test.

d. **Incubator or water bath:** BOD bottles are placed in an incubator or water bath set at a specific temperature, usually 20 or 25 °C, to promote microbial activity.

The BOD estimation procedure involves the following steps:

1. **Initial DO measurement:** Measure the initial DO concentration in the water sample using a DO metre. This value is noted as DO1.

2. **Preparation of BOD bottles:** Fill the BOD bottles with a specific volume of the water sample, usually 300 or 500 mL. Some bottles are also filled with a control solution without any organic matter for comparison.

3. **Sealing the BOD bottles:** Add a nutrient solution to each bottle to ensure that microorganisms have sufficient nutrients for decomposition. The bottles are then sealed with stoppers to prevent the exchange of gases with the surrounding air.

4. **Incubation period:** Place the sealed BOD bottles in an incubator or water bath at the specified temperature for a specific period of time, typically five days. During this time, microorganisms present in the water consume the organic matter, leading to the depletion of DO.

5. **Final DO measurement:** After the incubation period, remove the BOD bottles from the incubator and measure the DO concentration using a DO metre. This value is noted as DO2.

6. **Calculation of BOD:** The BOD is calculated using the following formula:

$$BOD = (DO1 - DO2) \times (Dilution\ factor)$$

The dilution factor is the ratio of the volume of the original water sample to the volume of the sample used in the BOD bottle. It accounts for the dilution of the organic matter during the test.

By comparing the initial and final DO concentrations along with the dilution factor, the BOD value can be determined. A higher BOD value indicates a greater level of organic pollution in the water sample.

It's important to note that the BOD test provides an estimate of the oxygen demand by microorganisms and may not reflect the actual conditions in a water body. It is a standardised method used for comparative purposes and regulatory compliance.

When nitrogen species are present in water, they can influence the BOD curves in several ways. Specific effects of nitrogen species on the BOD curves can vary depending on the characteristics of the water, the concentration of nitrogen species, the presence of other contaminants and the microbial communities involved:

1. **Nitrification:** Nitrification is the biological conversion of NH_3 to nitrite and then to nitrate (NO_3^-). This process consumes oxygen, contributing to the BOD. As NH_3 is oxidised to NO_3^-, the BOD curve may show a prolonged oxygen demand due to the additional oxygen consumption by nitrifying bacteria.

2. **Organic nitrogen decomposition:** Organic nitrogen compounds, such as proteins and amino acids, can be decomposed by microorganisms through microbial respiration. This decomposition process also consumes oxygen and contributes to the BOD. If nitrogen-rich organic matter is present in the water, the BOD curve may show a higher demand for oxygen compared to organic matter without nitrogen.

3. **Denitrification:** Denitrification is a process in which bacteria convert NO_3^- to nitrogen gas under anaerobic conditions. Denitrification can reduce the overall oxygen demand in water, as it removes NO_3^- from the system. However, it should be noted that denitrification occurs in oxygen-depleted zones, so the BOD curve may still exhibit a significant demand for oxygen before denitrification takes place.

4. **Ammonium toxicity:** High concentrations of ammonium (NH_4^+) can be toxic to aquatic organisms, leading to oxygen stress and affecting the BOD curve. When NH_4^+ levels are excessive, it can impair the metabolism of organisms, reducing their oxygen consumption and altering the BOD pattern.

Nitrogenous BOD (NBOD) is a measure of the amount of DO required for the biological oxidation of nitrogenous compounds in water. It is a variant of the

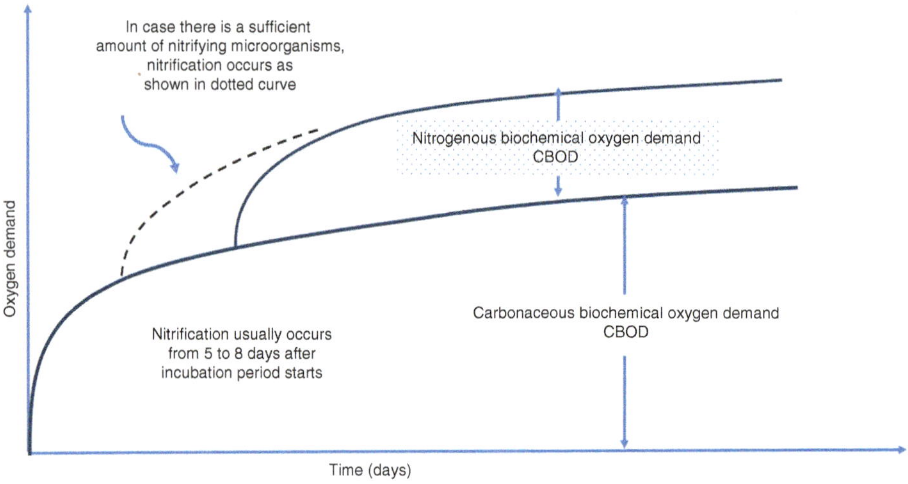

FIGURE 9.2 Curve of BOD in the presence of nitrogen compounds

traditional BOD test, which measures the oxygen consumption by microorganisms as they decompose organic matter in water.

NBOD specifically focuses on nitrogen-containing organic compounds, such as NH_3, urea and various organic nitrogen compounds. These compounds can be derived from sources like wastewater, agricultural runoff and industrial discharges. When nitrogenous compounds are present in water, microorganisms metabolise them, which leads to the consumption of DO. During incubation, microorganisms in the water consume the available nitrogenous compounds, resulting in the depletion of DO. The difference in DO concentration before and after incubation is then used to calculate the NBOD. An example of the curves of BOD in the presence of nitrogen compounds is given in Figure 9.2.

BOD is a measure of the amount of oxygen required by microorganisms to decompose organic matter in water. It is often used as an indicator of water quality and the level of pollution in aquatic ecosystems. Table 9.8 is a general guideline for interpreting BOD levels and their corresponding water quality.

BOD population equivalent (PE) is a way of expressing the amount of organic pollution that is being contributed to a body of water by a population of a certain size.

It is based on the assumption that each person generates a certain amount of organic matter in their waste that will contribute to the BOD of the water. The BOD

TABLE 9.8 **Water quality based on BOD**

BOD level (mg/L)	Water quality
Less than 2	Excellent
2–3.9	Good
4–6.9	Fair
7–10.9	Poor
Greater than 11	Very poor

TABLE 9.9 BOD population equivalent for different wastewater sources

Wastewater source	BOD population equivalent
Single-family household	4.5
Multi-family household	3.5
Restaurant	25
Hotel	20
Hospital	40
School	15
Shopping mall	10
Industrial wastewater	Variable
Municipal wastewater	Variable
Agricultural wastewater	Variable

PE is calculated by multiplying the population size by a factor that represents the average amount of organic matter generated per person. The units for BOD PE are typically expressed as PE, a relative measure that allows for the easy comparison of wastewater pollution loads to the equivalent number of people generating the same amount of organic matter. For example, if a certain wastewater source has a BOD PE of 100 PE, the organic pollution load of that wastewater is equivalent to what would be generated by 100 people.

The specific factor used can vary depending on the type of population being considered (e.g., residential, industrial or commercial) and the level of treatment provided to the wastewater before it is discharged into the water. In general, residential populations are assumed to generate less organic matter per person than industrial or commercial populations (Table 9.9).

The BOD PE is useful for assessing the impact of different populations on water quality and for designing and sizing wastewater treatment facilities. It is also commonly used in regulations and guidelines related to wastewater discharge and water quality standards.

Chemical Oxygen Demand (COD)

COD is a measure of the amount of oxidizable organic matter present in a water sample when it reacts with chemical oxidizers such as potassium dichromate or potassium permanganate under specific conditions of temperature and time. It is a commonly used parameter in environmental monitoring and wastewater treatment. COD is expressed in milligrams of oxygen per litre of water (mg/L). Higher COD values indicate that there is more organic matter present in the water sample. As BOD measures, they can be an indication of pollution or other sources of contamination. Common COD values for several types of waters are described in Table 9.10.

The relationship between BOD and COD can vary depending on the source and composition of the organic matter present in the water (Table 9.11). Generally, BOD

TABLE 9.10   Common COD values of several types of waters

Water type	COD range (mg/L)
Drinking water	1–10
Surface water	5–50
Municipal wastewater	100–400
Industrial wastewater	100–2000
Raw sewage	200–600
Treated wastewater	20–100
Seawater	5–50
Groundwater	1–10

TABLE 9.11   Comparison of the Biological Oxygen Demand (BOD) and Chemical Oxygen Demand (COD) for different types of waters

Water type	BOD range (mg/L)	COD range (mg/L)
Drinking water	0–2	2–10
Surface water	1–8	5–30
Municipal wastewater	150–300	250–600
Industrial wastewater	100–1,000	200–2,000
Sewage	200–600	400–1,200
Raw/untreated wastewater	500–1,000	1,000–3,000
Effluent (treated wastewater)	10–30	20–50

is lower than COD because not all organic matter in water is biodegradable. In some cases, the ratio of BOD to COD can provide information about the composition of organic matter in water. For example, a high BOD/COD ratio may indicate that the organic matter present is more biodegradable, whereas a low BOD/COD ratio may indicate the presence of non-biodegradable organic matter.

BOD and COD are often used in conjunction with other water quality parameters, such as total organic carbon (TOC) and nutrient levels, to provide a more complete picture of water quality. TOC refers to the amount of carbon in an organic compound or substance and it is commonly used as a measurement of the amount of organic matter present in water or soil samples. In water samples, TOC can be used as an indicator of water quality and can help identify sources of pollution or contamination. TOC measurements are typically made by analysing a sample for the amount of carbon present using specialised instruments such as a TOC analyzer. The results are expressed in terms of the concentration of carbon in the sample, typically in ppm or mg/L.

9.7   Water Pollutants: Priority Pollutants

Water pollutants refer to substances found in water bodies that can cause detrimental effects on the environment, human health and aquatic organisms. These harmful substances stem from a range of sources, such as industrial and agricultural activities, urban runoff, sewage discharges and even natural processes.

Water pollution is a grave concern as it can lead to significant ecological imbalances and endanger the well-being of both humans and aquatic life.

9.8   Priority Pollutants

Priority pollutants are substances that have been identified as potentially harmful to both human health and the environment. These pollutants are of particular concern due to their toxic nature, persistence in the environment and ability to accumulate in living organisms. While the specific list of priority pollutants may vary depending on the country or regulatory body, there are some commonly recognised examples.

Water quality standards are established to define the acceptable concentrations of pollutants in surface waters, including rivers, lakes and oceans, as well as in groundwater. These standards take into account the intended use of the water body, such as drinking water, recreational activities or the protection of aquatic life. They encompass criteria for various pollutants, including heavy metals, nutrients, organic chemicals and bacteria.

Regulatory standards for specific contaminants are based on two key parameters: maximum allowable levels (MALs) and maximum contaminant levels (MCLs).

a. **MALs:** MALs in water refer to the maximum concentrations of specific substances or parameters that are considered acceptable for human consumption or other designated purposes. These levels are established by regulatory bodies, such as government agencies or international organisations, to ensure the safety and quality of water.

b. **MCLs:** MCLs specifically refer to the maximum allowable concentrations of contaminants in drinking water. As MALs, MCL-specific levels are established by regulatory bodies to ensure that drinking water supplies are safe for human consumption. MCLs are typically based on extensive research, risk assessment and consideration of health effects associated with specific contaminants.

These levels are typically determined and enforced by government or regulatory agencies to safeguard public health and preserve the environment.

Topics of Interest: Guidelines for Drinking Water

The WHO has developed two primary sets of guidelines for drinking water quality: the Guidelines for Drinking Water Quality (GDWQ) and the International Standards for Drinking Water (ISDW). Dome's key points from these guidelines are:

1. **GDWQ:** The GDWQ provides a comprehensive framework for assessing and managing the risks associated with drinking water. They cover a wide range of potential contaminants

and establish both maximum acceptable levels and desirable levels for various parameters. The guidelines also provide guidance on water safety plans, monitoring and treatment technologies.

2. **ISDW:** The ISDW aims to harmonise the different national and regional drinking water standards and provide a basis for consistent and comparable water quality monitoring worldwide. They define basic parameters and indicators of water quality and outline the sampling and testing methods for these parameters.

3. **Key parameters:** The guidelines set limits for various physical, chemical, microbiological and radiological parameters in drinking water. Some important parameters include microbial contaminants (e.g., bacteria, viruses), chemical contaminants (e.g., heavy metals, pesticides, disinfection by-products) and physical characteristics (e.g., turbidity, colour, odour).

4. **Risk-based approach:** The WHO adopts a risk-based approach to drinking water safety, focusing on the prevention and control of hazards throughout the water supply chain. This includes source water protection, treatment processes, distribution systems and household practices.

5. **Water safety plans (WSPs):** The WHO encourages the implementation of WSPs by water suppliers. WSPs involve a comprehensive assessment and management of risks to ensure the safety of drinking water. They encompass all steps, from source to tap and are designed to be adaptable to local conditions.

It's important to note that specific regulations and standards for drinking water quality may vary between countries and regions. National or local authorities often adopt and adapt the WHO guidelines to suit their specific circumstances and priorities.

There are several regulatory bodies around the world that oversee and regulate water quality. Some examples of major regulatory bodies in different regions are:

1. **European Union:**
   a. **European Commission:** The European Commission establishes and implements policies and regulations related to water quality in the European Union (EU). It sets water quality standards and monitors compliance with those standards.

   b. **European Environment Agency (EEA):** The EEA provides information on water quality and supports the development and implementation of water-related policies in Europe.

2. **United States:**
   Environmental Protection Agency (EPA): The EPA sets and enforces water quality standards under the Clean Water Act. It regulates various aspects of water quality, including drinking water, surface water and wastewater.

3. **United Kingdom:**
   a. **Drinking Water Inspectorate (DWI):** The DWI is responsible for ensuring the quality of drinking water in England and Wales. It monitors water suppliers' compliance with standards and regulations.

   b. **Scottish Environment Protection Agency (SEPA):** SEPA is responsible for water quality regulation in Scotland. It monitors and enforces water quality standards for rivers, lakes and coastal waters.

4. **Canada:**
   a. **Health Canada:** Health Canada establishes guidelines and regulations for drinking water quality in Canada. It sets national standards and provides guidance to provinces and territories.

   b. **Environment and Climate Change Canada:** This federal department is responsible for monitoring and regulating water quality in Canada's aquatic ecosystems, including rivers, lakes and oceans.

5. **Australia:**
   a. **Australian Drinking Water Guidelines (ADWG):** The ADWG provides guidance on the management and regulation of drinking water quality in Australia. It sets health-based guidelines and quality standards.
   b. **National Health and Medical Research Council (NHMRC):** The NHMRC develops and maintains the ADWG and provides scientific advice on water quality issues.

9.9 Heavy Metal Pollution in Water

Heavy metal pollution in water is a significant environmental concern with potentially severe consequences for ecosystems and human health. Heavy metals are natural elements that have a high density and are toxic to living organisms at certain concentrations. They include substances such as lead, cadmium, arsenic, mercury, copper and chromium.

There are several sources of heavy metal pollution in water bodies. These sources can be natural or anthropogenic (human-related). Here are some common sources:

1. **Industrial discharges:** Industrial activities such as mining, smelting, metal processing and manufacturing can release heavy metals into water bodies. Effluents from industries may contain high concentrations of metals like lead, mercury, cadmium, chromium and arsenic.

2. **Municipal and industrial wastewater:** Untreated or inadequately treated sewage and wastewater from municipal treatment plants and industrial facilities can contain heavy metals. These metals can originate from household products, detergents and industrial processes.

3. **Agriculture:** The use of fertilisers, pesticides and animal manure in agricultural practices can contribute to heavy metal pollution in water bodies. Some fertilisers and pesticides contain metals such as mercury, cadmium and arsenic, which can enter water bodies through runoff and leaching.

4. **Mining and ore processing:** Mining activities can release heavy metals into water bodies through the disposal of mine tailings, which often contain high concentrations of metals. Additionally, the process of extracting metals from ores can result in the release of metals into water bodies.

5. **Atmospheric deposition:** Heavy metals can be deposited into water bodies through atmospheric deposition. Industrial emissions, vehicle exhaust and the burning of fossil fuels release heavy metals into the air, which can then be deposited into water bodies through rainfall or dry deposition.

6. **Landfills and waste disposal:** Improper disposal of waste, including electronic waste (e-waste), batteries and other products containing heavy metals, can lead to leaching of metals into groundwater and nearby water bodies.

7. **Urban runoff:** Stormwater runoff from urban areas can carry heavy metals from various sources, including roads, parking lots and rooftops. These metals can come from vehicle emissions, tyre wear, corrosion of infrastructure and other urban activities.

TABLE 9.12   **Maximum allowable levels for heavy metals (in mg/L)**

Heavy metal	WHO (mg/L)	EU (mg/L)	EPA (mg/L)
Arsenic	0.01	0.01	0.01
Cadmium	0.003	0.005	0.005
Chromium	0.05	0.05	0.1
Copper	2	2	1.3
Lead	0.01	0.01	0.015
Mercury	0.001	0.001	0.002
Nickel	0.02	0.02	0.1
Zinc	3	5	5

The regulated concentration of heavy metals in water varies depending on the country or region. In general, there are guidelines and standards set by organisations such as the WHO, United States EPA and EU that aim to protect public health by limiting exposure to harmful substances.

For example, the WHO has established GDWQ that sets MALs for heavy metals such as lead, arsenic, cadmium and mercury. In the case of lead, the WHO recommends a maximum concentration of $10\,\mu g/L$, while for arsenic, the recommended maximum is $10\,\mu g/L$.

Similarly, the EU and the EPA have established national primary drinking water regulations for heavy metals such as lead, copper and chromium. For lead, the MCL is set at $15\,\mu g/L$, while for copper, the MCL is set at $1.3\,mg/L$ (see Table 9.12).

9.10   Radionuclides and Radioactivity Pollution

Water pollution by radionuclides refers to the contamination of water bodies with radioactive substances.

There are several possible sources of radionuclide pollution in water bodies.

1. **Nuclear power plants:** Nuclear power plants generate electricity by utilising nuclear reactions. They release small amounts of radioactive materials, including radionuclides, into the environment through the discharge of cooling water or accidental leaks.

2. **Nuclear accidents:** Accidents at nuclear power plants or facilities that handle radioactive materials can result in the release of radionuclides into the surrounding environment. These releases can contaminate water bodies, leading to long-term pollution.

3. **Mining and milling operations:** Mining and milling of uranium, thorium and other radioactive minerals can release radionuclides into nearby water bodies.

These minerals often occur in association with other elements and the mining and processing activities can result in the release of radionuclides into water sources.

4. **Industrial activities:** Certain industrial processes, such as mining and processing ores, can lead to the release of radionuclides into nearby water bodies. Industries that handle radioactive materials or use radiation in their processes need to ensure proper disposal and treatment of wastewater to prevent contamination.

5. **Medical and research facilities:** Hospitals, research laboratories and other facilities that use radioactive materials for medical diagnosis, treatment or research purpose may release radionuclides into wastewater. Proper handling, storage and disposal of radioactive materials are essential to preventing contamination.

6. **Radioactive waste disposal sites:** Improperly managed or poorly designed radioactive waste disposal sites can contaminate groundwater and nearby surface water with radionuclides. Radioactive waste must be stored and disposed of in a controlled manner to prevent the migration of radionuclides into water sources.

7. **Natural sources:** Radionuclides also occur naturally in the environment. For example, radium-226 and radon-222 can be found in some groundwater sources due to the natural decay of uranium and thorium in rocks and soil.

The presence of radionuclides in water can have significant health and environmental implications. The radiation emitted by these substances can cause both short-term and long-term effects on human health, depending on the type of radionuclide, its concentration and the duration of exposure. Some radionuclides, such as iodine-131 and cesium-137, are known to accumulate in the body and can increase the risk of cancer and other diseases.

Nuclear accidents, such as the Chernobyl disaster in 1986 and the Fukushima Daiichi accident in 2011, have resulted in the release of large amounts of radionuclides into water bodies. These incidents have had severe consequences for the affected areas, leading to contamination of rivers, lakes and groundwater sources.

In addition to nuclear accidents, the improper disposal of radioactive waste can also contribute to water pollution by radionuclides. If not stored or disposed of correctly, radioactive materials can leach into the surrounding soil and groundwater, eventually reaching nearby water sources.

Monitoring and controlling water pollution by radionuclides is crucial to protecting public health and the environment. Regulatory bodies and environmental agencies establish guidelines and standards to limit the acceptable levels of radionuclides in drinking water and other water sources.

Regular monitoring of water sources, especially those near nuclear facilities or areas with a history of nuclear accidents, is essential to detect and mitigate radionuclide pollution promptly. Additionally, public awareness and education about the risks associated with radionuclides in water can help individuals make informed choices about their water sources and consumption habits.

MALs of radionuclides in drinking water can vary depending on the country and regulatory bodies involved. It's important to note that different countries may have different regulations and guidelines regarding the permissible levels of radionuclides in drinking water. A comparison of the MCLs for certain radionuclides in drinking water for the EPA in the United States and the EU is given in Table 9.13.

EU standards are given in Bq/L, which is a measure of radioactivity, while the EPA MCLs are given in pCi/L for alpha and beta emitters and in µg/L (micrograms per litre) for uranium.

TABLE 9.13	Comparison of MCLs for certain radionuclides in drinking water for the EPA and the EU	
Radionuclide	**EPA MCL (United States)**	**EU standard (Bq/L)**
Gross alpha emitters	15 pCi/L	0.1
Gross beta emitters	4 mrem/yr	1
Radium-226 + 228	5 pCi/L	0.5
Uranium	30 μg/L	2.0

9.11 Fertiliser Pollution

When fertilisers are not used properly or when they are applied in excessive amounts, they can enter water bodies through various pathways: runoff or leaching from agricultural fields, urban areas and industrial activities, causing a variety of adverse environmental and health effects:

1. **Algal blooms:** Elevated levels of nitrogen and phosphorus stimulate the rapid growth of algae in water bodies. This excessive algal growth, known as an algal bloom, blocks sunlight from reaching underwater plants and depletes oxygen levels in the water[1]. This process can harm or kill aquatic plants, fish and other organisms, leading to dead zones or fish kills. Oxygen-depleted waters cannot support the survival of many aquatic species, disrupting the entire ecosystem.

2. **Harm to aquatic life:** Elevated nutrient levels can directly harm aquatic organisms. High NH_3 concentrations resulting from nutrient pollution can be toxic to fish, causing deformities, reproductive issues and even death. Additionally, some algal blooms produce harmful toxins called harmful algal blooms (HABs), which can poison or suffocate fish, marine mammals and even humans if consumed.

3. **Disrupted food chain:** Nutrient pollution can alter the natural balance of food webs in aquatic ecosystems. Algal blooms can outcompete other plants, reducing biodiversity and negatively impacting the food sources of many species. This disruption can have cascading effects throughout the ecosystem.

4. **Drinking water contamination:** Excessive nutrient runoff can contaminate drinking water sources. Elevated NO_3^- levels in water can pose health risks as they can interfere with the oxygen-carrying capacity of their blood, leading to a condition called methemoglobinemia or *blue baby syndrome,* which can be fatal in infants. This condition occurs when the NO_3^- in the water are converted into nitrites in the body, which then reduce the ability of the blood to carry oxygen. Other health effects associated with NO_3^- pollution in drinking water include an increased risk of cancer and thyroid problems.

To mitigate the pollution of waters by fertilisers, several measures can be implemented, including:

a. Implementing best management practices in agriculture to reduce nutrient runoff, such as precision application of fertilisers, buffer zones and cover crops

---

[1] The process is known as eutrophication and occurs when a body of water, such as a lake, river or coastal area, becomes overly enriched with nutrients, particularly nitrogen and phosphorus.

b. Developing and enforcing regulations for fertiliser use and storage, especially near water bodies

c. Promoting sustainable farming practices that focus on soil health, nutrient management plans and efficient use of fertilisers

d. Educating farmers, landowners and the general public about the impacts of nutrient pollution and the importance of responsible fertiliser use

e. Investing in wastewater treatment systems and infrastructure upgrades to reduce nutrient discharges from urban and industrial sources

f. Monitoring water quality regularly helps identify problem areas and make appropriate resolutions.

To ensure the safety of drinking water, regulations have been established in many countries to limit the levels of NO_3^- and phosphate (PO_4^{3-}) in public water supplies. However, it's important to note that regulations can vary between countries and regions, so it's best to consult local guidelines for specific information.

In the United States, the EPA has set the MCL for NO_3^- at 10 mg/L or ppm. In the EU, the legal limit for NO_3^- in drinking water is 50 mg/L.

Different countries may have their own specific regulations and guidelines for NO_3^- in drinking water.

Regulations for PO_4^{3-} in drinking water are typically focused on controlling their discharge from wastewater treatment plants and industrial sources rather than setting specific limits for drinking water. In some areas, there may be guidelines or recommendations for PO_4^{3-} levels in drinking water, but they are not universally regulated like NO_3^-.

It's important to note that different countries and regions have their own specific regulations and guidelines for NO_3^- and PO_4^{3-} in drinking water. Local water authorities and health agencies are the best sources of information regarding the specific regulations and guidelines that apply to your area.

The WHO does not provide specific maximum allowable limits (MAL) for NO_3^- and PO_4^{3-} in drinking water. However, they do provide guidelines for these parameters. Here's a table summarising the WHO guidelines for NO_3^- and PO_4^{3-} in drinking water (Table 9.14).

The formation and concentration of these byproducts during chlorination can depend on various factors such as water chemistry, pH, temperature, contact time and chlorine dosage. Water treatment plants closely monitor and control these parameters to minimise the formation of both organic and inorganic byproducts and ensure compliance with regulatory standards for drinking water quality.

To reduce the formation of DBPs, water treatment plants may use alternative disinfectants, such as ozone or ultraviolet light or may use less chlorine and optimise the treatment process to minimise the amount of organic matter in the water. Additionally, individuals can reduce their exposure to DBPs by using a water filter or drinking bottled water.

TABLE 9.14 **The WHO guidelines for nitrates and phosphates in drinking water**

Parameter	Guideline value
Nitrates (NO_3^-)	50 mg/L
Phosphates (PO_4^{3-})	No specific guideline value provided

9.12 Organic Compounds Pollution in Waters

Organic compounds can be found in various sources and can contribute to water pollution, such as:

1. **Industrial discharges:** Industries such as chemical manufacturing, pharmaceuticals, oil refineries and pulp and paper mills release organic compounds into water bodies through their effluent discharges. These compounds can include solvents, pesticides, petroleum hydrocarbons and various synthetic chemicals.

2. **Agricultural runoff:** The use of pesticides, herbicides and fertilisers in agriculture can lead to the contamination of water bodies. These chemicals can be washed off the fields during rainfall or irrigation, entering nearby streams, rivers and lakes. Runoff from animal feedlots and manure storage areas can also introduce organic compounds into water.

3. **Municipal wastewater:** Wastewater from households, commercial establishments and sewage treatment plants contains organic compounds from human activities. These include detergents, personal care products, pharmaceuticals and various organic waste materials. While treatment plants are designed to remove many of these compounds, they may not remove all of them, leading to their release into water bodies.

4. **Stormwater runoff:** Urban areas generate stormwater runoff that can carry organic compounds from various sources. These include contaminants from road surfaces, parking lots, rooftops and other impervious surfaces. The runoff can transport motor oil, gasoline, pesticides and other pollutants into nearby water bodies.

5. **Leakage from underground storage tanks:** Underground storage tanks, such as those used for storing gasoline or industrial chemicals, can develop leaks over time. This can result in the release of organic compounds into the surrounding soil and groundwater, which can eventually find their way into nearby surface water bodies.

6. **Accidental spills and releases:** Accidental spills of chemicals during transportation, storage or industrial accidents can result in the direct release of organic compounds into water bodies. These events can have significant environmental impacts and pose risks to aquatic ecosystems.

7. **Natural sources:** Some organic compounds can enter water bodies through natural processes. For example, decaying organic matter from vegetation, algae blooms and natural oil seepages can introduce organic compounds into aquatic systems.

It's worth noting that the specific types and concentrations of organic compounds in water can vary depending on the location, the surrounding land use and the nature of the pollution sources. Water pollution monitoring and management practices aim to identify and mitigate the sources of organic compound contamination to protect water quality and ecosystem health.

9.13   Pesticide Pollution in Waters

A partical class of water pollution is caused by pesticides producing a significant environmental concern that can have detrimental effects on aquatic ecosystems and human health. Pesticides can enter water bodies through various pathways, including runoff from agricultural fields, leaching from soil and direct application into water bodies.

The pollution of waters with pesticides can have several negative consequences for both the environment and humans.

1. **Ecological impact:** Pesticides can be highly toxic to aquatic organisms, including fish, amphibians and invertebrates. Even at low concentrations, they can disrupt the balance of aquatic ecosystems, leading to a decline in biodiversity. Pesticides can harm not only the targeted pests but also beneficial organisms that play important roles in the ecosystem.

2. **Water quality:** Pesticides can contaminate water sources, making them unsafe for drinking, swimming and irrigation. Some pesticides can persist in water for long periods, increasing the risk of prolonged exposure to humans and wildlife.

3. **Human health concerns:** Exposure to pesticide-contaminated water can pose health risks to humans. Consuming water contaminated with pesticides may lead to acute or chronic health effects, such as nausea, vomiting, respiratory issues, neurological disorders and even cancer.

4. **Resistance and residues:** Over time, repeated pesticide use can lead to the development of pesticide-resistant pests, necessitating higher doses or more toxic pesticides. Pesticide residues can accumulate in the tissues of aquatic organisms, potentially entering the food chain and posing risks to organisms at higher trophic levels, including humans.

The regulation of residual pesticides in water is an important environmental and public health issue. In most countries, the regulation of pesticides in water is the responsibility of government agencies such as the EPA in the United States or the European Chemicals Agency (ECHA) in Europe. These agencies set standards for the maximum allowable concentration of pesticides in drinking water and monitor compliance with these standards through regular testing and reporting. In Table 9.15, some examples of the different maximum allowable concentration levels in the United States and Europe are collected.

The regulation of pesticides in water involves a number of different approaches. One approach is to require pesticide manufacturers to submit data on the

TABLE 9.15   Comparison of MAC levels for different pesticides in EEUU and EU

Pesticide	EPA (United States)	ECHA (European Union) (μg/L)
Glyphosate	No MCL established	0.1
Atrazine	3 μg/L	0.1
Chlorpyrifos	0.03 mg/L	0.1

environmental and health impacts of their products before they can be approved for use. This data is used to determine the appropriate level of regulation and develop guidelines for safe use.

Another approach is to regulate the use of pesticides through licensing and permit systems. This involves requiring farmers and other pesticide users to obtain a license or permit before they can use certain types of pesticides and fertilisers and requiring them to follow specific guidelines for their use.

9.14   Other Water Pollutants: Petroleum Derivatives, Fats, Oils and Residual Organic Matter, Plastics, and Microorganisms

Petroleum Products

These include oils and gasoline spills, as well as urban runoff from roads and parking lots, which can contain trace amounts of petroleum products.

An oil spill is a type of water pollution that occurs when crude oil or refined petroleum products are released into bodies of water, such as oceans, rivers or lakes. Oil spills can have serious environmental and economic consequences and can harm aquatic life and wildlife, as well as people who depend on the affected water for their livelihoods.

Oil spills can occur in a variety of ways, including from oil tanker accidents, pipeline leaks, offshore drilling accidents or even during the transportation of oil via trucks or trains. When an oil spill occurs, the oil can spread quickly across the surface of the water and may also sink to the bottom, contaminating the entire water column.

The impact of an oil spill on the environment and wildlife can be significant, as oil is toxic to many species of fish, birds and other aquatic animals. Oil can also coat the feathers or fur of animals, making it difficult for them to regulate their body temperature and leading to hypothermia. Additionally, oil spills can have long-term effects on the health of affected ecosystems, potentially causing long-term damage to food chains and habitats.

Efforts to clean up oil spills can be difficult and expensive, as the oil can spread quickly and become embedded in sediment or other materials. Techniques for cleaning up oil spills include the use of chemical dispersants, skimmers and absorbent materials, as well as manual labour to remove the oil from affected areas.

Preventing oil spills from occurring in the first place is the most effective way to avoid the negative impacts of water pollution caused by oil spills. This includes implementing and enforcing regulations to ensure the safe transportation and handling of oil, as well as investing in alternative energy sources to reduce the demand for fossil fuels.

Fats and Oils

Fats and oils can be significant water pollutants if they are released into waterways in large quantities. When these substances are discharged into water, they can create a thin layer on the surface that can impede the exchange of oxygen between the water and the atmosphere. This layer can also prevent sunlight from penetrating the water, which can impact photosynthesis in aquatic plants. Moreover, when fats and oils enter water, they can react with other chemicals and substances present in the water to form harmful compounds. For example, when fats and oils combine with nitrogen and phosphorus from agricultural and sewage runoff, they can create HABs that can lead to oxygen depletion in the water and create 'dead zones' that can harm aquatic life. Furthermore, fats and oils can also contribute to the accumulation of solids and debris in the water, which can clog up waterways and interfere with the functioning of aquatic ecosystems. Overall, the release of fats and oils into water can have significant negative impacts on aquatic environments and the organisms that inhabit them. It is essential to properly manage and dispose of fats and oils to prevent pollution and protect our water resources.

Residual Organic Matter

Residual Organic Matter (ROM) refers to a variety of sources, including agricultural runoff, manure, sewage discharge, industrial effluent and organic materials that remain after a process of decomposition of plants or animals. ROM can be a significant water pollutant if it is not properly managed. When organic matter accumulates in water, it can consume DO as it decomposes. This can lead to hypoxia or anoxia, which is a condition where there is not enough oxygen in the water to support aquatic life. Additionally, the decomposition of organic matter can release nutrients, such as nitrogen and phosphorus, which can cause eutrophication or excessive growth of algae and aquatic plants. This can result in the depletion of DO as the algae and plants consume it during their own respiration, leading to the death of fish and other aquatic organisms. ROM is a major contributor to BOD and COD.

Plastics

The name plastics, as it has been explained before, refers to synthetic polymer materials that can be moulded into various shapes and forms having an enormous variety of uses such as packaging, construction materials, textiles and consumer goods. Plastics can have negative impacts on the environment as happens with polyethylene (PE), polypropylene (PP), polyvinyl chloride (PVC), polystyrene (PS) and many others that are not biodegradable. So, they can persist in the environment for hundreds of years, leading to pollution and harm to wildlife. When plastic waste enters waterways, it can harm marine life by entangling animals or being mistaken for food, leading to ingestion and potential suffocation or death. Additionally, as plastics break down over time, they release harmful chemicals into the water, which can contaminate the food chain and affect the health of aquatic animals and humans who consume them. Plastic pollution can also have economic impacts on industries such as fishing and tourism, as well as on coastal communities that rely on a healthy marine environment for their livelihoods.

Microorganisms

Microbiological water pollution refers to the presence of harmful microorganisms in water bodies, such as bacteria, viruses and parasites. In untreated waters, these microorganisms can cause various waterborne diseases. Some common diseases borne by pathogens in water are as follows:

1. **Cholera:** Cholera is caused by the bacterium *Vibrio cholerae*. It spreads through contaminated water and food, primarily in areas with poor sanitation and hygiene practices.

2. **Typhoid fever:** Typhoid fever is caused by the bacterium *Salmonella enterica* serotype Typhi. It is transmitted through the consumption of contaminated water or food, usually in areas with inadequate sanitation.

3. **Hepatitis A:** Hepatitis A is a viral infection that affects the liver. It is transmitted through the ingestion of water or food contaminated with the faeces of an infected person.

4. **Giardiasis:** Giardiasis is caused by the parasite *Giardia lamblia*. It is spread through the ingestion of water contaminated with the parasite, often in settings where water sources are contaminated with human or animal faeces.

5. **Cryptosporidiosis:** Cryptosporidiosis is caused by the parasite *Cryptosporidium*. It can be contracted by ingesting water or food contaminated with the parasite, as well as through direct contact with infected individuals or animals.

6. **Legionnaires' disease:** Legionnaires' disease is caused by the bacterium *Legionella pneumophila*. It is typically contracted by inhaling aerosolized water droplets containing the bacteria, often from sources such as cooling towers, hot tubs or plumbing systems.

7. **Amoebic dysentery:** Amoebic dysentery is caused by the parasite *Entamoeba histolytica*. It is transmitted through the ingestion of water or food contaminated with the parasite, particularly in areas with inadequate sanitation.

8. **Shigellosis:** Shigellosis is caused by various species of the bacteria *Shigella*. It spreads through the ingestion of water or food contaminated with the bacteria, as well as through person-to-person contact.

9. ***E. coli* infection:** Certain strains of *E. coli* bacteria, such as enterotoxigenic *E. coli* (ETEC) or enterohemorrhagic *E. coli* (EHEC), can cause gastrointestinal infections when ingested through contaminated water or food.

10. **Norovirus infection:** Norovirus is a highly contagious virus that causes gastroenteritis. It can be transmitted through water contaminated with the virus, as well as through person-to-person contact and contaminated surfaces.

Sources of microbiological water pollution include untreated sewage, animal waste and runoff from farms and urban areas. It can also occur due to natural events such as heavy rainfall or floods that can wash pollutants into water sources.

To prevent and control microbiological water pollution, various measures can be taken, such as treating sewage and wastewater before discharging it into water bodies, improving sanitation and hygiene practices and regulating the use of pesticides and fertilisers in agriculture. Regular monitoring and testing of water sources for the presence of harmful microorganisms can also help identify potential risks and allow for prompt action to be taken to protect public health.

CHAPTER 10

Water Treatments

First Part: Drinking Water Treatments

10.1 Water Uses

Water uses encompass the diverse ways in which water is employed by humans and other living organisms. These applications can be broadly classified into several sectors, including domestic, agricultural, industrial and environmental uses.

1. **Domestic uses:** Water plays a vital role in households, serving multiple purposes such as drinking, cooking, washing clothes and dishes, bathing and toilet flushing.
2. **Agricultural uses:** Water holds indispensable significance in irrigation systems, essential for cultivating crops and providing water for livestock. It is also employed in aquaculture, facilitating fish farming activities.
3. **Industrial uses:** Various industries rely on water for their operations, including manufacturing, mining and energy production. Water is used for machinery cooling, cleaning purposes and, in some cases, even as a raw material.
4. **Recreational activities:** Water serves as a platform for leisure and recreational pursuits, such as swimming, boating and fishing.
5. **Environmental uses:** Water serves as a critical component in maintaining the health of ecosystems, fostering the growth of diverse aquatic plants and supporting a wide array of animal species.

10.2 Water Treatments

There is no single type of water treatment, as it depends on the type of water and its intended use. There is a vast diversity of treatment methods available for purifying or modifying the quality of water to make it suitable for specific purposes. The choice of water treatment method is determined by several factors, such as the source of water, whether it is surface water (from rivers, lakes or reservoirs) or groundwater (from wells or aquifers). On the other hand, wastewater treatment focuses on removing

Understanding the Chemistry of the Environment, First Edition. Francisco G. Calvo-Flores.
© 2025 John Wiley & Sons Ltd. Published 2025 by John Wiley & Sons Ltd.

pollutants and contaminants from domestic, industrial or agricultural wastewater before it is discharged back into the environment. Wastewater treatment methods can also vary widely and may include a set of unitary processes to attain the desired water quality, combining various physical, chemical and biological methods.

Water treatments play a crucial role in safeguarding public health, preserving the environment and supporting various human activities that rely on access to clean and safe water. By removing or reducing harmful substances, pathogens and pollutants, water treatment helps protect human health and prevent the transmission of waterborne diseases. Furthermore, it contributes to sustainable water management by conserving resources, reducing pollution and minimising the impact of human activities on water ecosystems.

It can be distinguished into several major groups of water treatments:

1. **Drinking water treatment:** This process ensures the safety of water for human consumption. It involves the removal of various contaminants, including bacteria, viruses, protozoa and chemicals. The treatment may include coagulation and flocculation, sedimentation, filtration, disinfection and pH adjustment.

2. **Swimming pool water treatment:** The goal of this treatment is to maintain safe and hygienic conditions in swimming pools. It involves the elimination of contaminants such as bacteria, algae and organic matter. Chemicals like chlorine or bromine are typically used to disinfect the water.

3. **Wastewater treatment:** Wastewater treatment aims to remove pollutants and contaminants from wastewater before it is discharged into the environment. The process incorporates physical, biological and chemical methods such as screening, sedimentation, activated sludge treatment and disinfection.

4. **Industrial water treatment:** This type of treatment is employed to remove contaminants from water used in industrial processes like manufacturing, cooling and boiler feedwater. Common methods include ion exchange, reverse osmosis (RO) and chemical treatment.

5. **Agricultural water treatment:** Agricultural water treatment focuses on removing contaminants from water used in irrigation, aquaculture and livestock farming. Processes like sedimentation, filtration and disinfection are employed.

These water treatment processes serve different purposes and ensure water quality and safety for their respective applications.

10.3   Freshwater and Drinking Water

Freshwater is a term used to describe water that contains low levels of dissolved salts and other minerals. It can come from a variety of sources, including lakes, rivers and underground aquifers. Freshwater is a limited and valuable resource on Earth. About 71% of the Earth's surface is covered by water, but only about 2.5% of that water is freshwater. Most of the freshwater is found in the form of glaciers, ice caps and underground aquifers.

Surface freshwater sources, such as rivers and lakes, only make up a small fraction of the freshwater available on Earth. These sources are important for human use and consumption, but they are vulnerable to pollution, climate change and overuse.

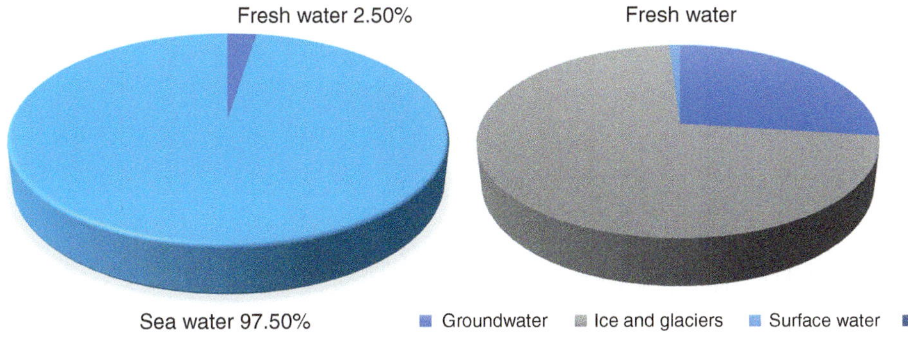

FIGURE 10.1   Freshwater average

The availability of freshwater varies across different regions of the world. Some areas, such as parts of Africa and Asia, experience water scarcity and drought, while others, such as parts of the Americas and Europe, have abundant freshwater resources (Figure 10.1).

Drinking water, on the other hand, is a type of water that has been specifically treated and purified to be safe for human consumption. It is free of harmful contaminants and pathogens that can cause illness or disease. Drinking water can come from freshwater sources, but it must go through a treatment process to remove impurities and ensure its safety.

10.4   Drinking Water Treatment

Drinking water must be free from harmful contaminants, such as pathogens, chemicals and impurities, that could cause illness or other adverse effects when ingested. The treatment of drinking water involves various unitary processes to ensure its safety and quality that are designed to remove or reduce contaminants and impurities, improve taste and odour and meet specific regulatory standards for that purpose. Such unitary processes can be grouped into: pre-treatments, disinfection and softening.

10.5   Pre-treatments: Screening, Sedimentation, and Coagulation-Flocculation

The first step in drinking water treatment is typically pre-treatment, which involves screening and filtering the water to remove large particles and debris from the water before it undergoes further steps.

The purpose of pre-treatment is to protect downstream treatment processes, such as filtration and disinfection, by removing larger particles that could clog or damage equipment or interfere with the effectiveness of subsequent treatment steps. There are several common pre-treatments including screening, sedimentation and coagulation-flocculation.

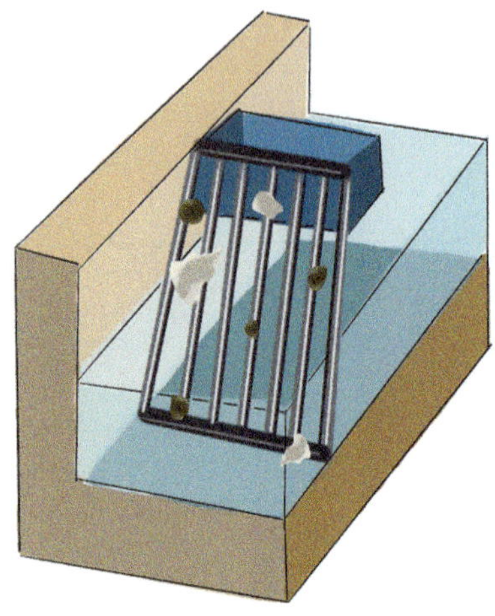

FIGURE 10.2 Screen

Screening

Screening is an essential procedure involving the removal of large debris, such as leaves, branches and other large particles before water enters the treatment plant to ensure that downstream equipment operates efficiently and effectively.

There are several types of screens used in water treatment, including coarse screens, fine screens and microscreens. Coarse screens are typically used to remove larger debris, (Figure 10.2), while fine screens and microscreens are used to remove smaller particles, such as sand and silt. Screening protects downstream equipment from damage, such as pumps and filters.

Sedimentation

Sedimentation is a commonly used process in water treatment to remove suspended solids from water that were not eliminated before by allowing them to settle out by gravity. The process works by reducing the velocity of water flow so that gravity can separate heavier particles from the water. It takes place in a sedimentation tank. In a typical sedimentation tank, the water enters the tank and slows down to a point where the suspended particles can settle out. The heavier particles will sink to the bottom of the tank, while the lighter particles will remain suspended in the water. Once the particles have settled, the clear water can be removed from the top of the tank and treated further (Figure 10.3).

Coagulation-Flocculation

In many cases, conventional sedimentation is not able to remove all the impurities, especially colloidal suspended particles. This is because they often carry an electrical charge on their surface, which interacts with the surrounding water molecules. Due to the polar nature of water molecules, they form a hydration layer around the charged colloidal particles that provide repulsion forces between particles, preventing them from aggregating or settling down. This phenomenon is known as electrostatic stabilisation.

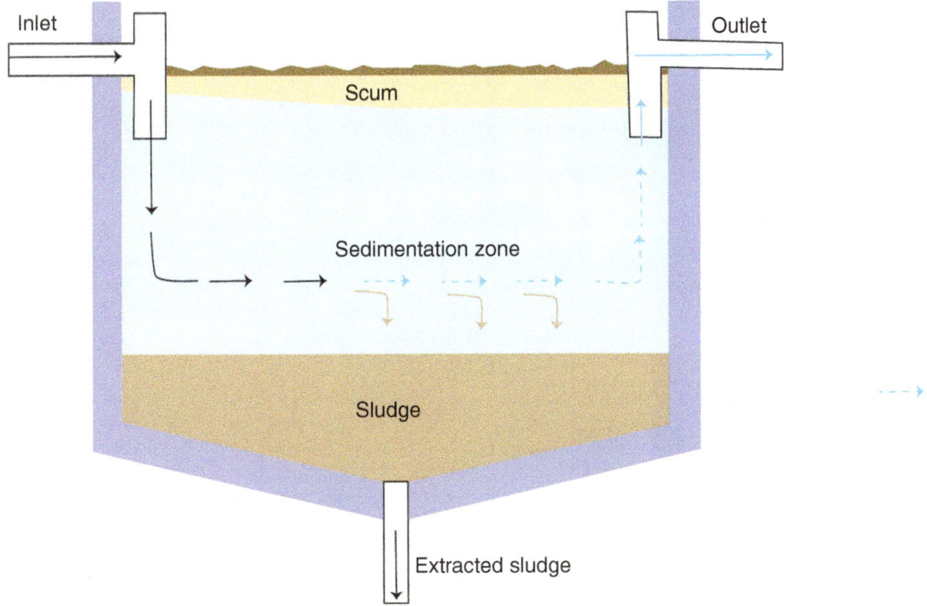

FIGURE 10.3   Sedimentation tank

At this point, the sedimentation process can be enhanced by adding chemicals such as alum, ferric chloride or polymers called flocculants that promote the coagulation of the suspended particles, making them easier to filtrate or settle out. They form a sticky precipitate that attracts and binds together suspended particles, colloids and other impurities. The process is commonly used both in drinking water and wastewater treatment (Figure 10.4).

The mechanism of action of flocculants involves several key steps (Figure 10.5):

1. **Adsorption:** The flocculant molecules adsorb onto the surface of the suspended particles through physical and chemical interactions between the flocculant and the charged particles present in the liquid.

2. **Bridging:** Once adsorbed onto the particle surface, the flocculant molecules can form bridges between adjacent particles. As a result, the particles become linked together, creating larger aggregates.

3. **Particle destabilisation:** The presence of the flocculant and the formation of bridges between particles destabilise the electrostatic repulsion forces that typically keep the particles dispersed in the liquid. The flocculant neutralises or reduces the charge on the particles, leading to decreased repulsion and increased attraction forces.

4. **Floccule formation:** As the flocculant continues to adsorb onto particles and form bridges, larger aggregates, known as floccules or flocs, begin to form. These flocs are much larger and denser than the individual particles and settle more rapidly under the influence of gravity or other separation mechanisms.

5. **Settling and separation:** The formed flocs settle down due to their increased size and weight, making it easier to separate them from the liquid. Sedimentation tanks or other separation processes are commonly used to allow the flocs to settle to the bottom, while the clarified liquid can be decanted or passed through filters for further purification.

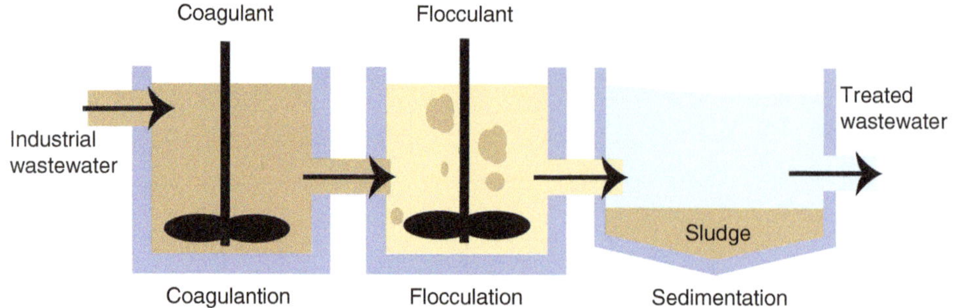

FIGURE 10.4 Sedimentation tank

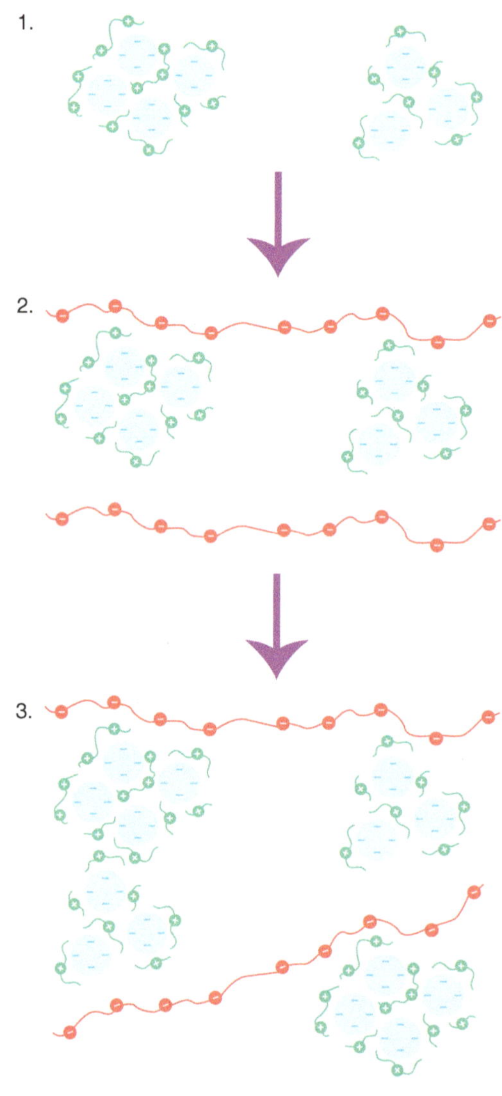

FIGURE 10.5 Flocculation steps

Topics of Interest: Types of Flocculants

There are various types of flocculants:

1. **Inorganic flocculants:** These are typically metal salts that contain aluminium or iron. Common examples include aluminium sulfate (alum), ferric chloride and poly-aluminium chloride. Inorganic flocculants work by neutralising the negative charges on particles and forming larger flocs through the processes of charge neutralisation and adsorption.

2. **Organic flocculants:** Organic flocculants are synthetic polymers or natural substances derived from plants. They are often used when inorganic flocculants are ineffective or environmentally undesirable. Organic flocculants can be further classified into the following categories:

   a. **Cationic flocculants:** These flocculants have positively charged functional groups, such as amino or quaternary ammonium groups. They are particularly effective in treating wastewater containing negatively charged particles.

   b. **Anionic flocculants:** These flocculants have negatively charged functional groups, such as carboxylate or sulfonate groups. They are commonly used for treating wastewater with positively charged particles.

   c. **Non-ionic flocculants:** These flocculants do not carry an electrical charge and are effective in a wide pH range. They are often used in combination with other flocculants to enhance flocculation efficiency.

   d. **Amphoteric flocculants:** These flocculants possess both positive and negative charges. They can adapt their charge depending on the pH of the solution and are suitable for treating wastewater with varying particle charges.

   e. **Natural flocculants:** Some naturally occurring substances can act as flocculants. For example, chitosan, a derivative of chitin found in the shells of crustaceans, is an effective natural flocculant. Other examples include starch, cellulose and certain plant-based extracts.

   f. **Composite flocculants:** Composite flocculants are a combination of inorganic and organic flocculants. By combining the advantages of both types, composite flocculants can offer enhanced flocculation performance and broader applicability.

The selection of a specific flocculant depends on factors such as the type of particles present, water chemistry, desired treatment efficiency and environmental considerations. It is essential to choose the appropriate flocculant for each application to achieve the desired.

After pre-treatment, the water is typically sent through a series of filters, which may include activated carbon, sand and other materials, to remove smaller particles and contaminants.

10.6 Disinfection with Chlorine

Untreated water sources, such as rivers, lakes and wells, may contain pathogens due to contamination from human or animal faecal matter, agricultural run-off or industrial pollutants. Disinfection of drinking water is the process of killing or inactivating harmful microorganisms such as bacteria, viruses and parasites that may be present in the water. The disinfection process is essential to protecting public health and preventing the spread of waterborne diseases. The most common method of disinfecting drinking water is chlorination.

Chlorination of water is the process of adding chlorine to drinking water to disinfect it and kill harmful bacteria, viruses and other pathogens that can cause illness. It has been used to treat drinking water for over 100 years.

When dissolved, chlorine actually reacts with water according to equilibrium (Eq. 10.1).

$$Cl_2 + H_2O \rightleftharpoons HOCl + HCl \tag{10.1}$$

Forming a mixture of chlorine, hypochlorous acid (HOCl) and hydrochloric acid (HCl). The average of every species is strongly dependent on the pH values of water.

HOCl undergoes a subsequent equilibrium reaction that involves the dissociation of the acid into hydrogen ions (H^+) and hypochlorite ions (OCl^-). The equilibria can be represented by the following chemical equation (Eq. 10.2):

$$HClO \rightleftharpoons H^+ + OCl^- \tag{10.2}$$

The value of K_a for HClO is around 3.5×10^{-8} at 25 °C.

This equilibrium reaction is also influenced by the pH of the solution. The concentration of HClO and ClO^- depends, therefore, on the pH.

At lower pH values (acidic conditions), the concentration of H^+ ions is high, favouring the formation of undissociated HOCl. As the pH increases (in alkaline conditions), the concentration of OH^- ions increases, leading to the formation of more OCl^- ions.

The pH-dependent equilibrium between HOCl and OCl^- has important implications for disinfection processes. As mentioned earlier, HOCl is a more potent disinfectant compared to ClO^-. Therefore, at lower pH values, where the concentration of HClO is higher, the disinfection efficacy of chlorine is increased.

Water treatment facilities often adjust the pH of the water to optimise the equilibrium between HClO and ClO^-, thereby minimising and enhancing disinfection. By maintaining the appropriate pH range, they can ensure an effective balance between the concentrations of the two species for efficient microbial control while minimising any potential adverse effects.

In acidic solutions, the major species are Cl_2 and HOCl, whereas in alkaline solutions, effectively only OCl^- is present.

HOCl kills microorganisms in water by disrupting their cell walls and denaturing their proteins.

The amount of chlorine required for effective disinfection depends on several factors, including the quality of the water source, the level of contamination and the desired level of disinfection. However, the standard range of chlorine concentration in drinking water is generally between 0.2 and 2.0 ppm. The United States Environmental Protection Agency (EPA) has set a maximum contaminant level (MCL) of 4.0 ppm for chlorine in drinking water. It is important to note that chlorine is added to drinking water as a disinfectant to kill harmful bacteria and viruses, but excessive amounts of chlorine in drinking water can be harmful to human health.

10.7 pH Dependence of Chlorine Concentration

The concentration of chlorine in water is influenced by the pH of the water (Figure 10.6). The most common form of chlorine present in disinfection with chlorine is HOCl and its conjugate base, hypochlorite ion (OCl^-). The ratio of HOCl to OCl^- is pH-dependent.

Disinfection is most effective at a pH level between 7.2 and 7.8. At this pH range, most of the chlorine is present in its active form, HOCl, which is able to penetrate bacterial cells and destroy them.

At lower pH values (acidic conditions), the concentration of HOCl is higher compared to OCl^-. This is because the dissociation of HOCl is favoured in acidic environments. HOCl is a more potent disinfectant than OCl^- and exhibits stronger antimicrobial activity.

As the pH increases (alkaline conditions), the concentration of OCl^- becomes dominant. This is because the dissociation of HOCl decreases and more OCl^- ions are present. OCl^- is a weaker disinfectant compared to HOCl and exhibits reduced antimicrobial activity.

If the pH level of the water is too low (acidic), the effectiveness of chlorine disinfection may be reduced. This is because, at low pH levels, chlorine is converted to its less effective form, the hypochlorite ion (OCl^-), which is less able to penetrate bacterial cells and destroy them. On the other hand, if the pH level is too high (alkaline), the effectiveness of chlorine may also be reduced because most of the chlorine is present in its inactive form, the chloride ion (Cl^-).

Therefore, it is important to monitor the pH level of water during chlorine disinfection to ensure that the water is within the recommended pH range of 7.2–10.8 for optimal disinfection. pH adjustment may be necessary in some cases to ensure effective disinfection. Additionally, it is important to ensure that the chlorine dosage is appropriate for the pH level of the water being treated to ensure that sufficient free chlorine is available to disinfect the water.

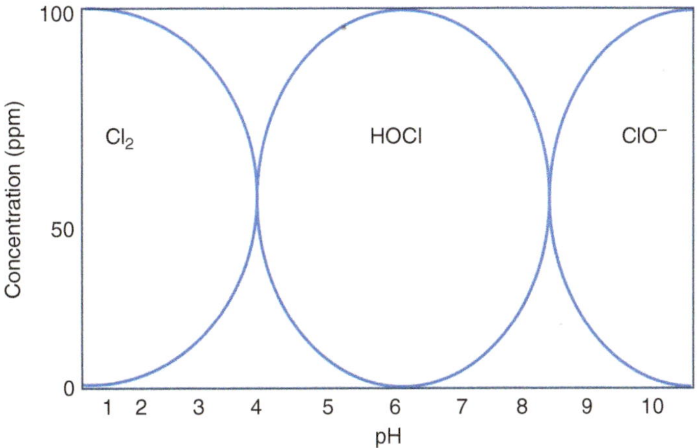

FIGURE 10.6 pH dependency of chlorine species

10.8 Chloramines

Chloramines are a byproduct that result from the reaction between chlorine and ammonia or organic nitrogen-containing compounds present in the water. The formation of chloramines can occur during various water treatment processes, such as drinking water disinfection or swimming pool maintenance.

When chlorine is added to water containing ammonia or organic nitrogen compounds, it undergoes a series of reactions that lead to the formation of different chloramine compounds. The most common types of chloramines include monochloramine (NH_2Cl), dichloramine ($NHCl_2$) and trichloramine (NCl_3). These chloramines have different disinfectant properties and characteristics.

The reaction between chlorine and ammonia can be summarised as follows:

When ammonia (NH_3) is present in water containing chlorine (Cl_2) or HOCl, the following reactions can occur:

$$\text{Formation of monochloramine}: \left(NH_2Cl\right): NH_3 + HOCl \rightarrow NH_2Cl + H_2O \qquad (10.3)$$

$$\text{Formation of dichloramine}: \left(NHCl_2\right): NH_2Cl + HOCl \rightarrow NHCl_2 + H_2O \qquad (10.4)$$

$$\text{Formation of trichloramine}: \left(NCl_3\right): NHCl_2 + HOCl \rightarrow NCl_3 + H_2O \qquad (10.5)$$

These reactions occur rapidly and can happen simultaneously. The relative amounts of each chloramine formed depend on various factors such as pH, temperature and chlorine-to-ammonia ratio.

The presence of chloramines in treated water can have both advantages and disadvantages. Chloramines are more stable than free chlorine and can provide a longer-lasting residual disinfectant effect. They are effective in controlling microbial growth and reducing the formation of disinfection byproducts (DBPs) such as trihalomethanes (THMs).

However, chloramines are weaker disinfectants compared to free chlorine. They may require longer contact times or higher concentrations to achieve the same level of microbial inactivation. Chloramines also have a characteristic odour that can be unpleasant and irritating to some people. In swimming pools, the presence of chloramines can cause eye and skin irritation and contribute to the 'chlorine smell' often associated with pools.

Water treatment facilities and pool operators carefully monitor and control the levels of chloramines to ensure effective disinfection while minimising the associated drawbacks. Techniques such as breakpoint chlorination, which involves adding sufficient chlorine to react with all the ammonia and organic nitrogen compounds, can help eliminate chloramines and restore free chlorine residual.

It's important to note that the formation and management of chloramines during disinfection are complex processes influenced by various factors such as chlorine dosage, contact time, pH, temperature and the initial concentrations of ammonia and organic nitrogen compounds in the water. Therefore, careful monitoring and control are essential to maintaining optimal disinfection and water quality.

10.9 Total, Free and Combined Chlorine

There are three main types of chlorine present in water: total chlorine, free chlorine and combined chlorine (Table 10.1).

1. **Total chlorine:** Total chlorine is the amount of both free chlorine and combined chlorine present in the water. It measures the total amount of chlorine in the water, regardless of its chemical state. Total chlorine is the sum of free chlorine and combined chlorine, while residual chlorine is the amount of free chlorine remaining in the water after the chlorine has reacted with organic and inorganic compounds.

    Total chlorine is typically measured using a colorimetric test that uses a reagent that reacts with all forms of chlorine in the water.

2. **Free chlorine:** Free chlorine is the amount of chlorine that is available to disinfect the water. It is the most effective form of chlorine for killing bacteria and viruses. Free chlorine reacts readily with contaminants in the water and breaks them down into harmless byproducts. Free chlorine is typically measured using a colorimetric test that uses a reagent that reacts specifically with free chlorine.

3. **Combined chlorine:** Combined chlorine is the amount of chlorine that has reacted with contaminants in the water and is no longer available to disinfect the water. Combined chlorine is less effective than free chlorine for disinfection purposes and can cause unpleasant odours and skin irritation. The most common form of combined chlorine is known as chloramines. Chloramines are chemical compounds that form when chlorine is added to water that contains ammonia or other nitrogen-containing compounds. Chloramines are commonly used as disinfectants in drinking water and swimming pools because they are more stable and last longer than free chlorine.

    As mentioned above, there are three types of chloramines: NH_2Cl, $NHCl_2$ and NCl_3. NH_2Cl is the most commonly used form of chloramine for water disinfection because it is the most stable and has the least potential for forming DBPs.

TABLE 10.1   Summary of chlorine species in drinking water

Type of chlorine	Definition
Total chlorine	The total amount of chlorine present in water, including both free chlorine and combined chlorine, represents the overall chlorine concentration and is often measured as a sum of free chlorine and combined chlorine.
Free chlorine	The form of chlorine that is not combined with other substances. It is the active and available form of chlorine used for disinfection. Free chlorine includes both HOCl and hypochlorite ion (OCl^-) species.
Combined chlorine	Chlorine that has reacted with organic or inorganic substances in water, typically resulting in the formation of chloramines. Combined chlorine is less effective as a disinfectant compared to free chlorine and can contribute to unpleasant odours and tastes.

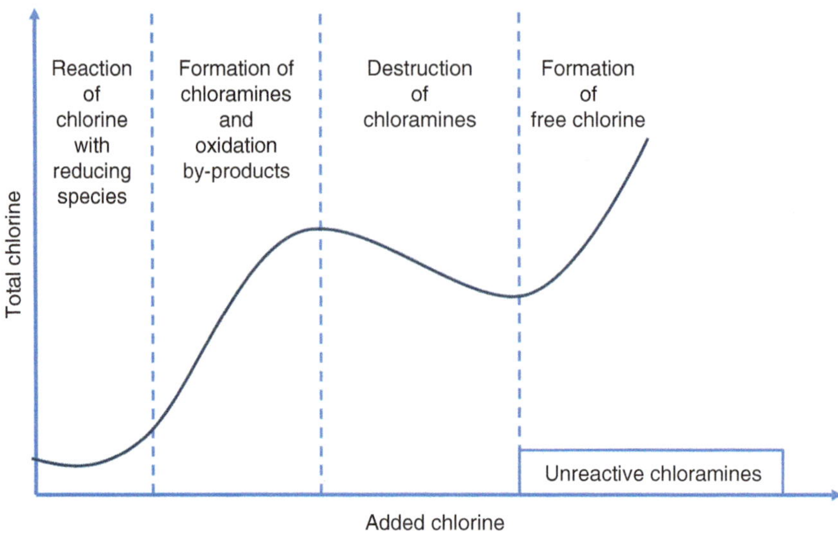

FIGURE 10.7   Breakpoint chlorination curve

The breakpoint chlorination curve (Figure 10.7) illustrates the behaviour of free and combined chlorine as the dosage of chlorine is increased within a system. The different stages of this process are described as follows:

a. **Initial reaction with reducing compounds:** When a small amount of chlorine is added, it initially reacts with reducing compounds present in the water, which are typically inorganic matter. These compounds could include metals, sulfides or other substances that can be oxidised by chlorine.

b. **Reaction with organic compounds and ammonia:** As the chlorine dosage is increased further, it starts to react with organic compounds and ammonia. Chlorine and ammonia combine to form NH_2Cl, which contributes to the total chlorine concentration. This stage is characterised by an increase in the measured total chlorine.

c. **Conversion of monochloramines:** Once all the ammonia in the water has reacted with chlorine, the NH_2Cl that were formed will undergo further reactions and convert into di- or NCl_3. These chloramines have a lower oxidation potential compared to NH_2Cl, resulting in a reduction of the measured total chlorine.

d. **Chlorine demand satisfied and free chlorine formation:** At this point, there is sufficient chlorine present to fulfil the chlorine demand of the system, including the oxidation of inorganic matter, organic matter and ammonia. As a result, free chlorine can begin to form, indicating that there is an excess of chlorine available beyond the demand.

While chlorination is an effective method of water disinfection, it can also have some drawbacks. For example, chlorination can produce DBPs such as THMs, which have been linked to cancer and other health problems or chloramines, which can be formed when chlorine reacts with ammonia or other organic nitrogen-containing compounds in water. Chloramines are commonly used as a secondary disinfectant in water treatment because they are more stable than free chlorine and provide longer-lasting disinfection. However, chloramines can have negative effects on water quality and human health. Long-term exposure to chloramines has been linked to various health problems, including eye and skin irritation, respiratory problems and an increased risk of cancer.

10.10 Water Disinfection By-Products with Chlorine

Water disinfection with chlorine is a common and effective method used to kill or inactivate various microorganisms, including bacteria, viruses and protozoa, in water. It's important to note that while chlorine is an effective disinfectant, it may react with certain organic and inorganic compounds in water to form DBPs (Table 10.2).

1. **Organic DBPs:** They are typically formed when the disinfectant reacts with organic matter in the water, such as algae, leaves, bacteria and organic molecules. Some examples of organic DBPs are:

   a. **Trihalomethanes (THMs):** THMs are one of the most common and well-known organic DBPs. They form when chlorine reacts with NOM and contain three halogen atoms (typically chlorine and bromine) and a carbon atom. Common THMs include chloroform ($CHCl_3$), bromodichloromethane ($CHCl_2Br$), dibromochloromethane ($CHClBr_2$) and bromoform ($CHBr_3$).

   b. **Haloacetic acids (HAAs):** HAAs are another significant class of organic DBPs. They are formed when chlorine or chloramine reacts with NOM.

TABLE 10.2   Water disinfection by-products with chlorine

Disinfection byproduct	Chemical formula	Health concerns
Trihalomethanes (THMs)	$CHCl_3$, $CHBrCl_2$, $CHBr_2Cl$, $CHBr_3$	Carcinogenic potential, hepatotoxins, reproductive effects
Haloacetic acids (HAAs)	$CH_2ClCOOH$, $CHCl_2COOH$ CCl_3COOH $CH_2BrCOOH$, $CHBr_2COOH$	Carcinogenic potential, liver and kidney issues
Haloacetonitriles (HANs)	$CHCl_2CN$, $CCl_3(CN)$ $CH_2Br(CN)$ $CHBrCl(CN)$	Carcinogenic potential, reproductive effects
Haloketones	CX_3COCH_3	Eye, skin and respiratory irritation
Chloroacetaldehyde	CCl_3CHO	Carcinogenic potential, nervous system effects
Chloramines	$ClNH_2$, Cl_2NH, Cl_3N	Allergic reactions, respiratory problems and gastrointestinal effects
Nitrosamines	$R_1R_2N\text{-}NO$	Carcinogenic potential
Chlorites	ClO_2^-	Affects thyroid function, nervous system effects
Chlorates	ClO_3^-	Infants and people with thyroid conditions
Bromates	BrO_3^-	Carcinogenic potential, kidney issues

Common HAAs include monochloroacetic acid (MCAA), dichloroacetic acid (DCAA), trichloroacetic acid (TCAA), monobromoacetic acid (MBAA) and dibromoacetic acid (DBAA).

c. **Haloacetonitriles (HANs):** HANs are formed when chlorine reacts with nitrogen-containing organic compounds present in water. They include compounds like dichloroacetonitrile (DCAN), trichloroacetonitrile (TCAN), bromochloroacetonitrile (BCAN) and dibromoacetonitrile (DBAN).

d. **Haloketones:** Haloketones are a group of DBPs that contain both halogen and carbonyl (ketone) functional groups. Examples include 1,1-dichloro-2-propanone and 1,1,1-trichloro-2-propanone.

e. **Haloacetaldehydes:** Haloacetaldehydes are formed when disinfectants react with natural organic matter. An example is chloroacetaldehyde.

f. **Chloramine byproducts:** Chloramines, formed by the combination of chlorine and ammonia, are another disinfection method used in water treatment. Chloramination can lead to the formation of DBPs such as nitrosamines, haloacetonitriles and haloacetic acids.

g. **Nitrosamines:** Nitrosamines are formed when nitrite, which can be present in water due to the reaction of disinfectants with naturally occurring nitrogen compounds, reacts with organic matter. Some examples include *N*-nitrosodimethylamine (NDMA) and *N*-nitrosodiethylamine (NDEA).

2. **Inorganic byproducts:** These inorganic byproducts typically arise from the reactions between chlorine and various inorganic substances present in the water. Some common inorganic byproducts include:

a. **Chlorites (ClO_2^-):** Chlorite is formed when chlorine reacts with naturally occurring or added inorganic chloride ions (Cl^-) in water. Chlorite is regulated in drinking water due to its potential health risks, particularly for infants.

b. **Chlorates (ClO_3^-):** Chlorate can form when chlorine reacts with HOCl or through other chemical reactions. High levels of chlorate in drinking water can have health implications, especially for infants and people with thyroid conditions.

c. **Bromates (BrO_3^-):** Bromate is produced when bromide ions (Br^-) naturally present in water react with chlorine. Bromate is a regulated contaminant due to its carcinogenic properties.

d. **Inorganic acids:** Chlorine can also react with certain inorganic compounds, such as sulfite ions (SO_3^{2-}) or sulfide ions (S^{2-}), leading to the formation of inorganic acids like sulfurous acid (H_2SO_3) or sulfuric acid (H_2SO_4).

Topics of Interest: Disinfection in Swimming Pools

Trichloroisocyanuric acid (TCCA) is a commonly used disinfectant that is effective against a wide range of microorganisms, including bacteria, viruses and fungi. It is often used in swimming pools, spas and other water treatment applications to kill harmful

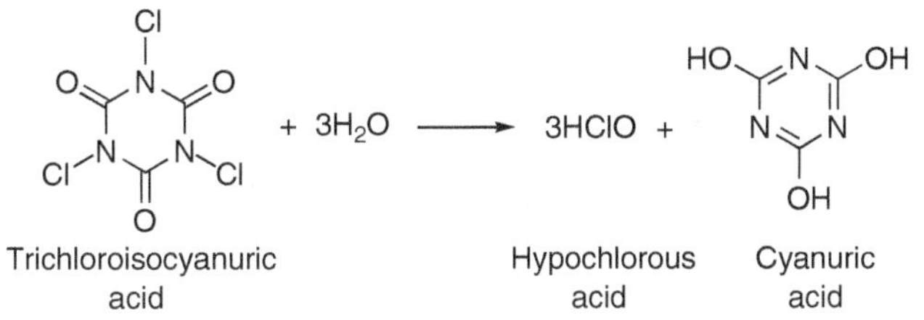

FIGURE 10.8 reaction:

Trichloroisocyanuric acid + 3H₂O → 3HClO + Cyanuric acid

$$\text{Trichloroisocyanuric acid} + 3H_2O \longrightarrow 3HClO + \text{Cyanuric acid}$$

FIGURE 10.8   Decomposition of TCCA

microorganisms and prevent the spread of disease. It is available in a variety of forms, including tablets, granules and powder and can be applied directly to water or dissolved in water before use.

When TCCA is added to water, it reacts and undergoes decomposition. The reaction can be represented as follows (Figure 10.8).

The decomposition of TCCA in water releases HOCl, hypochlorite ion (OCl^-) and cyanuric acid (CA).

HOCl and hypochlorite ions are both strong oxidising agents and are responsible for the disinfecting action of TCCA. They are effective at killing a wide range of microorganisms, including bacteria, viruses and fungi.

Cyanuric acid, on the other hand, is a stabiliser that is often added to TCCA to slow down the decomposition process and extend the effectiveness of the disinfectant. It helps prevent the loss of chlorine due to sunlight and other factors.

It reduces the rate at which chlorine is broken down by ultraviolet (UV) rays, allowing for more efficient use of chlorine over time.

Over time, especially with the continuous use of stabilised chlorine products, the concentration of CYA in the pool can gradually increase. Regular monitoring is essential to ensure levels remain within the optimal range (30–50 ppm).

Cyanuric acid is relatively stable in pool water. It does not break down or evaporate quickly and tends to accumulate unless actively managed. The most common method for reducing CYA levels is dilution. This involves partially draining the pool and refilling it with freshwater, which lowers the concentration of CYA.

In some cases, biological processes in the pool water can slightly degrade CYA, but this is typically minimal and not a reliable method for reducing levels.

However, it is important to note that if the concentration of CA in the water becomes too high, it can start to interfere with the disinfecting action of the HOCl and hypochlorite ion. This is because CA can form complexes with chlorine, which reduces the amount of free chlorine available to kill microorganisms.

Another key aspect of pools is maintaining an optimal pH value.

- **Ideal pH range:** The recommended pH range for a swimming pool is typically between 7.2 and 7.8. This range is slightly basic or alkalinain the water's clarity and prevent issues like corrosion, scaling and eye or skin irritation.

- **Low pH (below 7.2):** If the pH of a swimming pool falls below the recommended range, the water becomes more acidic. This can lead to several consequences, such as:

  - **Skin and eye irritation:** Acidic water can cause skin rashes, dryness and itching. It can also irritate the eyes, causing redness and discomfort.

  - **Corrosion:** Low pH levels can corrode metal fittings, pipes and pool equipment, leading to damage and potential leaks.

  - **Etching of pool surfaces:** Acidic water can etch and damage the pool surfaces, including the plaster or vinyl lining, tiles and grout.

- **High pH (above 7.8):** When the pH of a swimming pool exceeds the recommended range, the water becomes more alkaline. The consequences of high pH levels include:

- ○ **Reduced sanitiser efficiency:** A high pH can hinder the effectiveness of chlorine or other sanitisers, making it difficult to maintain proper disinfection levels in the pool.
- ○ **Cloudy water:** Alkaline water can cause cloudiness in the pool, making it less visually appealing.
- ○ **Scale formation:** High pH levels can lead to the formation of scale deposits on pool surfaces and equipment, including filters and heaters. This can affect their efficiency and longevity.
- ○ **Skin and eye discomfort:** Alkaline water may cause skin dryness, itchiness and eye irritation in swimmers.
- ○ **Reduced sanitiser effectiveness:** Chlorine or other sanitising agents become less effective at killing bacteria and controlling algae growth in acidic water. This can result in poor water quality and an increased risk of infection.

10.11 Disinfection Without Chlorine Gas

There are several methods for disinfecting water without using chlorine. The choice of method depends on various factors, such as the quality of the water source, available resources and the specific requirements of the situation.

Sodium Hypochlorite

Sodium hypochlorite is a chemical compound with the chemical formula NaClO. It is commonly known as bleach and is widely used as a disinfectant and cleaning agent. Sodium hypochlorite is a strong oxidising agent and has antimicrobial properties, which make it effective in killing bacteria, viruses and fungi.

Disinfection with sodium hypochlorite is a common and effective disinfectant that is widely used for water treatment and disinfection purposes. By adding hypochlorite to water, HOCl is formed (Eq. 10.6).

$$NaOCl + H_2O \rightarrow HOCl + NaOH \tag{10.6}$$

Due to the presence of sodium hydroxide in sodium, the pH of the water is increased. The pH of the water determines how much HOCl is formed. While sodium hypochlorite is used, HCl is used to lower the pH.

HOCl is a powerful oxidising agent that can destroy a wide range of microorganisms, including bacteria, viruses and protozoa.

Sodium hypochlorite is particularly effective at killing waterborne pathogens that can cause illnesses such as cholera, typhoid fever and dysentery.

However, it is important to use sodium hypochlorite carefully and in the correct concentrations when disinfecting water. Overuse of the chemical can lead to the formation of harmful DBPs.

Calcium Hypochlorite

Calcium hypochlorite ($Ca(ClO)_2$) is a white, crystalline salt highly soluble in water. $Ca(ClO)_2$ is a strong oxidising agent and releases chlorine gas when it comes in contact with acids. $Ca(ClO)_2$ is widely used as a disinfectant and sanitiser due to its ability to kill bacteria, viruses and other microorganisms. It is commonly used for water treatment in swimming pools, drinking water systems and wastewater treatment plants. The dosage typically ranges from 1 to 4 g/1,000 L (or 1 m^3) of water, depending on the level of contamination and the desired chlorine residual.

Chlorine Dioxide

Chlorine dioxide (ClO_2) is a highly reactive yellow-to-reddish gas at room temperature. ClO_2 is primarily used as an oxidising agent in various industrial applications and as a powerful disinfectant that is widely used as a disinfectant in various industries, including water treatment, healthcare facilities, food processing and more. Water disinfection involves adding ClO_2 to the water to kill bacteria, viruses and parasites. It is often used as an alternative to chlorine because it produces fewer disinfection by-products. The dosage for drinking water treatment can vary based on the source water quality, treatment objectives and regulatory requirements. Typically, ClO_2 is used at concentrations ranging from 0.5 to 2 ppm.

UV Radiation

This method involves exposing the water to UV light. UV radiation is effective against bacteria, viruses and parasites but does not remove any physical contaminants from the water. It is a non-chemical method of water treatment that is effective in removing microorganisms from water without altering its taste, colour or odour. It is a non-chemical and environmentally friendly method of disinfection, which makes it an attractive option, for example, for the drinks industry.

Water passes through a chamber that contains one or more UV lamps. These lamps emit UV-C light, which has a wavelength between 200 and 280 nm. Radiation penetrates the microorganisms present in the water and damages their DNA or RNA, preventing them from reproducing or becoming infectious. So, the microorganisms are effectively inactivated or killed, making the water safe for consumption (Figure 10.9).

UV disinfection is used in a wide range of industries that need safe water for various applications, including:

1. **Healthcare industry:** UV disinfection is commonly used in hospitals and healthcare facilities to sterilise equipment and surfaces, reducing the risk of healthcare-associated infections (HAIs).

2. **Bottled water industry:** UV disinfection is used to treat drinking water to kill harmful microorganisms, making the water safe for consumption and discharge into the environment.

3. **Food and beverage industry:** UV disinfection is used to sterilise food and beverage processing equipment, packaging materials and surfaces to prevent contamination and improve product safety.

4. **Pharmaceutical industry:** UV disinfection is used to sterilise equipment and surfaces in the production of pharmaceutical products to ensure product safety and purity.

5. **Agriculture industry:** UV disinfection is used to sterilise water used for irrigation and reduce the risk of plant diseases caused by harmful microorganisms.

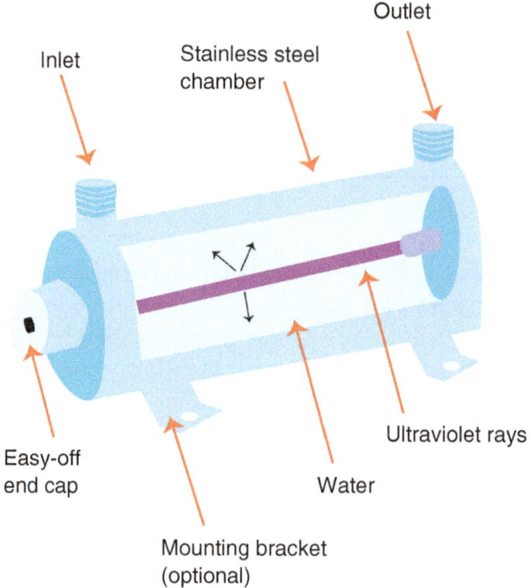

FIGURE 10.9   UV water disinfection device

Topics of Interest: Fluorination of Drinking Water Controversy

Fluorination of drinking water is the process of adding fluoride compounds, such as sodium fluoride, to public water supplies in order to prevent tooth decay. The practice of fluoridating water began in the United States in the 1940s and has since become widespread around the world.

Fluoridation is a safe and effective way to improve dental health, especially in areas where access to dental care is limited. Fluoride works by strengthening tooth enamel, making it more resistant to decay. It can also reverse early signs of decay and reduce the amount of acid that bacteria in the mouth produce.

While fluoridation is generally considered safe, there are some concerns about the potential health effects of long-term exposure to fluoride. In high concentrations, fluoride can cause dental fluorosis, a condition that affects the appearance of tooth enamel. In very high concentrations, it can also cause skeletal fluorosis, a condition that affects bone density.

Some opponents of water fluoridation argue that it is unethical to add substances to public water supplies without the consent of individuals and that individuals should have the right to choose whether or not they consume fluoride. Others argue that the benefits of fluoride can be obtained through other means, such as fluoridated toothpaste or dietary supplements and that adding fluoride to water is unnecessary.

Another concern is the potential health effects of long-term exposure to fluoride. Some studies have suggested that high levels of fluoride exposure may be associated with an increased risk of certain health problems, such as bone fractures, thyroid problems and neurological issues. However, other studies have not found a significant association between fluoride exposure and these health problems.

There have also been concerns about the potential impact of fluoride on the environment, particularly when high levels of fluoride are discharged into waterways.

However, the levels of fluoride added to public water supplies are carefully regulated and monitored to ensure that they are within safe limits. The World Health Organization recommends a fluoride concentration of between 0.5 and 1.5 mg/litre of water. Many countries, including the United States, follow this guideline.

It can be used in various stages of the drink production process, including water treatment, sterilisation of equipment and packaging. For example, UV disinfection systems can be installed in water treatment plants to ensure that the water used in the production of drinks is free from harmful microorganisms. Similarly, UV disinfection can be used to sterilise equipment used in the production process, such as tanks, pipes and filters.

The process of UV disinfection involves passing water through a chamber that contains UV lamps. The UV lamps emit UV-C light, which penetrates the cell walls of microorganisms and damages their DNA, preventing them from reproducing and rendering them harmless. The effectiveness of UV disinfection depends on several factors, including the intensity and duration of exposure to UV light, the quality of the water being treated and the design of the UV disinfection system.

Ozonation

Ozonation is a common method for water disinfection that involves the use of ozone, a powerful oxidising agent, to kill bacteria, viruses and other microorganisms in water. The process of ozonation involves the conversion of oxygen to ozone (O_3) through the application of high-energy electrical fields or UV radiation.

Ozone is a highly reactive gas that rapidly oxidises organic and inorganic compounds, including microorganisms, in water. When ozone is introduced into water, it reacts with the cell walls of microorganisms, causing them to rupture and die. Unlike other disinfection methods, such as chlorine, ozonation does not leave any harmful byproducts in the water.

Ozone is not a stable molecule in water and undergoes decomposition in a relatively short period of time. The half-life of ozone in water depends on various factors such as temperature, pH, the presence of organic matter and other water constituents.

At typical drinking water pH levels of around 7, the half-life of ozone in water is around 20–30 minutes. This means that after 20–30 minutes, half of the initial amount of ozone will have decomposed. At higher pH levels, the half-life of ozone decreases, meaning it decomposes more rapidly. Similarly, higher temperatures also increase the rate of ozone decomposition. For this reason, it is not applicable disinfection method for urban distribution networks due to their low permanence in water.

Ozonation is often used as a secondary treatment method for water purification, following primary treatment processes such as filtration and sedimentation. It is also used in the treatment of wastewater, swimming pools and industrial processes that require high levels of water purity.

One of the advantages of ozonation is that it is effective against a wide range of microorganisms, including bacteria, viruses and protozoa and it can remove some organic contaminants that are resistant to other disinfection methods. However, ozonisation requires careful monitoring and control to ensure that the proper levels of ozone are maintained and the process can be expensive due to the equipment and energy requirements.

Ozonation is used in a variety of industries for water disinfection and purification. Some of the industries that use ozonated water include:

1. **Drinking water treatment:** Many municipal water treatment plants use ozonation as a secondary treatment method to disinfect water before it is distributed to homes and businesses.

2. **Bottled water production:** The bottled water industry often uses ozonisation as a final step in the purification process to ensure that the water is free from harmful microorganisms.

3. **Food and beverage production:** The food and beverage industry uses ozonated water to disinfect equipment and surfaces, as well as to wash fruits and vegetables.

4. **Aquaculture:** Fish and shrimp farms use ozonated water to control the growth of algae and other microorganisms in the water and to prevent the spread of disease among the fish.

5. **Swimming pools and spas:** Ozonated water is used to disinfect swimming pools and spas, as well as to remove organic contaminants such as sweat and body oils.

6. **Healthcare facilities:** Hospitals and other healthcare facilities use ozonated water to disinfect medical equipment and surfaces, as well as to sterilise surgical instruments.

7. **Pharmaceutical production:** The pharmaceutical industry uses ozonated water to sterilise equipment and to clean and disinfect the production areas.

It is important to note that while ozone does not persist in water for an extended period, it is a powerful oxidising agent and can be used effectively for water treatment purposes. Ozone can effectively destroy various contaminants such as bacteria, viruses and organic compounds. However, it must be generated on-site and used immediately, as it cannot be stored for later use.

10.12 Water Softening

The purpose of water softening is to reduce the concentration of dissolved minerals, primarily calcium and magnesium ions, in water. This process aims to mitigate the negative effects of hard water and ensure the water is suitable for various domestic, industrial and commercial applications. While hard water is generally safe to drink, it can cause problems in household appliances and plumbing systems over time, such as clogging, corrosion and reduced efficiency. There are several methods for softening water. Each method has its advantages and disadvantages and the best method depends on the specific situation and needs of the user.

Distillation

Distillation involves boiling water and then condensing the steam back into water (Figure 10.10). This process removes the minerals that cause hardness, leaving behind pure water. It is inefficient and requires large amounts of energy, so its application is very limited.

Reverse Osmosis

Osmosis is the natural movement of solvent molecules (usually water) from an area of lower solute concentration to an area of higher solute concentration through a semipermeable membrane. In osmosis, water moves across the membrane to equalise the concentration of the solute on both sides. This process occurs spontaneously without the need for external pressure (Figure 10.11).

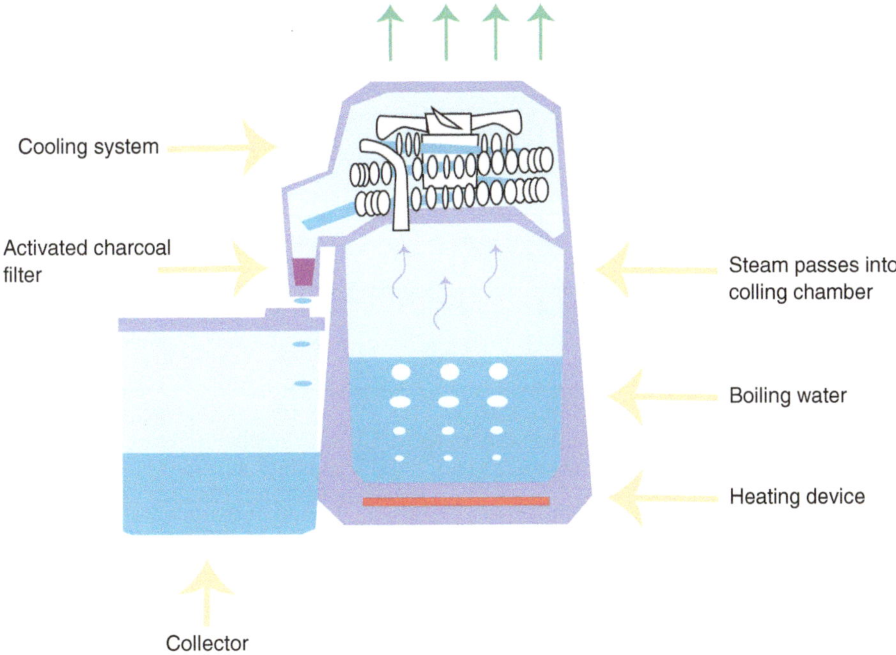

FIGURE 10.10 Water boiler

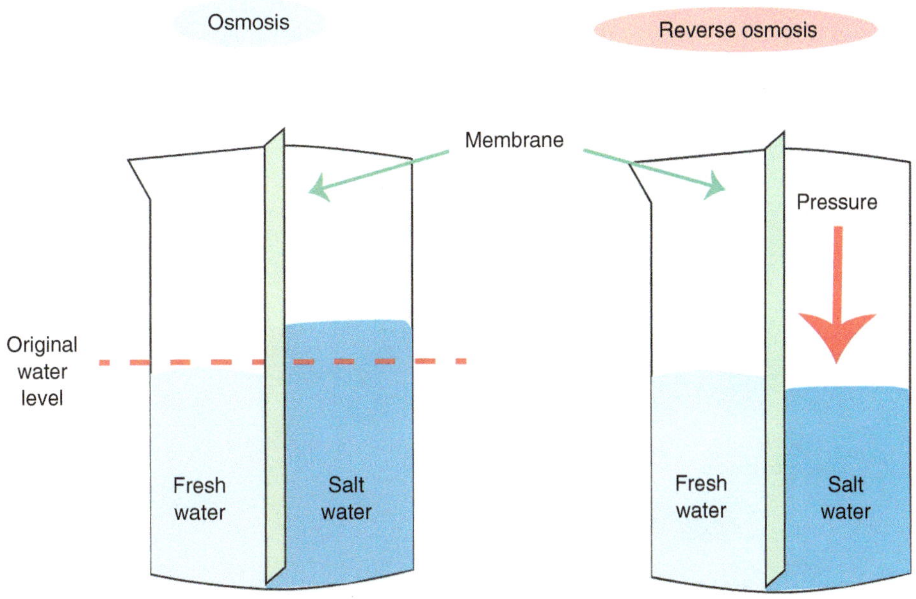

FIGURE 10.11 Osmosis and reverse osmosis

RO is a process that utilises external pressure to force water molecules from an area of higher solute concentration to an area of lower solute concentration through a semi-permeable membrane. The pressure applied is typically higher than the osmotic pressure of the solution, which results in the separation of water from solutes. RO is commonly used for water purification and desalination, where it can remove impurities, salts and other contaminants from water, such as bacteria and viruses (see Table 10.3).

TABLE 10.3   Osmosis versus reverse osmosis

Key facts	Osmosis	Reverse osmosis
Direction of water movement	From lower solute concentration to higher solute concentration	From higher solute concentration to lower solute concentration
Pressure requirement	No external pressure is required	External pressure is applied to overcome osmotic pressure
Purpose	The natural process for water balance in living organisms	Water purification, desalination and removal of contaminants
Membrane properties	A semipermeable membrane allows water to pass through	Semi-permeable membrane with smaller pores to effectively separate solutes
Efficiency	Not typically used for purification purposes	Can remove up to 99% of dissolved solutes

This process can also remove some minerals, including calcium and magnesium. The water is passed through the membrane under pressure, leaving behind minerals and other impurities and pollutants (Figure 10.12).

It is a popular method used to demineralise water in labs and households.

RO equipment can be used to produce deionised water in a laboratory setting. It is necessary to adjust the system settings, such as flow rate and pressure, to optimise performance. It is important to note that RO membranes can become fouled over

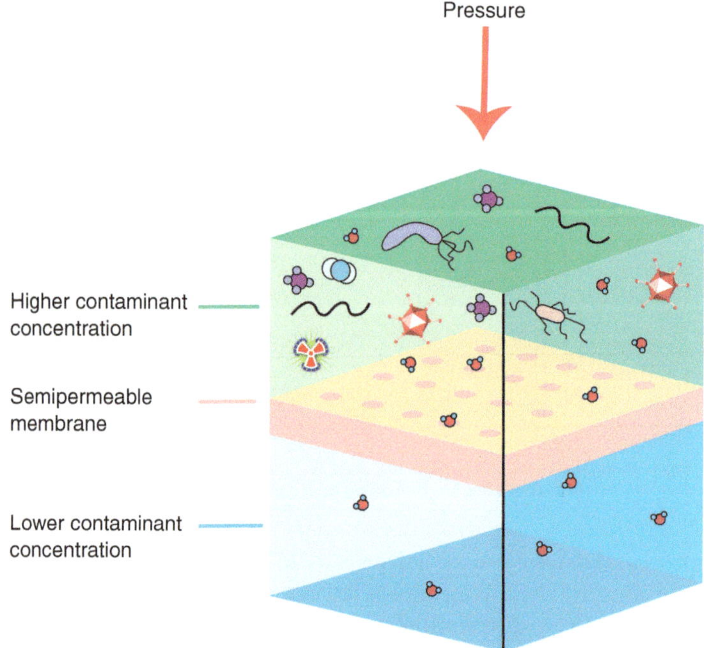

FIGURE 10.12   Reverse osmosis scheme

time, which can reduce their effectiveness in removing impurities from the water. Regular maintenance and cleaning of the RO system are necessary to ensure the consistent production of high-quality deionised water.

In households, RO is used to improve the taste and quality of drinking water. It can remove harmful contaminants like lead, arsenic and fluoride, as well as improve the taste of chlorine-treated municipal water.

Topics of Interest: Ion Exchange Resins

Ionic exchange resins are solid materials that are widely used as catalysts and for the purification, separation and recovery of different substances. They are typically composed of small beads or granules made of a polymer matrix, which provides mechanical stability and contains functional groups that are responsible for ion exchange. Depending on the nature of these functional groups, they have different uses and applications. In Table 10.4, the different types of resins are classified.

The resin beads are usually spherical and have a high surface area-to-volume ratio. They can be packed into a column or a tank, which is then filled with the solution to be treated. As the solution passes through the resin bed, ions in the solution are attracted to the functional groups on the resin beads and are exchanged with ions of the same charge that are already bound to the resin.

TABLE 10.4   **Exchange resin classification**

Exchange resin type	Uses
Strong acid	Removal of heavy metal ions from water Water softening and demineralisation Recovery of valuable metals
Weak acid	Removal of temporary hardness of water Purification of antibiotics and amino acids Recovery of valuable metals
Strong base	Removal of strong acids and anions from water Water demineralisation and dealkalisation Removal of nitrate and sulfate ions from water
Weak base	Removal of weak acids and anions from water Water demineralisation and dealkalisation Removal of organic matter and colour from water
Chelating resin	Removal of heavy metals and radionuclides from water Recovery of valuable metals Industrial wastewater treatment

Water Softening by Ion Exchange Resins

The general purpose of water softening is to reduce the concentration of dissolved minerals, primarily calcium and magnesium ions, in water. This process aims to mitigate the negative effects of hard water and ensure the water is suitable for various domestic, industrial and commercial applications.

Acidic ion exchange resins are a type of resin that contains acid functional groups such as sulfonic acid (—SO$_3$H), carboxylic acid (—COOH) and phosphonic acid (—PO$_3$H$_2$). These resins are used for cation exchange processes in various industries, including water treatment, pharmaceuticals and chemical manufacturing. Examples of acid resins and their trade names are described in Table 10.5.

Sulfonic resins are commonly used in water-softening applications. The resin is pre-loaded with a coating of positively charged sodium ions before they are packed into the water-softener resin bed.

These sulfonic acid groups are responsible for the ion exchange process, which removes hardness-causing ions from water and replaces them with sodium ions. Figure 10.13 shows how magnesium and calcium ions, responsible for water hardness, are removed from hard water and replaced with sodium ions, which do not cause scaling.

The water softening reaction occurs when hard water containing calcium (Ca^{2+}) and magnesium (Mg^{2+}) ions passes through a column filled with sulfonic resin. The sulfonic acid groups on the resin strongly bind with the calcium and magnesium ions, releasing sodium ions (Na$^+$) in exchange. The reaction can be represented as follows (Eqs. (10.7) and (10.8)):

$$\text{Sulfonic resin}\left(\text{Na}^+ \text{ form}\right) + \text{Ca}^{2+} \rightarrow \text{Sulfonic resin}\left(\text{Ca}^{2+} \text{ form}\right) + 2\text{Na}^+ \tag{10.7}$$

$$\text{Sulfonic resin}\left(\text{Na}^+ \text{ form}\right) + \text{Mg}^{2+} \rightarrow \text{Sulfonic resin}\left(\text{Mg}^{2+} \text{ form}\right) + 2\text{Na}^+ \tag{10.8}$$

Over time, the sulfonic resin becomes saturated with calcium and magnesium ions and needs to be regenerated to restore its ion exchange capacity. This is done

TABLE 10.5 Examples of acid resins

Resin type	Functional group	Examples
Sulfonic acid resin	-SO$_3$H	Amberlite IR-120, Dowex 50 and Purolite C100
Carboxylic acid resin	-COOH	Dowex 50W, Amberlite IRC-50
Phosphoric acid resin	-PO$_3$H$_2$	Amberlite IRP-69, Purolite C104

FIGURE 10.13 Ions interchange in sulfonic resin

by passing a concentrated solution of sodium chloride (brine) through the resin bed. The regeneration reaction involves displacing the calcium and magnesium ions from the resin with sodium ions from the brine solution. The reaction can be represented as follows (Eqs. (10.9) and (10.10)):

$$\text{Sulfonic resin}\left(Ca^{2+}\text{ form}\right)+2NaCl \rightarrow \text{Sulfonic resin}\left(Na^{+}\text{ form}\right)+CaCl_2 \qquad (10.9)$$

$$\text{Sulfonic resin}\left(Mg^{2+}\text{ form}\right)+2NaCl \rightarrow \text{Sulfonic resin}\left(Na^{+}\text{ form}\right)+MgCl_2 \qquad (10.10)$$

During the regeneration process, the calcium chloride ($CaCl_2$) and magnesium chloride ($MgCl_2$) formed are flushed out of the system along with the excess brine solution. The resin is then rinsed with water to remove any residual salts before it is ready for the next water-softening cycle.

All these reactions take place in a softener device (Figure 10.14), consisting of a tank filled with resin beads that are charged with sodium ions. As hard water flows through the resin bed, calcium and magnesium ions are attracted to the resin and are exchanged for sodium ions.

Eventually, the resin bed becomes saturated with calcium and magnesium ions and needs to be regenerated. This is typically done by backwashing the resin bed with saltwater, which replenishes the sodium ions in the resin and flushes out the accumulated calcium and magnesium ions.

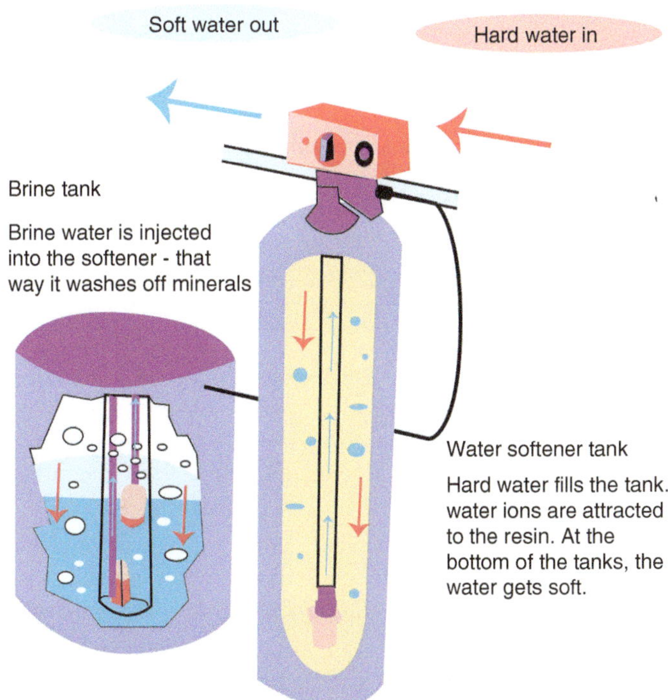

FIGURE 10.14   Ionic exchange softener

10.13 Desalination Plants

A desalination plant, also known as a desalination facility, is an installation that converts seawater or brackish water into freshwater suitable for various purposes, including drinking, irrigation and industrial processes, by RO. Desalination plants address water scarcity issues in regions where freshwater resources are limited or unavailable, particularly in arid coastal areas. The main objectives of desalination plants are as follows:

a. **Provide freshwater supply:** The main goal of desalination plants is to produce freshwater from saline water sources, thereby increasing the available water supply for human consumption, agriculture and industrial use.

b. **Address water scarcity:** Desalination helps mitigate water scarcity by tapping into abundant seawater resources that are otherwise unusable due to their high salinity. This is especially crucial in regions facing water shortages and drought conditions.

c. **Diversify water sources:** Desalination provides a supplementary or alternative water source to traditional freshwater sources such as rivers, lakes and groundwater. Diversifying water sources helps enhance water security and resilience to droughts and other environmental challenges.

d. **Support economic development:** Desalination facilitates economic development by ensuring a reliable water supply for residential, commercial and industrial sectors. It enables the growth of industries such as agriculture, tourism and manufacturing that rely on consistent access to freshwater.

e. **Emergency water supply:** Desalination plants can serve as a critical infrastructure component during emergencies, such as natural disasters or contamination events, by providing an alternative source of freshwater when traditional sources are compromised.

A simplified explanation of how a desalination plant works:

1. **Intake:** Seawater or brackish water is drawn into the plant from a nearby water source, typically an ocean, sea or underground well.

2. **Pre-treatment:** The water undergoes pre-treatment to remove large particles, debris and any organic matter that may be present. This process often involves screens, filters and settling tanks.

3. **Desalination process:** The most commonly used method is RO. In this process, the water is pressurised and forced through a set of semi-permeable membranes that allow water molecules to pass through while blocking salts and other impurities. The resulting freshwater is collected, while the concentrated saltwater, known as brine, is discharged.

4. **Post-treatment:** The desalinated water, whether goes through post-treatment processes to adjust its pH, mineral content and disinfection. This step ensures that the water meets quality standards and is safe for consumption or other specific uses.

5. **Product water distribution:** The treated freshwater is typically stored in reservoirs or distributed through pipelines to meet the local demand for drinking water or other purposes.

6. **Disposal of brine:** The concentrated brine generated during the desalination process contains a high concentration of salts and other dissolved substances. It needs to be carefully disposed of to minimise its impact on the environment. Common disposal methods include discharge into the ocean, deep well injection or evaporation ponds.

Desalination plants commonly employ spiral membrane devices as their filtration membranes. These membranes consist of a thin, semi-permeable layer wound around a perforated tube called the central permeate collection tube. The feedwater flows through the membrane in a spiral pattern (Figure 10.15), while the freshwater, known as permeate, is collected through the central tube. This design maximises the filtration surface area while occupying minimal space. The membranes are typically constructed using durable polymer materials like polyamide or cellulose acetate, which can withstand the harsh conditions encountered in desalination processes. Each membrane unit is referred to as a module and offers versatile functionality, allowing them to be used either in parallel or in series based on specific separation and concentration requirements.

These modules are housed in robust fibreglass enclosures known as 'vessels' and are equipped with plumbing connections at both the inlet and outlet points of the vessel. A pump assists in the fluid flow through the system before entering the filter module. To regulate the flow rate and create pressure within the module, a backpressure valve is installed on the concentrate outlet. Moreover, the pump speed at the inlet can be adjusted to achieve the desired pressure level.

Once the freshwater is separated, it is collected and undergoes further treatment if necessary. On the other hand, the concentrate, which contains concentrated salt and impurities, is discharged as brine.

It's important to note that desalination plants require a significant amount of energy to operate and the environmental impact of desalination, particularly the disposal of brine and energy consumption, are factors that need to be carefully managed and considered.

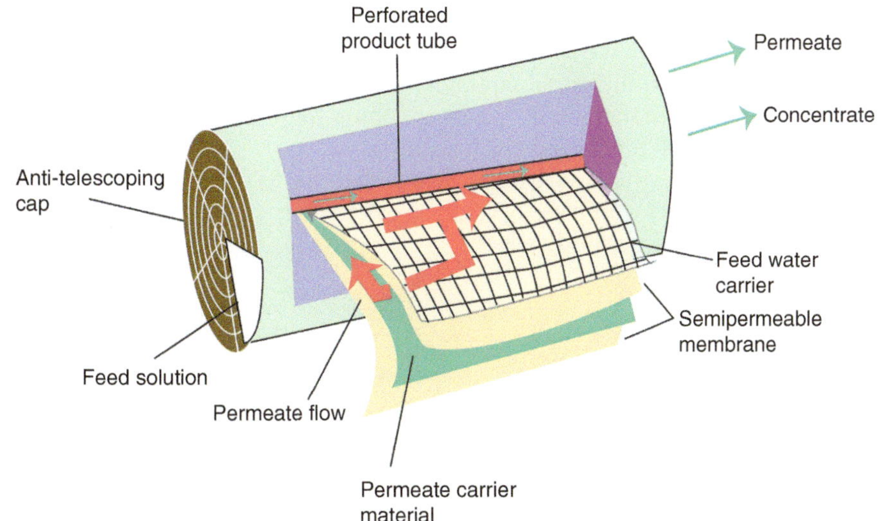

FIGURE 10.15 Spiral membrane device

10.14 Treatments of Water for Industrial Uses

Water is an essential resource in many industries and its uses vary depending on the specific industry. It is important to ensure that the water used is of good quality and meets specific standards. There are various treatments of water for industrial uses, which include similar processes to those applied in other procedures for water treatment but adjust to the specific requirements of every industry:

1. **Sedimentation:** This process involves the removal of suspended solids from water by allowing them to settle into the bottom of a tank.
2. **Filtration:** This process involves passing water through a filter to remove impurities such as sediment, bacteria and viruses.
3. **Disinfection:** This process involves the use of chemicals or physical processes such as UV light to kill microorganisms in the water.
4. **Softening and deionisation:** These processes involve the removal of ions responsible for water hardness such as calcium, magnesium and other ions from the water.
5. **Carbon active adsorption:** This process involves the removal of organic compounds such as pesticides and solvents by passing water through activated carbon filters.

10.15 Water Softening for Industrial Water Supply

Water softening for industrial water supply is a set process used to remove ions such as calcium and magnesium ions from water but on a large scale.

In case of hard water, it is important to remove those salts because they can cause damage and buildup in industrial equipment, which can lead to reduced efficiency and increased maintenance costs. Water softening helps prevent these issues and ensures that industrial processes run smoothly.

Together with methods described below such as RO and ion exchange, but adapted to great water volumes, two other, electrodialysis and lime softening are widely used in industry to purify water.

Electrodialysis

Electrodialysis is a separation process that uses an electric field to selectively transport ions through ion exchange membranes. It is commonly used to remove salts and other ions from solutions and is particularly useful in desalination processes (Figure 10.16).

In electrodialysis, a series of ion exchange membranes are placed between electrodes, creating a series of alternating compartments that contain the feed solution

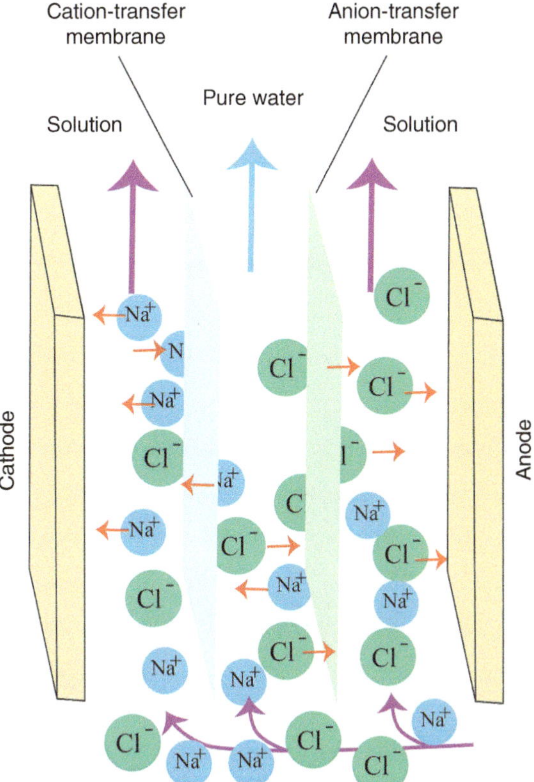

FIGURE 10.16 Electrodialysis device

and the target product solution. An electric field is applied across the membranes, causing ions in the feed solution to migrate towards the oppositely charged electrode and through the ion exchange membranes into the product solution.

The process is particularly effective for the removal of monovalent ions, such as sodium and chloride, from solutions. Electrodialysis is widely used in the food and beverage industry for the production of high-purity water, as well as in the chemical and pharmaceutical industries for the purification of process streams. It is also used in the production of salt from brine solutions.

Electrodialysis is an energy-intensive process and is typically used in conjunction with other separation technologies, such as RO, to increase overall efficiency.

Lime Softening

This is a chemical method involving the addition of lime (calcium hydroxide, $Ca(OH)_2$) to hard water. Lime softening is commonly used in large-scale water treatment facilities to form precipitates that can be removed through sedimentation or filtration.

The process is typically carried out by adding lime to the water and then allowing the mixture to settle for a period of time. The following chemical reactions take place:

When water is supplemented with slaked lime, any carbon dioxide present undergoes the following reaction (Eq. 10.11)

$$CO_2' + Ca(OH)_2 \rightarrow CaCO_3(s) + H_2O \qquad (10.11)$$

The lime reacts with carbonate hardness (Ca^{++} and Mg^{++} bicarbonates) as follows (Eqs. (10.12) and (10.13)):

$$Ca(HCO_3)_2 + Ca(OH)_2 \rightarrow 2CaCO_3(s) + 2H_2O \qquad (10.12)$$

$$Mg(HCO_3)_2 + Ca(OH)_2 \rightarrow CaCO_3(s) + MgCO_3(aq) + 2H_2O \qquad (10.13)$$

The magnesium carbonate formed in Eq. 10.13 is soluble. To remove it, more lime is added to give insoluble magnesium hydroxide and calcium carbonate (Eq. 10.14).

$$MgCO_3 + Ca(OH)_2 \rightarrow Mg(OH)_2(s) + CaCO_3(s) \qquad (10.14)$$

Furthermore, the elimination of non-carbonate hardness associated with magnesium, such as magnesium sulfate, is accomplished (Eq. 10.15).

$$MgSO_4 + Ca(OH)_2 \rightarrow Mg(OH)_2(s) + CaSO_4(s) \qquad (10.15)$$

At this point, the water contains the original calcium non-carbonate hardness and the calcium non-carbonate hardness produced in Eq. 10.15. Soda ash (Na_2CO_3) is added to remove calcium non-carbonate hardness (Eq. 10.16).

$$CaSO_4 + Na_2CO_3 \rightarrow CaCO_3(s) + Na_2SO_4 \qquad (10.16)$$

For $CaCO_3$ and $Mg(OH)_2$ precipitation, the pH must be 9.5 and 10.8, respectively; therefore, an excess of lime is required to raise the pH.

Lime softening has several benefits, including reducing the scaling and corrosion of pipes and appliances, improving the taste and appearance of water and reducing the amount of soap needed for cleaning. However, the process can also lead to an increase in pH levels and alkalinity in the water.

Second Part: Drinking Wastewater Treatment

10.16   Wastewater Treatments

Wastewater can be classified into several types based on its sources and characteristics:

1. **Domestic wastewater:** This type of wastewater comes from households, including kitchen sinks, bathrooms and toilets. It contains organic matter, human waste and detergents.

2. **Industrial wastewater:** This type of wastewater is generated from industries and manufacturing processes. It can contain toxic substances, heavy metals and organic compounds.

3. **Agricultural wastewater:** This type of wastewater comes from agricultural activities such as irrigation, animal husbandry and crop cultivation. It can contain fertilisers, pesticides and animal waste.

4. **Stormwater:** This type of wastewater comes from rainfall and other precipitation events. It can contain pollutants such as sediment, oil and debris.

5. **Medical wastewater:** This type of wastewater is generated from hospitals and clinics. It can contain infectious agents, pharmaceuticals and chemicals used in medical treatments.

6. **Leachate:** This type of wastewater is generated from landfills and can contain organic matter, heavy metals and other pollutants.

Wastewater treatments are the process of removing contaminants from wastewater to make it safe for disposal or reuse. The goal of wastewater treatment is to remove as many impurities and contaminants as possible to protect public health and the environment. Each type of wastewater requires different treatment processes to remove contaminants and make it safe for discharge or reuse.

10.17   Treatments of Domestic Wastewater

Domestic wastewater is the water that is discharged from households, commercial establishments and industries in urban areas. It contains a wide range of solid residues and pollutants, including organic matter, nitrogen and phosphorus compounds, heavy metals, pathogens and other contaminants. Proper management of urban wastewater is crucial for the prevention of water bodies pollution and the protection of human health and the environment. It is important to ensure that urban wastewater is treated to appropriate standards before being discharged into water bodies or reused for irrigation or other purposes.

There are several stages of wastewater treatment, including: Preliminary treatments and primary, secondary and tertiary treatments.

10.18   Preliminary Treatments

Preliminary treatments of wastewater are the first step in the treatment process. The purpose of the preliminary treatment is focused on the removal of large debris, grit and other coarse materials. Its primary goal is to protect downstream treatment processes and equipment from damage or clogging caused by these materials. It involves the removal of large and small solids and other materials that can cause blockages, damage equipment or interfere with subsequent treatment processes. The following are some of the common techniques used in preliminary treatment:

a. **Screening:** Screening is the process of removing large solids such as sticks, rags, plastics and other debris from the wastewater. This is typically done using bar screens or rotary drum screens (Figure 10.17).

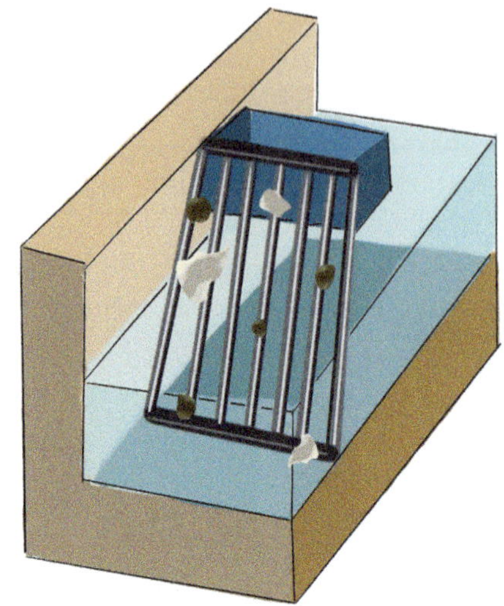

FIGURE 10.17 Bar screen

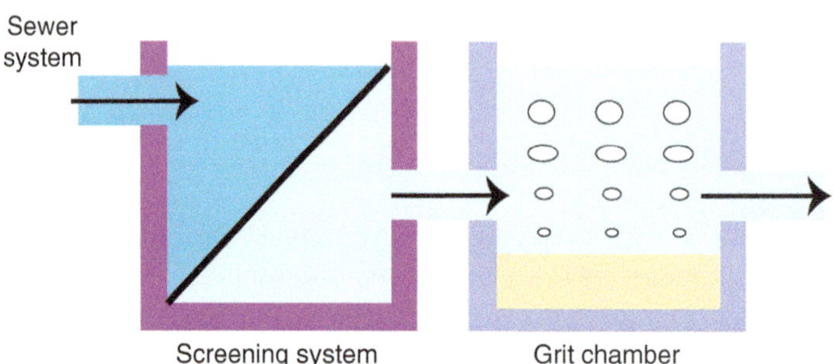

FIGURE 10.18 Grit removal equipment

b. **Grit removal:** Grit removal is an important process in wastewater treatment plants that aims to remove sand, gravel, grit and other heavy solid particles from the wastewater stream. These particles, collectively known as grit, can cause operational problems and damage to downstream equipment if not removed effectively. The grit removal process typically occurs at the preliminary treatment stage of wastewater treatment, combined with the initial screening process. Grit removal is often done using grit chambers or cyclones (Figure 10.18).

c. **Sedimentation:** Sedimentation is the process of allowing solids and heavy particles to settle out of the wastewater. This is done in sedimentation tanks or clarifiers.

d. **Flotation:** Flotation is a technique used to remove small and light solids such as fats, oils and grease from wastewater. This is typically done using dissolved air flotation or induced gas flotation (Figure 10.19).

e. **Equalisation:** Equalisation is the process of balancing the flow and composition of wastewater to ensure uniform treatment. This is typically done using so-called equalisation tanks or basins to equalise the flow rate and composition of incoming wastewater before it enters the main treatment process. In the context of wastewater treatment, an equalisation tank is typically used to manage wastewater that comes from various sources, such as residential, industrial or commercial facilities.

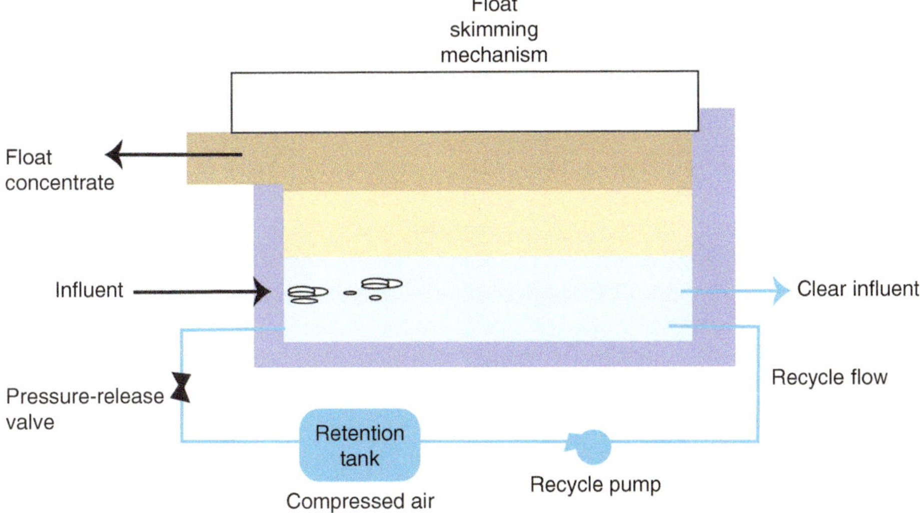

FIGURE 10.19 Dissolved air flotation unit

10.19 Primary Treatment

The purpose of primary treatment is focused on the physical removal of suspended solids and the separation of settleable organic and inorganic solids from wastewater. Its main objective is to reduce the overall pollutant load before proceeding to further treatment (Figure 10.20).

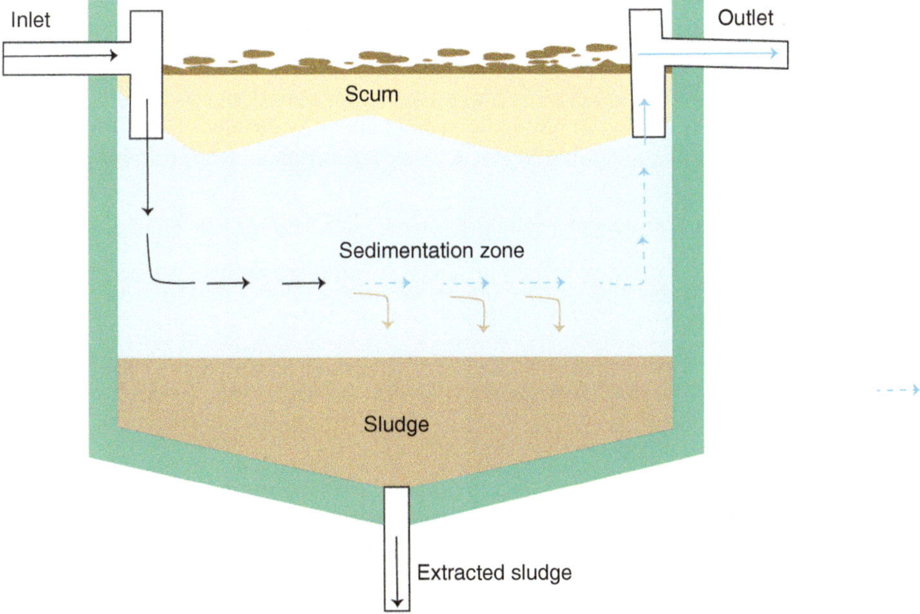

FIGURE 10.20 Sedimentation tank

Primary treatment relies on sedimentation or gravitational settling, to separate suspended solids and heavier particles from the wastewater. The wastewater is allowed to flow into large sedimentation tanks or clarifiers, where the velocity is reduced, enabling solids to settle to the bottom as sludge while floating debris and oils form a scum layer on the surface.

The wastewater that undergoes primary treatment is known as primary effluent and it still contains significant amounts of pollutants, including organic matter, suspended solids and nutrients.

The primary treatment process can reduce the biological oxygen demand (BOD) and total suspended solids (TSS) in the wastewater by up to 50%. The solid materials that are removed during primary treatment are known as primary sludge, which can be further processed and treated to produce biosolids that can be used as fertiliser or disposed of in a landfill.

10.20   Secondary Treatments: Aerobic and Anaerobic Treatments

Secondary treatment is a process used to treat wastewater after primary treatment to further remove contaminants and pollutants. The aim of secondary treatment is to produce effluent that is suitable for discharge into the environment or for reuse. The wastewater generated from households, industries and other sources undergoes secondary treatment, which involves aerobic and anaerobic biological processes.

Aerobic Biological Treatment

These treatments use living organisms such as bacteria, fungi and algae to break down organic matter and remove harmful substances from the water in both aerobic and anaerobic conditions.

There are several types of biological treatments in aerobic conditions used including: activated sludge, aerated lagoons, trickling filters and rotating biodiscs.

Activated Sludge Process   This process involves the use of microorganisms to break down organic matter in the wastewater. The microorganisms are suspended in a mixture of wastewater and sludge, which is constantly agitated to promote mixing and oxygen transfer.

The activated sludge process (ASP) is carried out in an aeration tank (Figure 10.21). Wastewater is first introduced into a large, oxygen-rich basin, typically equipped with mechanical aerators or diffusers that provide oxygen to support the growth of aerobic microorganisms.

To initiate the treatment process, a portion of activated sludge, which contains a high concentration of microorganisms, is added to the aeration tank. This seed sludge serves as a source of bacteria and other microorganisms needed for the breakdown of organic matter.

As the wastewater flows through the aeration tank, microorganisms in the activated sludge consume the organic pollutants present in the wastewater as their

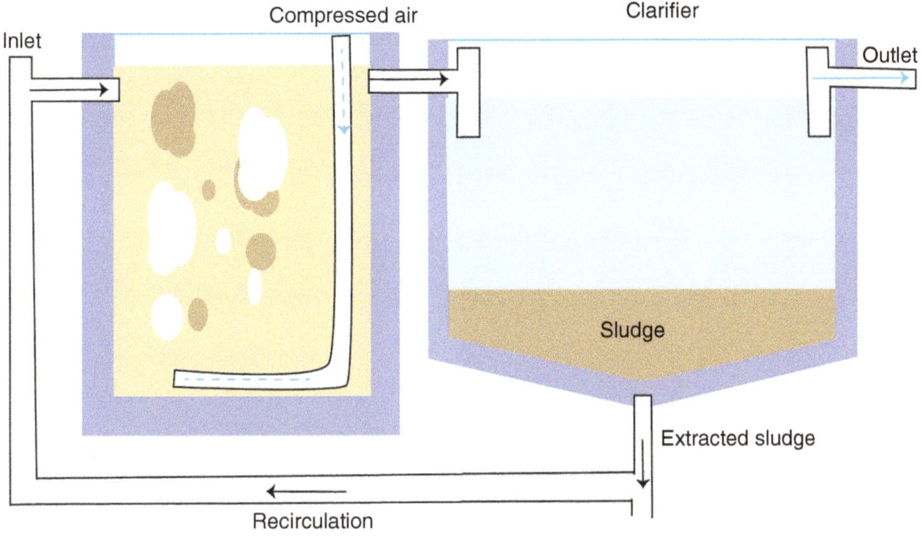

FIGURE 10.21   Activated sludge scheme

food source. These microorganisms break down the organic matter into carbon dioxide, water and additional microbial cells.

After the wastewater has been treated in the aeration tank, it flows into a settling tank or clarifier. In this tank, the mixed liquor is allowed to settle, allowing the microorganisms and other solids to form a sludge layer at the bottom of the tank. The clarified wastewater, also known as effluent, is discharged or further treated as required.

A portion of the settled sludge, known as the return activated sludge (RAS), is recycled back to the aeration tank. This helps maintain a sufficient concentration of microorganisms in the tank for effective treatment.

The excess sludge that accumulates in the settling tank is periodically removed and further processed. It can undergo processes such as thickening, dewatering and stabilisation to reduce its volume and convert it into a more manageable form for disposal or reuse.

The ASP is effective in removing organic pollutants, suspended solids and nutrients (such as nitrogen and phosphorus) from wastewater. It requires careful monitoring and control of various parameters, such as dissolved oxygen levels, nutrient concentrations and sludge age, to ensure optimal treatment efficiency.

This process typically removes 85%–95% of the BOD from the aeration influent.

Aerated Lagoons   Aerated lagoons are a type of wastewater treatment system that uses a combination of aerobic and anaerobic processes to treat wastewater (Figure 10.22). These lagoons are shallow basins or ponds designed to hold and treat wastewater for a specific period. In an aerated lagoon system, the treatment process typically involves two main zones: aerobic and anaerobic.

1. **Aerobic zone:** The aerobic zone of the lagoon is the area where oxygen is present and aerobic bacteria dominate the treatment process. These bacteria use oxygen to break down and consume organic matter in the wastewater. Oxygen is usually supplied to the aerobic zone through mechanical aeration devices, such

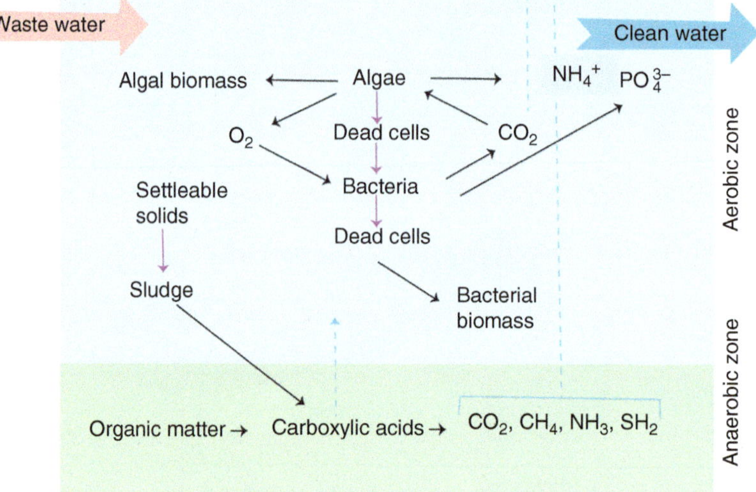

FIGURE 10.22   Aerated lagoon transformations

as diffusers or mechanical surface aerators. The aeration promotes the growth of aerobic bacteria, which metabolise organic pollutants such as carbonaceous compounds and ammonia. The aerobic zone is typically located near the surface of the lagoon, where oxygen levels are higher. Oxygen concentration is implemented by the presence of algae, which are photosynthetic microorganisms that utilise sunlight and nutrients, such as nitrogen and phosphorus, to carry out photosynthesis. In aerated lagoons, algae can flourish under favourable conditions, such as sufficient sunlight, appropriate temperature and availability of nutrients. The presence of algae in the lagoon helps absorb carbon dioxide and release oxygen through photosynthesis, contributing to the oxygenation of the water. The relationship between algae and microorganisms in aerated lagoons is symbiotic. Algae and microorganisms have a mutually beneficial relationship. The algae produce oxygen through photosynthesis, which is essential for the survival and growth of aerobic microorganisms. The microorganisms, in turn, contribute to the overall treatment process by consuming organic matter and nutrients present in the wastewater. This helps maintain a balanced ecosystem within the lagoon and improves the removal of contaminants from the wastewater.

Moreover, algae and microorganisms can compete for nutrients in the lagoon. Excessive algal growth can lead to the depletion of nutrients, limiting the growth of other microorganisms. Therefore, it is important to maintain a balance between algae and microorganisms to ensure optimal wastewater treatment performance in aerated lagoons.

2. **Anaerobic zone:** The anaerobic zone of the lagoon is the area where little to no oxygen is present, creating an oxygen-free environment. Anaerobic bacteria thrive in this zone and break down complex organic compounds, such as proteins and long-chain fatty acids. The anaerobic breakdown of organic matter produces gases such as methane, ammonia, hydrogen disulfide and carbon dioxide. The anaerobic zone is typically located at the bottom of the lagoon, where oxygen levels are lowest.

Initially, the population of microorganisms in an aerated lagoon is lower compared to an ASP due to the absence of sludge recycling. Consequently, a significantly longer residence time is necessary to achieve the same level of effluent quality. However, this extended residence time can be advantageous when dealing with complex organic chemicals that require degradation. Moreover, microorganisms in aerated lagoons demonstrate greater resilience to processing upsets caused by variations in feed, thanks to the larger tank volumes and longer residence times employed.

The key distinction between ASP and aerated lagoons is the absence of settling tanks and sludge recirculation arrangements in the latter. In some cases, an aerated lagoon water treatment system may include a maturation pond for further treatment of the effluent, as depicted below.

They also have some disadvantages.

1. **Land requirement:** Aerated lagoons require a significant amount of land compared to other wastewater treatment methods. The size of the lagoon is directly proportional to the volume of wastewater to be treated, making it unsuitable for areas with limited space availability and it must be properly impermeabilised to avoid pollution of soil and aquifers.

2. **Odour and aesthetics:** Aerated lagoons can produce unpleasant odours due to the decomposition of organic matter and the release of gases like hydrogen sulfide and ammonia. These odours can be a nuisance to nearby residents and businesses, affecting the aesthetics of the surrounding area.

3. **Climate sensitivity:** The performance of aerated lagoons can be influenced by weather conditions, especially temperature. Cold climates can slow down microbial activity and reduce treatment efficiency, while hot climates may lead to excessive evaporation and water loss.

4. **Sludge management:** Aerated lagoons produce a significant amount of sludge, which requires proper management. Sludge removal and disposal can be a costly and challenging process, requiring additional infrastructure and resources.

5. **Limited treatment efficiency:** While aerated lagoons can effectively remove organic matter and some nutrients from wastewater, they may have limitations in treating certain pollutants like heavy metals, pesticides and pharmaceuticals. These compounds may require additional treatment steps or alternative technologies.

6. **Water quality concerns:** If not properly designed or operated, aerated lagoons can pose a risk of nutrient enrichment in receiving water bodies. Excessive discharge of nitrogen and phosphorus from the lagoons can contribute to eutrophication, leading to algal blooms and water quality degradation.

Trickling Filters   These are beds of rocks, plastic or other media that are covered with a layer of microorganisms (Figures 10.23 and 10.24).

The influent wastewater is evenly distributed over the surface of the trickling filter bed. This can be done through a rotating distributor arm or by using a network of pipes with spray nozzles. The wastewater is then allowed to trickle down through the filter bed.

As the wastewater trickles down through the filter bed, a slimy, gelatinous layer of microorganisms called biofilm develops on the surface of the filter media. This biofilm contains various types of bacteria, fungi and protozoa, which are responsible for the biological degradation of organic pollutants present in the wastewater.

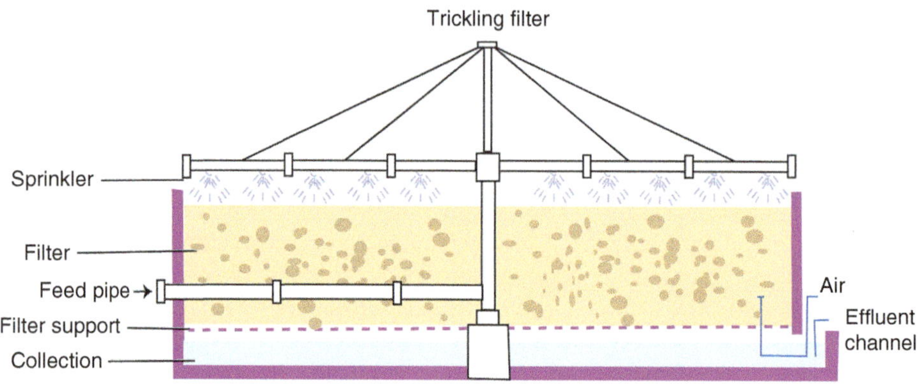

FIGURE 10.23 Trickling filter

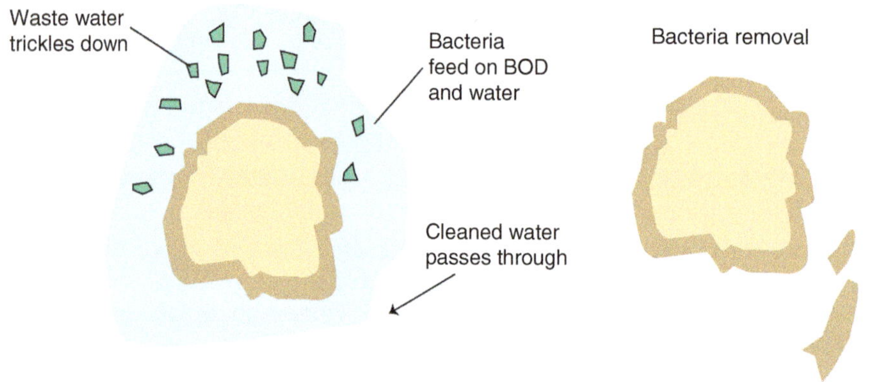

FIGURE 10.24 Detail of trickling filter

The organic pollutants present in the wastewater are biologically oxidised by the microorganisms in the biofilm. These microorganisms use organic matter as a carbon source and convert it into carbon dioxide, water and microbial biomass. This process helps remove organic pollutants and reduce the BOD of the wastewater. Trickling filters often require a constant supply of oxygen to support the aerobic microorganisms present in the biofilm. This is usually achieved by allowing air to flow through the filter bed. The air can be supplied either by a natural draft or by forced aeration using blowers.

The treated wastewater passes through the filter bed and collects at the bottom. It is then collected in an underdrain system or a perforated pipe network and removed for further treatment or discharge.

Trickling filters are commonly used in small- to medium-scale wastewater treatment plants as well as in decentralised systems. They are often combined with other treatment processes, such as sedimentation or secondary clarifiers, to achieve further removal of suspended solids and improve effluent quality.

Rotating Biodiscs Rotatory biodiscs, also known as rotating biological contactors (RBCs), are a type of wastewater treatment technology that utilises a series of rotating discs to facilitate the treatment process.

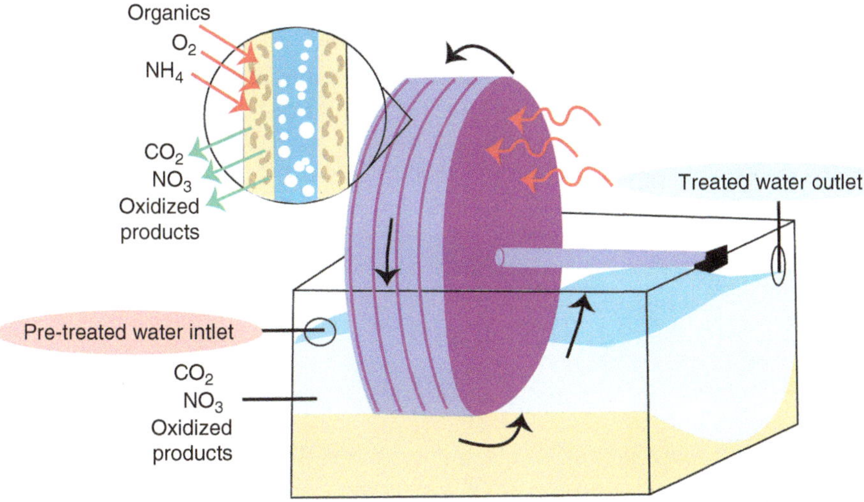

FIGURE 10.25   Rotating biological contactors (RBCs)

The rotatory biodisc system consists of a number of plastic discs that are mounted on a horizontal shaft and arranged in a series. These discs are partially submerged in the wastewater, allowing for the growth of a biofilm on their surface. The biofilm is composed of microorganisms such as bacteria, fungi and protozoa (Figure 10.25).

As the discs rotate, they come into contact with the wastewater, allowing the microorganisms in the biofilm to consume and break down organic compounds. This process helps in the removal of organic matter, suspended solids and some types of nutrients like nitrogen and phosphorus.

The rotating motion of the discs provides an oxygen-rich environment, ensuring the availability of oxygen for the microorganisms. The oxygen is essential for the aerobic biological processes that occur within the biofilm, enabling the microorganisms to degrade and metabolise the organic contaminants effectively.

The wastewater flows continuously over the rotating discs, ensuring a constant supply of wastewater to the biofilm for treatment. The treated water is then separated from the biofilm and discharged or subjected to further treatment processes, depending on the specific requirements. Rotatory biodiscs offer several advantages in wastewater treatment:

1. **High treatment efficiency:** The rotating discs provide a large surface area for the growth of the biofilm, allowing for a high population of microorganisms and efficient treatment of organic matter.

2. **Compact design:** The compact nature of rotatory biodisc systems makes them suitable for installations in areas with space limitations.

3. **Low energy requirements:** The rotating motion of the discs is often driven by low-power motors, resulting in lower energy consumption compared to other wastewater treatment technologies.

4. **Robust and reliable:** Rotatory biodiscs are known for their durability and ability to withstand variations in wastewater composition and flow rates.

However, biodiscs may have limitations in treating certain types of wastewater, such as those with high levels of toxic substances or excessive amounts of solids. In such cases, additional treatment steps or alternative technologies may be required.

Each system has its advantages and disadvantages and is suited for different applications depending on the volume of wastewater, type and concentration of contaminants and available resources. Biological treatments are often used in combination with other treatment methods, such as physical and chemical treatments, to ensure the water is safe for discharge into the environment.

Anaerobic Biological Treatment

Secondary treatment in anaerobic conditions typically involves the use of anaerobic bacteria to further break down organic matter in wastewater. This process is called *anaerobic digestion* and it can occur in various types of reactors in the absence of oxygen. It relies on the activity of anaerobic microorganisms to break down organic matter and convert it into simpler compounds such as methane and carbon dioxide.

The treatment takes place in an oxygen-free environment, typically in a sealed tank (Figure 10.26). The digester provides the ideal conditions for anaerobic bacteria, archaea and other microorganisms to thrive and carry out the degradation process.

During anaerobic treatment, complex organic substances are hydrolyzed into simpler compounds by acid-forming bacteria. These compounds, such as sugars, amino acids and fatty acids, are then converted into volatile fatty acids (VFAs) by

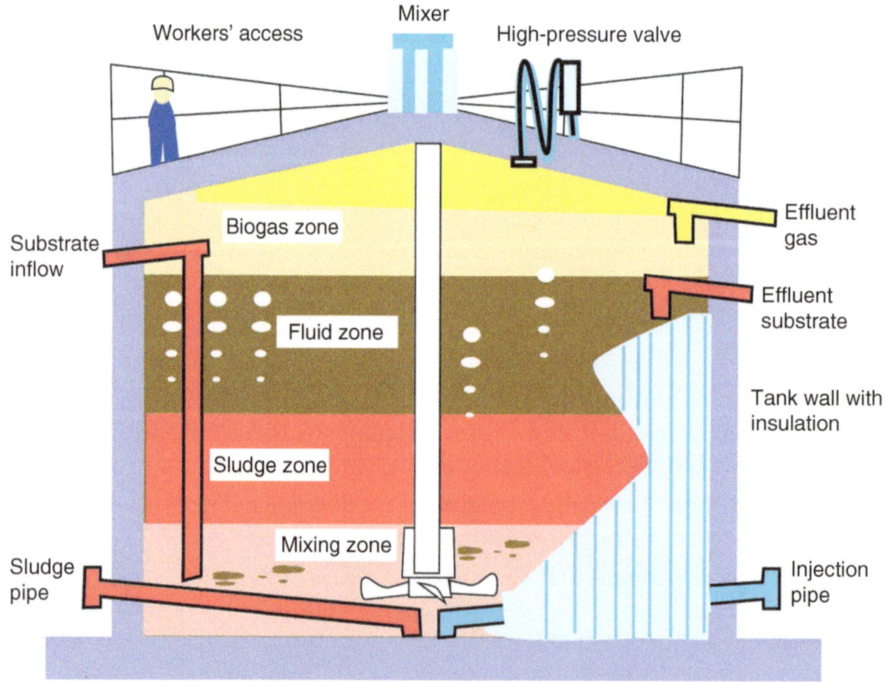

FIGURE 10.26 Anaerobic digestion tank

acidogenic bacteria. The VFAs serve as a food source for methanogenic archaea, which further metabolise them to produce methane gas and carbon dioxide.

The methane produced during anaerobic digestion can be captured and used as a renewable energy source, typically for generating heat and electricity. This process is known as anaerobic digestion for biogas production. Additionally, the remaining sludge or digestate from the treatment process can be further processed and used as fertiliser or soil amendment due to its nutrient-rich content.

Anaerobic biological treatment offers several advantages over aerobic treatment methods. It requires less energy input since it doesn't rely on aeration and it produces methane, a valuable biogas. It also generates less sludge compared to aerobic treatment processes. However, it has limitations in treating certain types of wastewater with high concentrations of toxic or inhibitory substances.

During anaerobic digestion, bacteria consume organic matter in the absence of oxygen and produce biogas (a mixture of methane and carbon dioxide as a byproduct). The biogas can be captured and used as a source of renewable energy. The remaining sludge can be further treated in aerobic conditions, such as in activated sludge systems, to remove any remaining organic matter and nutrients.

Anaerobic treatment usually requires lower energy requirements and reduced sludge production than aerobic transformations. A combination of anaerobic and aerobic treatments may be used to achieve the desired level of treatment.

10.21   Tertiary Treatment

Tertiary treatment of wastewater is the final stage of the wastewater treatment process, which aims to remove any remaining contaminants and improve the quality of the effluent before it is discharged into the environment or reused for various purposes.

The primary and secondary treatments of wastewater focus on removing organic matter. Tertiary treatment, on the other hand, focuses on removing nutrient residues (nitrates and phosphates), dissolved and suspended solids, pathogens and other harmful substances that may not have been removed through the previous treatments.

Some common methods of tertiary treatment of wastewater include:

a. Filtration is one of the key processes used in tertiary treatment. It involves passing the wastewater through various filtration media to remove suspended particles, such as sand. In this method, the wastewater is passed through a bed of sand or sand-like media, which traps and removes suspended solids and fine particles. Sand filters can effectively remove particles down to the micron level.

b. Membrane filtration is a separation process that utilises a thin, semi-permeable material that allows certain components to pass through while retaining others based on their size, molecular weight or other characteristics. It is a widely used technique in various industries and applications, including food and beverage processing, pharmaceutical manufacturing and biotechnology. This technique plays a crucial role in wastewater treatment processes, particularly in the advanced treatment and polishing stages. It helps remove contaminants, particles and microorganisms from wastewater, resulting in cleaner water that can be discharged or reused.

The membrane acts as a physical barrier, selectively allowing the passage of molecules or particles smaller than its pore size. The retained components are typically larger particles, suspended solids, colloids, microorganisms or macromolecules.

There are several types of membrane filtration processes, including:

i. **Microfiltration (MF):** This process uses membranes with relatively large pore sizes, typically in the range of 0.1–10 µm. It is commonly used for the removal of suspended solids, bacteria and some larger viruses.

ii. **Ultrafiltration (UF):** UF membranes have smaller pore sizes, typically ranging from 0.001 to 0.1 µm. They are effective in removing smaller particles, colloids, proteins, macromolecules and some viruses.

iii. **Nanofiltration (NF):** NF membranes have even smaller pore sizes, typically between 0.001 and 0.01 µm. NF is capable of removing divalent ions, organic matter and larger organic molecules.

iv. **Reverse osmosis (RO):** RO membranes have the smallest pore sizes, typically below 0.001 µm. RO can remove ions, dissolved solids, organic compounds and almost all types of molecules, providing high-quality water purification.

The selection of the membrane and the filtration process depends on the specific application and the desired separation requirements (Figure 10.27).

c. **Adsorption:** This involves using adsorbents, such as activated carbon, to remove organic and inorganic pollutants from the effluent. Activated carbon filters are used to remove organic compounds, odours and residual disinfectants from the wastewater. The activated carbon has a high surface area that can adsorb a wide range of organic contaminants, such as pesticides, pharmaceuticals and many other pollutants improving the water quality. Figure 10.28 shows an active carbon absorption tank.

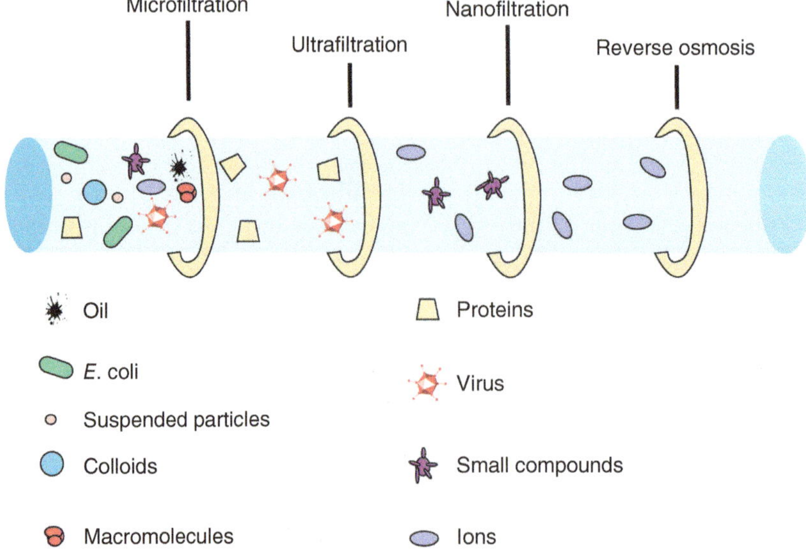

FIGURE 10.27 Comparison of diverse filtration membranes

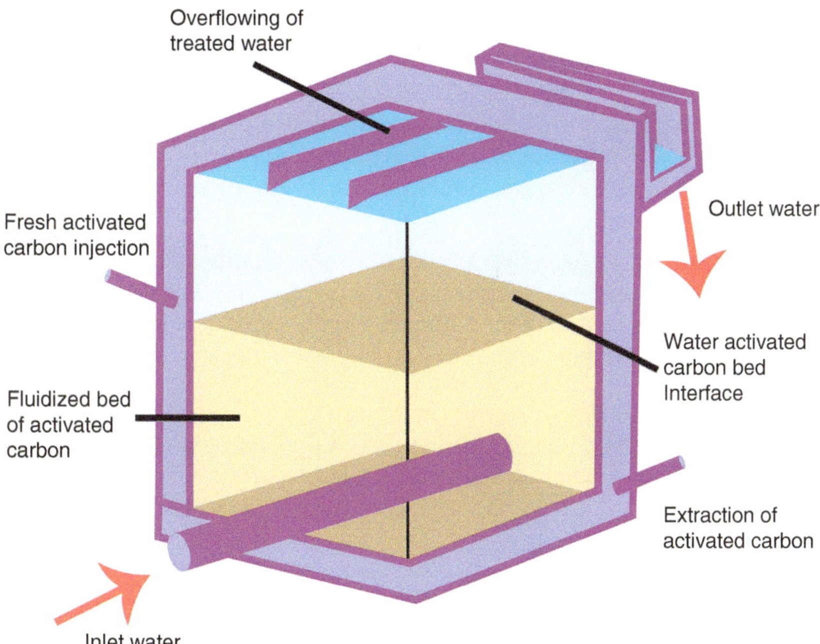

FIGURE 10.28   Active carbon absorption tank

d. Nutrient removal: This involves using chemical or biological processes, such as biological denitrification or chemical precipitation, to remove nitrogen and phosphorus from the effluent. There are several methods commonly used for nutrient removal from wastewater:

  i. Biological nutrient removal (BNR): This process uses microorganisms to biologically convert nitrogen and phosphorus compounds into harmless gases or solids. BNR typically involves two stages: the removal of nitrogen through nitrification and denitrification processes and the removal of phosphorus through biological uptake by bacteria.

  ii. Enhanced biological phosphorus removal (EBPR): EBPR is a variation of the BNR process specifically designed for phosphorus removal. It involves creating an environment that promotes the growth of specific bacteria capable of accumulating phosphorus in their cells. The excess bacteria, containing phosphorus, are then removed from the system.

  iii. Chemical precipitation: Chemicals such as aluminium or iron salts are added to wastewater to precipitate phosphorus as insoluble compounds. This process, known as phosphorus precipitation or chemical dosing, forms solid particles that can be easily separated from the water.

  iv. Membrane filtration: Membrane filtration techniques such as MF, UF and RO can effectively remove nutrients from wastewater by physically separating the dissolved and suspended particles, including nitrogen and phosphorus compounds.

  v. Wetlands and natural systems: Constructed wetlands and natural systems like lagoons can be used to remove nutrients from wastewater. These systems rely on a combination of physical, chemical and biological processes, including plant uptake, microbial degradation and sedimentation, to remove nutrients.

e. **Advanced oxidation:** Advanced oxidation processes (AOPs) involve the use of powerful oxidants, such as ozone, oxone[1] or hydrogen peroxide, to degrade persistent organic pollutants that are resistant to conventional treatment methods.

f. **Disinfection** is the last step in a wastewater treatment plant. This involves using disinfectant agents, such as chlorine, ozone or UV light, to kill any remaining bacteria, viruses and other pathogens in the effluent before discharging it to protect the environment.

Tertiary treatment of wastewater is essential to protect public health and the environment by ensuring that the effluent discharged or reused is safe and of high quality with special attention to the presence of emerging pollutants that primary and secondary treatment use to be inefficient for removing these pollutants.

Topics of Interest: Anaerobic Ammonium Oxidation (Anammox©)

Anammox is a type of wastewater treatment technology that specifically targets the removal of ammonia from wastewater. Ammonia is a common pollutant found in industrial and municipal wastewater and its removal is important to protect the environment and ensure compliance with regulatory standards.

Anammox wastewater treatment typically involves a two-step process: nitrification and denitrification, as shown in Figure.

1. Nitrification is the biological conversion of ammonia to nitrate through the action of nitrifying bacteria. This process occurs in an aerobic (oxygen-rich) environment. The nitrification step is usually accomplished in an ASP, where wastewater is mixed with a culture of nitrifying bacteria in an aeration tank. The bacteria oxidise ammonia to nitrite and then to nitrate.

2. Denitrification is the subsequent step where nitrate is converted to nitrogen gas, which is released into the atmosphere. Denitrification is an anaerobic (oxygen-free) process that requires the presence of denitrifying bacteria. In this step, the wastewater is transferred to an anoxic tank, where the denitrifying bacteria consume the nitrate and convert it to nitrogen gas.

The initial step involves partial nitrification, also known as nitritation, wherein approximately half of the ammonium is converted to nitrite by the activity of ammonia-oxidising bacteria (Eq. 10.17).

$$2NH_4^+ + 3O_2 \rightarrow 2NO_2^- + 4H^+ + 2H_2O \qquad (10.17)$$

---

[1] Oxone, also known as potassium peroxymonosulphate, is a powerful oxidising agent commonly used as a disinfectant, sanitiser and cleaner.

Ammonium and nitrite are then converted into nitrogen and nitrate (not shown) by anammox bacteria (Eq. 10.18):

$$NH_4^+ + NO_2^- \rightarrow N_2 + 2H_2O \tag{10.18}$$

The global reaction can be represented in Eq. 10.19.

$$NH_4^+ + NO_2^- \rightarrow N_2 + 2H_2O \tag{10.19}$$

10.22 Sludge Treatments

The waste generated during the wastewater treatment consists mainly of sludges, which are a mixture of organic and inorganic solids. The sludges must be treated and disposed of properly to prevent pollution. There are several methods for treating and disposing of sludge, including:

a. **Thickening:** This involves removing water from the sludge to reduce its volume.

b. **Digestion:** This is a biological process in which microorganisms break down the organic matter in the sludge, producing a stabilised product that can be used as a soil conditioner.

c. **Dewatering:** This process involves removing water from the sludge to produce a solid cake that can be transported for disposal.

d. **Incineration:** This process involves burning the sludge to reduce its volume and produce ash that can be disposed of in a landfill.

e. **Land application:** This involves using the stabilised sludge as a fertiliser or soil conditioner.

10.23 Treatments of Industrial Effluents

Industrial effluents refer to the wastewater or liquid waste that is discharged from industrial processes. These effluents may contain a wide range of pollutants, including chemicals, heavy metals, organic matter and other substances that can harm the environment and public health.

Industrial effluents can be generated from a variety of industries, such as manufacturing, mining, agriculture and energy production. The composition of these effluents depends on the type of industry, the raw materials used and the manufacturing processes involved.

To minimise the harmful effects of industrial effluents on the environment and public health, industries are required to treat their effluent before discharging it into the environment. Treatment processes may involve physical, chemical and biological methods to remove pollutants and contaminants from the effluent.

The treatment of industrial effluents is an important process to minimise the negative impact of industrial waste on the environment. There are several methods of treating industrial effluents, depending on the specific characteristics of the waste and the desired level of treatment and each must be designed specifically in every case.

1. **Physical treatment:** This involves the use of physical processes such as sedimentation, filtration and flotation to remove suspended solids and other contaminants from the effluent.

2. **Chemical treatment:** This involves the use of chemicals such as coagulants, flocculants and oxidising agents to remove dissolved contaminants from the effluent.

3. **Biological treatment:** This involves the use of microorganisms to break down organic matter in the effluent. This can be done through processes such as aerobic and anaerobic digestion, activated sludge treatment and biofiltration.

CHAPTER 11

Stratospheric Pollution

11.1 Introduction

Stratospheric pollution refers to the presence of harmful substances in the stratosphere. The most well-known example of stratospheric pollution is the depletion of the ozone layer, a specific type of stratospheric pollution that refers to the thinning of the ozone layer in the stratosphere caused by the release of certain chemicals.

Other types of stratospheric pollution include the presence of particulate matter from volcanic eruptions, which can have a cooling effect on the Earth's surface and the presence of soot and other particles from aircraft emissions, which can contribute to climate change. In addition, the release of nitrogen oxides from high-altitude aircraft can lead to the formation of ozone in the stratosphere, which can exacerbate the depletion of the ozone layer.

Stratospheric pollution is a significant environmental issue because it can have far-reaching impacts on human health, ecosystems and the Earth's climate. Efforts to reduce stratospheric pollution typically focus on reducing emissions of ozone-depleting substances (ODSs) and other pollutants from human activities, such as industrial processes, transportation and energy production.

11.2 Ozone Layer

Ozone is a colourless gas made up of three oxygen atoms (O_3) that are found in different layers of the Earth's atmosphere.

Tropospheric ozone, also known as ground-level ozone, is formed by the reaction of sunlight with pollutants. Stratospheric ozone, on the other hand, is found in the stratosphere. It is formed naturally by the interaction of sunlight with oxygen molecules and is responsible for absorbing harmful ultraviolet radiation from the sun. This helps to protect life on Earth from the harmful effects of UV radiation, such as skin cancer and cataracts.

In the stratosphere ozone is formed when oxygen molecules (O_2) are broken down by high-energy ultraviolet radiation in the atmosphere and the resulting oxygen atoms then react with other oxygen molecules to form ozone.

Understanding the Chemistry of the Environment, First Edition. Francisco G. Calvo-Flores.
© 2025 John Wiley & Sons Ltd. Published 2025 by John Wiley & Sons Ltd.

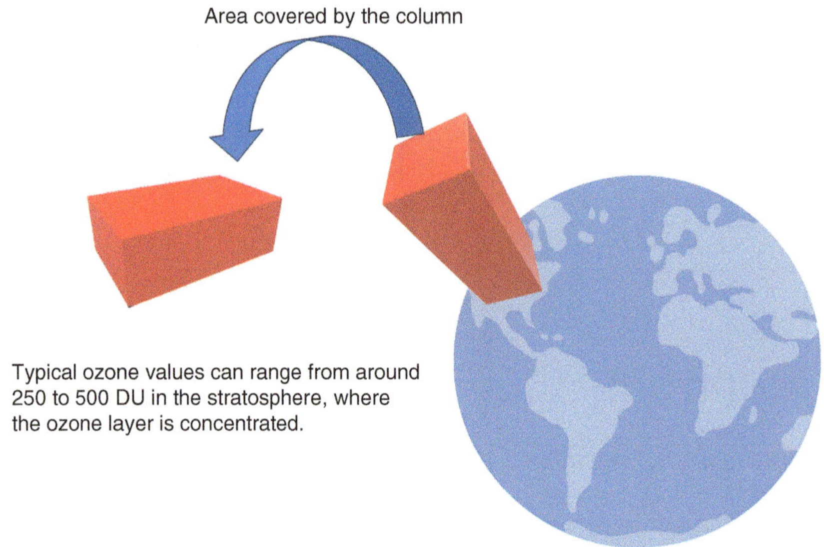

FIGURE 11.1   Ozone layer and Dobson units

The zone in the Earth's stratosphere that contains the highest concentration of ozone (O_3) molecules is named the ozone layer. It is located approximately 10–50 km above the Earth's surface. The ozone layer is important because it absorbs a significant portion of the sun's ultraviolet (UV) radiation, which is harmful to life on Earth.

In the stratosphere, the molecules (O_3) are constantly formed and destroyed by photochemical reactions, keeping the concentration approximately stable in the absence of pollutants.

The ozone layer thickness in the atmosphere is expressed in Dobson units (DU). One DU is the number of molecules of ozone that would be required to create a layer of pure ozone of 0.01 mm thick at a temperature of 273 K and a pressure of 1 atm (Figure 11.1). A column of air with an ozone concentration of 1 DU would contain about 2.69×10^{16} ozone molecules for every square centimetre of area at the base of the column.

Over the Earth's surface, the ozone layer's average thickness is about 300 DU or a layer that is 3 mm thick.

11.3   Chapman Mechanism

In the 1930s decade of the 20th century, S. Chapman proposed a series of reactions to explain why ozone is formed and how it is destroyed in the stratosphere. According to this proposal, ozone is formed from atmospheric oxygen by absorption of UV radiation (100–240 nm) to produce two oxygen radicals (Eq. 11.1)

$$O_2 + UV\left(< 242\,\text{nm}\right) \rightarrow O\cdot + O\cdot \qquad (11.1)$$

Every radical reacts with a molecule of oxygen to give two molecules of ozone (Eq. 11.2).

$$2O\cdot + 2O_2 + M \rightarrow 2O_3 + M*$$

(11.2)

M is any inert molecule, generally N_2 and O_2, that participates in the process, only removing the excess energy.

Simultaneously, ozone molecules are destroyed by absorption of UV radiation (240–235 nm) by a non-catalyzed process to give one oxygen molecule and one oxygen radical (Eq. 11.3)

$$O_3 + UV \rightarrow O_2 + O\cdot$$

(11.3)

Oxygen radicals are able to react with other ozone molecules to give two oxygen molecules (Eq. 11.4) or interact with another oxygen radical to give a new molecule of oxygen (Eq. 11.5).

$$O_3 + O\cdot \rightarrow 2O_2$$

(11.4)

$$O\cdot + O\cdot \rightarrow O_2$$

(11.5)

Based only on the Chapman hypothesis, stratospheric O_3 levels should be higher than observed. It is recognised that there is another way for ozone destruction to occur through catalytic cycles that involve radical species present in the atmosphere (Figure 11.2).

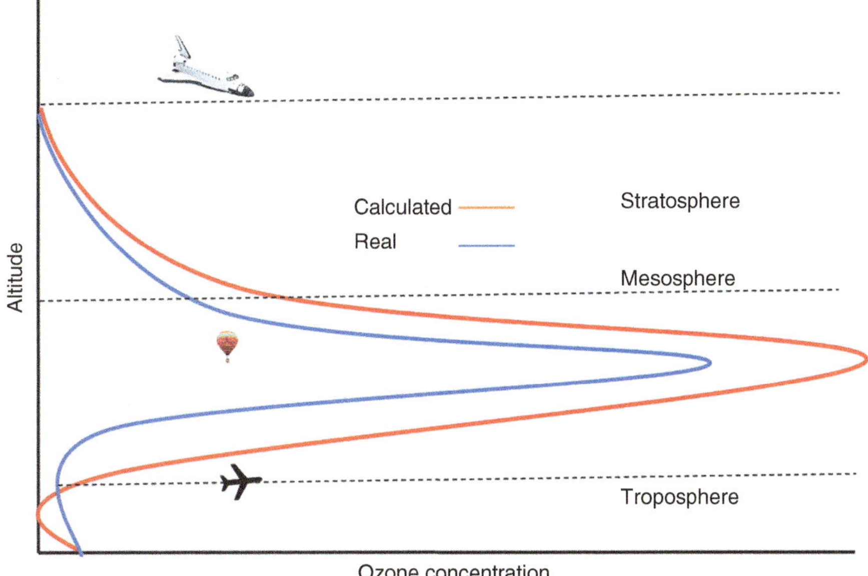

FIGURE 11.2   Calculated and real ozone concentrations in the atmosphere

11.4   Catalytic Mechanism of Ozone Destruction

In the stratosphere, there are different radical species (molecules, X.) that are able to react with ozone molecules, removing an oxygen atom (Eq. 11.6). This is a two-step mechanism that takes place in the middle and high stratosphere:

The four main families of radicals responsible for ozone destruction are hydroxyl ($OH^\cdot$), nitric oxide ($NO^\cdot$), chlorine radical ($Cl^\cdot$) and bromine radical ($Br^\cdot$).

There are two accepted catalytic mechanisms for ozone depletion:

$$X\cdot + O_3 \rightarrow XO\cdot + O_2$$

$$XO\cdot + O \rightarrow X\cdot + O_2$$

$$\text{------------------}$$

$$O_3 + O\cdot \rightarrow 2O_2 \tag{11.6}$$

In the global reaction, $O_3 + O^\cdot \rightarrow 2O_2$, X acts as a catalyst because it participates in the reaction but is not consumed in the global process. Every radical can be responsible for the destruction of hundreds of molecules of ozone, despite their very low concentrations.

When concentrations of $X^\cdot$ species are very high. A second mechanism of significance takes place (Eq. 11.7).

$$X\cdot + O_3 \rightarrow XO\cdot + O_2$$

$$XO\cdot + O_3 \rightarrow X\cdot + 2O_2$$

$$\text{------------------}$$

$$2O_3 \rightarrow 3O_2 \tag{11.7}$$

11.5   Reservoir Species

The so-called reservoir species are relatively stable molecules that can exist in the atmosphere for an extended period without reacting with other molecules. They act as temporary storage for reactive radicals, preventing them from immediately reacting with other compounds and allowing for the transport of these radicals to different regions (Figure 11.3). For example, $Cl^\cdot$ and $ClO^\cdot$

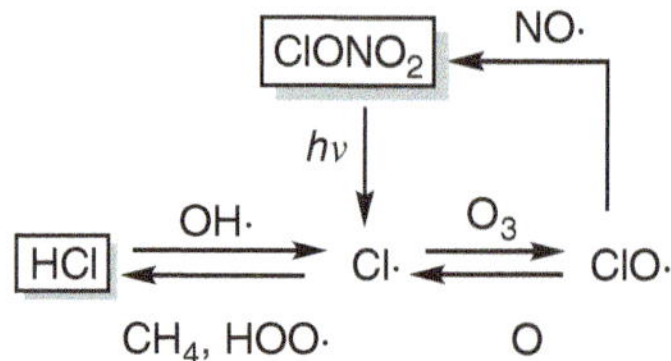

FIGURE 11.3   Reservoir species

radicals can combine with other molecules such as NO_x (Eqs. 11.8 and 11.9) or methane (Eq. 11.10) to form several reservoir species (bold), which do not react with ozone.

$$ClO \cdot + NO_2 + M^* \rightarrow \mathbf{ClONO_2} + M^* \tag{11.8}$$

$$ClO \cdot + NO \rightarrow Cl \cdot + NO_2 \tag{11.9}$$

$$Cl \cdot + CH_4 \rightarrow \mathbf{HCl} + CH_3 \tag{11.10}$$

Chlorine nitrate ($ClONO_2$) and hydrogen chloride (HCl) can be removed by deposition or regeneration of radicals. The role of NO_2 is complex; it acts by itself as an ozone-depleting agent and on the other hand, it is capable of sequestering Cl· in reserve molecules.

A dynamic equilibrium has been established, so the ozone concentration has remained relatively stable during the decades. Satellite observations of the ozone layer began in the 1970s, when the possibility of ozone depletion was just becoming an environmental concern. The atmosphere has maintained its natural balance, but the presence of man-made substances has implemented X· concentrations in the stratosphere and as a consequence, the ozone layer has been damaged over the last few decades.

11.6   Ozone Depletion

Ozone depletion is a specific type of stratospheric pollution that refers to the thinning of the ozone layer in the stratosphere due to the stratospheric pollutants that are released into the Earth's atmosphere and react with ozone molecules, therefore diminishing ozone concentration.

Ozone molecules absorb and filter out most of the harmful ultraviolet (UV) radiation from the sun. This is important because UV radiation can cause skin cancer and other health problems, as well as damage crops and ecosystems.

Mario Molina and Frank Sherwood Rowland discovered in 1974 that CFCs[1] were capable of destroying ozone in the stratosphere. Their investigations led to their groundbreaking publication in the journal 'Nature' in 1974. This discovery was significant because CFCs were widely used in many industrial applications and were thought to be safe.

Their research findings led to increased public awareness of the dangers of CFCs and other halogenated compounds and eventually to the adoption of the Montreal Protocol in 1987, an international treaty designed to phase out the production and use of CFCs and other ODSs. For their work on the ozone layer, Molina and Rowland were awarded the Nobel Prize in Chemistry in 1995, along

---

[1] CFCs, or chlorofluorocarbons, are organic compounds that contain carbon, chlorine and fluorine atoms. They were widely used as refrigerants, solvents and propellants in aerosol cans until the late 20th century.

with chemist Paul Crutzen. Their discovery and subsequent advocacy helped prevent further damage to the ozone layer and protect the Earth from the harmful effects of UV radiation.

As mentioned above, in the stratosphere, ozone is formed from oxygen molecules and decomposes to give back oxygen molecules. Catalysed by radicals, the decomposition of ozone takes place.

Radicals Responsible of Ozone Depletion
Hydroxyl Radical

One of the radicals responsible for this transformation is the hydroxyl radical, which is present in the stratosphere and is formed mainly in two processes:

1. **Photolysis of ozone:** Ozone (O_3) in the atmosphere can be dissociated by ultraviolet (UV) radiation to produce O_2 and an excited oxygen atom (O^*). This excited oxygen atom can then react with water vapour to produce two hydroxyl radicals (Eqs. 11.11 and 11.12):

$$O_3 + UV\ radiation \rightarrow O_2 + O\cdot \tag{11.11}$$

$$O\cdot + H_2O \rightarrow 2OH\cdot \tag{11.12}$$

2. **Photolysis of nitrous oxide:** Nitrous acid (HNO_2) is a compound that can be found in the Earth's atmosphere, particularly in the lower stratosphere. HONO can be directly emitted into the troposphere by combustion processes such as biomass burning, vehicle exhaust, domestic heating and industrial burning, but it is also formed through the heterogeneous reaction of nitric oxide (NO) and water (Eq. 11.13).

$$NO + H_2O \rightarrow HNO_2 \tag{11.13}$$

When exposed to ultraviolet (UV) radiation, nitrous acid decomposes into nitric oxide and hydroxyl radicals (OH·), which can react with ozone to form oxygen (O) and molecular oxygen (O_2), thus reducing the concentration of ozone in the stratosphere (Eq. 11.14).

$$HNO_2 + hv \rightarrow NO\cdot + OH\cdot \tag{11.14}$$

The natural equilibrium of these processes is modified by extensive agriculture and its massive use of fertilisers. Fertilisers can contribute to the formation of nitrogen oxides in the atmosphere through nitric oxide (NO) emission, which occurs when nitrogen in the fertiliser is converted into nitric oxide (NO) by soil microbes and then released into the atmosphere, where it can react with other pollutants to form nitrogen oxides (NO_x).

Halogen Radicals

But the stronger impact on stratospheric ozone is caused by halogen radical producers. All of these volatile substances with halogen atoms are stable enough to not degrade in the troposphere and reach the stratosphere unchanged.

Halogen radicals are formed in the stratosphere through the photo dissociation of halogen-containing molecules, where they are broken down by ultraviolet radiation from the sun (Eq. 11.15).

$$R - C - X + h\nu \rightarrow R - C \cdot + X \cdot \qquad (11.15)$$

This process releases halogen radicals, which can then react with ozone (O_3) to form oxygen (O_2) and halogen oxide (e.g., ClO· or BrO·). The halogen oxide radicals can also interact with other molecules of ozone, causing a chain reaction that ultimately leads to the depletion of the ozone layer. This is a serious problem, as the ozone layer plays a critical role in protecting life on Earth from harmful ultraviolet radiation.

Examples of volatile substances with halogen atoms are:

- **Bromomethane:** Used as a fumigant for soil and stored commodities.
- **Carbon tetrachloride and dichloromethane:** Used as a solvent in industry and laboratories.
- **Trichloroethylene:** Used as a solvent in metal degreasing and dry cleaning.
- **Chlorofluorocarbons (CFCs):** CFCs are a group of synthetic organic compounds that contain chlorine, fluorine and carbon atoms. They were widely used in the past as refrigerants, solvents, propellants and foam-blowing agents due to their stability, non-toxicity and non-flammability.
- **Hydrochlorofluorocarbons (HCFCs):** HCFCs are chemical compounds that contain hydrogen, chlorine, fluorine and carbon. They were developed as a replacement for CFCs, which were found to be contributing to the depletion of the Earth's ozone layer.
- **Halons:** Halons are a family of chemicals that contain the elements fluorine, chlorine, bromine or iodine and carbon. They were commonly used as fire extinguishing agents in the past due to their effectiveness in suppressing fires. However, they have been largely phased out due to their negative impact on the environment, specifically their ability to deplete the ozone layer and contribute to global warming.

To measure the ability of a substance to destroy the ozone layer in the atmosphere, scientists have developed the Ozone Depletion Potential (ODP) scale. ODP measures the relative ability of a substance to destroy ozone molecules compared to CFC-11.

For example, the ODP of CFC-11 is defined as 1.0. Substances with higher ODP values are more destructive to the ozone layer than CFC-11, while substances with lower ODP values are less destructive.

ODP is a useful tool for assessing the environmental impact of chemicals and substances on the ozone layer. Some of the ways in which ODP is used are:

1. **Scientific research:** ODP is a key parameter in atmospheric chemistry models used to study the ozone layer and its depletion. Understanding the behaviour of

ODSs in the atmosphere and their impact on the ozone layer requires accurate measurements of ODP.

2. **Regulation of ozone depleting substances:** ODP is used by governments and international organisations to regulate the production, import and export of ODSs, such as CFCs and halons. The Montreal Protocol, an international agreement to protect the ozone layer, uses ODP as a key metric to determine which chemicals are subject to control.

3. **Development of alternatives:** The ODP of a substance can be used to guide the development of less harmful alternatives. For example, the replacement of CFCs with HCFCs was seen as a step forward in reducing the impact on the ozone layer, even though HCFCs still have an ODP (although significantly lower than CFCs). Since then, efforts have been made to phase out HCFCs as well and replace them with even less harmful substances.

4. **Environmental impact assessments:** ODP can be used to evaluate the environmental impact of a product or process, especially in industries that use ODSs. This can help identify opportunities for reduction, substitution or elimination of such substances.

Topics of Interest:　The Case of Thomas Midgley

Thomas Midgley Jr. (1889–1944) was an American engineer and chemist who played key roles in the development of two significant but ultimately controversial technologies: leaded gasoline and chlorofluorocarbons (CFCs).

Leaded gasoline: Midgley's work with General Motors in the 1920s led to the development of tetraethyl lead as an additive to gasoline. As explained in Chapter 3, this compound was done to reduce engine knocking and improve engine performance. However, leaded gasoline was later found to be highly toxic, leading to widespread environmental and health problems. It was eventually phased out in most countries due to its detrimental effects on human health and the environment.

Chlorofluorocarbons (CFCs): Midgley also invented CFCs, which were initially hailed as safe alternatives to toxic substances used in refrigeration and other applications. However, it was later discovered that CFCs contributed significantly to the depletion of the ozone layer, which protects Earth from harmful ultraviolet radiation. This led to the implementation of international agreements, such as the Montreal Protocol, to phase out the production and use of CFCs.

Midgley's contributions to these technologies were significant at the time but have since been heavily criticised due to their negative impacts on public health and the environment. He tragically died in 1944 due to complications from polio.

Table 11.1 summarises the ODP values of some representative halogenated substances, their residence time in the atmosphere and their Global Warming Potential (GWP).

The depletion of the ozone layer has several significant consequences for the environment, human health and ecosystems. Some of the key consequences are as follows:

1. **Increased UV radiation:** The ozone layer acts as a shield, absorbing most of the harmful ultraviolet (UV) radiation from the sun. The first consequence

TABLE 11.1 Ozone depletion potential of some ozone depleting substances

Substance	Residence time (years)	ODP
CFC-11 (trichlorofluoromethane)	50–80	1.0
CFC-12 (dichlorodifluoromethane)	100	1.0
CFC-113 (trichlorotrifluoroethane)	90–210	0.8
HCFC-22 (chlorodifluoromethane)	12.1	0.05–0.2
HCFC-141b (1,1-dichloro-1-fluoroethane)	9.2	0.11
HCFC-142b (1-chloro-1,1-difluoroethane)	18.9	0.06
Halon-1211 (bromochlorodifluoromethane)	14	3.0
Halon-1301 (bromotrifluoromethane)	65	10.0
Halon-2402 (dibromotetrafluoroethane)	21	6.0
Methyl chloroform (1,1,1-trichloroethane)	5.3	0.11

is that when the ozone layer is depleted, more UV radiation reaches the Earth's surface.

2. **Impact on marine life:** Marine ecosystems are also affected by increased UV radiation. Phytoplankton, which form the base of the marine food chain, can be particularly vulnerable to UV radiation. Damage to phytoplankton can disrupt the entire aquatic food web, affecting fish populations and other marine organisms.

3. **Crop damage:** Increased UV radiation can harm agricultural crops, leading to reduced yields and lower- quality produce. UV radiation can inhibit photosynthesis, stunt plant growth and cause physiological damage to plants, reducing agricultural productivity.

4. **Disruption of ecosystems:** UV radiation can also affect terrestrial ecosystems. It can inhibit the growth of plants, particularly in sensitive areas such as forests and wetlands, affecting the biodiversity and stability of these ecosystems. Changes in vegetation can also impact the habitat and food sources of many animal species.

5. **Materials damage:** Ozone depletion can have significant effects on various materials, such as

   a. **Plastics and polymers:** UV radiation can cause the degradation of plastics and polymers. Ozone depletion leads to increased UV exposure, which accelerates the breakdown of polymer's molecular structure. This degradation can result in discoloration, brittleness, cracking and reduced mechanical strength in plastic materials, making them less durable and prone to failure.

   b. **Rubber:** Ozone is known to react with rubber, leading to a process called ozone cracking. Ozone molecules can penetrate rubber materials, such as tyres, belts and seals and cause chemical reactions that weaken the rubber's molecular bonds. This results in the formation of cracks on the surface of the rubber, reducing its elasticity, flexibility and overall lifespan.

   c. Paints and coatings: UV radiation can also degrade paints and coatings. Ozone depletion allows more UV radiation to reach painted surfaces, leading to fading, discoloration and deterioration of the protective coatings. This can result in the need for more frequent repainting and maintenance of structures and surfaces exposed to sunlight.

   d. Fabrics and textiles: UV radiation can have damaging effects on fabrics and textiles. Ozone depletion can cause fading, discoloration and weakening of fibres in clothing, curtains, upholstery and outdoor fabrics. This can result in reduced colorfastness, loss of strength and decreased lifespan for textile materials.

   e. Building materials: Ozone depletion can impact building materials such as plastics, sealants and coatings used in construction. Increased UV radiation can cause degradation and deterioration of these materials, leading to reduced structural integrity and performance. For example, sealants and adhesives used in windows, roofs and joints can be susceptible to cracking and failure, compromising the building's waterproofing and insulation capabilities.

6. **Human health effects:** Increased UV radiation due to ozone depletion can have direct effects on human health, such as skin cancer, cataracts, weakened immune system and DNA damage. It can also indirectly impact health by affecting ecosystems and disrupting food chains, leading to changes in food availability and nutritional content.

Topics of Interest: Skin Cancer and UV Radiation

UV-B radiation is a known environmental risk factor for various skin conditions, including sunburn, premature skin ageing and skin cancer.

Skin cancer, particularly the two most common types – basal cell carcinoma (BCC) and squamous cell carcinoma (SCC) – is strongly associated with exposure to UV radiation. Prolonged or excessive exposure to UV-B radiation can damage the DNA in skin cells, leading to genetic mutations and the development of cancerous cells. Additionally, UV radiation can weaken the immune system's ability to control the growth of cancer cells, further contributing to the development of skin cancer.

It is important to note that not all UV radiation is blocked by the ozone layer. UV-A radiation, which is not significantly affected by ozone depletion, also contributes to skin damage and skin cancer development. Therefore, even with a healthy ozone layer, it is essential to take precautions to protect the skin from UV radiation.

Efforts to address ozone depletion have been made through international agreements, such as the Montreal Protocol, which aims to phase out the production and use of ODSs. These efforts have led to a gradual recovery of the ozone layer, but it may take several more decades for a full recovery to occur.

To protect yourself from the harmful effects of UV radiation and reduce the risk of skin cancer, it is recommended to:

1. Limit exposure to the sun during peak hours when UV radiation is strongest (usually between 10 a.m. and 4 p.m.).

2. Seek shade and use protective clothing, such as wide-brimmed hats, long-sleeved shirts and sunglasses that block UV rays.

3. Apply a broad-spectrum sunscreen with a high sun protection factor (SPF 30 or higher) to exposed skin, even on cloudy days.

4. Avoid indoor tanning beds, as they emit concentrated UV radiation and increase the risk of skin cancer.

5. Regular skin self-examinations and professional skin screenings are also crucial for early detection and treatment of any potential skin abnormalities or cancers.

6. **Climate change:** ODSs, such as CFCs, are also potent greenhouse gases. The phase-out of these substances under the Montreal Protocol has contributed to reducing their impact on global warming. However, the interactions between ozone depletion and climate change are complex and some studies suggest that ozone depletion has had a cooling effect on the atmosphere.

As the ozone layer recovers, this Ozone depletion affects the entire planet to some degree, but certain regions have experienced more significant impacts due to a combination of factors such as geography, climate and human activities. Some of the more affected countries or regions are:

a. **Antarctica:** The ozone hole, a severe depletion of the ozone layer, was first discovered over Antarctica in the 1980s. It forms every year during the Southern Hemisphere's spring (September–October) due to a combination of extremely cold temperatures and human-made CFCs. The ozone hole over Antarctica has had significant consequences for the region's ecosystems and has affected countries like Argentina, Chile and Australia, which have research stations in Antarctica.

b. **Arctic region:** While not experiencing an ozone hole like Antarctica, the Arctic region has also seen significant ozone depletion, particularly during the winter and early spring. The depletion is influenced by unique atmospheric conditions and the presence of ODSs. Indigenous communities in Arctic regions are particularly vulnerable to increased UV radiation due to their traditional lifestyles and reliance on outdoor activities.

c. **Australia and New Zealand:** These countries experience higher levels of UV radiation due to their proximity to the Antarctic ozone hole. Increased UV radiation poses risks to human health, including higher rates of skin cancer. Australia and New Zealand have implemented extensive public awareness campaigns and policies to protect their populations from the harmful effects of UV radiation.

d. **Highly populated countries in general:** Countries with large populations, particularly in tropical and subtropical regions, can experience higher levels of UV radiation due to ozone depletion. Some examples include India, Brazil, China and countries in Southeast Asia. The effects can be particularly pronounced in densely populated urban areas with limited access to protective measures like shade and sunscreen.

Efforts to address ozone depletion have been largely successful, particularly through the implementation of the Montreal Protocol, which has resulted in a gradual recovery of the ozone layer. However, continued vigilance and adherence to international agreements are necessary to ensure the long-term protection of the ozone layer and mitigate its consequences.

11.7 Ozone Hole

The ozone hole refers to a region in Earth's stratosphere where the concentration of ozone (O_3) molecules becomes significantly depleted.

The ozone hole typically forms during the Antarctic spring (September–November), when sunlight returns to the region after the polar winter. Cold temperatures, unique wind patterns and the presence of a special kind of cloud that occurs in the Earth's polar stratosphere during winter are the origins of such phenomena.

The ozone hole was first discovered in the 1980s, primarily over Antarctica and later observed in a milder form over the Arctic. In the early 1980s, scientists started to notice a significant decline in the ozone layer over Antarctica during the Southern Hemisphere spring. This was detected through ground-based measurements, satellite observations and data from weather balloons. This severe depletion of the ozone layer in the Earth's stratosphere over the polar regions was nominated as the ozone hole.

The groundbreaking discovery of the Antarctic ozone hole was credited to British Antarctic Survey scientists Joe Farman, Brian Gardiner and Jonathan Shanklin. In 1985, they published a paper in the scientific journal Nature reporting a dramatic decline in ozone concentrations over Antarctica. They had been studying data collected at the British Antarctic Research Station in Halley Bay since the 1950s and their findings showed an unprecedented loss of ozone during the spring months.

The ozone hole is usually depicted in scientific reports and public awareness campaigns using colour scales that show the extent of ozone depletion. Commonly used colours are shades of blue and purple.

In these representations, darker shades of blue or purple typically indicate larger ozone depletion or a more significant ozone hole, while lighter shades represent smaller depletion. The colour scale is often accompanied by numerical values to provide a quantitative understanding of the severity of the ozone hole. Figure 11.4 shows a typical representation of the ozone hole in the Antarctic zone[2].

Colour coding ozone hole representations is crucial for visualising the extent and severity of ozone depletion in the Earth's stratosphere.

The following are common colour schemes used in such visualisations, along with their typical interpretations:

- **Blues and purples:** Represent areas with low ozone concentration (severe depletion).
- **Greens:** Represent moderate ozone concentration.
- **Yellows and oranges:** Represent higher ozone concentrations, indicating healthier levels.
- **Reds:** Represent very high ozone concentrations, which are typically seen outside the ozone hole region.

Table 11.2 provides the colour scale for ozone hole representations.

To follow the values and monitor the ozone hole, it is possible to use various online resources and organisations that provide real-time data and updates because

---

[2] https://creativecommons.org/licenses/by/2.0/.

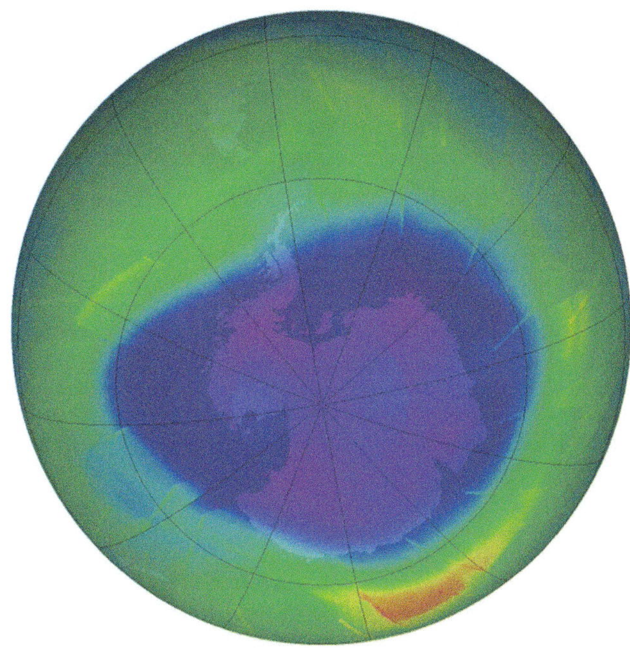

FIGURE 11.4   Ozone hole

TABLE 11.2   **Colour scale for ozone hole representations**

Colour	Ozone concentration (DU)
Dark purple	100–150
Light purple	150–200
Blue	200–250
Green	250–300
Yellow	300–350
Orange	350–400
Red	>400

the ozone layer is continuously monitored by scientific institutions and they publish their findings regularly. Here are the steps to follow the values of the ozone hole:

1. **Visit the official websites:** Go to the websites of organisations responsible for monitoring the ozone layer. The primary authorities in this area are the World Meteorological Organisation (WMO) (https://public.wmo.int/en) and the United Nations Environment Programme (UNEP) (https://www.unep.org/). They often collaborate on initiatives like the Scientific Assessment Panel for the Montreal Protocol. Keep an eye on their websites for updates, reports and data.

2. **NASA ozone watch:** NASA's ozone watch website (https://ozonewatch.gsfc. nasa.gov/) is a valuable resource for tracking the ozone layer's status. They

provide up-to-date information, data visualisations and educational materials related to ozone depletion and recovery.

3. **National meteorological agencies:** Many countries have their own meteorological agencies that monitor and report on atmospheric conditions, including the ozone layer. Check your country's meteorological agency website for relevant information.

4. **NOAA's Earth System Research Laboratory (ESRL):** The NOAA ESRL (https://www.esrl.noaa.gov/) provides data on various atmospheric components, including ozone. They have a dedicated section for ozone monitoring that you can explore.

5. **European Space Agency (ESA):** ESA (https://www.esa.int/) also has missions and projects focused on monitoring the ozone layer. Check their website for relevant data and updates.

6. **Mobile apps:** There might be mobile applications available that provide real-time ozone data such as Ozone Data Hub. Look for apps developed by reputable organisations or government agencies specialising in environmental monitoring.

7. **Scientific journals:** Keep an eye on scientific journals that publish research related to the ozone layer and atmospheric conditions. Many studies on ozone depletion and recovery are published in peer-reviewed journals, which can give you valuable insights into the topic.

The discovery of the ozone hole serves as a testament to the importance of scientific research and international cooperation in addressing global environmental challenges. It has led to increased awareness about the need to protect the ozone layer and has had a significant impact on policies and regulations regarding the use of ODSs.

The primary cause of the ozone hole is the release of human-made chemicals known as ODS, combined with two important facts produced in polar zones by the very special climate conditions of such regions. They are the **polar stratospheric clouds (PSCs)** and the so-called **polar vortex**.

11.8 Polar Stratospheric Clouds (PSCs)

PSCs, also known as nacreous clouds or mother-of-pearl clouds, are a type of cloud that forms in the stratosphere at high latitudes, typically over the polar regions during winter. PSCs are composed of tiny crystals of ice and other components that are suspended in the stratosphere and are characterised by their beautiful colours (Figure 11.5).

PSCs form when temperatures in the stratosphere drop below $-78\,°C$ $(-108\,°F)$, which is cold enough to freeze water vapour into ice crystals. These clouds form at high altitudes where the air is very dry and the pressure is low. A persistent low-pressure system traps cold air in the polar stratosphere, leading to the formation of PSCs.

According to their composition, PSCs can be classified into two types:

1. **Type I clouds** contain water, nitric acid and/or sulfuric acid. When temperatures fall to $-78\,°C$ or lower, they form:
   - **Type Ia clouds**, which are made up of large, aspherical particles made up of nitric acid trihydrate (NAT).

FIGURE 11.5 Polar stratospheric clouds

- **Type Ib clouds**, which contain small, non-depolarising spherical particles of a liquid supercooled ternary solution (STS) of sulfuric acid, nitric acid and water.

- **Type Ic clouds**, which are composed of metastable, water-rich nitric acid in solid form.

2. **Type II clouds**

   - They are made up entirely of water ice and are extremely rare in the Arctic.

   - They are formed at temperatures of −83 °C or lower.

Type I clouds, like any other cloud, can be iridescent under certain conditions, whereas Type II clouds are necessarily nacreous.

Table 11.3 details the composition, temperature and size of different types of PSCs.

Ground-based observations of PSCs and research on their formation processes date back several years before the scientific community recognised their role in polar ozone destruction. However, a complete understanding of the geographical and altitude extent of PSCs in both polar regions was not available until satellite instruments observed them in the late 1970s. The scientific community's comprehension

TABLE 11.3 **Properties of polar stratospheric clouds**

PSC type	Composition	Temperature formation	Size
Type Ia	Nitric acid trihydrate	Below −78 °C (−108 °F)	Micrometre-sized particles
Type Ib	Supercooled ternary solution (STS)	Below −78 °C (−108 °F)	Micrometre-sized particles
Type Ic	Nitric acid dihydrate	Below −78 °C (−108 °F)	Micrometre-sized particles
Type II	Ice crystals	Below −78 °C (−108 °F)	Larger particles than type I

of the chemical role of PSC particles in converting reactive chlorine gases to active halogen radicals emerged only after the discovery of the Antarctic ozone hole in 1985. Further development in the knowledge of the chemical role of PSC particles came from laboratory studies of their surface reactivity, coupled studies of polar stratospheric chemistry and measurements that directly sampled particles and reactive chlorine gases such as Cl˙ and ClO˙ in the polar stratosphere.

11.9 Polar Vortex

The other key fact for explaining characterised ozone hole generation is the so-called polar vortex (Figure 11.6).

The Antarctic polar vortex is a persistent low-pressure system that forms over the Antarctic region during the winter months. It is a large-scale circulation pattern that is characterised by a strong, counter-clockwise flow of air around the South Pole, with winds reaching speeds of up to 400 km/h (250 mph). Such strong winds prevent the exchange of matter between the interior and exterior. The polar vortex is a natural occurrence that is driven by the temperature difference between the cold air in the polar region and the warmer air in the mid-latitudes. During the winter months, the polar region is almost completely dark and the surface temperature can drop to as low as −90 °C. This creates a large temperature gradient between the polar region and the surrounding oceans, which drives the formation of the polar vortex. The polar vortex plays an important role in the Earth's climate system, as it helps regulate the exchange of heat and moisture between the polar region and the mid-latitudes. However, in recent years, there has been concern that the polar vortex may be weakening due to global warming, which could have significant implications for weather patterns and sea ice in the Southern Hemisphere.

The Antarctic vortex plays a crucial role in the formation of the ozone hole by creating a stable, cold environment in which the chemical reactions that destroy

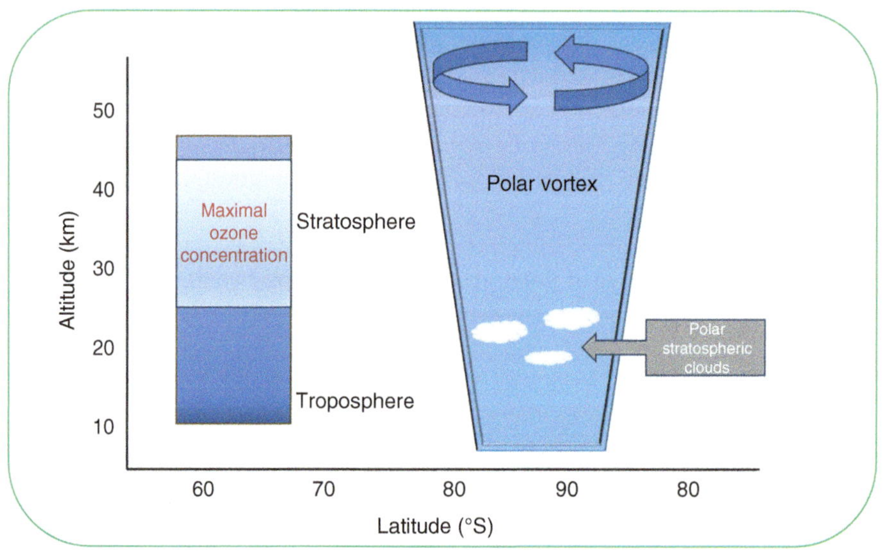

FIGURE 11.6 Schematic representation of the polar vortex

ozone can occur more efficiently. The combination of low temperatures, high-altitude clouds and the presence of CFCs leads to a chain reaction that rapidly destroys ozone molecules, resulting in the depletion of the ozone layer over Antarctica.

At the North Pole, a similar phenomenon occurs: a persistent circulation pattern centred near the North Pole. Nevertheless, it is formed due to the temperature difference between the cold Arctic air and the warmer air at lower latitudes. Nevertheless, there are significant differences in addition to the geographic location:

1. **Size and strength:** The Arctic vortex is generally smaller and weaker compared to the Antarctic vortex. The Arctic vortex can undergo significant variations in size and strength, with a core that can be disrupted or split during certain conditions. The Antarctic vortex tends to be more stable and powerful, displaying a well-defined circular shape with a strong, consistent circulation.

2. **Seasonal variation:** Both vortices exhibit seasonal variations, but their patterns differ. The Arctic vortex experiences more pronounced seasonal changes due to the presence of landmasses surrounding the Arctic Ocean. During winter, the Arctic vortex becomes stronger and more tightly organised. In contrast, the Antarctic vortex undergoes less variability throughout the year and has a relatively stable and well-structured circulation.

3. **Ozone depletion:** One of the most significant differences between the two vortices is related to the presence of the ozone hole. The Antarctic vortex experiences the formation of an ozone hole during the Southern Hemisphere's spring (September–November). Cold temperatures and unique chemical reactions within the vortex lead to the depletion of ozone molecules. The Arctic vortex, on the other hand, does not typically develop an ozone hole as the temperature and chemical conditions are not as conducive to ozone depletion.

4. **Interactions with mid-latitude weather:** Both vortices can influence weather patterns beyond the polar regions, but the interactions differ due to the contrasting geography and circulation patterns. The Arctic vortex has a greater potential to interact with mid-latitude weather systems, such as the polar jet stream, due to the proximity of landmasses and the greater variability in its structure. The Antarctic vortex, being relatively isolated by the Southern Ocean, has a lesser impact on mid-latitude weather systems.

The ozone hole is produced by a complex series of chemical reactions involving ODSs, primarily human-made CFCs and halons, that are released into the atmosphere. These chemicals rise into the stratosphere and are broken down by the intense ultraviolet radiation from the sun. As a result, chlorine and bromine atoms are released, which then react with ozone molecules, breaking them down into oxygen molecules.

11.10 Steps in the Formation of the Ozone Hole

Pre-hold period: Before the ozone hole forms, the ozone concentration in the Antarctic stratosphere is relatively stable throughout the year, with some seasonal variations. During the pre-hole period, the concentration of ozone is maintained by a balance between production and destruction processes, as occurs in the rest of the planet's ozone layer.

a. Ozone is produced by the photodissociation of molecular oxygen (O_2) by UV radiation, followed by the reaction of the resulting oxygen radicals ($O^{\cdot}$) with molecular oxygen.

b. Ozone is destroyed according to Chapman's mechanism and the catalysed reactions because of man-made ODS.

Reactions in darkness: A series of heterogeneous phase reactions take place in the PSCs, leading to the formation of $Cl^{\cdot}$ and $HOCl$ from the reservoir species HCl and $ClONO_2$.

1. **First step:** The nitrogen oxides (NO and NO_2), which help convert $ClO^{\cdot}$ into HCl, are removed from the gaseous phase of the stratosphere through the following reactions (Eq. 11.16 i–iv):

$$\textbf{i. } NO + O_3 \rightarrow NO_2 + O_2$$

$$\textbf{ii. } NO_2 + O_3 \rightarrow NO_3 + O_2$$

$$\textbf{iii. } NO_2 + NO_3 + M^* \rightarrow N_2O_5 + M$$

$$\textbf{iv. } N_2O_5 + H_2O \rightarrow 2HNO_3 \qquad (11.16)$$

Nitric acid is incorporated into particles in PSCs.

2. **Second step:** On the surface of PSCs, the ice particles of the 'reservoir species' of non-reactive chlorine, HCl and $ClONO_2$, react with each other (Eq. 11.17).

$$HCl + ClONO_2 \rightarrow Cl_2 + HNO_3 \qquad (11.17)$$

The nitric acid is immediately incorporated into the particles of PSCs.

Hole period: End of the polar night

3. **Third step:** When daylight returns at the end of the polar night, the photolysis of Cl_2 produces two $Cl^{\cdot}$ radicals (Eq. 11.18).

$$Cl_2 + h\nu \rightarrow 2Cl\cdot \qquad (11.18)$$

4. **Fourth step:** The chlorine radicals initiate a chain of catalytic reactions, leading to the destruction of ozone, as long as there are no nitrogen oxides available to remove them (Eq. 11.19 1–5).

$$\textbf{1. } 2Cl\cdot + \mathbf{2O_3} \rightarrow 2ClO\cdot + \mathbf{2O_2}$$

$$\textbf{2. } ClO\cdot + ClO\cdot + M \rightarrow Cl_2O_2 + M^*$$

$$\textbf{3. } Cl_2O_2 + h\nu\cdot \rightarrow Cl\cdot + ClO_2$$

$$\textbf{4. } Cl\cdot + ClO_2 \rightarrow 2Cl\cdot + \mathbf{O_2}$$

$$\text{-------------------------}$$

$$\textbf{5. } 2O_3 \rightarrow 3O_2 \qquad (11.19)$$

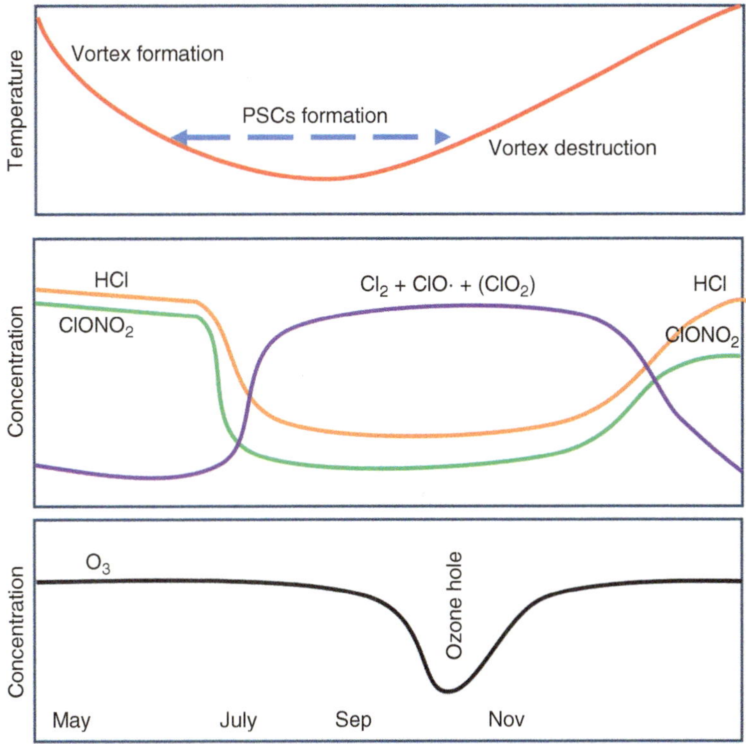

FIGURE 11.7   Resume of ozone and other species' concentrations in the Antarctic

The most common chlorine species, such as Cl·, ClO· and Cl$_2$O$_2$, are formed and concentrated mainly in the upper stratosphere, while ozone is found in the lower stratosphere. The polar vortex brings the destructive chlorine species into contact with the ozone.

Hole recovery: As the Antarctic Spring ends and the sun's intensity decreases, the chemical reactions that cause ozone depletion slow down. As a result, the ozone hole begins to shrink and the concentration of ozone in the atmosphere over Antarctica begins to increase.

During the recovery period, the concentration of CFCs and other ODSs in the atmosphere decreases, the rate of ozone depletion slows down and the ozone layer begins to recover. However, the recovery of the ozone layer is also influenced by other factors, such as changes in atmospheric circulation patterns and temperature.

Figure 11.7 resumes the evolution along the year of the temperature and the species responsible for the ozone depletion in the Antarctic.

11.11   Montreal Protocol

The Montreal Protocol is an international treaty designed to protect the ozone layer by phasing out the production and consumption of substances that deplete it. The protocol was signed on September 16, 1987 in Montreal, Canada and came into effect on January 1, 1989. The substances targeted by the Montreal Protocol are primarily CFCs and other ODSs, The Montreal Protocol targets a number of substances that deplete the ozone layer, including: CFCs, halons, carbon tetrachloride, methyl chloroform, HCFCs, hydrobromofluorocarbons (HBFCs) and bromochloromethane.

Under the Montreal Protocol, participating countries agreed to gradually phase out the production and consumption of ODSs, with specific targets and timelines for each substance. The treaty has been highly successful, with most countries having already met their targets for the phase-out of these substances. As a result, the ozone layer is now showing signs of recovery and it is expected to fully recover by the middle of the 21st century.

The Montreal Protocol is widely regarded as one of the most successful international environmental agreements and its success has been attributed to its strong scientific foundation, its use of market-based mechanisms to drive compliance and its ability to adapt to changing scientific and technological developments.

In addition to the Montreal Protocol, there have been several other international agreements and initiatives aimed at protecting the ozone layer, including:

1. **The Vienna Convention for the Protection of the Ozone Layer (1985):** This treaty provided the framework for the development of the Montreal Protocol and called for international cooperation to address the issue of ozone depletion.

2. **The Copenhagen Amendment to the Montreal Protocol (1992):** This amendment accelerated the phase-out of ODSs and included additional chemicals, such as halons and carbon tetrachloride, in the list of substances to be phased out.

3. **The London Amendment to the Montreal Protocol (1990):** This amendment established a financial mechanism, the Multilateral Fund for the Implementation of the Montreal Protocol, to help developing countries phase out ODSs.

4. **The Beijing Amendment to the Montreal Protocol (1999):** This amendment banned the production and consumption of methyl bromide, a potent ODS.

5. **The Kigali Amendment to the Montreal Protocol (2016):** This amendment aims to phase down the production and consumption of hydrofluorocarbons (HFCs), which are not ozone-depleting but are potent greenhouse gasses that contribute to climate change.

These agreements and initiatives demonstrate the international community's commitment to protecting the ozone layer and the environment as a whole.

CHAPTER 12

Tropospheric Pollution

12.1 Introduction

Tropospheric pollution, occurring in the Earth's lower atmosphere, is primarily caused by the presence of harmful substances. These substances have adverse effects on human health, ecosystems and the climate. While some instances of tropospheric pollution can be attributed to natural causes, the predominant source is human activity.

Tropospheric pollution is not only contemporary; evidence of air pollution can be traced back to ancient Rome and China. In Rome, the combustion of wood and coal in homes and public baths emitted smoke, leading to respiratory issues and raising concerns among city officials. Similarly, during the Han Dynasty in China (206 BCE–220 CE), the use of coal for heating and cooking resulted in severe air pollution. Although the sources and types of pollution in the Middle Ages differed from modern times, the Industrial Revolution in the 19th century witnessed widespread air pollution in UK cities like London and Manchester due to the extensive use of coal in factories and steam engines. The resultant smog, a combination of smoke and fog, caused respiratory problems and even fatalities for numerous individuals.

Following are some of the sources of tropospheric pollution:

1. **Natural sources:** Volcanic eruptions, wildfires and dust storms can release large amounts of particulate matter (PM), carbon monoxide (CO) and other pollutants into the atmosphere.

2. **Transportation:** Emissions from vehicles, aeroplanes and ships, release pollutants such as CO, nitrogen oxides (NO_x), PM and sulfur dioxide (SO_2) into the air.

3. **Industrial activities:** Fossil fuel combustion in power plants, refineries and factories releases large amounts of pollutants, including SO_2, NOx, CO and PM.

4. **Agricultural activities:** Farming practices like the use of fertilisers, pesticides and herbicides, as well as livestock management, can release pollutants like ammonia (NH_3), methane (CH_4) and nitrous oxide (N_2O).

5. **Burning of biomass and waste:** The burning of wood, charcoal and other biomass for cooking and heating purposes can release large amounts of PM and other pollutants. Open burning of waste, such as garbage and crop residue, can also lead to significant emissions of pollutants.

Understanding the Chemistry of the Environment, First Edition. Francisco G. Calvo-Flores.
© 2025 John Wiley & Sons Ltd. Published 2025 by John Wiley & Sons Ltd.

6. **Residential and commercial heating:** The use of fossil fuels, such as natural gas and oil, for heating in residential and commercial buildings can also contribute to tropospheric pollution.

7. **Indoor activities:** Indoor air pollution can also be a significant source of tropospheric pollution, particularly in developing countries where cooking and heating are done using solid fuels like wood, charcoal and coal.

Common pollutants include PM, NOx, SO_2, CO, carbon dioxide (CO_2) and ozone (O_3).

Exposure to tropospheric pollution can cause respiratory problems such as asthma and bronchitis, increase the risk of cardiovascular disease, lung cancer and stroke and harm ecosystems, crops and wildlife.

To reduce tropospheric pollution, efforts can be made to reduce emissions from transportation, industry and energy production, as well as promote cleaner technologies and renewable energy sources. Additionally, individuals can take actions to reduce their own contribution to air pollution, such as driving less, using public transportation and conserving energy.

12.2 Classification of Atmospheric Pollutants

Atmospheric pollutants are classified into two major categories:

1. **Primary pollutants** are those that are directly emitted into the atmosphere from a source, such as a car or a factory. Examples of primary pollutants include CO and CO_2, SO_2, NOx, volatile organic compounds (VOCs) and PM.

2. **Secondary pollutants** are those that are formed in the atmosphere through chemical reactions between primary pollutants and other substances. These reactions can be caused by sunlight, heat or other chemical reactions. Examples of secondary pollutants include O_3, which is formed through a reaction between NOx and VOCs and acid rain, which is formed through a reaction between SO_2, NOx and water vapour (H_2O). Some pollutants can be found in both groups. Secondary pollutants are formed when primary pollutants react with other substances in the atmosphere. Smog is a type of air pollution characterised by a mixture of smoke, fog and other pollutants, often occurring in urban areas with high population density and heavy industrial activity. It is typically formed when emissions from various sources, such as vehicle exhaust, industrial emissions and chemical reactions in the atmosphere, combine with atmospheric moisture and sunlight. The resulting smog is a visible haze that can hang over cities, reducing visibility and causing a range of adverse effects on human health and the environment.

   There are two primary types of smog:

   Industrial smog: Also known as 'grey smog' or 'London smog', industrial smog is typically associated with areas where industries burn fossil fuels, particularly coal. It consists of SO_2, PM and other pollutants emitted by industrial processes. Industrial smog is often characterised by a thick, greyish, fog-like appearance.

Photochemical smog: Photochemical smog, also referred to as 'brown smog' or 'Los Angeles smog', is prevalent in areas with high levels of automobile traffic and abundant sunlight. It forms when pollutants, primarily NO_x and VOCs, react in the presence of sunlight. Photochemical smog often appears as a brownish haze and contains harmful pollutants such as O_3.

Smog is a significant environmental and public health concern as it can lead to respiratory problems, eye irritation, reduced lung function and an increased risk of cardiovascular diseases. It also contributes to the deterioration of ecosystems, damages crops and vegetation and contributes to climate change. Efforts to reduce smog include implementing emission controls, promoting cleaner technologies and implementing policies to reduce air pollution from various sources.

12.3   Primary Carbon Pollutants: Carbon Monoxide and Dioxide

Carbon Monoxide

Carbon monoxide (CO) is a colourless, odourless and toxic gas that can be found in the Earth's atmosphere, including the troposphere. CO is produced by natural and human activities such as volcanic eruptions, forest fires and the incomplete combustion of fossil fuels from transportation, power generation and industrial processes.

CO is produced in combustion reactions when there is not enough oxygen available to fully oxidise carbon (C) to CO_2.

A fuel (such as coal, oil, gas or wood) reacts with oxygen (O_2) to release energy in the form of heat and light. The basic chemical equation for the combustion of hydrocarbons (compounds made up of hydrogen and carbon) is (Eq. 12.1):

$$Fuel + O_2 \rightarrow CO_2 + H_2O \tag{12.1}$$

However, if there is insufficient oxygen available, the carbon in the fuel may not fully react with oxygen to form CO_2. Instead, it may react with only one oxygen molecule to form CO, which is less stable than CO_2 and more likely to react further to form other compounds. The chemical equation for the incomplete combustion of hydrocarbons can be represented in (Eq. 12.2):

$$Fuel + O_2 \rightarrow CO + CO_2 + H_2O \tag{12.2}$$

In the troposphere, CO can have significant impacts on air quality and human health. It can react with other atmospheric compounds to form ground-level O_3, a harmful pollutant that can cause respiratory problems and other health issues.

CO is sometimes referred to as the 'silent killer' because it is a colourless, odourless gas that can be deadly in a home.

CO poisoning can occur when CO builds up in an enclosed space, such as a home or car. When inhaled, CO binds with haemoglobin in the blood, preventing oxygen from being delivered to the body's organs and tissues.

The effects of CO on health can vary depending on the level and duration of exposure. Even low levels of CO exposure over an extended period can cause chronic health problems such as heart disease and cognitive impairment.

CO binds with haemoglobin in red blood cells, reducing the amount of oxygen that can be carried by the blood. This can lead to a range of health effects, including:

1. **Headaches:** CO exposure can cause headaches, which may range from mild to severe.

2. **Dizziness:** Dizziness is a common symptom of CO poisoning and it may be accompanied by nausea or vomiting.

3. **Fatigue:** CO poisoning can cause fatigue, weakness and difficulty concentrating.

4. **Chest pain:** Exposure to high levels of CO can cause chest pain and shortness of breath.

5. **Confusion:** CO poisoning can lead to confusion, disorientation and memory loss.

6. **Coma:** In severe cases of CO poisoning, a person may slip into a coma.

7. **Death:** In the most severe cases, CO poisoning can be fatal.

In indoor spaces, CO poisoning is preventable through proper ventilation of fuel-burning appliances, regular maintenance of heating systems and the installation of CO detectors, which can help minimise the risk of exposure to CO.

In addition to its effects on human health, CO can also contribute to climate change. It is a greenhouse gas, meaning that it can trap heat in the atmosphere and contribute to the warming of the planet.

Carbon Dioxide

CO_2 is considered a pollutant when it is present in excessive amounts in the Earth's atmosphere. While CO_2 is a naturally occurring gas and is necessary for life on Earth, human activities such as burning fossil fuels, deforestation and industrial processes have significantly increased the levels of CO_2 in the atmosphere.

CO_2 is formed as a product of combustion when a fuel (such as gasoline, coal or natural gas) reacts with oxygen in the air. During the combustion process, carbon in the fuel reacts with oxygen from the air to form CO_2, H_2O and other byproducts. For example, when gasoline is burned in a car engine, the hydrocarbons in the gasoline react with oxygen from the air to produce CO_2 and H_2O according to Eq. (12.3 (assuming for gasoline an average formula of C_8H_{18}):

$$C_8H_{18} + 25/2 O_2 \rightarrow 8CO_2 + 9H_2O \tag{12.3}$$

This means that for every molecule of gasoline (C_8H_{18}) burned, 8 molecules of CO_2 and 9 molecules of H_2O are produced.

The amount of CO_2 formed during combustion depends on the type of fuel being burned and the efficiency of the combustion process. In general, burning fossil fuels such as coal, oil and natural gas produces more CO_2 per unit of energy than burning renewable fuels such as biomass or wind and solar power.

While CO_2 is not directly harmful to human health at normal atmospheric concentrations, high levels of exposure can cause several adverse effects.

1. **Headaches and dizziness:** Exposure to high levels of CO_2 can cause headaches, dizziness and feelings of fatigue. These symptoms may be more pronounced in enclosed spaces with poor ventilation.

2. **Respiratory problems:** CO_2 can displace oxygen in the air, which can make it harder for people to breathe. This is especially true for individuals with pre-existing respiratory conditions, such as asthma or chronic obstructive pulmonary disease (COPD).

3. **Cardiovascular problems:** Elevated CO_2 levels can also cause an increase in heart rate and blood pressure, which can lead to cardiovascular problems.

4. **Cognitive impairment:** High levels of CO_2 can impair cognitive function, including memory, attention and decision-making ability.

5. **Sleep disturbances:** Exposure to high levels of CO_2 during sleep can cause disruptions in sleep patterns, leading to reduced sleep quality and daytime fatigue.

It is worth noting that these effects are generally only observed at very high levels of CO_2 exposure, such as those found in industrial settings or poorly ventilated indoor spaces. In general, the levels of CO_2 found in the ambient outdoor air are not high enough to cause significant health problems. However, it is important to maintain good indoor air quality by ensuring adequate ventilation, particularly in enclosed spaces where people spend a lot of time, such as homes and workplaces.

In the atmosphere, when the concentration of CO_2 becomes too high, it contributes to global climate change by trapping heat from the sun and causing the Earth's temperature to rise. This can lead to a variety of negative impacts, such as rising sea levels, more frequent and severe weather events and harm to ecosystems and human health.

The CO_2 concentration in 2023 could be around 419 ppm, based on measurements taken at the Mauna Loa Observatory in Hawaii. This is the highest atmospheric CO_2 concentration in at least the past 800,000 years and is primarily due to human activities such as burning fossil fuels and deforestation.

However, it's important to note that CO_2 levels continue to increase each year. This ongoing rise in CO_2 levels is contributing to climate change, as CO_2 is a greenhouse gas that traps heat in the Earth's atmosphere and causes the planet's temperature to rise.

12.4 Primary Sulfur Pollutants: Sulfur Oxides and Hydrogen Sulfide

Sulfur Oxides

Sulfur oxides (SO_x) are a group of air pollutants that are formed when sulfur-containing fuels are burned. The most common sulfur oxide pollutants are SO_2 and sulfur trioxide (SO_3).

Natural sources of sulfur oxides are volcanic eruptions and forest fires. The burning of biomass, including forest fires, can also release sulfur oxides.

Anthropogenic sources of sulfur oxides are primarily the combustion of fossil fuels such as coal, oil and gas in industrial processes and transportation. Some specific sources of sulfur oxides include:

1. **Burning of biomass**
2. **Power plants:** Coal-fired power plants are one of the largest sources of sulfur oxide emissions.
3. **Industrial processes:** Sulfur oxides are produced by a variety of industrial processes, such as the smelting of ores, the refining of petroleum and the production of sulfuric acid (H_2SO_4).
4. **Transportation:** Vehicles that use diesel fuel can produce significant amounts of sulfur oxides, particularly if the fuel is not low-sulfur.
5. **Heating:** The combustion of heating oil and natural gas can also produce sulfur oxides.
6. **Marine transportation:** Ships that use heavy fuel oil can emit significant amounts of sulfur oxides.

Once SO_2 is released into the air, it can be transported by the wind and atmospheric circulation patterns. The transport of SO_2 in the atmosphere can occur in both the short term and the long term. In the short term, local winds can transport SO_2 from its source to nearby areas. In the long term, SO_2 can be carried over long distances by global wind patterns. So, this pollutant can affect areas far away from the point of emission.

Transformation of SO_2 to SO_3

In the atmosphere, SO_2 can react with oxygen and other atmospheric chemicals to form SO_3 through a series of chemical reactions involving sunlight and other compounds (Eq. 12.4). Once SO_3 is formed, it can react with H_2O to form H_2SO_4, which can then contribute to the formation of sulfate aerosols, tiny particles of sulfate compounds that are suspended in the atmosphere.

$$SO_2 + 1/2O_2 \rightarrow SO_3 + H_2O \rightarrow H_2SO_4 \qquad (12.4)$$

Sulfate aerosols are typically removed from the atmosphere through precipitation, but they can also be transported long distances before being deposited on the ground or into bodies of water.

SO_x pollutants can have harmful effects on human health and the environment. When SO_2 and SO_3 are released into the air, they can react with other chemicals to form fine PM that can penetrate deep into the lungs and cause respiratory problems. SO_x emissions can also contribute to acid rain, which can damage crops, forests and aquatic ecosystems.

Hydrogen Sulfide

Hydrogen sulfide (H_2S) is a colourless gas with a strong, unpleasant odour that is often described as being similar to that of rotten eggs. It is naturally produced by the decay of organic matter, such as in swamps, sewers and volcanic areas. It is also released as a byproduct of various industrial processes, including petroleum refining, natural gas extraction, the paper industry and certain chemical manufacturing operations.

Exposure to H_2S can have harmful effects on human health. At low concentrations, it can cause irritation of the eyes, nose and throat. Higher concentrations can lead to more severe symptoms, including headaches, dizziness, nausea and difficulty breathing. In extreme cases, exposure to very high concentrations can be fatal.

It's important to note that H_2S is present in the atmosphere at very low concentrations, typically in the parts per billion (ppb) or parts per trillion (ppt) range and the atmospheric lifetime of H_2S is relatively short. The exact lifetime depends on factors such as atmospheric conditions, concentration levels and the presence of reactive species. On average, H_2S has a lifetime of a few hours to a few days before it undergoes the aforementioned chemical reactions and removal processes.

12.5   Primary Nitrogen Pollutants: Nitrogen Oxides and Ammonia

Nitrogen Oxides

NO_x is a term used to refer to two oxides of nitrogen: nitrogen oxide (NO) and nitrogen dioxide (NO_2). They are both produced from the combustion of fossil fuels, such as in vehicles and power plants, but also from natural sources, including wildfires and microbial activity in soils.

NO_x emissions contribute to the formation of smog and acid rain, as well as being harmful to human health. Some common anthropogenic sources of NO_x emissions are as follows:

1. **Transportation**: NO_x is produced by burning fossil fuels in engines, including those in cars, trucks, ships and planes. The highest NO_x emissions come from diesel engines.

2. **Industrial processes:** Many industrial processes, such as metal refining, cement production and glass manufacturing, release NO_x emissions into the atmosphere.

3. **Energy production:** Power plants that burn fossil fuels, such as coal and natural gas, are a significant source of NO_x emissions.

4. **Residential sources:** Indoor combustion sources such as gas stoves, heaters and fireplaces can also contribute to NO_x emissions.

5. **Agricultural sources:** Fertilisers used in agriculture contain nitrogen, which can be converted into NO_x emissions through microbial activity in soil.

Specifically, NO_x formation in the combustion process is primarily due to the reaction between nitrogen and oxygen molecules. Nitrogen is quite an inert molecule, but during combustion, the high temperature causes the nitrogen and oxygen molecules in the air to react and form various NOx, including nitrogen monoxide (NO) and NO_2, according to Eqs. (12.5 and (12.6 respectively.

$$N_2 + O_2 \rightarrow 2NO \tag{12.5}$$

$$NO + \tfrac{1}{2}O_2 \rightleftharpoons NO \tag{12.6}$$

The second step is insignificant at high temperatures and the reaction reverts.

The effects of NO_x on human health can be significant. Short-term exposure to NO_2 can cause respiratory irritation, coughing and wheezing, while long-term exposure can lead to chronic respiratory problems such as asthma and bronchitis. NO_x can also contribute to the formation of ground-level O_3, which can exacerbate respiratory problems and cause eye and throat irritation.

In addition to its direct effects on human health and the environment, NO_x also contributes to environmental problems such as acid rain, eutrophication of water bodies and the formation of low-level atmospheric O_3, which is a greenhouse gas that contributes to global warming.

Dinitrogen Monoxide (Nitrous Oxide, N_2O)

Dinitrogen monoxide, also known as N_2O, is a colourless and odourless gas with the chemical formula N_2O, but it is not a significant contributor to acid rain, as it happens with NO_x.

It also has various industrial, medical and recreational uses.

N_2O is primarily produced by both natural and human activities.

Some of the major sources of N_2O emissions are as follows:

1. **Natural sources** such as microbial activity in soils, oceans and wetlands, as well as volcanic activity and lightning.

2. **Agriculture:** as a byproduct of the nitrogen cycle, which is an essential part of agriculture. Fertilisers, animal manure and crop residues all contribute to N_2O emissions from agricultural activities.

3. **Industrial processes** including the production of nitric acid (HNO_3), adipic acid and nylon. N_2O is also used as a propellant in aerosol cans, which can release the gas into the atmosphere.

4. **Combustion** of fossil fuels in vehicles, power plants and industrial processes. It is also a byproduct of biomass burning.

5. **Wastewater treatment** particularly when biological processes are used.

In medicine, N_2O is used as an anaesthetic, particularly for dental surgery and minor surgical procedures. It is also used as a pain reliever and as a sedative for patients with anxiety disorders. In the food industry, dinitrogen oxide is used as a propellant in whipped cream dispensers.

However, N_2O can also have negative effects on human health and the environment. When inhaled in large quantities, it can cause dizziness, headaches and even unconsciousness.

This gas is also used recreationally for its euphoric and dissociative effects. In this context, it is usually inhaled from small canisters, often referred to as 'whippets', which are typically used to charge whipped cream dispensers. When inhaled, N_2O can produce a sense of lightheadedness, relaxation and laughter, hence the name 'laughing gas'.

The legal status of N_2O varies across jurisdictions. In many countries, it is legal to possess and use N_2O for legitimate purposes, such as medical or dental procedures. However, recreational use may be regulated or prohibited in some places due to safety concerns and potential misuse.

N_2O has other negative impacts as a contributor to global warming and, additionally, can deplete the O_3 layer when it is released into the atmosphere.

Ammonia

NH_3 is a colourless gas with a pungent odour composed of one nitrogen and three hydrogen atoms (NH_3). It is produced naturally by various sources, such as the decomposition of organic matter and bacterial processes in soil and water.

However, human activities also contribute significantly to NH_3 emissions. The primary sources of anthropogenic NH_3 emissions include:

1. **Agriculture**: The largest contributor to NH_3 emissions is agricultural practices, particularly livestock farming and the use of synthetic fertilisers. NH_3 is released from animal waste, including urine and manure, as well as from the application of nitrogen-based fertilisers.

2. **Industrial processes:** Certain industries, such as chemical manufacturing, petroleum refining and the production of explosives and refrigerants, can release NH_3 as a byproduct or during handling and storage.

3. **Waste management:** Landfills and wastewater treatment plants can emit NH_3 due to the decomposition of organic waste and the breakdown of nitrogen-containing compounds in wastewater.

When NH_3 is released into the atmosphere, it undergoes a series of chemical reactions and transformations that determine its fate and impact.

12.6 Primary Volatile Organic Pollutants

VOCs are a group of primary pollutants that have a high vapour pressure at room temperature, meaning they easily evaporate into the air. Those directly emitted into the atmosphere from various sources, natural and anthropogenic, are the primary pollutants. They can be divided into two groups:

1. *Biogenic VOCs (BVOCs)* are organic compounds that are produced by living organisms, including plants, animals and microbes. These compounds are volatile, meaning they can evaporate easily into the air and can have important effects on the atmosphere, including affecting air quality and climate. Plants are the largest source of BVOCs and they produce a wide variety of these compounds for different purposes, such as attracting pollinators, defending against herbivores and communicating with other plants. BVOCs can also have indirect effects on the atmosphere by influencing the formation of clouds and aerosols,

which can in turn affect climate. Some examples of BVOCs include isoprene (C_5H_8), terpenes and methanol (CH_3OH).

a. Biogenic CH_4 comes from natural sources, such as wetlands, rice paddies and the digestive systems of animals. Wetlands are the largest natural source of biogenic CH_4 emissions. They occur when organic matter decomposes in an anaerobic environment, such as in flooded rice paddies or swamps. CH_4 is also emitted by livestock, such as cows and sheep, during the process of digestion, as well as by other animals such as termites and wild ruminants.

b. Isoprene is the most abundant VOC, which is primarily produced by trees and other woody plants.

c. Terpenes are produced by a wide variety of plants and can have a range of different functions, including attracting pollinators, repelling herbivores and acting as antimicrobial agents.

d. Biogenic CH_3OH is produced by both plants and microbes and can be a significant contributor to atmospheric chemistry.

e. Biogenic carbonyl compounds are a specific subset of BVOCs that contain carbonyl functional groups ($C=O$). These compounds are primarily emitted by living organisms, including plants and certain microorganisms, as part of their metabolic processes. Examples include:

    i. Formaldehyde (HCHO): HCHO is a simple and highly reactive carbonyl compound that is emitted by both plants and animals. It can contribute to the formation of secondary organic aerosols (SOAs) and atmospheric reactions that affect air quality.

    ii. Acetaldehyde (CH_3CHO): CH_3CHO is a VOC that can be found in various natural sources and processes such as:

    Fermentation: CH_3CHO is an intermediate in the fermentation process. This can occur in various natural settings, such as the fermentation of sugars in fruits, grains and other organic matter. It's responsible for the fruity and sometimes alcoholic aroma of ripening fruits.

    Decomposition of organic matter: CH_3CHO can be released during the decomposition of organic matter, including plant material, soil organic matter and decaying vegetation. Microorganisms involved in the breakdown of organic material can produce CH_3CHO as part of their metabolic processes.

    Forest fires: CH_3CHO is one of the VOCs released during forest fires. The combustion of vegetation and organic matter in wildfires can lead to the production of CH_3CHO, contributing to air pollution and the characteristic smoky odour associated with wildfires.

    Microbial metabolism: Some microorganisms, including bacteria and yeast, can produce CH_3CHO as a metabolic byproduct during their growth and fermentation processes. This can occur in various natural environments, such as soil and aquatic ecosystems.

    Ripening of fruits: CH_3CHO is produced naturally during the ripening of certain fruits, such as apples, bananas and tomatoes. It contributes to the flavour and aroma of these fruits.

    Plant metabolism: As mentioned earlier, CH_3CHO can be generated in plants during specific metabolic processes, such as alcohol fermentation under anaerobic conditions.

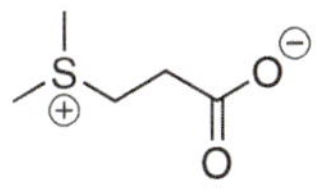

FIGURE 12.1   Dimethylsulfoniopropionate

Seawater: CH_3CHO has also been detected in seawater. It can be produced through various biological and chemical processes in marine environments.

    iii. Acetone (CH_3COCH_3): CH_3COCH_3 is a ketone that is emitted by plants and microorganisms. It plays a role in atmospheric chemistry and contributes to the production of VOCs, which can have various atmospheric implications.

    iv. Butanone ($CH_3CH_2COCH_3$): $CH_3CH_2COCH_3$ is another ketone compound emitted by vegetation. It can undergo chemical reactions in the atmos.

f. Dimethyl sulfide (DMS) (CH_3-S-CH_3): DMS (CH_3-S-CH_3) is a colourless gas produced by marine phytoplankton and from the breakdown of dimethylsulfoniopropionate, DMSP[1] (Figure 12.1), which is a sulfur-containing compound found in marine algae and other organisms.

g. Ethyl mercaptan (CH_3CH_2SH): This is a colourless gas with a strong, unpleasant odour that is often described as being similar to that of rotten cabbage or garlic. Methyl mercaptan is produced by the breakdown of organic matter in seawater.

2. *Anthropogenic VOCs (AVOCs)* are organic chemicals that are released into the atmosphere as a result of human activities. These compounds include a wide range of chemicals, such as benzene, toluene, HCHO and other organic chemicals that are commonly used in industrial processes, transportation and consumer products (See chapter IV).

12.7   Primary Aerosols

Atmospheric aerosols are tiny solid or liquid particles suspended in the Earth's atmosphere. These particles can range in size from a few nanometers to tens of micrometres. They can come from both natural and human sources.

Natural aerosols may be emitted from several sources, such as volcanic eruptions, dust storms, sea spray and wildfires.

a. Volcanic eruptions are a major source of natural aerosols. They release SO_2 and other gases, which react with H_2O in the atmosphere to form sulfate aerosols. These aerosols can stay in the atmosphere for years and affect the Earth's climate by reflecting incoming solar radiation back into space, cooling the planet.

b. Dust storms are another natural source of aerosols. They occur when winds blow over dry, arid areas, picking up dust and sand particles and carrying them into the atmosphere. These particles can affect air quality, visibility and human health.

c. Seaspray is a natural source of aerosols that occur when waves break on the surface of the ocean, releasing tiny droplets of water into the air. These

---

[1] DMSP is produced by phytoplankton and other marine organisms as a response to stress and it can be converted into DMS by bacteria and other microorganisms in the ocean, plays a role in the global sulfur cycle.

droplets contain salts, organic matter and other particles that can affect cloud formation and precipitation.

d. Wildfires also produce natural aerosols, which are mainly composed of smoke particles. These particles can affect air quality and human health and can also contribute to climate change by absorbing solar radiation and warming the atmosphere.

Table 12.1 summarises some common atmospheric natural aerosols, origins, compositions and sizes:

Anthropogenic aerosols refer to tiny particles or droplets produced by any human-made activity. Some common examples of anthropogenic sources of aerosols include:

a. Industrial processes: Many industrial processes, such as combustion, mining and manufacturing, release large amounts of PM into the atmosphere.

b. Transportation: The burning of fossil fuels in transportation, such as cars, trucks and aeroplanes.

c. Agricultural activities: The use of fertilisers, pesticides and other chemicals in agriculture can also contribute to the release of aerosols.

d. Construction and demolition: Dust and other particles generated during construction and demolition activities can also contribute to aerosol emissions.

e. Burning of solid waste: The burning of solid waste, such as trash and biomass, also releases aerosols into the atmosphere.

f. Residential sources: Residential sources of aerosols include cooking, heating and burning of solid fuels, such as wood and coal, in homes.

g. Aerosol sprays: Certain consumer products, such as hair spray.

Their composition is diverse:

a. Inorganic aerosols: These aerosols consist of particles composed primarily of inorganic substances, such as sulfates (SO_4^{-2}), nitrates (NO_3^-), ammonium (NH_4^+), mineral dust and sea salt.

b. Organic aerosols: These aerosols contain primarily carbonaceous compounds, such as soot, organic carbon and SOAs formed from the oxidation of VOCs.

TABLE 12.1   Natural aerosols

Aerosol type	Origin	Composition	Size range
Dust	Natural sources (e.g., deserts, soil erosion)	Mineral particles (e.g., silicates)	Coarse ($>10\,\mu m$)
Sea salt	Ocean surface (sea spray)	Sodium chloride (NaCl)	Coarse ($>1\,\mu m$)
Soot	Combustion processes (e.g., vehicles, industry)	Carbonaceous particles, organic compounds	Fine ($<1\,\mu m$)
Pollen	Plants (trees, flowers)	Biological particles (pollen grains)	Coarse ($>10\,\mu m$)
Volcanic ash	Volcanic eruptions	Silicate minerals (e.g., silica, feldspar)	Variable (micrometres to millimetres)

   c. Mixed aerosols: These aerosols are composed of both inorganic and organic components.

Another relevant characteristic is size. According to their size, aerosols can be classified into three groups:

1. **Coarse particulate matter (PM10):** These are aerosols with a diameter between 2.5 and 10 µm. They are larger and tend to settle more quickly than PM2.5.
2. **Fine particulate matter (PM2.5):** These are aerosols with a diameter less than or equal to 2.5 µm. They can be inhaled deep into the respiratory system and have significant health impacts.
3. **Ultrafine particles (UFP):** These are aerosols with a diameter of less than 0.1 µm. They are generally smaller and can remain suspended in the air for longer periods.

The classification systems mentioned above provide a general framework, but there can be overlaps and variations depending on specific research fields or applications.

The impact of aerosols on human health depends on their size, composition and concentration.

Some potential effects of atmospheric aerosols on human health are as follows:

1. **Respiratory problems:** Inhalation of fine PM (PM2.5) can cause respiratory problems such as asthma, bronchitis and COPD. These particles can penetrate deep into the lungs and cause inflammation, leading to decreased lung function and increased risk of respiratory infections.
2. **Cardiovascular problems:** Exposure to high levels of PM2.5 has been linked to an increased risk of cardiovascular disease, including heart attacks and strokes. This is because these particles can enter the bloodstream and cause inflammation, leading to the formation of plaques in the arteries.
3. **Cancer:** Some types of atmospheric aerosols, such as diesel exhaust particles and certain types of industrial emissions, have been classified as carcinogenic by the World Health Organization (WHO). Long-term exposure to these particles may increase the risk of lung cancer and other types of cancer.
4. **Neurological problems:** Fine PM has also been linked to neurological problems such as cognitive decline and dementia. These particles can enter the brain and cause inflammation, leading to damage to brain cells and impaired cognitive function.
5. **Other health effects:** Exposure to atmospheric aerosols has also been linked to a range of other health effects, including reproductive problems, low birth weight and premature death.

12.8 Secondary Pollutants

Secondary pollutants are not directly emitted as a result of a specific activity, such as burning of fossil fuels, but they are formed through chemical reactions that occur in the atmosphere involving primary pollutants, such as NOx, SO_2 and VOCs. These chemical reactions can happen through a variety of pathways, including photochemical reactions, oxidation and hydrolysis.

12.9 Tropospheric Oxidants

Most of the transformations of primary pollutants into secondary pollutants are responsible for the tropospheric oxidants hydroxyl (OH·) and hydroperoxyl (H—O—O·) radicals.

Both are very reactive species that are formed in the atmosphere and contribute to removing the majority of gas pollutants, giving other molecules considered secondary pollutants the ability to act as natural 'cleaners' by reacting with and breaking down pollutants.

Hydroxyl Radical (OH·)

OH· is formed in the atmosphere by several reactions. It has a very short lifetime of only a few seconds, but during this time, it can reoxidise and oxidise a wide range of compounds, including NO_x, CO and VOCs.

The most common pathway for the formation of OH· in the atmosphere is through the photodissociation of O_3 by ultraviolet (UV) radiation. This reaction generates an oxygen radical (O·) that is able to attack a molecule of vaporised water to give two OH· radicals (Eqs. 12.7 and (12.8).

$$O_3 + h\nu \rightarrow O_2 + O\cdot \tag{12.7}$$

$$O\cdot + H_2O \rightarrow 2OH\cdot \tag{12.8}$$

This process occurs primarily in the upper atmosphere, where UV radiation is most intense.

Hydroperoxyl Radical (H-O-O·)

H-O-O· is another highly reactive and unstable species that is found in the Earth's atmosphere. It is formed by the reaction of O_3 with an OH· radical, as shown below (Eq. 12.9):

$$O_3 + OH\cdot \rightarrow H\text{-}O\text{-}O\cdot + O_2 \tag{12.9}$$

The H-O-O· radical is also formed from the hydrogen radical (H·)[2] in the presence of molecular oxygen (O_2). The reaction pathway for the formation of the H-O-O· radical from the hydrogen radical can be represented in (Eq. 12.10):

$$H\cdot + O_2 \rightarrow H\text{-}O\text{-}O\cdot \tag{12.10}$$

This reaction is exothermic and typically occurs in the gas phase, such as in the Earth's atmosphere.

H-O-O· radicals are highly reactive and can participate in a wide range of chemical reactions. In particular, it can react with other radicals, such as the OH· radical, to form hydrogen peroxide (H_2O_2).

H-O-O· is an important player in the chemistry of the atmosphere, as it helps control the concentrations of both O_3 and other pollutants. It also plays a key role in

---

[2] See oxidation of CO to CO_2.

the self-cleansing mechanism of the atmosphere as it participates in the oxidation of pollutants and helps remove them from the air.

The concentration of H-O-O· in the atmosphere is dependent on a number of factors, including the levels of O_3, NO_x and VOCs, as well as the availability of sunlight. Understanding the behaviour of HO_2 in the atmosphere is an important area of research, as it can help improve our understanding of air pollution and climate change.

Ground-Level O_3

Ground-level O_3 is a secondary pollutant formed by a set of complex chemical reactions between NO_x and VOCs in the presence of sunlight.

Formation of NO_2: NO_2 is formed by the reaction of nitrogen oxide (NO) with oxygen (O_2) in the atmosphere (Eq. 12.11).

$$NO + O_2 + sunlight \rightarrow NO_2 + O· \tag{12.11}$$

Photodissociation of NO_2: NO_2 is broken down by sunlight into NO and an oxygen atom (O) (Eq. 12.12).

$$NO_2 + sunlight \rightarrow NO + O· \tag{12.12}$$

Formation of O_3: Oxygen atoms (O) react with O_2 to form O_3 (Eq. 12.13)

$$O· + O_2 + M \rightarrow O_3 + M^* \tag{12.13}$$

O_3 is capable of oxidising new molecules of NO. The reaction is much faster than oxidation with oxygen (Eq. 12.14):

$$NO + O_3 \rightarrow NO_2 + O_2 \tag{12.14}$$

As the concentration of NO_2 increases, the production of O_3 accelerates. Therefore, NO_2 is a secondary pollutant.

While tropospheric oxidants such as O_3 and OH· radicals can help clean the air by breaking down pollutants, they can also have harmful effects on human health and the environment. O_3 at ground level can cause respiratory problems.

Secondary CO_2

In the atmosphere, CO reacts with OH· radicals to form CO_2 through the following reaction (Eq. 12.15):

$$CO + OH· \rightarrow CO_2 + H· \tag{12.15}$$

The OH· radical is a highly reactive molecule that is naturally present in the atmosphere and acts as a natural 'cleaner' by reacting with and breaking down pollutants such as CO.

12.10 Oxidation Products from Hydrocarbons: Aldehydes and Carbon Dioxide

The oxidation of hydrocarbons in the atmosphere is a complex process that involves a series of chemical reactions. The first step is typically the reaction of hydrocarbons with OH· radicals, which are highly reactive and abundant in the atmosphere. This reaction produces an alkyl radical and H_2O. The alkyl radical can then react with oxygen to form peroxy radicals (RO_2·). RO_2 can react with NOx to form additional secondary pollutants such as O_3 and organic nitrates.

The most abundant hydrocarbon in the atmosphere is CH_4. The photochemical transformation of CH_4 occurs primarily through reactions with OH· radicals in the presence of sunlight and other atmospheric oxidants.

The oxidation of CH_4 in the atmosphere occurs through a complex series of reactions involving a variety of chemical species.

First, CH_4 reacts with the OH· radical, which is produced by the photolysis of O_3 and H_2O in the presence of sunlight (Eq. 12.16).

$$CH_4 + OH\cdot \rightarrow CH_3\cdot + H_2O \tag{12.16}$$

The resulting methyl radical (CH_3·) can then react with molecular oxygen (O_2) to form a peroxyl methyl radical ($CH_3O\text{-}O\cdot$) (Eq. 12.17):

$$CH_3\cdot + O_2 + M \rightarrow CH_3O\text{-}O\cdot + M^* \tag{12.17}$$

Peroxymethyl radical reacts with NO to give NO_2 and methoxy radical ($CH_3O\cdot$) (Eq. 12.18).

$$CH_3OO\cdot + NO \rightarrow NO_2 + CH_3O \tag{12.18}$$

Methoxy radical reacts with a molecule of oxygen to give HCHO and H-O-O· radical (Eq. 12.19).

$$CH_3O + O_2 \rightarrow HCHO + H\text{-}O\text{-}O\cdot \tag{12.19}$$

HCHO and other aldehydes present in the atmosphere, such as CH_3CHO and pro-pionaldehyde, are known to be respiratory irritants and can cause a range of health effects, such as eye, nose and throat irritation, coughing and difficulty breathing.

HCHO is oxidised to CO_2 and new molecules of O_3 as follows (Eqs. (12.20–(12.22).

$$HCHO + HO \rightarrow HCO\cdot + H_2O \tag{12.20}$$

$$HCO\cdot + O_2 \rightarrow H\text{-}O\text{-}O\cdot + CO \tag{12.21}$$

$$CO + 2O_2 + h\nu \rightarrow CO_2 + O_3 \tag{12.22}$$

The global process of CH_4 oxidation is resumed in Figure 12.2.

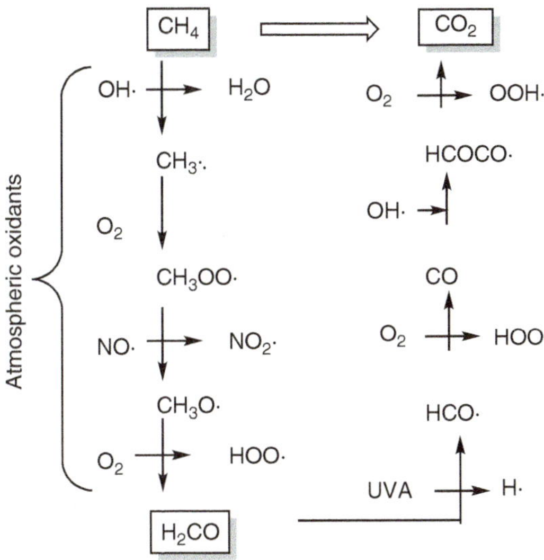

FIGURE 12.2   Oxidation of methane (CH$_4$) in the atmosphere

12.11   Peroxyacetyl Nitrates

Peroxyacetyl nitrates (PANs) are a group of secondary pollutants that are formed in the atmosphere through a series of chemical reactions involving VOCs and NO$_x$.

The formation of PANs begins with the emission of VOCs, such as C$_5$H$_8$ and other hydrocarbons, from natural and anthropogenic sources. These VOCs react with OH· radicals in the atmosphere, producing peroxy radicals (ROO·).

For example, CH$_3$CHO reacts with OH· and oxygen to give an acetyl radical (Eq. 12.23).

$$CH_3CHO + OH\cdot \rightarrow CH_3CO + H_2O \tag{12.23}$$

Acetyl radical reacts with oxygen to give peroxyacetyl radical (CH$_3$C(O)OO·) (Eq. 12.24), which is combined with NO$_x$, typically in the form of NO$_2$, to give PAN (Eq. 12.25).

$$CH_3CO + O_2 \rightarrow CH_3C(O)OO \tag{12.24}$$

$$CH_3C(O)OO\cdot + NO_2\cdot \rightleftarrows CH_3C(O)OONO_2 \tag{12.25}$$

The last reaction is reversible and constitutes an important reservoir of radicals.

PANs are relatively stable and can persist in the atmosphere for several hours, allowing them to be transported over long distances. Exposure to high levels of PANs can cause symptoms such as eye irritation, nose and throat irritation, coughing, shortness of breath and chest pain. Long-term exposure to PANs can cause chronic respiratory problems and may contribute to the development of asthma.

12.12 Secondary Sulfur and Nitrogen Pollutants: Sulfuric and Nitric Acids and Related Compounds

Sulfuric and Nitric Acids

H_2SO_4 and HNO_3 are two of the most common acids present in the atmosphere. H_2SO_4 is formed primarily from SO_2 emissions and HNO_3 is formed primarily from NO_x emissions. So, they are secondary pollutants.

Transformation of SO_2 to SO_3

In the atmosphere, SO_2 can react with oxygen and other oxidants, such as O_3 and H_2O_2, to form SO_3 to form SO_3 (Eq. 12.26) through a series of chemical reactions involving sunlight and other compounds. Once SO_3 is formed, it can react with H_2O to form H_2SO_4, which can then contribute to the formation of sulfate aerosols, tiny particles of sulfate compounds that are suspended in the atmosphere (Eq. 12.27):

$$SO_2 + Oxidants \rightarrow SO_3 \tag{12.26}$$

$$SO_2 + 1/2O_2 \rightarrow SO_3 + H_2O \rightarrow H_2SO_4 \tag{12.27}$$

Sulfate aerosols are typically removed from the atmosphere through precipitation, but they can also be transported long distances before being deposited on the ground or into bodies of water.

SO_x pollutants can have harmful effects on human health and the environment. When SO_2 and SO_3 are released into the air, they can react with other chemicals to form fine PM that can penetrate deep into the lungs and cause respiratory problems. SO_x emissions can also contribute to acid rain, which can damage crops, forests and aquatic ecosystems.

Similarly, NO reacts with other oxidants, such as O_3, to give NO_2 (Eq. 12.28), which, when combined with H_2O, forms HNO_3 (Eq. 12.29).

$$NO + oxidants \rightarrow NO_2 \tag{12.28}$$

$$NO_2 + H_2O \rightarrow HNO_3 \tag{12.29}$$

The tiny particles of H_2SO_4 and HNO_3 derivatives suspended in the air are named sulfate and nitrate aerosols. Typically, fine aerosol sulfate and nitrate exist in the form of ammonium salts.

Secondary Pollutants from Hydrogen Sulfide

H_2S released into the atmosphere undergoes various chemical reactions and processes that determine its fate. The primary mechanisms involved in the transformation and removal of H_2S in the atmosphere are as follows:

a. **Chemical oxidation:** H_2S can react with atmospheric oxidants such as OH·
   radicals, O_3 and NO_2. These reactions result in the formation of SO_2 and
   water (H_2O). The reaction with OH· radicals is the most dominant process for
   removing H_2S from the atmosphere (Eq. 12.30).

$$H_2S + OH \rightarrow SO_2 + H_2O \qquad (12.30)$$

b. **Deposition:** As with other acidic compounds, H_2S can be deposited from
   the atmosphere onto surfaces such as vegetation, soil or bodies of water. This
   occurs through dry deposition, where H_2S particles settle directly onto surfaces or through wet deposition, where H_2S is dissolved in rainwater and falls
   to the ground as acid rain.

Secondary Pollutants from Ammonia

The fate of NH_3 pollution depends on several factors, including its release source,
environmental conditions and the presence of other compounds in the surrounding
environment. It involves a combination of physical, chemical and biological
processes that transform and remove NH_3 from the environment. The primary
pathways through which NH_3 pollution can be transformed or removed from the
environment are:

a. **Atmospheric reactions:** NH_3 released into the atmosphere can undergo various chemical reactions. It can react with acidic compounds such as H_2SO_4 or
   HNO_3 to form ammonium sulfate $(NH_4)_2SO_4$ or ammonium nitrate (NH_4NO_3),
   respectively. These reactions contribute to the formation of PM, which can be
   a component of air pollution.

b. **Deposition:** NH_3 can be deposited from the atmosphere onto land or water
   surfaces through a process known as dry or wet deposition. Dry deposition
   occurs when NH_3 gas directly settles onto surfaces, while wet deposition
   involves NH_3 being dissolved in rain or other forms of precipitation and then
   deposited onto the Earth's surface.

c. **Soil absorption:** NH_3 can be absorbed by soils, particularly those with acidic
   properties. In the soil, NH_3 can undergo nitrification, a process where it is converted to nitrate ions (NO_3-) by nitrifying bacteria. These NO_3- can be taken
   up by plants as a nutrient source or can further react in the environment.

d. **Water bodies:** When NH_3 is deposited into water bodies, it can have various
   effects depending on the concentration and pH of the water. In aquatic environments, NH_3 can be toxic to aquatic organisms, particularly fish and other
   species sensitive to NH_3. However, some bacteria present in water can convert
   NH_3 to nitrate through a process called nitrification, which reduces its toxicity.

e. **Volatilisation:** NH_3 can also undergo volatilisation, where it returns to the
   atmosphere from water or soil surfaces. This process occurs more readily in
   alkaline conditions. Volatilised NH_3 can then be transported over long distances
   and contribute to the formation of secondary pollutants, such as fine PM.

f. **Biological uptake:** Plants and algae can take up NH_3 from the soil or water,
   utilising it as a nutrient for growth. This process helps reduce the NH_3
   concentration in the environment and incorporates it into the biological cycle.

12.13 Acid Rain

Acid rain is a type of precipitation that is significantly more acidic than normal. The pH of normal rainwater is slightly acidic, typically ranging from 5.0 to 5.5, due to the natural presence of CO_2 in the atmosphere, which combines with water to form carbonic acid. However, this acidity is not considered harmful to the environment. In contrast, acid rain refers to rainwater that has a pH lower than 5.0, typically ranging from 4.2 to 4.4. This is caused by the presence of pollutants such as SO_2 and NO_x in the atmosphere, which react with water to form H_2SO_4 and HNO_3, respectively, as explained before. Acid rain can be deposited on the ground and other surfaces through two processes: dry deposition and wet deposition.

1. **Dry deposition** occurs when acidic particles and gases in the atmosphere come into contact with surfaces such as soil, plants and buildings. The acidic particles and gases can then stick to the surface, causing damage over time.

2. **Wet deposition** occurs when acidic particles and gases are dissolved in rain, snow, sleet or other forms of precipitation. The acidic precipitation can then fall onto the ground, bodies of water and other surfaces, causing damage and harm to ecosystems (Figure 12.3).

The main consequences of acid rain are:

1. **Forests:** Acid rain can damage trees and other plants, making them more vulnerable to disease and insect infestations.

2. **Lakes and rivers:** Acid rain can make bodies of water more acidic, which can harm fish, aquatic plants and other organisms that live in the water.

3. **Soil:** Acid rain can leach important minerals and nutrients from soil, making it less fertile and less able to support plant life.

4. **Buildings and monuments:** Acid rain can corrode building materials such as limestone and marble, causing damage to historical monuments and buildings.

5. **Human health:** Acid rain can release toxic metals such as lead and mercury into the air, which can be harmful to human health when inhaled.

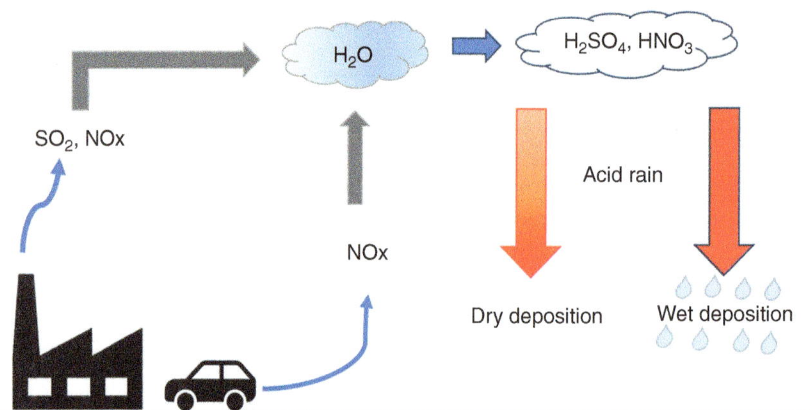

FIGURE 12.3 Acid rain

Many zones of the world have been affected by acid rain, particularly in regions with high levels of industrialisation and emissions of these pollutants. Some of the most affected regions include:

a. **Europe:** Acid rain has been a significant problem in many European countries, particularly in the 1980s and 1990s. Countries such as Germany, the United Kingdom, Poland and Scandinavian countries have experienced significant damage to forests and lakes due to acid rain.

b. **North America:** The eastern United States and eastern Canada have also been affected by acid rain, particularly in the Appalachian Mountains and the Great Lakes region. Acid rain has caused damage to forests, lakes and other ecosystems in these areas.

c. **Asia:** Many parts of Asia, particularly China and India, have experienced significant problems with acid rain due to their high levels of industrialisation and emissions of pollutants.

d. **Africa:** Acid rain has been reported in some parts of Africa, including South Africa, due to emissions from power plants and industrial activities.

Topics of Interest: The Black Triangle

The 'Black Triangle' is a term used to describe a heavily industrialised and polluted region in Central Europe, primarily located in parts of Poland, the Czech Republic and Germany. It earned its name due to the extensive coal mining and heavy industrial activities that took place in the area during the 20th century, which led to widespread environmental degradation and pollution.

During the 1970s and 1980s, acid rain was affecting this region so badly that it was given its name (Figure 12.4). In areas falling under the black triangle, environmental damages produced by acid rain caused severe harm to forests, lakes, rivers and soils, leading to a hard

FIGURE 12.4   The black triangle in central Europe

forest decline and loss of biodiversity. Acid rain corroded buildings, monuments, bridges and infrastructure made of limestone, marble and metal, such as railroad tracks and concrete, leading to deterioration and structural damage. In addition, pollution contributed to improved respiratory problems and cardiovascular diseases in the area.

Efforts to mitigate acid rain in the Black Triangle and other regions have included the implementation of emission controls, such as installing pollution control technologies in industrial facilities, transitioning to cleaner energy sources and international agreements aimed at reducing sulfur and nitrogen emissions. These measures have led to enhanced air quality, although challenges remain in addressing historical pollution legacies.

12.14 Secondary Aerosols

Secondary aerosols refer to airborne particles that form in the atmosphere through chemical reactions involving precursor gases emitted by various sources. Unlike primary aerosols, which are directly emitted into the atmosphere (such as dust, sea salt or soot), secondary aerosols are formed from precursor gases such as SO_2, NO_x, VOCs, NH_3 and other gases that are transformed into secondary pollutants as described before. These gases undergo complex chemical reactions in the atmosphere, leading to the formation of new particles or the growth of existing ones, through processes like gas-to-particle, nucleation, condensation coagulation and evaporation (see Figure 12.5).

1. **Nucleation:**
   Aerosol nucleation is the process by which tiny clusters of gas molecules come together to form the initial building blocks of aerosol particles. This process is fundamental to the formation of new aerosols in the atmosphere and plays a crucial role in determining the number, concentration and size distribution of aerosols. Nucleation occurs through two main mechanisms: homogeneous nucleation and heterogeneous nucleation.

   - **Homogeneous nucleation:** In regions with high concentrations of certain precursor gases (e.g., SO_2, NH_3), new aerosol particles can form through homogeneous nucleation. This process involves the spontaneous clustering of gas molecules to form small clusters or nuclei. These nuclei serve as the initial building blocks for aerosol particles.

   - **Heterogeneous nucleation:** In heterogeneous nucleation, aerosol formation occurs on pre-existing particles or surfaces, such as dust, soot or other aerosols. These surfaces provide sites for gas molecules to accumulate and form new aerosol particles more readily than in the surrounding air.

2. **Condensation:**
   Aerosol condensation is the process by which gas-phase molecules in the atmosphere adhere to the surfaces of existing aerosol particles, leading to the growth of these particles in size concentration and distribution. In this step, atmospheric aerosol particles typically show sizes between 10 and 100 nm in diameter and represent the smallest end of the accumulation mode of atmospheric aerosols.

   Aitken mode particles can originate from various sources, including emissions from combustion processes (e.g., vehicle exhaust, biomass burning),

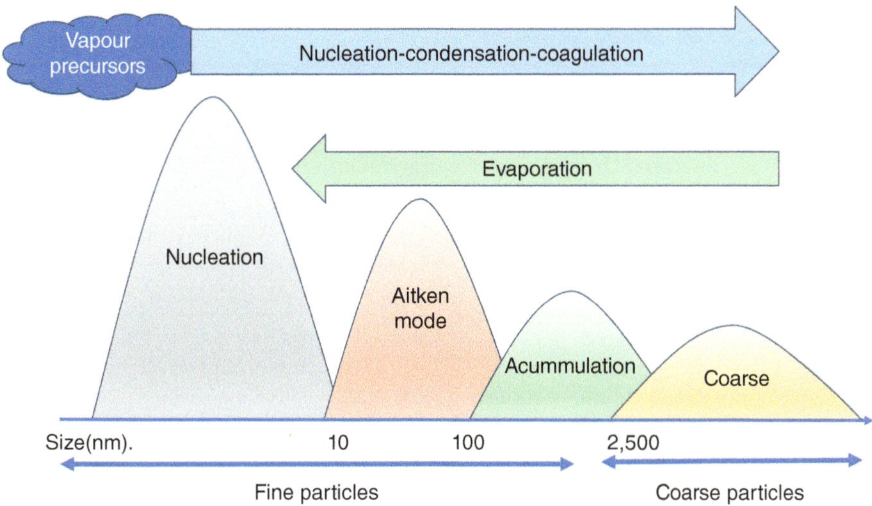

FIGURE 12.5   Like gas-to-particle, nucleation, condensation, coagulation and evaporation

biogenic emissions (e.g., from plants), sea spray and nucleation events triggered by atmospheric gases such as SO_2, NH_3 and VOCs.

3. **Coagulation:**
   - Coagulation refers to the process by which aerosol particles collide and merge with each other to form larger particles. This occurs as a result of Brownian motion and turbulent mixing in the atmosphere, causing aerosol particles to collide and aggregate. Coagulation is more significant for larger aerosol particles and can lead to the formation of coarse-mode particles through the aggregation of smaller particles.

4. **Evaporation:**
   - In addition to growth processes, aerosol particles can also undergo evaporation, where volatile components evaporate from the particle surface back into the gas phase. Evaporation can occur due to changes in temperature, humidity or exposure to other atmospheric constituents. This process can affect the size distribution and composition of aerosols in the atmosphere.

The formation and properties of secondary aerosols have significant implications for air quality, visibility, climate and human health. They can contribute to haze, smog and PM pollution, affecting respiratory health and visibility (Table 12.2).

12.15   Photochemical Smog: Evolution During the Day

Photochemical smog is a type of air pollution that is formed by the interaction of sunlight with certain chemical pollutants and typically occurs in urban areas with high levels of automobile traffic and industrial activity. The primary components of photochemical smog include:

TABLE 12.2 **Secondary aerosols**

Type	Source	Composition	Size range
Sulfate aerosols	Oxidation of sulfur dioxide (SO_2) emissions	Sulfuric acid (H_2SO_4), ammonium sulfate ((NH_4)$_2SO_4$)	Fine (<1000 nm)
Nitrate aerosols	Oxidation of nitrogen oxides (NO_x) emissions	Nitric acid (HNO_3), ammonium nitrate ((NH_4)NO_3)	Fine (<1000 nm)
Ammonium aerosols	Ammonia (NH_3) emissions, neutralisation reactions	Ammonium sulfate ((NH_4)$_2SO_4$), ammonium nitrate ((NH_4)NO_3)	Fine (<1000 nm)
Organic aerosols	VOCs emissions, biogenic emissions	Organic compounds (e.g., hydrocarbons, alcohols)	Variable

1. **Nitrogen oxides (NO_x):** These are a group of highly reactive gases, including nitric oxide (NO) and NO_2, that are produced during combustion processes. NO_x reacts with other pollutants to form O_3 and other photochemical oxidants.

2. **VOCs:** These are a group of organic chemicals that have high vapour pressure at room temperature. They are emitted by sources such as motor vehicles, industrial processes and solvents. VOCs can react with NO_x in the presence of sunlight to form O_3 and other photochemical oxidants.

3. **Ozone (O_3):** O_3 is a highly reactive gas that is formed by the reaction of NO_x and VOCs in the presence of sunlight. It is a major component of photochemical smog and can cause respiratory problems and other health effects.

4. **Peroxyacetyl nitrate (PAN):** This is a secondary pollutant that is formed by the reaction of NO_x and VOCs. It is a major component of photochemical smog and can cause eye and respiratory irritation.

5. **Aldehydes:** These are a group of organic chemicals that are emitted by sources such as motor vehicles, industrial processes and solvents. They can cause eye and respiratory irritation, as well as other health effects.

6. **Particulate matter (PM):** These are tiny particles of solid or liquid matter that are suspended in the air. They can be emitted directly from sources such as motor vehicles and industrial processes or formed by the reaction of other pollutants in the atmosphere. PM can cause respiratory and cardiovascular problems, as well as other health effects.

It typically forms on hot and sunny days and its formation and evolution during the day can be divided into several stages.

a. **Morning:** In the morning, the air is relatively clean and the concentration of NO_x and VOCs is low. As the sun rises and the temperature increases, these pollutants begin to evaporate from sources such as cars, factories and trees.

b. **Late morning/early afternoon:** As the day progresses, the concentration of NO_x and VOCs continues to rise and the sunlight becomes more intense. These pollutants begin to react with each other to form a mixture of O_3, NO_2 and other secondary pollutants.

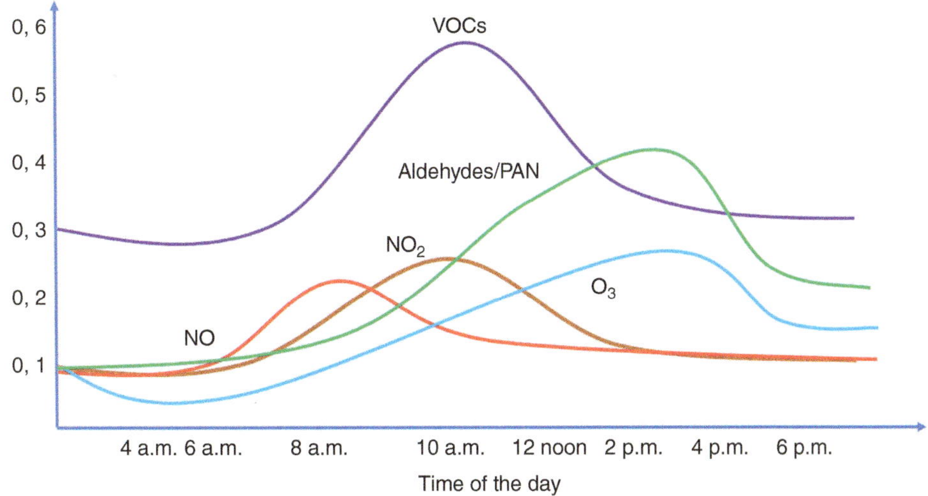

FIGURE 12.6 Evolution of photochemical smog components along the day

c. **Afternoon:** By the afternoon, the concentration of pollutants has reached its peak and the smog is often visible as a brownish haze. This is the most dangerous time for people with respiratory issues, as the pollutants can cause lung irritation and other health problems.

d. **Evening:** As the sun begins to set and the temperature cools, the reaction between NO_x and VOCs slows down and the concentration of pollutants begins to decrease. However, the smog can still linger in the air, particularly in urban areas with high levels of traffic and industry.

Figure 12.6 shows the typical evolution of the different components of photochemical smog along the day.

12.16 Thermal Inversion

Thermal inversion, also known as temperature inversion, is a meteorological phenomenon where the typical temperature gradient of the troposphere is inverted. Normally, air temperature decreases with altitude, allowing warm air near the surface to rise and cool as it ascends. However, during a thermal inversion, the temperature increases with altitude over a specific layer, trapping a cooler air layer near the surface beneath a warmer layer above. This situation can significantly impact air quality, exacerbating pollution-related problems, especially in urban areas and altering weather patterns (Figure 12.7).

Thermal inversions can occur under various conditions:

1. **Radiative cooling of the surface:** During clear nights, the ground loses heat rapidly through radiation, cooling down the surface air layer. If a warmer air layer moves in above this cooler air, a temperature inversion occurs.

2. **Topographic effects:** In valleys and basins, cold air can get trapped under a layer of warmer air. The topography of such areas can prevent the mixing of air layers, sustaining the inversion.

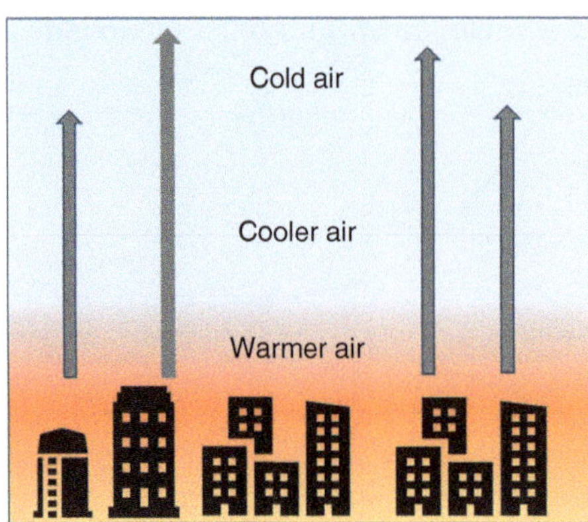

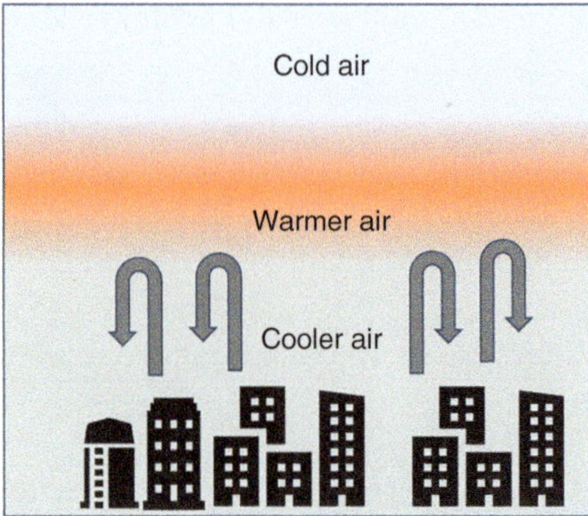

FIGURE 12.7   Normal conditions versus thermal inversion

3. **Frontal systems:** Warm air can be forced over a layer of cooler air during the movement of weather fronts, particularly warm fronts, creating an inversion.

4. **Subsidence:** In high-pressure systems, air descending from higher altitudes warms adiabatically (compression heating), creating a warmer air layer over cooler surface air.

When this phenomenon takes place, some effects are produced such as:

1. **Air pollution:** Thermal inversions trap pollutants close to the ground. Pollutants from vehicles, industry and other sources accumulate near the ground because they cannot rise and disperse vertically. This can lead to high concentrations of smog and other pollutants. Cities surrounded by mountains or valleys are particularly susceptible to this effect, showing high levels of PM, NO_x, SO_2 and ground-level O_3 that can exacerbate respiratory conditions like asthma, bronchitis and other lung diseases. On the other hand, long-term exposure to polluted air during inversion periods can lead to chronic health conditions, cardiovascular diseases and premature death.

2. **Fog formation:** Inversions can trap moisture close to the surface, leading to the formation of fog or exacerbating existing fog conditions.

3. **Temperature effects**: Inversion layers can create a marked difference in temperature between the surface and higher altitudes, affecting local climate conditions and weather patterns.

4. **Impact on aviation:** Temperature inversions can cause turbulence and wind shear, posing risks to aviation during takeoff and landing.

12.17   Permanency of Pollutants in the Atmosphere

The persistence of pollutants in the atmosphere depends on several factors. A key measure to evaluate the permanence of pollutants in the atmosphere is their half-life. In general, the half-life of a pollutant is the amount of time it takes for half of the initial amount of the pollutant to degrade, break down into other compounds or be removed from the environment through natural processes such as decay or chemical reactions. The half-life can vary depending on the specific pollutant and environmental conditions. Understanding the half-life of pollutants is important for predicting their behaviour in the environment and developing effective strategies for pollution control and remediation.

Pollutants in the atmosphere can have varying half-lives depending on their chemical properties and the environment in which they exist. Here are some examples:

1. **Carbon monoxide (CO):** The half-life of CO in the atmosphere is about 1–2 months.

2. **Sulfur dioxide (SO_2):** The half-life of SO_2 in the atmosphere can vary from a few minutes to several days, depending on the concentration and environmental factors such as temperature, humidity and wind speed.

3. **Nitrogen oxides (NO_x):** The half-life of NO_x in the atmosphere can range from hours to several days, depending on the specific compound.

4. **VOCs:** The half-life of VOCs in the atmosphere can vary widely depending on the specific compound, ranging from a few hours to several months.

5. **Particulate matter (PM):** The half-life of PM in the atmosphere can range from a few hours to several days, depending on the size of the particles and environmental conditions.

The longer the average lifespan of a pollutant in the atmosphere, the greater its environmental impact will be and the larger the area or region where its effects can be observed.

Figure 12.8 shows an estimation of the residence time of some well-known pollutants in the atmosphere and the scale of the problem.

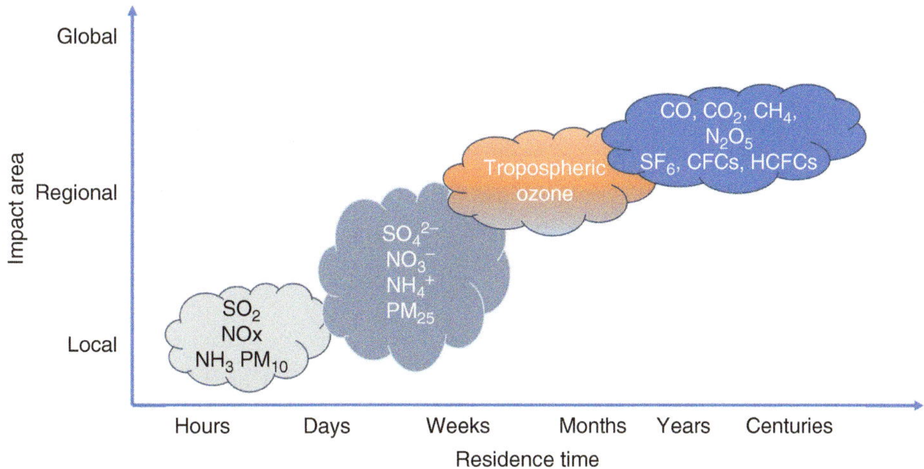

FIGURE 12.8   Residence time of some pollutants in the atmosphere

12.18 Indoor Pollution

Indoor pollution is a sort of tropospheric pollution caused by the presence of harmful pollutants inside buildings or enclosed spaces that can have negative impacts on the health and well-being of the occupants. In general, indoor air can be more polluted than outdoor air.

In modern cities, indoor pollution occurs in tightly sealed buildings where air exchange is limited. On the other hand, in many regions of the world, indoor air pollution is mainly caused by cooking methods. This is because traditional cooking methods involve using open fires or inefficient stoves that burn solid fuels such as wood, charcoal or dung. The smoke and fumes generated by these fuels contain a range of harmful pollutants, including PM, CO, NOx and VOCs. These pollutants can have serious health effects, particularly for women and children, who spend the most time indoors and are exposed to the highest levels of pollution.

According to the WHO, indoor air pollution was responsible for an estimated 3.2 million deaths per year in 2020, including about 237,000 deaths of children under the age of 5. The combined effects of indoor and outdoor pollution are associated with 6.7 million premature deaths annually[3]. The majority of these deaths occur in low- and middle-income countries. This is because indoor sources of pollution can accumulate over time without proper ventilation, leading to higher concentrations of pollutants.

Common sources of indoor pollutants include combustion products from:

1. Cooking methods mentioned above

2. Gas appliances and heating equipment may cause indoor pollution, which can be a serious health hazard if they are not properly vented or maintained. Gas appliances, such as heaters, stoves and water heaters, produce combustion products when they burn fuel to produce heat or hot water. These combustion products can include CO, NO_x and PM.

3. Tobacco smoke pollution occurs when someone smokes cigarettes, cigars or pipes inside a building or enclosed space. Tobacco smoke is a complex mixture of more than 7,000 chemicals, including both gases and particles. Here are some of the most common chemicals found in tobacco smoke:

    a. Nicotine is a highly addictive stimulant that increases heart rate and blood pressure.

    b. Tar is a mixture of particles that contains numerous harmful chemicals, such as polycyclic aromatic hydrocarbons (PAHs), benzene and HCHO.

    c. CO is a toxic gas that reduces the amount of oxygen that the body can absorb.

    d. Nitrogen oxides can contribute to respiratory irritation and damage.

    e. NH_3 is a corrosive gas that can irritate the respiratory system.

    f. Hydrogen cyanide is a poisonous gas that can cause headaches, nausea and dizziness.

    g. Acrolein is a strong irritant that can damage the lungs and cause respiratory problems.

    h. CH_3CHO is a toxic chemical that can cause cancer and contribute to cardio-vascular disease.

[3] https://www.who.int/data/gho/data/themes/air-pollution

4. Indoor VOCs come from household products, building materials or furniture. Some common sources of VOCs in indoor environments include paints, solvents, cleaning products, air fresheners, adhesives, carpets and furniture.

5. Building materials and furnishings, cleaning products and outdoor pollutants that enter through ventilation systems.

Indoor pollution can cause a range of health problems, including respiratory issues, headaches, allergies and other symptoms. Long-term exposure to indoor pollutants has been linked to more serious health problems, such as cancer and heart disease.

12.19 Sick Building Syndrome

Sick building syndrome (SBS)[4] is a term used to describe a sort of indoor pollution in which people experience a range of health problems and discomfort when they spend time in a particular building. The main symptoms associated with SBS are summarised in Table 12.3 but they may vary from person to person.

The exact causes of SBS are not completely understood, but they may be related to a variety of factors and can include the following:

1. **Poor indoor air quality:** This is the most common cause of SBS. It can be caused by inadequate ventilation, poor maintenance of heating, ventilation and air conditioning (HVAC) systems and the presence of contaminants such as mould, bacteria and VOCs from building materials, cleaning agents and office equipment.

TABLE 12.3 **Main symptoms associated with sick building syndrome (SBS)**

Symptom	Description
Headache	Persistent or recurrent headaches, often dull or throbbing in nature.
Fatigue	Constant tiredness or lack of energy.
Dizziness	Feeling lightheaded or unstable.
Eye irritation	Redness, itching or dryness of the eyes.
Nasal congestion	Blocked or stuffy nose.
Sore throat	Discomfort or pain in the throat.
Coughing	Frequent or persistent cough.
Skin rashes	Unexplained skin irritations or rashes.
Difficulty breathing	Breathlessness or shortness of breath.
Nausea	Feeling of queasiness or an upset stomach.
Concentration issues	Difficulty focusing or concentrating on tasks.
Memory problems	Trouble remembering things or recalling information.

[4] Quah, S.R. (2017). *International Encyclopedia of Public Health Reference Work*. 2: Academic Press.

2. **Chemical contaminants:** These can include HCHO, pesticides and VOCs from building materials, cleaning agents and office equipment.

3. **Biological contaminants:** These can include mould, bacteria and viruses.

4. **Poor lighting**: Inadequate or harsh lighting can cause eye strain and headaches.

5. **Inadequate acoustics:** Too much noise or poor sound quality can cause stress and fatigue.

6. **Psychological factors:** Stressful work environments, lack of control and poor morale can contribute to SBS.

7. **Personal factors:** Allergies, asthma and other underlying health conditions can increase the risk of developing SBS.

Certain types of buildings may be more susceptible to SBS due to their construction, ventilation or other factors. Some of the buildings that may be more likely to experience SBS include:

a. **Office buildings:** Because people spend a significant amount of time in offices and often in closed spaces, they may be more likely to experience symptoms of SBS, such as headaches, fatigue and dry eyes.

b. **Schools:** Children and teachers may be more vulnerable to SBS due to the large number of people in the building and the potential for poor ventilation.

c. **Hospitals:** Healthcare facilities may be more susceptible to SBS due to the high volume of people, the use of chemicals and the need for controlled temperatures and humidity levels.

d. **Residential buildings:** Houses and apartments if they have poor ventilation, mould or mildew or use chemicals in cleaning or construction.

e. **Industrial buildings:** Factory or warehouse workers may be at risk of SBS due to poor air quality, chemical exposure and other occupational hazards.

Some steps that can help avoid SBS are

1. **Improve ventilation:** Proper ventilation can help circulate fresh air throughout the building, reducing the concentration of indoor air pollutants that can cause SBS symptoms.

2. **Keep humidity levels in check:** High humidity levels can promote the growth of mould and other microorganisms, which can cause SBS. Maintaining indoor humidity levels between 30% and 60% can help prevent this.

3. **Use indoor plants:** Plants can help remove indoor air pollutants and increase oxygen levels, which can improve indoor air quality and reduce the risk of SBS.

4. **Use natural cleaning products:** Many conventional cleaning products contain harsh chemicals that can release toxic fumes into the air. Using natural cleaning products can help reduce the risk of SBS.

5. **Avoid smoking:** Smoking indoors can release a variety of harmful chemicals into the air, which can cause SBS symptoms.

6. **Regular maintenance:** Regular maintenance of HVAC systems, air filters and other building systems can help prevent the buildup of indoor air pollutants and reduce the risk of SBS.

CHAPTER 13

Methods for Controlling Tropospheric Pollution

13.1 Introduction

The management of atmospheric pollution aims to eliminate or reduce to acceptable levels of gases, suspended particles, biological agents, etc., in the atmosphere that can cause adverse effects on the population in particular, and the environment in general. Emissions are produced by:

a. Static sources of pollution which are stationary sources that emit pollution from a fixed location. Examples of static sources include factories, power plants and industrial facilities.

b. Mobile sources of pollutants which are sources that move and emit pollution while in motion. Examples of mobile sources include cars, trucks, aeroplanes and ships.

The level of pollution they produce can vary depending on the type of source and its location. For example, a large industrial plant may produce more pollution than a single car, but a densely populated city with many cars can have a higher overall level of pollution than a rural area with a single large factor.

To achieve this, the application of technologies is necessary, such as:

1. **Treatment of emissions**
2. **Development of new methods to prevent emissions**

13.2 Treatment of Emissions Static Sources: Aerosols and Gases

Static sources such as factories, power plants and industrial facilities produce two generic types of emissions: aerosols and gases.

Understanding the Chemistry of the Environment, First Edition. Francisco G. Calvo-Flores.
© 2025 John Wiley & Sons Ltd. Published 2025 by John Wiley & Sons Ltd.

Aerosols

Aerosols are tiny particles suspended in the air and they can be both natural and man-made. Man-made aerosols such as smoke, soot and various chemicals are considered air pollutants. These aerosols can come from a variety of industrial activities, including manufacturing, power generation and transportation.

One common source of aerosol pollutants from industry is particulate matter. This can come from sources such as the burning of fossil fuels in power plants, manufacturing processes that create dust and debris and transportation activities that kick up dust and other particles from roads and construction sites.

There are several ways to eliminate particulate matter in industrial effluents and the choice of method will depend on the specific nature and characteristics of the particulate matter and the industrial process involved. Following are some of the most popular tools and techniques for removing particulate matter:

1. **Filters and baghouses:** These are large fabric filters that capture particulate matter by passing the effluent through a series of bags made of woven or felted fabrics (Figure 13.1). Fabric filters are the most commonly used type of dust filter and work by capturing dust particles on the surface of a fabric filter bag. Baghouse dust collectors are large-scale air-material separators that utilise fabric filters to remove particulates from industrial and manufacturing operations. These devices are designed to prevent dust and other solid particles from entering the workplace or being released into the environment, thereby reducing

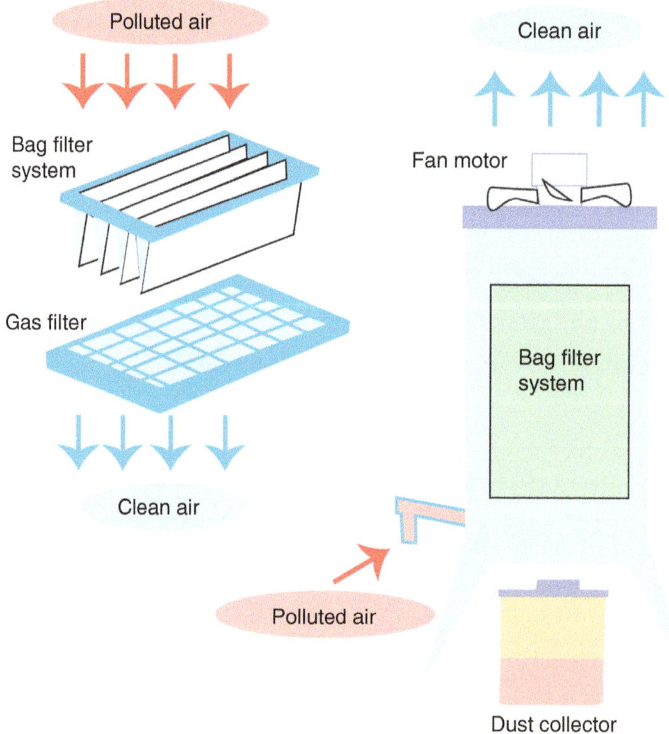

FIGURE 13.1 Dust filters and baghouse collectors

air pollution. Essentially, baghouses function as industrial-grade fabric filter systems and are critical components of air pollution control strategies.

2. **Sedimentation chambers:** Effluent passes through the device, allowing the particles to settle out and accumulate at the bottom of the chamber. The gas flow enters a chamber, where the speed decreases and particulate matter falls by gravity to the collector's hoppers. It is applicable to particles larger than or equal to 50 μm. (Figure 13.2.)

3. **Electrostatic precipitators:** These are devices that use an electric field to charge the particulate matter, causing it to be attracted to the charged plates or surfaces, where it is collected and removed (Figure 13.3).

4. **Cyclones:** Cyclone separators use centrifugal force to separate particulate matter from the effluent by directing it through a spiral path, which causes the heavier particles to be flung outward and collected in a separate chamber. There are two types: dry and wet cyclones.

   Dry cyclones (Figure 13.4) are used for the removal of particulate matter from a gas stream using centrifugal force to separate particles from the gas

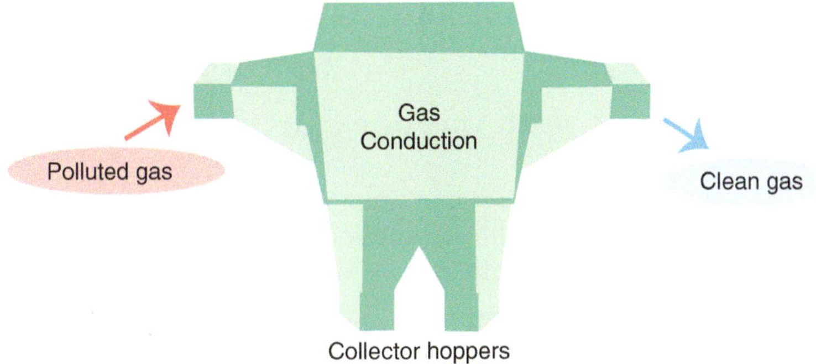

FIGURE 13.2 Sedimentation chamber

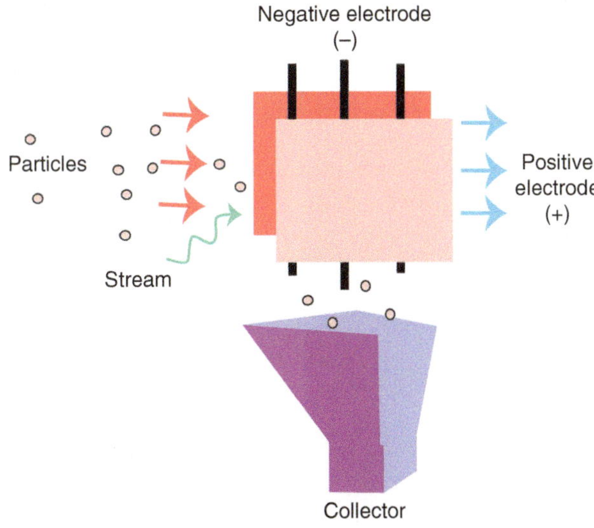

FIGURE 13.3 Electrostatic precipitator

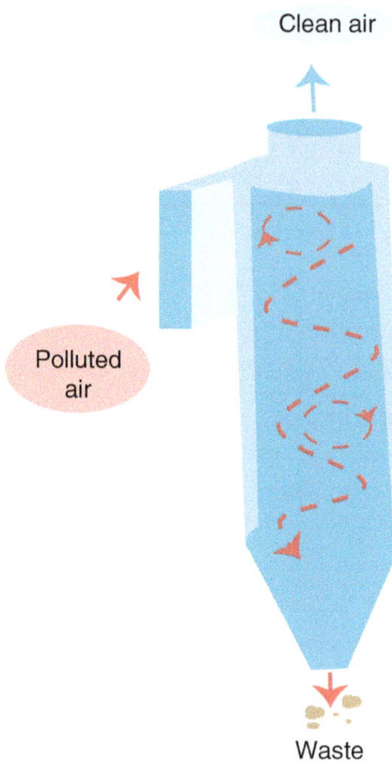

FIGURE 13.4 Dry cyclone

stream. As the gas stream enters the cyclone, it is forced to spin around the inside of the cyclone. This spinning motion creates a centrifugal force that pushes the heavier particles to the outside of the cyclone, where they are collected in a hopper or bin for disposal.

Wet cyclones (Figure 13.5), on the other hand, use liquid droplets to capture particles from a gas stream, combining the centrifugal force with the drag effect of the liquid. Liquid droplets are introduced into the cyclone and they capture the particles as they pass through the cyclone. The liquid droplets then exit the cyclone with the captured particles and are sent to a downstream system for separation.

5. **Wet scrubbers:** These devices use a liquid spray to capture and remove particulate matter from the effluent through wetting (Figure 13.6).

The use of wet scrubbers with chemical reactions is an effective way to control air pollution, as it removes harmful pollutants from the exhaust gas stream before they are released into the atmosphere.

The most effective method for eliminating particulate matter will depend on the specific characteristics of the industrial effluent and the particulate matter present. It may be necessary to use a combination of these methods to achieve optimal results.

Gas Emissions

There are several methods for treating gas emissions from static sources. The appropriate method will depend on the type of gas being emitted and the quantity of emissions. The most appropriate treatment method will depend on the specific

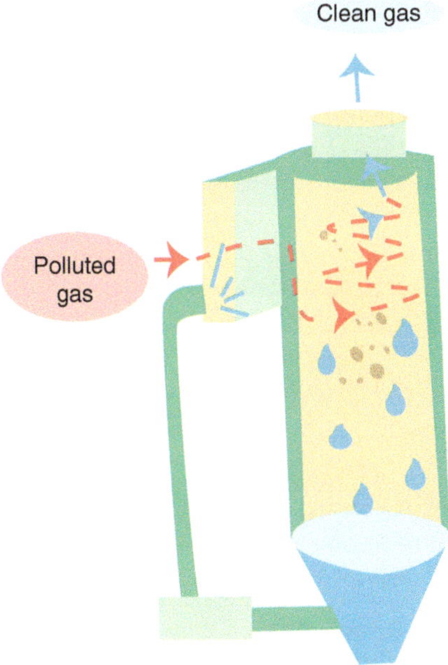

FIGURE 13.5   Wet cyclone

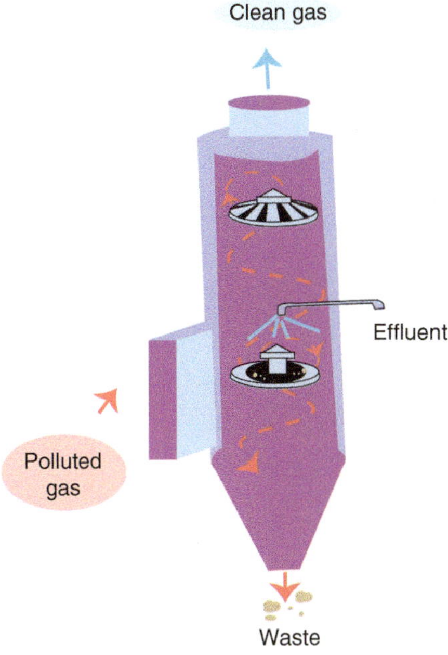

FIGURE 13.6   Wet scrubber

type of gas emissions and the regulations in place in the region where the factory is located. Here are a few methods commonly used:

1. **Gas condensers** are devices used for emission treatment in various industrial processes, such as chemical manufacturing, oil refining and power generation.

The main purpose of a gas condenser is to cool down and condense the vaporised gas or liquid contaminants in the exhaust gas stream, allowing them to be separated and removed from the gas stream. Gas condensers (Figure 13.7) are commonly used in air pollution control systems to reduce the emissions of volatile organic compounds (VOCs), hazardous air pollutants (HAPs) and other toxic gases from industrial processes By condensing the gases, the condenser can capture and recover valuable chemicals, reduce the emission of harmful pollutants and prevent damage to downstream equipment such as scrubbers and filters. There are different types of gas condensers available, including air-cooled condensers, water (H_2O)-cooled condensers and hybrid condensers that combine both cooling methods. The selection of a particular type of gas condenser depends on the specific process conditions, such as the gas flow rate, temperature and composition, as well as the desired level of efficiency and cost-effectiveness.

2. **Venturi gas purifier:** A venturi gas purifier is a type of gas cleaning device that uses the venturi effect to remove impurities from a gas stream (Figure 13.8). The venturi effect is a physical phenomenon where the velocity of a fluid increases

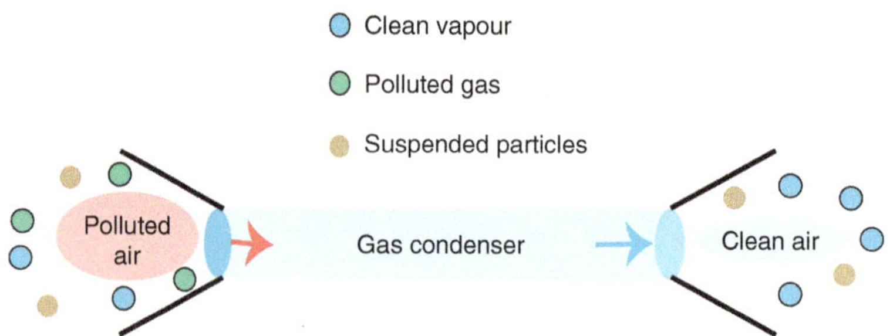

FIGURE 13.7 Gas condenser

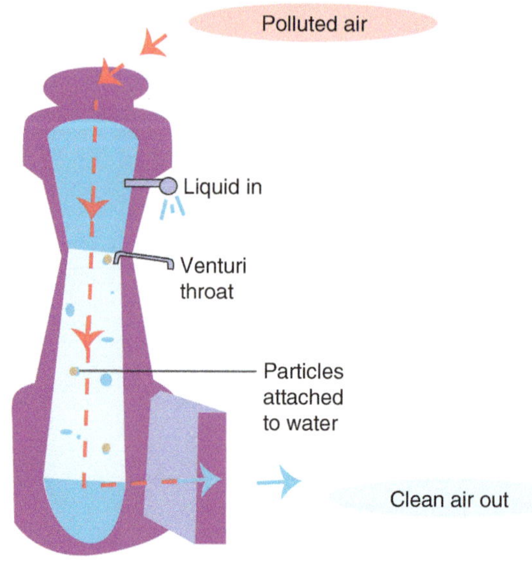

FIGURE 13.8 Venturi gas purifier

as it flows through a narrow section of a pipe and the pressure decreases. In a venturi gas purifier, the gas stream is forced through a constricted section of a pipe called the venturi throat. This creates a region of low pressure in the throat, which causes the gas to accelerate and mix with a liquid or solid material (usually a scrubbing liquid or solid particles). This mixture of gas and liquid/solid particles passes through a cyclone separator, where the liquid/solid particles are separated from the gas stream.

The purified gas stream then exits the cyclone separator, while the separated liquid/solid particles are collected in a separate chamber or container. Venturi gas purifiers are commonly used in industries such as chemical processing, power generation and environmental remediation to remove pollutants such as sulfur dioxide (SO_2), nitrogen oxides (NO_x) and particulate matter from exhaust gases.

3. **Air scrubbers:** This involves using chemicals or H_2O to remove pollutants from the air. This method is effective for removing acidic gases such as SO_2 and NO_x. Desulfurisation with calcium oxide, also known as lime, is a widely used method to remove sulfur from flue gas emissions in industrial processes such as power generation and cement production. The process involves injecting calcium oxide into the flue gas stream, where it reacts with SO_2 to form calcium sulfite ($CaSO_3$), which can then be further oxidised to calcium sulfate ($CaSO_4$) by introducing air or oxygen (Figure 13.9). The reaction between calcium oxide and SO_2 can be represented by the following equations (Eqs. (13.1) and (13.2)):

$$CaO + SO_2 \rightarrow CaSO_3 \tag{13.1}$$

$$2CaSO_3 + O_2 \rightarrow 2CaSO_4 \tag{13.2}$$

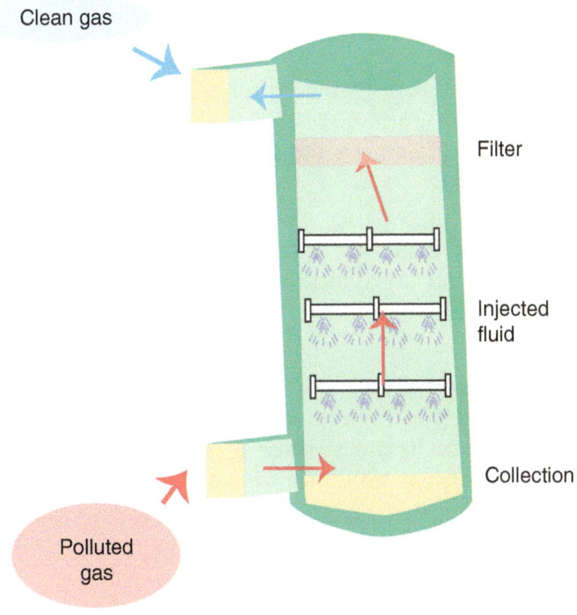

FIGURE 13.9   Air scrubber

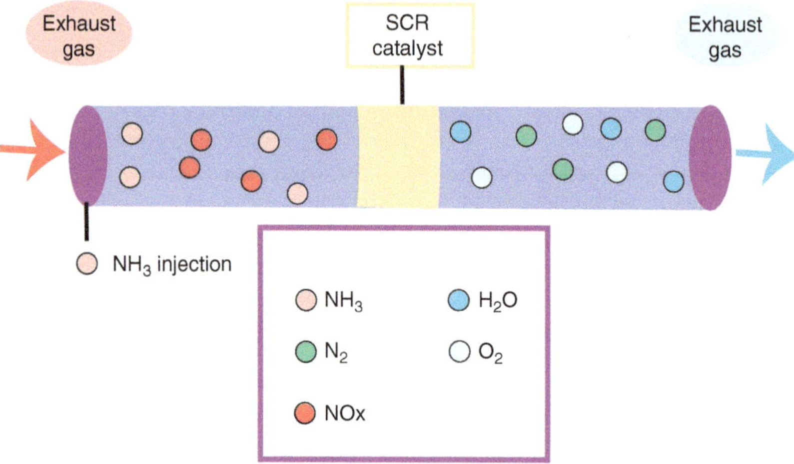

FIGURE 13.10   SCR scrubber

The resulting $CaSO_4$ is a solid byproduct that can be collected and disposed of.

Desulfurisation with calcium oxide is a cost-effective and widely used method for reducing sulfur emissions from industrial processes. However, it is important to properly handle and dispose of the resulting $CaSO_4$ to prevent environmental contamination.

In the case of NO_x emissions, the most common method for treating them in air scrubbers is through a process called selective catalytic reduction (SCR) (Figure 13.10).

SCR involves injecting ammonia (NH_3) or urea into the exhaust gas stream before it enters the scrubber. The NO_x reacts with the NH_3 or urea over a catalyst, typically made of vanadium oxide or titanium oxide, to form harmless N_2 and H_2O. The resulting gas is then passed through the scrubber, where other pollutants are removed.

SCR is a highly effective method for treating NO_x in air scrubbers, with removal efficiencies of up to 90%. However, it requires careful control of the injection rate of the NH_3 or urea and the temperature of the catalyst and can be expensive to implement.

Other methods for treating NO_x in air scrubbers include non-catalytic reduction (NCR) and oxidation (NO_x scrubbing), but these methods are less effective than SCR and are typically used only when SCR is not feasible or cost-effective.

4. **Adsorption:** This process involves passing the gas through a material that absorbs the pollutants. Commonly used adsorbents are activated carbon, zeolite, silica gel or clay. When a pollutant comes into contact with the adsorbent, it attaches to the surface of the solid through Van der Waals forces. The pollutants are held onto the surface of the adsorbent until they can be removed by a process called desorption, which involves releasing the pollutants from the adsorbent into a different medium, such as H_2O or air (Figure 13.11).

5. **Thermal oxidisers:** Thermal oxidisers are air pollution control devices used to remove VOCs and HAPs from industrial air emissions. They work by breaking

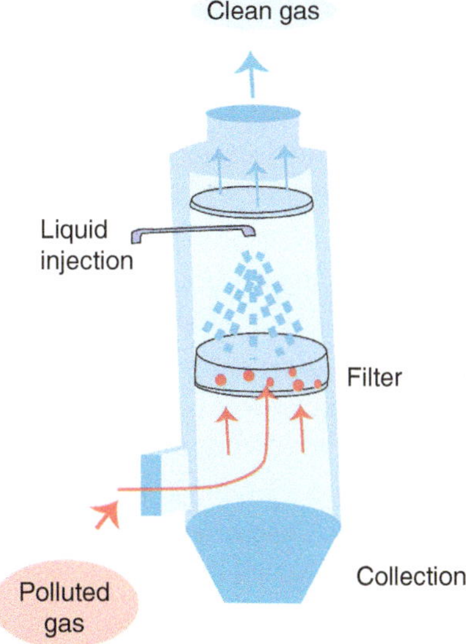

FIGURE 13.11   Absorption column

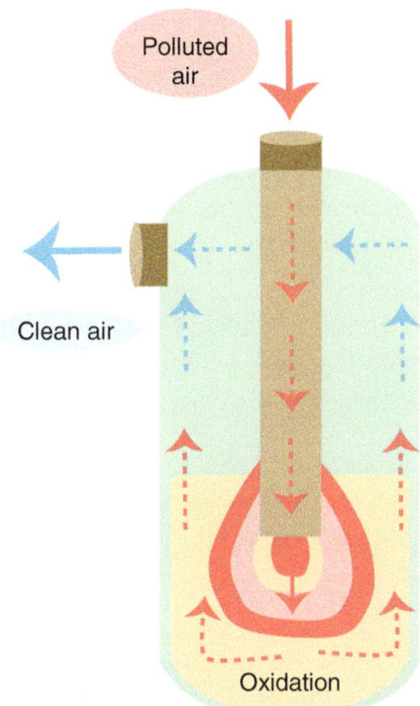

FIGURE 13.12   Thermal oxidiser

down pollutants at high temperatures in the presence of oxygen, converting them into carbon dioxide (CO_2) and H_2O vapour (Figure 13.12).

There are several types of thermal oxidisers, including regenerative thermal oxidisers (RTOs), catalytic oxidisers, direct-fired thermal oxidisers (DFTOs) and recuperative thermal oxidisers (RTOs). Each type of thermal oxidiser uses a different method to heat and destroy the pollutants, but all of them rely on high temperatures to break down the pollutants.

Thermal oxidisers are commonly used in industries such as chemical manufacturing, pharmaceuticals, automotive manufacturing and printing environments where VOCs are generated during the manufacturing process. They are an effective and efficient way to reduce air pollution and protect the environment. These systems burn the pollutants in the gas stream, converting them into carbon dioxide and water. This method is effective for treating VOCs.

6. **Gas separation:** Gas separation devices (Figure 13.13) are used to separate various gases from mixtures, including pollutants from air or other gases. Some common pollutants that can be separated using gas separation devices are:

   a. **Carbon dioxide (CO_2):** Gas separation devices can be used to separate CO_2 from industrial processes or power plants, which can then be captured and stored underground to reduce greenhouse gas emissions.

   b. **Sulfur dioxide (SO_2):** SO_2 is produced by burning fossil fuels, such as coal and oil. Gas separation devices can be used to remove SO_2 from power plant emissions before they are released into the atmosphere.

   c. **Nitrogen oxides (NO_x):** NO_x are pollutants that are produced by burning fossil fuels as well as by industrial processes. Gas separation devices can be used to remove NO_x from emissions before they are released into the atmosphere.

   d. **Methane (CH_4):** CH_4 is a potent greenhouse gas that is produced by a variety of sources, including natural gas production and agriculture. Gas separation devices can be used to separate CH_4 from other gases, which can then be captured and used as fuel or stored underground to reduce greenhouse gas emissions.

   e. **VOCs:** VOCs are chemicals that are emitted from a variety of sources, including industrial processes and vehicle emissions. Gas separation devices can be used to separate VOCs from other gases, which can then be captured and either reused or safely disposed of to reduce air pollution.

7. **Biofiltration:** This method uses microorganisms to break down the pollutants in the gas stream. It is commonly used for treating odorous emissions. In biofiltration, a bed of organic material, such as compost or peat moss, is inoculated with microorganisms that break down the pollutants into less harmful substances (Figure 13.14).

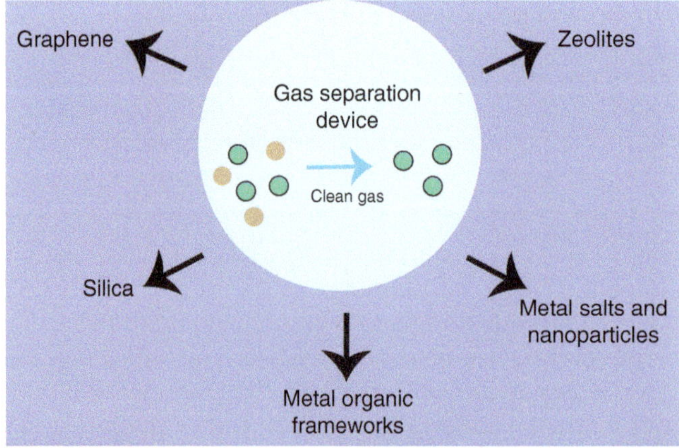

FIGURE 13.13 Gas separation device

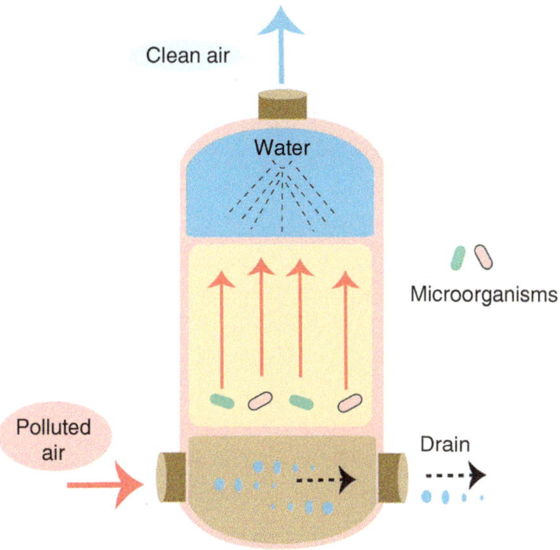

FIGURE 13.14   Biofilter

In the case of air pollution, the contaminated air is drawn through the bed of organic material and the microorganisms in the bed use the pollutants as a food source. As the air passes through the bed, the pollutants are removed and the clean air is released back into the environment.

Biofiltration is an effective and sustainable way to remove pollutants from the environment, as it does not rely on harsh chemicals or energy-intensive processes. It is used in a variety of industries, including wastewater treatment, air pollution control and agriculture. However, the efficiency of biofiltration depends on a variety of factors, such as the type of pollutants being treated, the size and design of the biofilter and the environmental conditions in which it is operating.

13.3 Environmental Impact of Combustion Engines

Combustion engines have a significant environmental impact. Here are some of the ways in which combustion engines contribute to environmental problems:

a. **Air pollution:** Combustion engines emit various air pollutants, which can have negative effects on human health and the environment. The emissions from combustion engines include:

   i. Carbon monoxide (CO) is a colourless, odourless gas that is toxic to humans and animals. It is formed when there is incomplete combustion of the fuel.

   ii. NO_x are gases that contribute to the formation of smog and acid rain. They are formed when the N_2 in the air reacts with oxygen at high temperatures, which occurs during the combustion process.

iii. **Particulate matter (PM):** Tiny particles that are released into the air and can be inhaled into the lungs. PM can cause respiratory problems, particularly in vulnerable populations such as children, the elderly and people with pre-existing respiratory conditions.

iv. VOCs are gases that contribute to the formation of ground-level ozone, a major component of smog. They are released during the combustion of gasoline and other fuels.

b. **Greenhouse gases,** including CO_2, CH_4 and nitrous oxide (N_2O), contribute to climate change by trapping heat in the Earth's atmosphere. These pollutants can cause respiratory problems, contribute to smog formation and damage crops and forests.

c. **Climate change:** Combustion engines also emit greenhouse gases, primarily carbon dioxide, which contribute to climate change. The burning of fossil fuels in combustion engines is one of the largest sources of carbon dioxide emissions, which trap heat in the atmosphere and cause the Earth's temperature to rise.

With some responsible actions, there are several ways to reduce the environmental impact of combustion engine emissions, such as:

1. **Use cleaner fuels:** Switching to cleaner fuels, such as ethanol, biodiesel or natural gas, can significantly reduce emissions. Biofuels are renewable and produce lower emissions compared to fossil fuels.

2. **Improve engine efficiency:** Increasing the efficiency of the engine can reduce the amount of fuel burned and, thus, lower emissions. This can be achieved by improving the combustion process, reducing friction and optimising engine design.

3. **Maintenance of the vehicle:** Regular maintenance, such as changing the oil, replacing air filters and keeping tyres properly inflated, can help ensure that your vehicle is running efficiently and emitting fewer pollutants.

4. **Use catalytic converters:** Catalytic converters reduce harmful emissions by converting pollutants into less harmful compounds. They are a standard feature on most modern vehicles and are required by law in many countries.

13.4 Catalytic Converters

Catalytic converters are devices used in the exhaust systems of vehicles to reduce the emissions of harmful pollutants. The catalytic converter works by using a catalyst to convert the pollutants in the exhaust into less harmful substances.

The catalyst typically used in catalytic converters is a mixture of precious metals such as platinum, palladium and rhodium. When the exhaust gases pass over the catalyst, chemical reactions occur that convert the pollutants into less harmful substances like carbon dioxide, water vapour and N_2.

Catalytic converters are essential for reducing harmful pollutants emitted by vehicles. They are required by law in most countries and failure to have a working catalytic converter can result in fines or even the impounding of the vehicle.

In addition to their use in vehicles, catalytic converters are also used in other applications, such as industrial processes to reduce emissions and in some types of power plants.

The catalytic converter contains a ceramic honeycomb or a bead-filled monolith coated with metals that act as catalysts to speed up the chemical reactions that occur in the converter.

When the exhaust gas from the engine passes through the catalytic converter, it comes into contact with the precious metals on the surface of the honeycomb or monolith. The pollutants in the exhaust gas, such as carbon monoxide, nitrogen oxides and hydrocarbons (HCs), react with the metals to form less harmful compounds such as carbon dioxide, nitrogen and water vapour.

The chemical reactions that occur in the catalytic converter depend on the specific metals and the chemical composition of the exhaust gas. Different types of catalytic converters are designed to work with different types of engines and fuels.

There are three main types of catalytic converters:

Two-way Catalytic Converter

A two-way catalytic converter is called so because it can only transform two types of pollutants: CO and HCs. It cannot reduce NO_x, which require a different type of catalyst called a three-way catalytic converter. In a two-way catalytic converter, the exhaust gases pass through the honeycomb and undergo the following reactions:

a. Oxidation of carbon monoxide (Eq. 13.3)

b. Oxidation of hydrocarbons (Eq. 13.4)

$$CO + \tfrac{1}{2}O_2 \rightarrow CO_2 \tag{13.3}$$

$$\left(HC\right) + O_2 \rightarrow CO_2 + H_2O \tag{13.4}$$

The reactions occur at high temperatures, typically between 300 and 800 °C, which is why the two-way catalytic converter is located in the exhaust system where the exhaust gases are at their hottest. By converting the pollutants into harmless substances, the two-way catalytic converter helps reduce air pollution and protect the environment.

Three-way Catalytic Converter (TWC)

In a three-way catalyst (Figure 13.15), three reactions take place simultaneously: the oxidation of both HC and CO (Eqs. 13.5 and 13.6) and the reduction of oxides (NO_x) into the harmless components, H_2O, N_2 and CO_2 (Eq. 13.7).

$$\left(CH\right) + O_2 \rightarrow CO_2 + H_2O \tag{13.5}$$

$$CO + O_2 \rightarrow CO_2 \tag{13.6}$$

$$NO_x + CO \rightarrow N_2 + CO_2 \tag{13.7}$$

In modern automobiles, the process is controlled through a lambda[1] sensor, which is located in the exhaust system. The lambda value refers to the air-to-fuel ratio

---

[1] The lambda value refers to the air-to-fuel ratio (AFR) of the engine, which is the ratio of the mass of air to the mass of fuel in the combustion mixture.

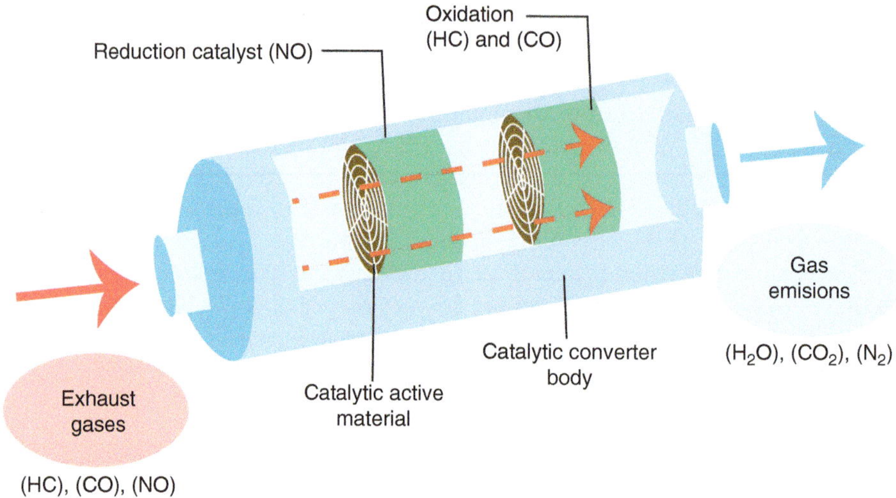

FIGURE 13.15   Three-way catalytic converter

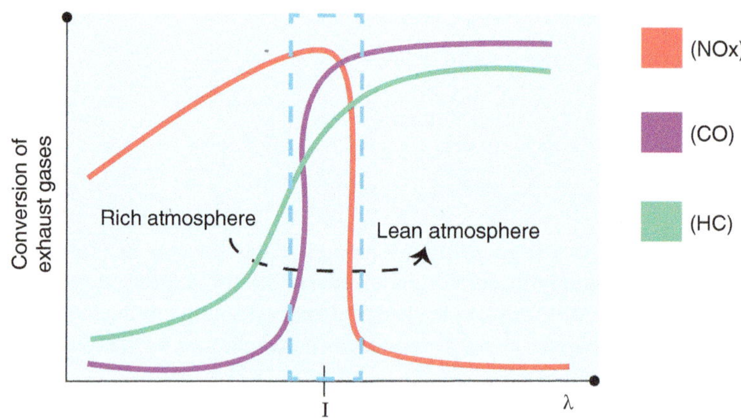

FIGURE 13.16   Conversion of exhaust gases versus lambda value

(AFR) of the engine, which is the ratio of the mass of air to the mass of fuel in the combustion mixture (Figure 13.16).

Lambda-controlled TWCs use a sophisticated feedback control system to adjust the AFR of the engine in real-time, ensuring that the catalytic converter is operating within its optimal temperature range and that the exhaust gases are being converted as efficiently as possible.

Depending on the operating conditions of the engine and the exhaust gas composition, conversion rates upwards of 98% can be achieved at close to stoichiometric (lambda one) conditions. The necessary reaction conditions can be reached in less than a minute by the introduction of special cold-start measures, especially a fast heat-up of the exhaust gas after engine cranking. This is especially important for city driving, which is characterised by frequent start-stop events.

The three-way catalytic converter is designed to operate within a narrow temperature range (typically 200–400 °C) for maximum efficiency. This temperature range is achieved by placing the catalytic converter close to the engine, where it can quickly reach operating temperature and by using engine management systems that optimise the AFR to reduce emissions.

Diesel Oxidation Catalyst (DOC)

This type of converter is specifically designed for diesel engines, which produce higher levels of PM and NO_x than gasoline engines. The DOC oxidises the PM and CO in the exhaust gas to reduce emissions, but it does not reduce NO_x. Diesel engines may also use a separate SCR system to reduce NO_x emissions.

AdBlue is a registered trademark for a urea-based solution used in SCR systems in diesel engines. The purpose of AdBlue is to reduce NO_x emissions that are harmful to the environment.

AdBlue is injected into the SCR catalyst (Figure 13.17) where it reacts with the NO_x in the exhaust gas to form harmless nitrogen and water vapour. The SCR catalyst is a crucial component of the system as it provides the surface on which the following chemical reactions take place.

a. When urea solution is injected into the hot exhaust gas stream, the water evaporates and the urea $[(NH_2)_2CO]$ thermally decomposes to form NH_3 and isocyanic acid (HNCO) (Eq. 13.8).

$$\left(NH_2\right)_2 CO \rightarrow NH_3 + HNCO \tag{13.8}$$

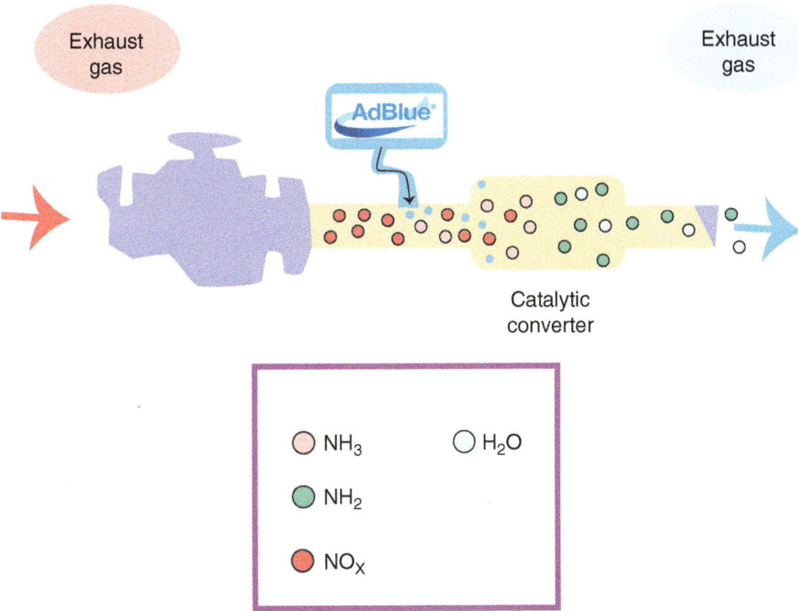

FIGURE 13.17   Diesel catalytic converter

b. HNCO reacts with the H_2O to give CO_2 and NH_3 (Eq. 13.9).

$$HNCO + H_2O \rightarrow CO_2 + NH_3 \tag{13.9}$$

c. Ammonia, in the presence of oxygen and a catalyst, reduces two different NO_x (Eqs. 13.10 and 13.11):

$$4NO + 4NH_3 + O_2 \rightarrow 4N_2 + 6H_2O \tag{13.10}$$

$$NO_2 + 8NH_3 \rightarrow 7N_2 + 12H_2O \tag{13.11}$$

PM emissions from diesel engines refer to the particulate matter that is released into the air as a result of the combustion of diesel fuel. These emissions can have harmful effects on human health and the environment and as such, there are regulations in place to limit their release.

PM emissions are made up of very small particles, typically less than $10\,\mu m$ in diameter, that can penetrate deep into the lungs and cause respiratory problems. They can also contribute to the formation of smog and acid rain.

To reduce PM emissions, diesel engines can be fitted with particulate filters that capture the particles before they are released into the air. In addition, low-sulfur diesel fuel can also help reduce PM emissions.

There are also emission standards in place, such as the Euro 6 standard in the European Union, which sets limits on the amount of PM emissions that diesel engines are allowed to release. Particulate filters are a type of exhaust after-treatment device that is commonly used in diesel engines to reduce the emission of harmful PM into the atmosphere. PM refers to microscopic particles that are small enough to be inhaled by humans and can cause respiratory problems and other health issues.

The particulate filter works by trapping the PM in the exhaust gases before they are released into the atmosphere. The filter is made up of a ceramic honeycomb structure that has tiny channels through which the exhaust gases pass. The walls of these channels are coated with a special material that traps the PM particles as they pass through.

Over time, the PM accumulates in the filter, which can reduce engine performance and fuel economy. To prevent this, the filter needs to be periodically cleaned through a process called regeneration. During regeneration, the accumulated PM is burned off at high temperatures, leaving the filter clean and ready to trap more PM.

There are two types of regeneration: passive and active. Passive regeneration occurs naturally when the exhaust temperature is high enough to burn off the PM. Active regeneration, on the other hand, is initiated by the engine control unit (ECU) when it detects that the filter is getting clogged. The ECU injects extra fuel into the engine to raise the exhaust temperature and initiate regeneration.

CHAPTER 14

Greenhouse Effect and Climate Change

14.1 Absorption of Radiation in the Atmosphere

The Sun emits a wide range of electromagnetic radiation. The electromagnetic spectrum refers to the entire range of electromagnetic radiation, which includes all types of radiation that travel through space at the speed of light.

Radiation can be classified according to its wavelength (Table 14.1 and Figure 14.1), which determines its position on the electromagnetic spectrum. The different types of radiation, ordered from shortest to longest wavelength, are:

Absorption of radiation in the atmosphere is the process by which gases and particles in the Earth's atmosphere absorb energy from incoming solar radiation and outgoing thermal radiation from the Earth's surface. This absorption process plays a critical role in determining the temperature and composition of the Earth's atmosphere.

Different types of electromagnetic waves are filtered in the Earth's atmosphere depending on their wavelengths and the Earth's atmosphere acts as a natural filter, absorbing or reflecting certain types of electromagnetic waves while allowing others to pass through. The types of waves that are filtered by the atmosphere are:

1. **X-rays and gamma rays:** These high-energy electromagnetic waves are mostly absorbed in the atmosphere, which is a good thing since they can be harmful to living organisms.

2. **Ultraviolet (UV) radiation:** Earth's atmosphere absorbs most of the harmful UV radiation from the sun, particularly in the range of wavelengths known as UV-C and most of UV-B. O_3 molecules in the stratosphere are particularly effective at absorbing these types of radiation, which is why the ozone layer is so important.

    However, some UV radiation still reaches the Earth's surface and can cause sunburn, skin cancer and other health problems.

Understanding the Chemistry of the Environment, First Edition. Francisco G. Calvo-Flores.
© 2025 John Wiley & Sons Ltd. Published 2025 by John Wiley & Sons Ltd.

TABLE 14.1   Types of radiation

Type of radiation	Wavelength range	Frequency range	Energy range
Radio waves	>1 m	<300 MHz	<1.24 μeV
Microwaves	1 m–1 mm	300 MHz–300 GHz	1.24 μeV–1.24 meV
Infrared	1 mm–700 nm	300 GHz–430 THz	1.24 meV–1.7 eV
Visible light	700–400 nm	430–750 THz	1.7–3.3 eV
Ultraviolet	400–10 nanometers	750 THz–30 PHz	3.3–124 eV
X-rays	10–0.01 nm	30 PHz–30 EHz	124 eV–124 keV
Gamma rays	<0.01 nm	>30 EHz	>124 keV

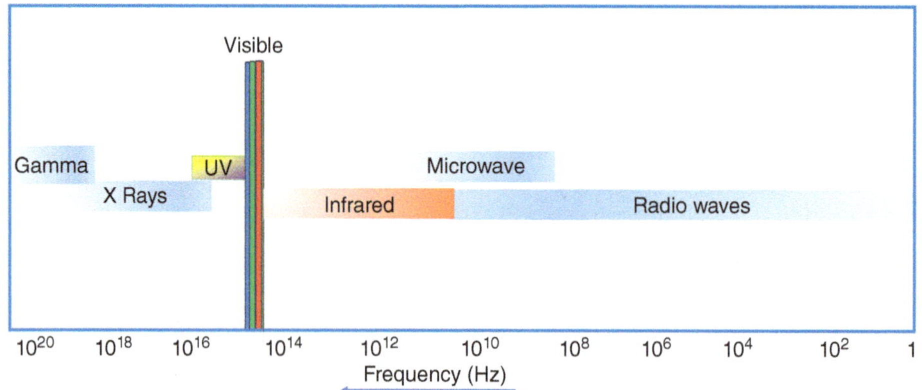

FIGURE 14.1   Types of electromagnetic radiation

3. **Visible light:** Most visible light is able to pass through the atmosphere, although some shorter-wavelength light is scattered by the atmosphere, which is why the sky appears blue.

4. **Infrared (IR) radiation:** IR radiation is able to pass through the atmosphere. In fact, a significant portion of the Sun's energy that reaches the Earth's surface is in the form of IR radiation. This energy is absorbed by the Earth's surface and then re-emitted back into the atmosphere as IR radiation. Most of the IR radiation emitted by the Earth's surface is absorbed by the atmosphere, particularly by water vapour and carbon dioxide (CO_2). This is what causes the greenhouse effect and contributes to the warming of the planet.

5. **Microwaves:** Most microwave frequencies are able to pass through the atmosphere without much attenuation, although some specific frequencies are absorbed by water molecules in the atmosphere, such as those used by satellite-based remote sensing of the Earth's surface.

6. **Radio waves:** Radio waves with long wavelengths (greater than 10 m) are mostly unaffected by the atmosphere and can travel long distances. However, shorter-wavelength radio waves (less than 10 m) are absorbed by the atmosphere, which limits their range.

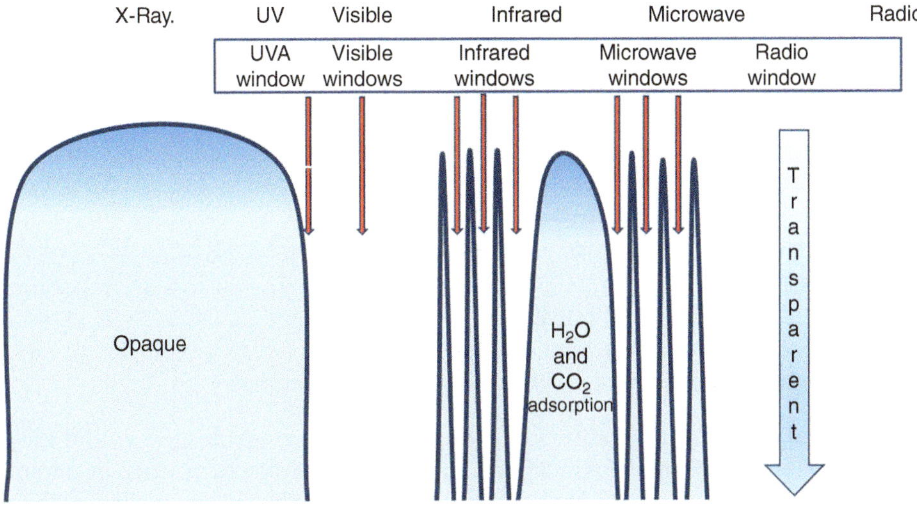

FIGURE 14.2   Schematic representation of wavelength windows in the atmosphere

Most of this radiation is filtered out by the Earth's atmosphere, which protects us from harmful levels of radiation, but some of them are able to break through. The term 'windows for radiation' refers to specific wavelength ranges in the electromagnetic spectrum that are able to pass through the Earth's atmosphere without being significantly absorbed or scattered. These wavelength ranges are also referred to as atmospheric windows or atmospheric transmission windows (Figure 14.2).

14.2   Natural and Anthropogenic Greenhouse Effect

The Earth receives energy from the Sun in the form of solar radiation, visible light, UV and IR radiation. About 30% of the incoming solar radiation is reflected back into space by clouds, ice and the Earth's surface. The remaining 70% is absorbed by the Earth's surface, oceans and atmosphere.

Once absorbed, the energy is converted into heat, which is radiated back into the atmosphere as IR radiation. This radiation can be absorbed by certain gases in the atmosphere, which trap the heat and prevent it from escaping into space. This process is known as the greenhouse effect and it helps keep the Earth's temperature within a range that is suitable for life. This warming effect is essential for life on Earth to exist.

Greenhouse gases are gases in the Earth's atmosphere that absorb and emit radiation within the thermal IR range, playing a critical role in regulating the Earth's climate. Some of the key properties of greenhouse gases include:

1. **Absorption of radiation:** Greenhouse gases have the ability to absorb and re-emit IR radiation. This property allows them to trap heat in the Earth's atmosphere, which contributes to the greenhouse effect.

2. **Wavelength selectivity:** Different greenhouse gases absorb radiation at different wavelengths. This means that they are more effective at trapping certain types of radiation than others. For example, carbon dioxide is particularly effective at trapping long-wavelength radiation.

3. **Concentration dependence:** The greenhouse effect depends on the concentration of greenhouse gases in the atmosphere. As the concentration of these gases increases, their ability to trap heat also increases, leading to an overall warming of the planet.

4. **Lifetime:** The lifetime of a greenhouse gas in the atmosphere is an important factor in determining its impact on the climate. Some gases, such as carbon dioxide, can remain in the atmosphere for hundreds or even thousands of years, while others, such as methane (CH_4), have shorter lifetimes.

Greenhouse gases are produced by natural and anthropogenic sources and they can also be removed from the atmosphere through natural processes, such as photosynthesis and ocean uptake, so that a thermal balance is maintained over time, as long as there is no excessive contribution of these anthropogenic gases, as has been happening in recent years, with scientific evidence that such a balance is being disrupted.

The mechanism of the greenhouse effect can be explained as follows:

1. The sun emits radiation, which passes through the Earth's atmosphere and reaches the surface of the planet.

2. The Earth's surface absorbs some of this energy and reflects the rest back into the atmosphere as longwave radiation.

3. Greenhouse gases in the atmosphere, such as carbon dioxide, methane and water vapour, absorb some of this longwave radiation and trap it in the atmosphere.

4. This trapped heat causes the atmosphere to warm, which in turn warms the Earth's surface.

The mechanism of the greenhouse effect is summarised in Figure 14.3.

Greenhouse gases absorb radiation because of their molecular structure, which gives them a unique ability to absorb certain wavelengths of energy, which include radiation from the Sun and the Earth's surface. When these greenhouse gases absorb radiation, they become excited and vibrate. This vibration, in turn, heats up the surrounding air molecules, causing the Earth's atmosphere to warm up. This is commonly referred to as the **greenhouse effect** and it is a natural process that helps keep our planet warm enough to support life.

The natural greenhouse effect is a critical process that keeps the Earth's climate warm enough to support life. It involves greenhouse gases such as carbon dioxide, methane, water vapour and nitrous oxide (N_2O) naturally present in the atmosphere. These gases allow sunlight to enter the atmosphere and reach the Earth's surface, where it is absorbed and then radiated back as heat. Some of this heat is trapped by greenhouse gases, maintaining the Earth's average temperature at around 15 °C.

Without the natural greenhouse effect, the Earth's average temperature would be too cold to support life as we know it.

The anthropogenic greenhouse effect refers to the enhancement of the natural greenhouse effect caused by human activities. This process leads to increased concentrations of greenhouse gases in the Earth's atmosphere, which trap more heat and result in global warming and climate change.

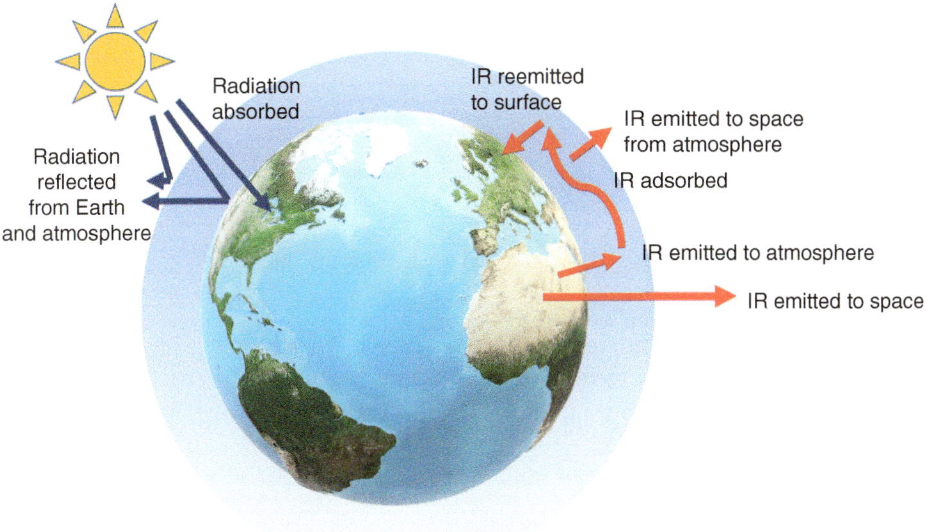

FIGURE 14.3   The greenhouse effect

14.3   Main Greenhouse Gases

Greenhouse gases are components of the Earth's atmosphere that can trap IR radiation. They are produced by natural and anthropogenic sources. The following describes the main greenhouse gases and their most notable characteristics.

a. Water Vapour

Water vapour is a natural component of the Earth's atmosphere and is constantly present in varying amounts depending on location, temperature and weather conditions. The primary natural sources of water vapour in the atmosphere are:

1. **Evaporation:** This is the process by which liquid water changes into water vapour due to the absorption of heat energy from the sun. This process occurs naturally on the surface of oceans, lakes, rivers and other bodies of water.

2. **Transpiration:** This is the process by which water vapour is released from plants through tiny openings called stomata on the leaves. This process is also known as evapotranspiration.

3. **Sublimation:** This is the process by which ice and snow directly change into water vapour without first melting into liquid water. This occurs naturally in areas with cold temperatures and low humidity, such as at high altitudes or in polar regions.

4. **Respiration:** This is the process by which living organisms release water vapour into the atmosphere during the process of breathing.

Water molecules absorb IR radiation in several ways due to their unique vibrational modes (Figure 14.4). The IR radiation causes the water molecule to vibrate and rotate, leading to changes in its dipole moment and absorption of the radiation.

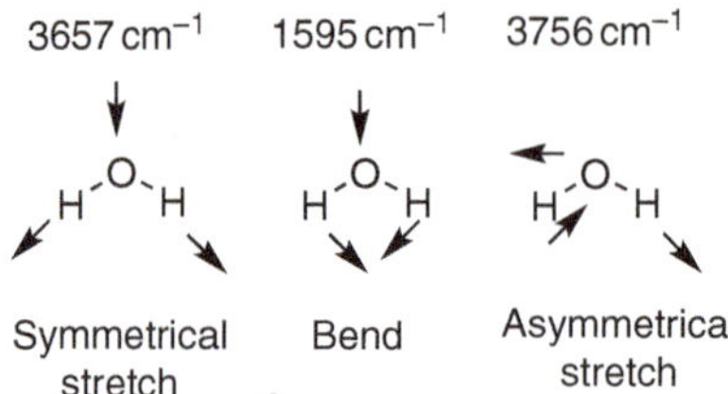

FIGURE 14.4 IR vibrational modes of water molecules

The main vibrational modes of a water molecule are the stretching and bending modes. The stretching modes involve the symmetric and asymmetric stretching of the O-H bond, while the bending modes involve the rocking, wagging and twisting of the H-O-H bond angle.

The strongest IR absorption of water molecules occurs in the 3-μm region, which corresponds to the stretching mode of the O-H bond. This absorption band is often used in IR spectroscopy to identify the presence of water molecules in a sample.

Water molecules also absorb in the 6-μm region, which corresponds to the bending mode of the H-O-H bond angle. This absorption band is weaker than the O-H stretching band but still contributes to the overall absorption of IR radiation by water molecules.

Water vapour is the most abundant greenhouse gas in the atmosphere and it is responsible for about 60% of the Earth's natural greenhouse effect.

b. Carbon Dioxide

There are several natural sources of carbon dioxide including:

1. **Volcanic eruptions:** When volcanoes erupt, they release large amounts of carbon dioxide into the atmosphere.

2. **Respiration:** Both plants and animals respire, which means they release carbon dioxide into the atmosphere.

3. **Decomposition:** When organic matter such as dead plants and animals decays, it releases carbon dioxide into the atmosphere.

4. **Ocean release:** The ocean contains dissolved carbon dioxide and as water temperatures rise, the ocean releases more carbon dioxide into the atmosphere.

5. **Forest fires:** When forests burn, they release large amounts of carbon dioxide into the atmosphere.

6. **Geothermal activity:** Geothermal vents on the ocean floor and hot springs on land can release carbon dioxide into the atmosphere.

7. **Weathering of rocks:** The natural weathering of rocks can release small amounts of carbon dioxide into the atmosphere.

Natural sources of carbon dioxide have been occurring for millions of years and the Earth has a natural way of balancing them out. However, human activities such as burning fossil fuels, deforestation, land-use changes and agriculture have caused an imbalance in the carbon cycle, leading to increased levels of carbon dioxide in the atmosphere and contributing to global climate change.

Carbon dioxide absorbs IR radiation in a characteristic manner, which has important implications for Earth's climate (Figure 14.5). Specifically, carbon dioxide

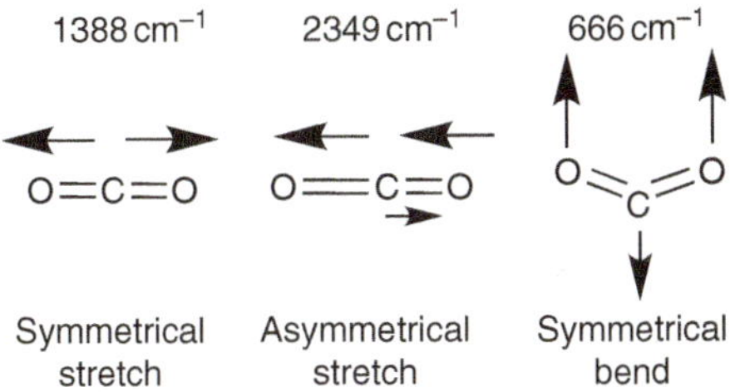

FIGURE 14.5   IR vibrational modes of carbon dioxide molecules

molecules have two vibrational modes, which involve the bending and stretching of the molecule's bonds.

The bending and stretching modes of carbon dioxide correspond to wavelengths of approximately 15 and 4.3 μm, respectively, which fall within the range of IR radiation that is absorbed by Earth's atmosphere. This means that when IR radiation emitted by the Earth's surface or lower atmosphere encounters carbon dioxide molecules in the upper atmosphere, it can be absorbed and re-emitted in all directions, including back towards the Earth's surface.

c. Methane

Overall, natural sources account for about 40% of methane emissions, with the remainder being attributed to human activities such as agriculture, fossil fuel production and waste management. Natural sources of methane in the environment are:

1. **Wetlands:** Wetlands are one of the largest natural sources of methane emissions. Microorganisms in the soil produce methane as a byproduct of their metabolism, which is then released into the atmosphere.

2. **Termites:** Termites are another significant source of methane emissions. They produce methane as a byproduct of their digestion.

3. **Oceans:** Methane is produced in the sediments at the bottom of the ocean and is released into the water column and eventually into the atmosphere.

4. **Ruminants:** Ruminants, such as cows and sheep, produce methane as a byproduct of their digestive processes.

5. **Permafrost:** Methane is also trapped in permafrost, which is frozen soil in the Arctic and Antarctic regions. As the permafrost thaws, methane is released into the atmosphere.

6. **Wildfires:** Methane can be released during wildfires from burning organic matter.

7. **Geologic sources:** Methane can be found in natural gas deposits and is released during oil and gas production as well as coal mining.

Anthropogenic sources contribute significantly to methane emissions, including those emitted during the extraction, transportation and use of fossil fuels such as coal, oil and natural gas. Additionally, Methane is produced during the digestion of food by livestock like cows and sheep, as well as during the decomposition of organic matter in soils like rice paddies. Landfills and wastewater treatment plants also release methane during the decomposition of organic waste. Moreover, methane is formed during the production of chemicals like plastics and other organic materials.

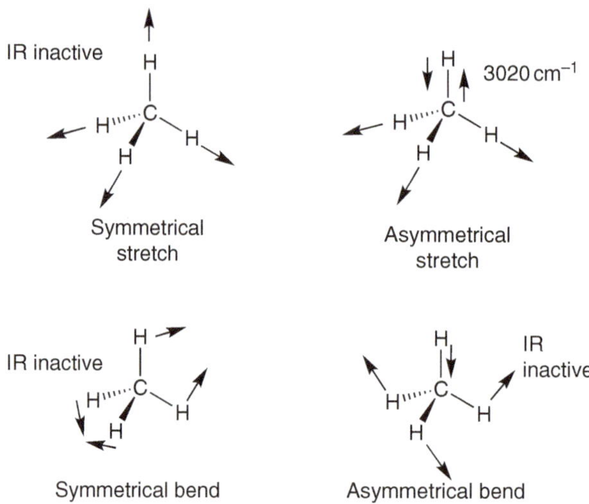

FIGURE 14.6   Bending and stretching vibrations of methane

Methane absorbs IR radiation in several regions of the electromagnetic spectrum. The strongest and most well-known absorption occurs in the mid-IR region, specifically at wavelengths around 3.3 microns. This absorption band is known as the 'fundamental band' or the v3 band and it arises from the stretching vibration of the C—H bonds in methane.

In addition to the fundamental band, methane also absorbs IR radiation in several other regions, including the overtone and combination bands at shorter wavelengths and the hot band at longer wavelengths. These absorption bands correspond to various bending and stretching vibrations of the molecule (Figure 14.6).

The absorption of IR radiation by methane is of particular interest because it is a major component of natural gas and a potent greenhouse gas. methane is responsible for about 20% of the greenhouse effect on Earth, despite being present in much lower concentrations than carbon dioxide. The absorption of IR radiation by methane contributes to its warming effect on the planet, which is why reducing methane emissions is an important goal in mitigating climate change.

d. Nitrous Oxide

Nitrous oxide is a naturally occurring gas that is produced naturally in a number of ways, including:

1. **Microbial processes:** Nitrous oxide is produced by a variety of microbial processes in soil, oceans and freshwater systems. Bacteria, fungi and archaea play a significant role in producing and consuming nitrous oxide in these environments.

2. **Lightning:** Nitrous oxide is also produced naturally during lightning strikes. When lightning occurs, it causes nitrogen and oxygen in the air to combine and form nitrous oxide.

3. **Oceanic processes:** Nitrous oxide is also produced in the ocean through a variety of processes, including denitrification by bacteria, nitrification by microorganisms and the breakdown of organic matter.

4. **Volcanic activity:** Volcanic activity can also release nitrous oxide into the atmosphere. This is because nitrogen is a common element found in volcanic rocks and the intense heat and pressure of volcanic eruptions can cause nitrogen to react with other elements and form nitrous oxide.

Although nitrous oxide can be found naturally in the atmosphere, human activities have led to a significant increase in its concentration through anthropogenic emissions. One of the main sources of these emissions is the use of nitrogen fertilisers in agricultural practices such as farming and livestock production, as well as land-use changes. In addition, nitrous oxide is also released during the production of certain chemicals, including nitric acid and adipic acid and from the combustion of fossil fuels.

Waste management practices, such as the treatment of wastewater solid waste and animal manure, also contribute to the release of nitrous oxide. Furthermore, combustion in vehicles, particularly diesel engines, is a significant source of nitrous oxide emissions in the transportation sector.

Nitrous oxide has several IR absorption bands in the mid-IR region (Figure 14.7).

The fundamental stretching vibration of the N=O bond in nitrous oxide is observed at a wavenumber of around $2280\,cm^{-1}$. This is a very strong absorption band and the most intense peak in the IR spectrum of nitrous oxide.

In addition to the fundamental N=O stretching vibration, there are also several combination and overtone bands in the IR spectrum of nitrous oxide. These include a combination band at around $4580\,cm^{-1}$, an overtone band at around $4430\,cm^{-1}$ and another combination band at around $3650\,cm^{-1}$.

e. Ozone

Ozone can act in the troposphere as both a greenhouse gas and an air pollutant. As a greenhouse gas, ozone absorbs and emits radiation in the IR portion of the electromagnetic spectrum, which can contribute to warming the Earth's atmosphere. However, the role of ozone as a greenhouse gas is generally considered to be small compared to other greenhouse gases such as carbon dioxide, methane and water vapour. This is because the concentration of ozone in the atmosphere is much lower than that of these other gases. In addition, ozone's lifetime in the atmosphere is

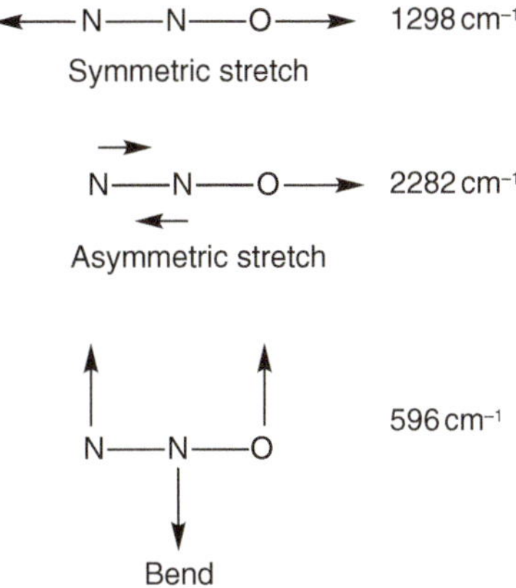

FIGURE 14.7   IR vibrational modes of nitrous oxide

TABLE 14.2	Relative percentages of some greenhouse gases in Earth's atmosphere	
Greenhouse gas	**Percentage in Earth's atmosphere**
Carbon dioxide (CO_2)	Approximately 0.0415%
Methane (CH_4)	Approximately 0.00017%
Nitrous oxide (N_2O)	Approximately 0.00003%
Ozone (O_3)	Varies depending on altitude (0.000001–0.0008%)
Water vapour (H_2O)	Highly variable, typically around 1–4%

relatively short, typically on the order of a few weeks. This means that ozone concentrations can vary greatly depending on local conditions and are strongly influenced by human activities such as industrial emissions and transportation.

Relative percentages of some greenhouse gases in Earth's atmosphere are summarised in Table 14.2.

f. Fluorinated Greenhouse Gases (Also Known as F-Gases)

Fluorinated greenhouse gases (f-gases) are synthetic gases that are used in a variety of applications. F-gases are potent greenhouse gases that can remain in the atmosphere for very long periods of time.

Fluorinated gases have a strong ability to absorb IR radiation in the range of 8–14 μm, which is known as the 'atmospheric window'. This means that they can trap heat in the atmosphere and contribute to the greenhouse effect. In fact, fluorinated gases are among the most potent greenhouse gases.

These gases include chlorofluorocarbons (CFCs), hydrofluorocarbons (HFCs), perfluorocarbons (PFCs) and sulfur hexafluoride (SF_6).

a. **Chlorofluorocarbons (CFCs)** are a group of synthetic organic compounds that contain carbon, chlorine and fluorine atoms. As commented in Chapter 10, the gaseous CFCs have a strong ozone-depleting effect but they are also very potent greenhouse gases. CFCs have been banned or heavily restricted in most countries around the world, but despite this fact, their persistence in the atmosphere makes them still a subject of attention as greenhouse gases.

b. **Hydrofluorocarbons (HFCs)** with several technical applications were developed as a replacement for CFCs. They do not contain chlorine, which means they do not contribute to the depletion of the ozone layer. However, HFCs are potent greenhouse gases. Examples of HFCs are HFC-134a, HFC-410a and HFC-23 (Table 14.3).

One of the most significant regulations regarding HFCs is the Kigali Amendment to the Montreal Protocol, which was agreed upon in 2016. The Kigali Amendment aims to phase down the use of HFCs globally and replace them with alternative refrigerants that have lower global warming potentials (GWPs). As of 2021, more than 100 countries have ratified the Kigali Amendment. The amendment sets out a timeline for the phase-down of HFCs, with developed countries starting to reduce their use of HFCs in 2019 and developing countries following suit in 2024 or 2028, depending on their level of development.

TABLE 14.3   **Examples of HFCs**

Compound	Formula	Common name	Uses
HFC-134a	CH_2FCF_3	1,1,1,2-Tetrafluoroethane	Refrigerant, propellant and blowing agent for foam insulation
HFC-410a	$CH_2F_2CHF_2CF_3$	R-410A	Air conditioning and refrigeration systems
HFC-23	CHF_3	Trifluoromethane	Production of refrigerants and fluoropolymers

c. **Perfluorocarbons (PFCs)** are a class of man-made chemicals that contain only carbon and fluorine atoms. PFCs are very stable and do not break down easily, so they persist in the environment for a long time and they are also considered to be potent greenhouse gases. PFCs are used in the manufacturing of semiconductors and are produced as a byproduct of aluminium production. PFCs generally have long atmospheric lifetimes. Examples of such substances are: Tetrafluoromethane (CF_4), Hexafluoroethane (C_2F_6), Perfluoropropane (C_3F_8) and Perfluorocyclohexane (C_6F_{12}).

Several countries and international organisations have taken steps to regulate and reduce PFC emissions. The United Nations Framework Convention on Climate Change (UNFCCC) has included PFCs as one of the six greenhouse gases covered under the Kyoto Protocol and has set emission reduction targets for PFCs.

d. **Sulfur hexafluoride (SF_6)** is a versatile gas with multiple important applications across various industries. It is a colourless, odourless, non-flammable and chemically stable gas that exhibits high dielectric strength and excellent arc-quenching properties. The primary use of SF_6 is as an insulating gas in electrical equipment, including circuit breakers, transformers and switchgear. Moreover, it finds extensive use in the semiconductor industry as a plasma-etching agent. In the medical field, SF_6 serves as a contrast agent in ultrasound imaging, where it is injected into the bloodstream to enhance the visibility of organs and blood vessels during medical procedures. Additionally, SF_6 is used in different industrial processes, such as the production of magnesium and aluminium. It can also be used as a cover gas in the semiconductor industry during the production of electronic components. Despite its relatively small atmospheric concentration, SF_6 is a very powerful greenhouse gas and is a significant contributor to anthropogenic climate change. According to the Intergovernmental Panel on Climate Change (IPCC)[1], SF_6 is the most potent greenhouse gas regulated under the Kyoto Protocol, accounting for around 10% of total greenhouse gas emissions from the industrial sector.

---

[1] The Intergovernmental Panel on Climate Change (IPCC) is an international body established in 1988 by the United Nations Environment Programme (UNEP) and the World Meteorological Organization (WMO). Its primary purpose is to provide policymakers with regular scientific assessments on climate change, its implications and potential future risks, as well as to propose adaptation and mitigation strategies.

Efforts are underway to reduce SF_6 emissions, including through the use of alternative insulating gases in high-voltage equipment and the development of technologies to detect and repair SF_6 leaks. The European Union has implemented regulations to phase out the use of SF_6 in most non-electrical applications by 2022 and to reduce its use in the electrical sector by two-thirds by 2030.

14.4 Consequences of the Anthropogenic Greenhouse Effect

The enhanced greenhouse effect due to human activities has several significant impacts:

1. **Global warming:** Increased greenhouse gas concentrations lead to higher global average temperatures. Since the late 19th century, the Earth's average surface temperature has risen by approximately 1.2 °C.

2. **Climate change:** Global warming drives changes in climate patterns, resulting in more frequent and severe weather events such as hurricanes, droughts, heatwaves and heavy precipitation.

3. **Melting ice and rising sea levels:** Higher temperatures cause polar ice caps and glaciers to melt, contributing to sea level rise. This threatens coastal communities and ecosystems.

4. **Ocean acidification:** Increased carbon dioxide levels in the atmosphere lead to higher concentrations of carbon dioxide being absorbed by oceans, causing ocean acidification. This affects marine life, particularly organisms with calcium carbonate shells or skeletons.

Addressing the anthropogenic greenhouse effect requires both mitigation and adaptation strategies:

a. **Mitigation:** Efforts to reduce or prevent the emission of greenhouse gases include transitioning to renewable energy sources (e.g., wind, solar, hydro), enhancing energy efficiency, protecting and restoring forests and developing carbon capture and storage technologies.

b. **Adaptation:** Societies must also adapt to the changes already set in motion. This includes building resilient infrastructure, developing sustainable agriculture practices and planning for sea-level rise

---

Topics of Interest: Ocean Acidification

When carbon dioxide dissolves in seawater, it reacts with water molecules to form carbonic acid, which lowers the pH of the water and makes it more acidic. This process is known as ocean acidification.

Ocean acidification is a significant concern because it has the potential to disrupt marine ecosystems and threaten the survival of many species. As the ocean becomes more acidic, it can make it more difficult for organisms like corals, mollusks and plankton to build and maintain their shells and skeletons. This can have cascading effects throughout the food chain, impacting everything from tiny plankton to larger fish and marine mammals.

In addition, ocean acidification can also affect the chemistry of seawater, which can have implications for nutrient cycling and the overall health of marine ecosystems. The impacts of ocean acidification are not yet fully understood, but they are likely to be significant and far-reaching.

Here are some key highlights about ocean acidification:

1. **pH levels are decreasing:** The ocean's pH levels have been decreasing due to the increase in atmospheric carbon dioxide, making it more acidic.

2. **Impacts on marine life:** Ocean acidification can negatively impact marine life, especially those that rely on calcium carbonate to build their shells or skeletons. This includes corals, mollusks and some types of plankton.

3. **Impacts on food webs:** The impacts of ocean acidification can also ripple through the food web, as organisms that rely on these affected species for food or habitat may also be impacted.

4. **Decreased biodiversity:** As species struggle to adapt to the changing conditions, there is a risk of decreased biodiversity and even extinctions.

5. **Economic impacts:** Many coastal communities rely on marine resources for their livelihoods and ocean acidification can threaten these industries, including fishing and tourism.

6. **Mitigation is possible:** While the impacts of ocean acidification are concerning, reducing greenhouse gas emissions and pursuing other mitigation strategies can help slow the rate of change and reduce the severity of impacts.

14.5 Radiative Forcing and Global Warming Potential

Radiative forcing is the difference between the amount of solar energy absorbed by the Earth's atmosphere and the amount of energy that is radiated back to space. It is a measure of the impact that human activities, such as burning fossil fuels, have on the Earth's climate system[2].

Radiative forcing is typically expressed in units of watts per square metre (W/m^2) and is used to quantify the amount of energy that is added or subtracted from the Earth's energy balance due to changes in atmospheric composition.

Radiative forcing may be calculated from the sum of several components that contribute positive or negative signs.

Positive radiative forcing occurs when there is a net increase in the amount of energy absorbed, which can lead to warming. Positive radiative forcing is the main greenhouse gases responsible for positive radiative forcing are carbon dioxide, methane and nitrous oxide. These glasses absorb and trap heat in the atmosphere, which contributes to global warming.

There are several factors that can produce positive radiative forcing, including:

1. **Solar radiation:** Increases in the amount of solar radiation that reaches the Earth's surface, such as during periods of high solar activity, can also produce a positive radiative forcing.

---

[2]https://climate.mit.edu/explainers/radiative-forcing

2. **Greenhouse gases:** Certain gases in the Earth's atmosphere, such as carbon dioxide, methane, VOCs and water vapour, trap heat and prevent it from escaping into space, resulting in a positive radiative forcing.

3. **Tropospheric ozone:** Ozone in the lower atmosphere is a greenhouse gas and can contribute to a positive radiative forcing. However, ozone in the upper atmosphere can absorb harmful UV radiation from the sun, reducing the amount of incoming energy and producing a negative radiative forcing.

4. **Some aerosols.**

Negative radiative forcing occurs when there is a net decrease in the amount of energy absorbed by the Earth's atmosphere, which can lead to cooling.

Some factors that produce negative radiative forcing are:

1. **Reduced solar radiation:** Variations in the output of the sun can lead to reduced solar radiation, which can result in negative radiative forcing.

2. **Stratospheric ozone:** Ozone in the stratosphere is formed when UV radiation from the sun reacts with oxygen molecules. Ozone absorbs solar radiation and therefore has a warming effect on the Earth's surface. However, in the lower stratosphere, ozone can be destroyed by chemical reactions with human-made substances such as CFCs, which produce a negative radiative forcing effect.

3. **Albedo:** Albedo is a measure of the reflectivity of a surface, with higher albedo surfaces reflecting more incoming solar radiation back into space. Some surfaces have the ability to reflect incoming solar radiation back into space, resulting in a net cooling effect on the plane, as occurs with cloud cover or snow and ice.

4. **Some aerosols.**

14.6 The Role of Aerosols on the Greenhouse Effect

Aerosols play a complex and multifaceted role in the greenhouse effect and Earth's climate system.

Aerosols are critical to understanding and predicting climate change due to their dual role in both cooling and warming the Earth's climate. While some aerosols reflect sunlight and cool the surface, others absorb heat and contribute to atmospheric warming. Their interactions with clouds further complicate their overall impact, making aerosols a key area of study in climate science. Reducing certain types of aerosols, particularly black carbon, can provide a quick win for climate mitigation by reducing warming and improving air quality simultaneously.

Their main effects are as follows:

1. **Direct radiative effects.**

   Aerosols interact directly with solar and terrestrial radiation (Figure 14.8):

   a. **Scattering:** Many aerosols scatter sunlight back into space, which tends to cool the Earth's surface. This scattering effect is most pronounced with sulfate aerosols, which are highly reflective.

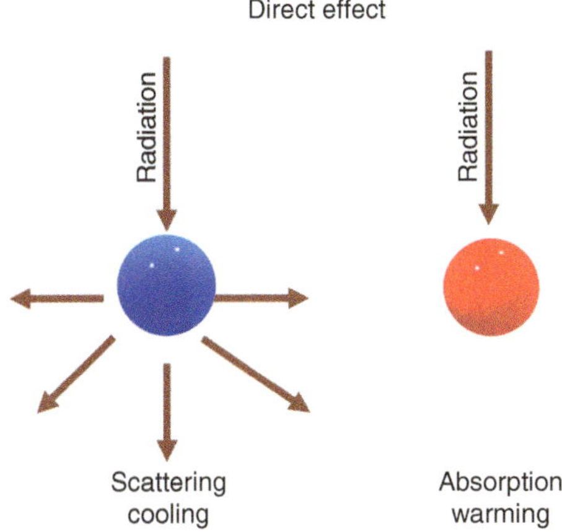

FIGURE 14.8   Interaction of aerosols with solar radiation

   b. Absorption: Some aerosols, like black carbon (soot), absorb sunlight and heat the atmosphere. This absorption can contribute to warming, both locally and globally.

2. **Indirect radiative effects.**

   Aerosols also influence cloud properties and behaviour:

   a. Cloud condensation nuclei (CCN): Aerosols act as nuclei around which cloud droplets can form. An increase in aerosols can lead to the formation of more but smaller cloud droplets, making clouds more reflective (a phenomenon known as the Twomey effect).

   b. Cloud lifetime: By modifying cloud microphysics, aerosols can affect the lifetime of clouds. Some studies suggest that increased aerosols can extend the lifetime of clouds, enhancing their cooling effect.

   c. Precipitation: Aerosols can influence precipitation patterns. For instance, more numerous but smaller droplets in clouds can suppress rainfall, potentially altering regional hydrological cycles.

3. **Semi-direct effects.**

   When aerosols like black carbon absorb sunlight, they heat the surrounding air, which can lead to changes in cloud formation and dissipation:

   a. Cloud dissipation: Heating the atmosphere can reduce cloud cover by promoting the evaporation of cloud droplets.

   b. Albedo effect on snow and ice: When deposited on snow and ice, dark aerosols like black carbon reduce their albedo (reflectivity), leading to increased absorption of sunlight and accelerated melting.

4. **Surface and atmospheric effects.**

   Aerosols can affect both surface and atmospheric temperatures:

   a. Surface cooling: Reflective aerosols, such as sulfates, cool the surface by reflecting incoming solar radiation.

   b. Atmospheric heating: Absorptive aerosols, like black carbon, can warm the atmosphere by absorbing radiation.

5. **Overall impact on climate.**

   The net effect of aerosols on climate is complex and depends on their type, concentration and distribution:

   a. **Cooling effect:** On balance, many aerosols, particularly sulfates, tend to have a net cooling effect by reflecting sunlight.

   b. **Warming effect:** Absorptive aerosols, such as black carbon, contribute to warming, although their overall impact is generally smaller than that of greenhouse gases.

   Therefore, aerosols can have both a warming and cooling effect on the climate, depending on their properties and interactions with the atmosphere. For example, some aerosols, such as black carbon particles, absorb incoming solar radiation and warm the Earth's atmosphere, resulting in a positive radiative forcing.

   On the other hand, other types of aerosols, such as sulfate particles, reflect incoming solar radiation back into space, which can have a cooling effect on the Earth's surface by providing a negative radiative forcing.

   The net effect of aerosols on radiative forcing is complex and depends on a variety of factors, including their size, composition, location and time of day or year.

14.7 Calculation of Radiative Forcing

It is important to point out that the calculation of radiative forcing has evolved over time as our understanding of climate science has advanced and as more comprehensive data and models have become available. Radiative forcing calculations have evolved according to several factors:

1. **Simple energy balance models (early approach):** In the early stages of understanding climate change, scientists used simplified energy balance models to estimate radiative forcing. These models considered the energy inputs from the sun and the energy radiated back into space by Earth. However, these models did not take into account the complex interactions between different greenhouse gases, aerosols and other factors.

2. **Consideration of greenhouse gases:** As our understanding of the role of greenhouse gases in influencing Earth's climate grew, scientists started to focus on calculating radiative forcing due to specific gases like CO_2, CH_4 and N_2O. The radiative forcing due to these gases is calculated by comparing the energy balance before and after accounting for their increased concentrations.

3. **Climate models and comprehensive calculations:** With advancements in computing power, scientists developed more sophisticated climate models that simulate the interactions between various factors, including greenhouse gases, aerosols, solar variability and more. These models take into account feedback mechanisms and complex interactions, providing a more comprehensive understanding of radiative forcing.

4. **Improved data collection:** Accurate radiative forcing calculations require precise data on factors like greenhouse gas concentrations, aerosol properties and land use changes. Over time, improvements in data collection through satellite observations, ground-based measurements and global monitoring networks have enhanced the accuracy of radiative forcing estimates.

5. **Accounting for aerosols and complex interactions:** Calculating radiative forcing due to aerosols (small particles suspended in the atmosphere) is challenging due to their diverse properties and effects. Researchers have worked to better understand aerosol-cloud interactions and their impact on radiative forcing. These interactions can either amplify or mitigate the effects of radiative forcing from other sources.

6. **Scenario-based assessments:** Radiative forcing calculations are often performed for different emission scenarios, reflecting different potential trajectories of greenhouse gas emissions and other factors. These scenarios help policymakers understand the potential impacts of different policy choices on radiative forcing and subsequent climate change.

7. **Ongoing research and refinement:** Radiative forcing calculations continue to evolve as new research emerges, computational capabilities improve and our understanding of climate processes deepens. Ongoing efforts are made to refine models, incorporate new data and account for emerging factors that influence radiative forcing.

Table 14.4 shows radiative forcing values for various greenhouse gases and other factors influencing Earth's energy balance, expressed in watts per square metre (W/m^2).

The values are approximate and they can vary slightly depending on the source and the specific methodology used in their determination. The numbers above are

TABLE 14.4   Radiative forcing data

Greenhouse gases	Radiative forcing (W/m^2)
Carbon dioxide (CO_2)	+1.82
Methane (CH_4)	+0.48
Nitrous oxide (N_2O)	+0.17
Halocarbons (CFCs, HCFCs, HFCs)	+0.36
Aerosols	
Sulfate aerosols	−0.40
Black carbon on snow	+0.04
Organic carbon aerosols	−0.12
Ozone	
Tropospheric ozone	+0.40
Stratospheric ozone	−0.05
Other factors	
Land use (Albedo changes)	−0.15
Solar irradiance	+0.05
Total (all factors combined)	+2.29

TABLE 14.5 **Global warming potential (GWP) for common greenhouse gases**

Gas	Formula	GWP (AR1)	GWP (AR2)	GWP (AR3)	GWP (AR4)	GWP (AR5)
Carbon dioxide	CO_2	1	1	1	1	1
Methane	CH_4	21	23	25	28	34
Nitrous oxide	N_2O	310	296	298	298	265
Fluorinated gases (HFCs)	Various	N/A	N/A	N/A	N/A	Varies by gas
Sulfur hexa-fluoride	SF_6	5,000	23,900	22,800	22,800	23,500

Note: The values for HFCs vary widely by gas and the AR5 uses a different method for calculating their GWP, so a specific value cannot be given in this table.
Adapted from https://archive.ipcc.ch/publications_and_data/ar4/wg1/en/ch2s2-10-2.html

drawn from the IPCC Fifth Assessment Report (AR5) and other relevant scientific literature[3].

- **Positive radiative forcing** indicates a warming effect (more energy retained in the Earth's system).
- **Negative radiative forcing** indicates a cooling effect (less energy retained in the Earth's system).
- Values are based on scientific assessments and represent changes relative to the pre-industrial era (around 1750).

The contribution of every greenhouse gas to the greenhouse effect is not the same. GWP is a measure of how much a particular greenhouse gas contributes to global warming over a given period of time, usually 100 years. It is a relative measure that compares the warming potential of a particular gas to that of carbon dioxide, which is the reference gas with a GWP of 1 (Table 14.5).

The GWP of greenhouse gas is calculated based on its ability to trap heat in the atmosphere compared to that of carbon dioxide over a specific period of time. The GWP calculation takes into account several factors, including:

1. Greenhouse gas's molecular structure.
2. Its absorption spectrum.
3. Lifetime.

The basic formula (Eq. 14.1) for calculating the GWP of a greenhouse gas is:

$$GWP = \frac{Q(1-r)}{5.35.10^{-3}} \tag{14.1}$$

[3]https://www.ipcc.ch/site/assets/uploads/2018/03/TAR-06.pdf

where:

- Q is the total radiative forcing caused by a given amount of greenhouse gas over a specific time period (usually 100 years).
- r is the fraction of the gas that is removed from the atmosphere over the same time period (i.e., its atmospheric lifetime).
- 5.35×10^{-3} is the radiative forcing of a unit mass of carbon dioxide over a 100-year time frame.

The value of Q is determined by calculating the total radiative forcing caused by the greenhouse gas over the specified time period.

Radiative forcing is a measure of the imbalance between the amount of energy that the Earth receives from the sun and the amount of energy that is radiated back into space. Greenhouse gases increase radiative forcing by trapping more heat in the atmosphere and preventing it from being radiated back into space.

The value of r is determined by the atmospheric lifetime of the gas, which is the amount of time it takes for the gas to be removed from the atmosphere through chemical or physical processes. For example, the atmospheric lifetime of methane is much shorter than that of carbon dioxide, so even though methane has a higher GWP, it has a lower overall impact on the climate system.

By using the GWP formula, scientists can compare the overall climate impact of different greenhouse gases and determine the most effective ways to reduce emissions and slow down the rate of global warming (Table 14.5).

The calculation of GWP has evolved over time to better reflect the complex processes of climate change. GWP is a measure of the ability of greenhouse gases to trap heat in the Earth's atmosphere, compared to that of carbon dioxide, which is assigned a GWP of 1.

The initial method for calculating GWP was developed in the 1990s by the IPCC and was based on the radiative forcing of a greenhouse gas over a given period of time (usually 100 years) compared to carbon dioxide. This approach used simple formulas that assumed all greenhouse gases had the same atmospheric lifetime and that the warming effect of a gas was linearly proportional to its concentration.

WP stands for 'GWP' and AR1, AR2, AR3, AR4 and AR5 refer to the different Assessment Reports produced by the IPCC.

Each Assessment Report represents a comprehensive review of the state of knowledge on climate change science, impacts and potential solutions and is produced by hundreds of experts from around the world.

The first assessment report (AR1) was published in 1990, followed by the AR2 in 1995, the AR3 in 2001, the AR4 in 2007 and the AR5 in 2014. Each report builds on the previous one, incorporating new research findings and updating previous assessments.

GWP is a measure of how much a given greenhouse gas contributes to global warming relative to carbon dioxide, which is assigned a GWP of 1. For example, methane has a GWP of 28–36 over a 100-year time horizon, which means that it contributes 28–36 times more to global warming than an equal mass of carbon dioxide over that time period. The different ARs assess the GWP of different greenhouse gases and their potential impact on the climate.

However, as understanding of the Earth's climate system has improved, so too has the calculation of GWP. In 2007, the IPCC introduced a new method for calculating GWP that took into account the varying lifetimes of different greenhouse gases in the atmosphere as well as the effects of indirect radiative forcing. This approach is known as GWP 100 or fourth assessment report (AR4) GWP.

In 2013, the IPCC released the fifth assessment report (AR5), which included an updated GWP methodology, known as GWP 20 or AR5 GWP. This approach further refined the calculation of GWP, taking into account the impacts of changes in short-lived climate pollutants (SLCPs) such as black carbon and methane on the climate system.

The most recent IPCC report, published in 2021, has continued to refine the GWP methodology, incorporating the latest research on the impacts of SLCPs and improving the treatment of non-carbon dioxide emissions from land-use change and forestry.

The long-term increase in Earth's average surface temperature is defined as global warming. Here are some facts about global warming: The Earth's average temperature has increased by about 1 °C (1.8 °F) since the late 1800s, with two-thirds of that warming occurring since 1980. This increase has been linked to human activities such as burning fossil fuels, deforestation and agriculture, which have led to an increase in the concentration of greenhouse gases in the atmosphere, trapping heat and causing global temperatures to rise. The consequences of global warming include more frequent and intense heat waves, droughts, floods and hurricanes; melting glaciers and ice caps; rising sea levels and ocean acidification. Climate scientists agree that global warming is happening and that human activities are the primary cause. Multiple lines of evidence, including temperature records, ice core data and computer models, support this conclusion.

The Paris Agreement is an international treaty signed in 2015 that aims to limit global warming to well below 2 °C (3.6 °F) above pre-industrial levels and pursue efforts to limit the temperature increase to 1.5 °C (2.7 °F). Despite global efforts to reduce greenhouse gas emissions, especially carbon dioxide, the temperature continues to rise and the Earth is on track to exceed the Paris Agreement's temperature goals. Mitigating the effects of global warming will require a global effort to reduce greenhouse gas emissions, transition to renewable energy sources and adapt to the changing climate.

14.8 Relationship Between ODP and GWP

Ozone depletion potential (ODP) and GWP are two different metrics used to assess the environmental impact of various substances. They are related but measure different aspects of the substances' effects on the Earth's atmosphere.

While ODP and GWP both relate to the environmental impact of substances, they are distinct and measure different effects. ODP focuses on the depletion of the ozone layer, which leads to increased levels of harmful UV radiation reaching the Earth's surface. On the other hand, GWP measures the ability of a substance to trap heat in the Earth's atmosphere over a specific time frame relative to carbon dioxide. Carbon dioxide is used as the reference gas with a GWP of 1 and other greenhouse gases are compared against it. The GWP takes into account the IR radiation-absorbing properties of different gases and their atmospheric lifetimes and its value depends on the time horizon considered (usually 20, 100 or 500 years). For example, methane has a

GWP of about 25 over a 100-year period, which means it is about 25 times more effective at trapping heat than carbon dioxide on a per-molecule basis over that time frame.

In some cases, there can be substances that have both a high ODP and a high GWP. For instance, some hydrochlorofluorocarbons (HCFCs) not only contribute to ozone depletion (although to a lesser extent than CFCs) but also have a significant GWP. Therefore, it's essential to consider both ODP and GWP when evaluating the environmental impact of substances, especially those used in various industrial applications, refrigeration and air conditioning systems. This is why international agreements like the Montreal Protocol and the Kyoto Protocol have been established to regulate the production and use of such substances to mitigate their impact on the environment.

14.9 Variation of Temperature and Carbon Dioxide Concentration Along Earth's History

The Earth's surface temperature has varied significantly over geological timescales, spanning millions to billions of years. These variations are primarily driven by a variety of natural factors, including changes in solar radiation, atmospheric composition, volcanic activity and continental drift. Here are some key temperature variations in Earth's history:

1. **Hadean aeon (4.6 billion–4 billion years ago):** During this time, the Earth was still cooling down from its formation and surface temperatures were extremely high due to intense meteorite impacts and residual heat from the planet's formation.

2. **Archean aeon (4 billion–2.5 stabilisation billion years ago):** The Earth's surface began to cool down and the formation of oceans and the emergence of early continents led to some stabilisation of temperatures. However, the presence of higher levels of greenhouse gases like methane and carbon dioxide in the atmosphere likely resulted in warmer conditions than today.

3. **Proterozoic aeon (2.5 billion–541 million years ago):** This era saw several instances of significant glaciations, often referred to as 'Snowball Earth' events. During these events, much of the planet's surface was covered in ice, dramatically lowering global temperatures.

4. **Palaeozoic era (541 million–252 million years ago):** This era began with the Cambrian Explosion, a period of rapid diversification of life forms. Temperatures during the early Palaeozoic were generally warmer than today and there were no polar ice caps. However, by the late Palaeozoic, the planet had experienced a series of glaciations.

5. **Mesozoic era (252 million–66 million years ago):** The Mesozoic is often referred to as the 'Age of Dinosaurs'. During much of this era, the Earth experienced generally warm conditions, with higher atmospheric carbon dioxide levels contributing to a greenhouse effect. The mid-Cretaceous period, around 100 million years ago, is believed to have been one of the warmest periods in Earth's history.

6. **Cenozoic era (66 million years ago–present):** The Cenozoic began with the extinction event that wiped out the dinosaurs. Since then, the Earth has seen a general cooling trend with the development of polar ice caps and periods of glaciation. The current Ice Age, which began around 2.6 million years ago, has seen repeated cycles of glaciations (ice ages) and interglacial periods (warmer periods).

7. **Quaternary period (2.6 million years ago–characterised present-characterised):** The Quaternary is the most recent period in Earth's history and includes the Pleistocene epoch, characterised by the repeated advance and retreat of ice sheets across continents. We are currently in the Holocene epoch, an interglacial period that began around 11,700 years ago.

The relationship between solar activity and Earth's surface temperature over geological timescales is a complex and debated topic among scientists. Solar activity refers to the variations in the Sun's energy output, primarily driven by changes in sunspots and solar radiation. On the other hand, Earth's surface temperature has experienced natural variations over millions of years due to factors such as continental drift, greenhouse gas concentrations, volcanic activity and more.

Some key points to consider when exploring the relationship between solar activity and Earth's surface temperature over geological time:

1. **Milankovitch cycles:** The Earth's climate has been influenced by changes in its orbit and axial tilt known as Milankovitch cycles. These cycles lead to variations in the amount and distribution of solar radiation received by the Earth, which can impact global temperatures. However, these variations operate on much longer timescales (tens of thousands of years) compared to the shorter-term solar activity variations.

2. **Solar variability:** The Sun's energy output does vary over time due to factors like sunspot cycles, solar flares and changes in solar magnetic activity. These variations can influence Earth's climate, but the magnitude of their impact on long-term temperature trends is still debated. The Little Ice Age (roughly the 14th–19th centuries), for example, coincided with a period of reduced solar activity known as the Maunder Minimum. However, the exact causal relationship between the two is not well understood.

3. **Climate feedbacks:** Earth's climate is also influenced by feedback mechanisms, where changes in temperature can trigger additional processes that amplify or dampen the initial change. For example, changes in temperature can affect the distribution of greenhouse gases like carbon dioxide and water vapour, which in turn influence the greenhouse effect and temperature.

4. **Proxy data:** Scientists use various proxy data, such as ice cores, sediment layers and tree rings, to reconstruct past climate conditions. These records provide insights into temperature changes and can sometimes be correlated with variations in solar activity.

5. **Greenhouse gas concentrations:** Over geological timescales, the concentration of greenhouse gases in the atmosphere has varied due to processes like volcanic activity, changes in ocean circulation and the weathering of rocks. These variations in greenhouse gas concentrations have played a significant role in shaping Earth's climate over millions of years.

On the other hand, the concentration of carbon dioxide in the Earth's atmosphere has varied significantly over geological timescales. These variations have played a crucial role in shaping the planet's climate and environment. Here are some key points about the variation of carbon dioxide concentration over geological time:

1. **Geological history:** Over the past 600 million years, the Earth has experienced periods of higher and lower carbon dioxide concentrations. These variations are associated with different geological events, such as volcanic activity, changes in ocean circulation and the carbon cycle.

2. **Ancient Earth:** During the early history of the Earth, carbon dioxide concentrations were much higher than they are today. For instance, during the Cambrian Period (around 500 million years ago), carbon dioxide levels were several times higher than present levels. This high carbon dioxide content contributed to a warmer climate.

3. **Carboniferous period:** Roughly 359–299 million years ago, during the Carboniferous and Permian periods, carbon dioxide levels were relatively high. These periods were marked by lush vegetation growth, which led to the accumulation of organic matter that eventually became coal deposits.

4. **Mesozoic Era:** During the Mesozoic Era (about 252–66 million years ago), carbon dioxide concentrations varied. They were generally higher than present levels during the early part of this era, contributing to a greenhouse climate and supporting the growth of dinosaurs. However, there were also periods of lower carbon dioxide levels, particularly during the Jurassic Period.

5. **Cretaceous Period:** Carbon dioxide concentrations during the Cretaceous Period (about 145–66 million years ago) were significantly higher than they are today. This era had a much warmer climate and there was relatively little ice at the poles.

6. **Cenozoic Era:** The last 66 million years, known as the Cenozoic Era, have seen a general decline in carbon dioxide levels. The transition from the Eocene to the Oligocene (around 34 million years ago) marked a period of cooling and a decrease in carbon dioxide levels. The Miocene (about 23–5.3 million years ago) saw further cooling and the growth of ice sheets in Antarctica.

7. **Quaternary Period:** The most recent period, the Quaternary, began around 2.6 million years ago. This period includes the Pleistocene Epoch, marked by ice ages and interglacial periods. During ice ages, carbon dioxide levels were lower and during warm interglacial periods, carbon dioxide levels were higher.

8. **Recent carbon dioxide levels:** Prior to the Industrial Revolution, carbon dioxide levels remained relatively stable for thousands of years, hovering around 280 parts per million (ppm). But since the onset of industrial times, atmospheric carbon dioxide concentration has been growing with small fluctuations. In the NASA web (https://climate.nasa.gov/vital-signs/carbon-dioxide/), a counter shows the updated carbon dioxide values.

14.10   Climate Change and Global Warming

Climate change refers to long-term changes in the Earth's climate, such as changes in weather patterns. These changes are largely caused by human activities, such as burning fossil fuels and deforestation, which release greenhouse gases into the atmosphere and contribute to the warming of the planet. The effects of climate change are far-reaching and include rising sea levels, more frequent and severe natural disasters, shifts in agricultural patterns and the extinction of species. Addressing climate change is a critical global challenge that requires coordinated efforts to reduce greenhouse gas emissions and develop sustainable practices for energy, transportation and land use.

Part of these long-term changes are caused by global warming. It refers to the increase in the Earth's average surface temperature, primarily caused by the buildup of greenhouse gases (such as carbon dioxide, methane and nitrous oxide) in the atmosphere.

Some of the most notable consequences of climate change include:

1. **Rising temperatures:** Global temperatures have been rising steadily, leading to heat waves, droughts and wildfires. This can have negative impacts on agriculture, water resources and public health.

2. **Melting glaciers and rising sea levels:** The Earth's glaciers are melting, causing sea levels to rise. This can lead to flooding, erosion and loss of coastal habitats.

3. **Ocean acidification:** As carbon dioxide levels rise, the oceans are becoming more acidic. This can harm marine life and damage coral reefs, which support a variety of species.

4. **Changing weather patterns:** Climate change is causing more frequent and severe weather events, such as hurricanes, tornadoes and floods.

5. **Extinctions:** Many species are struggling to adapt to the rapidly changing climate, leading to a loss of biodiversity and potential extinctions.

6. **Health impacts:** Climate change can lead to the spread of disease, as warmer temperatures can increase the range of disease-carrying insects. It can also worsen air quality, which can lead to respiratory problems.

7. **Economic impacts:** Climate change can have significant economic impacts, such as crop losses and property damage from extreme weather events.

The predictions about long-term changes in Earth's climate are based on Climate models, a set of mathematical representations that use data and scientific principles to simulate past, present and future climate conditions. These models help scientists understand how the Earth's climate system works and how it may respond to changes in greenhouse gas concentrations, solar radiation, volcanic eruptions and other factors.

Climate models have evolved significantly over the years (Figure 14.9), with improvements in computational power, data availability and scientific understanding. Here is a brief overview of the evolution of climatic models:

1. **Early climate models (1960s–1970s):** The first climate models were developed in the 1960s and 1970s. They were relatively simple and relied on simplified representations of the Earth's atmosphere. These models aim to simulate large-scale climate patterns and understand the fundamental processes driving climate variability.

2. **General circulation models (GCMs) (1980s–1990s):** In the 1980s and 1990s, GCMs became the primary tool for climate modelling. GCMs are complex computer models that simulate the atmosphere, oceans, land surface and sea ice. They divide the Earth into a three-dimensional grid and solve mathematical equations to represent the physical processes affecting the climate. GCMs

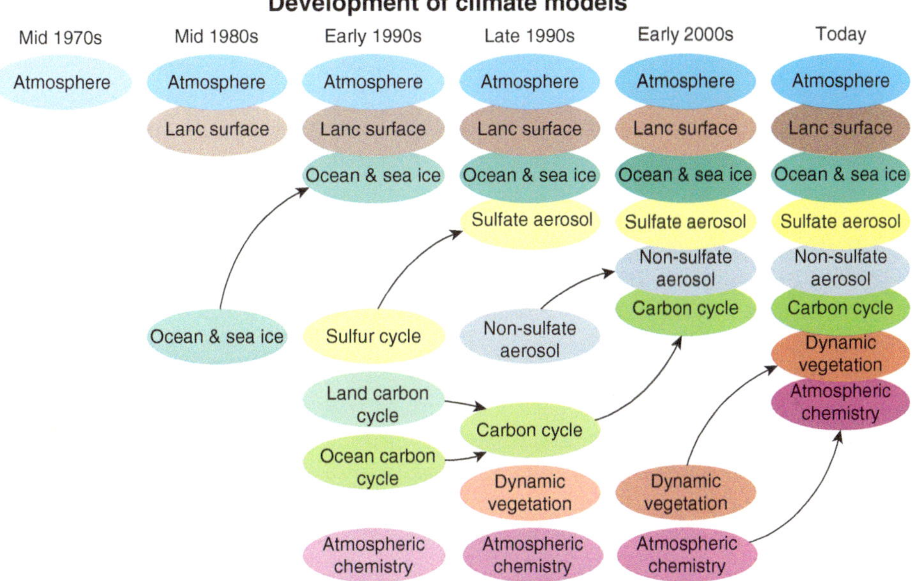

FIGURE 14.9   Evolution of climate models

incorporate more detailed representations of atmospheric dynamics, radiative transfer and the interaction between the atmosphere and the oceans.

3. **Improved resolution and regional models (2000s):** As computing power increased, climate models started to be run at higher spatial resolutions. Higher resolution allows for better representation of small-scale features such as clouds, mountain ranges and regional climate patterns. Regional climate models (RCMs) emerged, which focused on simulating climate at a smaller scale, typically a region or a single country. RCMs are nested within GCMs to provide more detailed information on regional climate changes.

4. **Earth system models (ESMs) (2000s–2010s):** ESMs expanded the scope of climate modelling by incorporating additional components, such as the carbon cycle, vegetation dynamics and biogeochemical processes. By including these components, ESMs can simulate the interactions between the climate system and the Earth's ecosystems. This allows for a more comprehensive understanding of climate change and its impacts on the biosphere.

5. **Improved parameterisations and uncertainty quantification:** Over time, scientists have developed better parameterisations, which are mathematical representations of physical processes that occur at scales smaller than the grid resolution. Parameterisations are necessary because the computational power required to resolve all processes explicitly is still beyond our capabilities. Additionally, efforts have been made to quantify uncertainties in climate models, considering uncertainties in initial conditions, model structure and future emission scenarios.

6. **Advancements in data assimilation and observations:** The availability of more comprehensive and accurate observational data, including satellite observations, weather stations and ocean buoys, has improved model initialisation, calibration and validation. Data assimilation techniques have been developed to integrate observations into models, improving their accuracy and reliability.

7. **High-resolution and Earth system models (present and future):** Recent advancements in computing power have allowed for the development of high-resolution climate models, which can simulate climate processes at fine scales. These models can capture local climate phenomena, extreme events and regional climate patterns with greater fidelity. ESMs continue to be refined, incorporating more detailed representations of various Earth System components and their interactions.

It is important to note that climate models are continually evolving as new research emerges, computational capabilities improve and observational data becomes available. These advancements help scientists gain a better understanding of past and future climate changes and their potential impacts on society and the environment.

Nevertheless, climate models are important tools for predicting future climate change and assessing the potential impacts of climate change on ecosystems, human health and the economy. They can also be used to test the effectiveness of different climate mitigation and adaptation strategies.

Topics of Interest: IPCC

The IPCC is a scientific body established by the United Nations (UN) and the World Meteorological Organization (WMO) in 1988. The IPCC's mandate is to provide policymakers with regular assessments of the scientific basis of climate change, its impacts and options for adaptation and mitigation.

The IPCC produces assessment reports that are based on the latest scientific evidence from thousands of peer-reviewed scientific papers and are produced by hundreds of scientists from around the world. These reports are typically published every 5–7 years. The most recent report, the sixth assessment report (AR6), was released in 2021. These reports are widely recognised as the most authoritative source of information on climate change.

The IPCC's assessment reports are divided into three working groups:

1. **Working group I: The physical science basis:** This group assesses the physical scientific aspects of the climate system and climate change.

2. **Working group II: Impacts, vulnerability and adaptation:** This group assesses the vulnerability of socio-economic and natural systems to climate change, negative and positive consequences of climate change and options for adapting to it.

3. **Working group III: Mitigation of climate change:** This group assesses options for mitigating climate change through limiting or reducing greenhouse gas emissions and enhancing activities that remove them from the atmosphere.

The IPCC's most recent assessment report, the AR6, was released in 2021 and provides the most up-to-date scientific understanding of climate change. The report confirms that climate change is unequivocally caused by human activities and that urgent and ambitious action is needed to address the problem.

Index